Pipeline Integrity Management Systems

A Practical Approach

RAFAEL G. MORA
PHIL HOPKINS
EDGAR I. COTE
TAYLOR SHIE

Library of Congress Cataloging-in-Publication Data

Names: Mora, Rafael G., author. | Hopkins, Phil (Pipeline Integrity Engineer), author. | Cote, Edgar I., author. | Shie, Taylor, author.
Title: Pipeline integrity management systems : a practical approach / Rafael G. Mora, Phil Hopkins, Edgar I. Cote, Taylor Shie.
Description: New York : ASME Press, 2016. | "ASME Order No. 860472." | Includes bibliographical references and index.
Identifiers: LCCN 2016012084 | ISBN 9780791861110
Subjects: LCSH: Petroleum pipelines--Safety measures. | Petroleum pipelines--Reliability. | Petroleum pipelines--Inspection. | Petroleum industry and trade--Risk management.
Classification: LCC TN879.59 .M67 2016 | DDC 665.5/440289--dc23 LC record available at https://lccn.loc.gov/2016012084

DEDICATION

This book is dedicated to the new generation of technicians, technologists, engineers, and professionals working in the pipeline industry and their mentors that guide them on their journey to achieve a safe and reliable pipeline operation protecting our environment.

The authors would also like to dedicate the book to their families, Cristal and Alexandra Shie, Marujita Sinning, Liliana Ortegon, David Esteban and Nicolas Mora, Benny Omaña and Josie Hopkins for accompanying us in the journey of writing the book for the next generation of pipeline industry professionals.

ACKNOWLEDGMENTS

ASME STAFF

The authors would like to acknowledge and thank the ASME staff specifically Ms. Mary Grace Stefanchik, Ms. Tara Smith, Ms. Kimberly Miceli, and Mr. Timothy Graves for their kind and professional support in the timely and quality preparation and publication of this book.

We would also like to acknowledge the contribution, guidance and support from the ASME Pipeline Systems Division (PSD).

BOOK REVIEWERS: ENCOURAGEMENT, KNOWLEDGE, AND EXPERIENCE

This book also benefited from many colleagues from multiple pipeline industry sectors such as pipeline operators, consultants, regulators, and services around the globe from their encouragement and knowledge and experience reflected in the technical review of the chapters of this book (alphabetical order by last name):

Nader Al-Otabi
Oswaldo Baron
Robert Basaraba
Colby Bell
Thomas Beuker
Tom Bubenik
Alex Cook
Graham Emmerson
James Ferguson
Colin Gagne
Adrian Luhowy
Robin Magelky
German Melendrez
Parimal More
Husain M. Muslim
Xavier Ortiz
Joe Paviglianiti
Jonathan Prescott
Debbie Price
Jerry Rau
Miguel Antonio Ramirez
Gabriela Rosca
Thushanthi Senadheera
Nauman Tehsin
Howard Wallace
Hong Wang
Adnan Zaheer
Dario Zapata

CONTRIBUTIONS TO INTEGRITY KNOWLEDGE AND EDUCATION

Following our distinguished Mo Mohitpour's efforts in the Pipeline Integrity Assurance book, the authors would like to continue the list of professionals (not comprehensive) that, over the years, have relentlessly, through their professional dedication, made technical contributions to pipeline integrity knowledge and education (alphabetical order by last name):

Jose V. Amortegui
Tom Barlo
John Beavers
William R. Byrd
Iain Colquhoun
Aaron Dinovitzer
Ray Fessler
Manuel Garcia Lopez
Ron Hugo
Dave Jones
Shahani Kariyawasam
Roger King
Don Lefevre
BJ Lowe
Michael McManus
Jim Marr
Giancarlo Massucco
Oliver Moghissi
Alan Murray
Maher Nessim
Ken Paulson
Andrew Palmer
Moness Rizkalla
N. Daryl Ronsky
Wilson Santamaria
Bill Santos
William J. Shaw
Ed Seiders
Mark Stephens
John Tiratsoo
Carlos Vergara
Patrick Vieth
David Weir
Mike Yoon

PERMISSIONS

ASME and the authors would like to thank all the individuals and organizations who have granted their permission to use and reproduce figures, tables, data, and other material in this book. These are referenced along the book, accordingly.

We apologize in advance for any reference or attributions that we may have overlooked and will be pleased to remedy those brought to our attention as well as properly recognize them in any subsequent editions of the book.

Baker Hughes Incorporated, www.bakerhughes.com
British Standards Institute, BSI[1]
Chemical Institute of Canada, www.cheminst.ca
Colpet al Dia: Colombian Petroleum Company, *La Historia de Coveñas* in Facebook
Diakont Advanced Technologies, www.diakont.com
EnhanceCo, www.enhanceco.net
Great Southern Press, www.gs-press.com.au
Gus Gonzalez-Franchi
Institution of Gas Engineers and Managers, IGEM, www.igem.org.uk
JW's Pipeline Integrity Services, LLC, www.jwspiservices.com
M. E. Alonso, E. C. Vasquez [2]
Manuel Aguirre, Mexico Maxico, www.mexicomaxico.org
Mechling Bookbindery, www.mechlingbooks.com

Moness Rizkalla, Visitless Integrity Assessment Ltd, www.via-plus.net
Museum Rosneft, www.russianmuseums.info/M3115
NACE International, www.nace.org
National Energy Board|Office national de l'énergie, www.neb-one.gc.ca
Neil McElwee, Oil Creek Press, www.manta.com/c/mmj9sfd/oil-creek-press
New Century Integrity Plus, www.ncintegrityplus.com
Petroleum History Institute, www.petroleumhistory.org, Samuel T. Pees
Petrolia Heritage, www.Petroliaheritage.com
Phil Hopkins, phil.hopkins.work@gmail.com
PII Pipeline Solutions, a GE Oil & Gas and Al Shaheen Joint Venture, www.geoilandgas.com/pipeline-storage/pipeline-integrity-services
Pipe Line Contractors Association of Canada, www.pipeline.ca
ROSEN Group, www.rosen-group.com
Russell NDE Systems. Inc. and PICA Corp, www.russelltech.com
Thinkstock, www.thinkstockphotos.ca
TRC Pipeline Services Sector, www.trcsolutions.com
USA National Transportation Safety Board, www.ntsb.gov
Wikimedia Commons, https://commons.wikimedia.org/wiki/Main_Page
World Oil, www.worldoil.com

---

[1] Permission to reproduce extracts from PAS 551:2008 is granted by BSI. British Standards can be obtained in PDF or hard copy formats from the BSI online shop: www.bsigroup.com/Shop or by contacting BSI Customer Services for hardcopies only: Tel: +44 (0)20 8996 9001, Email: cservices@bsigroup.com.

CONTENTS

PREFACE

By the end of 2000s, the idea of an integrity management book came about from students attending the University of Calgary's Pipeline Integrity Management courses. An increasing need to develop combined engineering and management practices for the day-to-day became important for achieving integrity management effectiveness. Then, creating a resource or material with integrated and multi-disciplinary knowledge was one of many steps needed to contribute technicians, technologists, engineers and managers with knowledge and a practical approach for responding to the increasing society expectations for safe and reliable pipelines.

In the summer of 2014, ASME Publishing Committee notified their approval for publication of the book. This initiated the authors journey to collectively share their engineering and management experiences from four (4) interlinked pipeline industry sectors (i.e., pipeline transmission, engineering integrity consulting, technology services, and regulatory oversight) and their multiple geographical exposures (i.e., Arctic, Canada, USA, South America, Europe, Asia Pacific, and Australia). However, the authors' interest was focused on making the reader the "Hero of the Story" not the writers by engaging them with core, relevant, and attractive knowledge needed at work. Figures and real case examples were made generic for transferring knowledge and experience to the *Hero* for easier applicability and building-block innovation.

The book outline portrays a management system (MS) approach that enables management in providing direction, guidance, support, and evaluation to sustain continuous improvement, while it connects to and explains the Integrity Management Program (IMP) elements providing the engineering integrity core for sustaining pipeline risk reduction. The key for success resides in the linkage between MS and IMP, named hereafter as *Pipeline Integrity Management System (PIMS)*. PIMS approach enables organizations to achieve state-of-the-art adequacy, timely implementation, and measured effectiveness of the relentless integrity goals, objectives, and targets toward the *safety of employees and the public, the protection of the environment, and a reliable service.*

The table of contents follows the PIMS structure detailing within each management system element the engineering elements; however, they can be read individually as each applicable chapter has a PLAN-DO-CHECK-ACT built-in process for enabling the reader to become the *Hero*.

During the 2015 winter, the review of the book chapters became an insightful experience for all authors discovering the richness and value added by 28 reviewers from multiple engineering and management backgrounds and experience levels located around the globe. The review was conducted relaxing the reviewers by not knowing whose materials they were commenting on and empowering them by applying their need for changes in the final book: *their review guided the writers for reader's benefit.*

In the early 2016, when the time for writing the acknowledgments came, the authors realized that there were not enough pages and expressive words of gratitude for all individuals who reviewed the book and have contributed with pipeline integrity knowledge and education. In the Acknowledgments page, we continued the list initiated by Dr. Mo Mohitpour in the Pipeline Integrity Assurance book with the extent possible. We all know that *our reward at the end of the day* is enjoying that the *people and environment along the pipeline are safe and protected.*

With appreciation,
Rafael G. Mora, Taylor Shie, Edgar I. Cote, and Phil Hopkins
The Authors

If you had any questions, comments, clarifications or improvement for the next edition, please email us at asme.pims@outlook.com. Please indicate your phone number in the email, if you would like to be contacted as soon as possible.

LIST OF FIGURES

LIST OF TABLES

PIPELINE INTEGRITY MANAGEMENT (IM): HISTORY AND FRAMEWORK

This Chapter 1 defines the core purpose of pipeline integrity management recommended for building the foundation of a pipeline integrity management system (PIMS). It also provides key integrity management definitions differentiating pipeline and pipeline systems, pipeline integrity attributes, integrity hazards and threats, types of pipeline systems with their own integrity challenges, integrity management programs (IMP), management systems, the pipeline life cycle stages related to integrity as well as the elements of and participants in an integrity management framework in today's society.

This chapter briefly walks you through the first transmission pipelines listed chronologically, but not comprehensive (e.g., *Canada, USA, Azerbaijan—formerly Russia, Mexico, Iran, Colombia, Saudi Arabia, Brazil, Europe, China, Australia, and Peru*) and a context related to the initial reservoir discoveries and legislation, considering the publicly and readily available information to the authors. The pipeline legislation during those beginnings is also briefly described capturing some successes and challenges faced by those pioneering countries in building hydrocarbon pipelines.

Subsequently, the chapter outlines the evolution of Integrity Management (IM) approach around the world. The IM evolution started with a pipeline maintenance approach evolving into IMP and, later on expanding into PIMS. The IM evolution is explained chronologically going over countries and/or regions.

The regulatory framework was classified in four (4) approaches used to achieve compliance with the regulations defining intent, advantages and limitations of each approach. Pipeline regulatory agencies introducing some key pipeline integrity changes in Canada, United States of America (USA), United Kingdom, Argentina, Colombia, Peru, and China are described and referenced.

Pipeline integrity industry standards from CSA Z662, ASME B31.8S, API 1160, BS PD 8010, BS PAS-55, IGEM TD/1, ICONTEC NTC 5747 and 5901, and AS 2885 are briefly described related to their pipeline integrity focus and coverage. Industry associations such as CEPA, CAPP, INGAA, AOL, API PPTS, ARPEL, MARCOGAZ, UKOPA, and APIA; and industry research entities such as NRCan, PRCI, EPRG, and APIA-RSC are briefly described highlighting some of their great efforts in developing best industry practices to support the pipeline integrity management framework.

1.1 INTEGRITY MANAGEMENT (IM) CORE PURPOSE

The core purpose of Integrity Management is to *protect people and environment along and around pipeline systems providing a reliable service to customers and shippers*. This core purpose is the reason for the existence of pipeline integrity enabling organizations to create, modify, or improve processes and procedures, technology and organizational culture. Thus, organizations would walk the extra mile in developing, customizing and implementing the pipeline integrity management system (PIMS) to systemically, systematically and proactively achieve this core purpose. This core purpose drives the need for continuous improvement to increase and sustain the adequacy, implementation and effectiveness of PIMS.

1.2 KEY INTEGRITY MANAGEMENT TERMINOLOGY

As pipeline Integrity has and continues to be built from multiple disciplines (e.g., engineering and management), industry standards, and regulatory jurisdictions, this section provides you with the key integrity management definitions being used in this book.

1.2.1 Pipeline and Pipeline Systems

The word "*Pipeline*" has been differently defined in some technical context (e.g., CSA-Z662) and in some regulatory context (e.g., NEB Act, PHMSA CFR and UK HSE). The industry standards typically define *Pipeline* as the components through which oil and gas industry fluids are conveyed including pipe, isolating valves and other appurtenances, but excluding pumping, compression and metering stations. This definition makes the pipeline starting and ending at the station fence excluding components within the stations or facilities. Simply described, pipeline definition would include pipe and accessories from fence to fence.

Regulations typically define *Pipeline* as the pipeline definition in industry standards plus the facilities such as tanks, reservoirs, pumps, compressors, and loading. This regulatory definition equates to a *Pipeline System* definition in some industry standards, which includes all components of a pipeline and facilities required to move oil, gas and products including measurement, non-formation storage (e.g., tank farms), transportation, and distribution.

Furthermore, some industry standards typically define *Pipeline Systems* within their standard applicability flowcharts having in common the exclusion of production, underground formations, bulk plant, steam generation, tanker and barge loading/unloading, gas processing, and refinery facilities.

1.2.2 Pipeline Integrity Attributes: Structural Reliability, Availability and Safe Conditions

Pipeline Integrity is the status of a pipeline defined by its *structural reliability and availability to transport a fluid under safe conditions*.

Structural Reliability of a pipeline, in the context of pipeline integrity, reflects its net resistance at a given time to avoid a release (i.e., leak and/or rupture) or sustain a damage in order to safely transport a fluid (function) under maximum operating conditions. Structural reliability quantifies and qualifies the condition of the pipeline in terms of its capacity to "resist" damage, leaks or ruptures from any of the pipeline integrity threats. The net result between the effects of every applicable integrity threat on the pipeline and its resistance at a given time determines the structural reliability.

In the pipeline integrity context, structural reliability can be deterministic or probabilistic. The deterministic structural reliability is typically defined by worst threat's characteristics such as anomaly dimensions (e.g., depth, length, and width), their frequency, safety margins (e.g., operation factor of safety, landslide failure potential) and material properties. They may account for the deterministic tolerance (e.g., +/−10% wall thickness) of the measurement equipment. Furthermore, current integrity threat characteristics shall be predicted to future conditions considering degradation and growth mechanisms over a period of time under maximum operating and/or environmental (e.g., flooding) conditions since first reported or last time measured.

In the pipeline integrity context, probabilistic structural reliability is typically defined by probability density functions represented by its mean and standard deviation. They account for both tolerance and certainty and are typically expressed in non-dimensional (i.e., chances to fail) or failures per distance-time (e.g., 1/mile-year, 1/km-year).

Availability of a pipeline, in the context of pipeline integrity, reflects the estimated time needed to "fail" to operate under maximum operating pressure or alternatively, under pressure reduction conditions including the safety margin. Availability is a deterministic measure typically expressed in years, instead of percentage, providing an estimated safe remaining life of the pipeline with the required or exceeded structural reliability.

Safe conditions for pipeline integrity is the expected outcome from the structural reliability and availability processes providing the barriers that protect employees, society, and general public as well as the environment from any pipeline release or damage.

1.2.3 Integrity Hazards and Threats

In the pipeline industry the terms hazard and threat are either used interchangeably or independently of each other to represent different or same meanings. However, the term *hazard* has been extensively defined in other disciplines (e.g., safety) prior to the birth of IMP; whereas, the term *threat* has been further developed in recent years (e.g., security). This book differentiates terms integrity *hazard* and *threat* defining their action and relationship within pipeline integrity.

Hazard is defined by Webster's dictionary as "*a source of danger*" and danger is explained as the *exposure or liability to harm or loss*. Thus, an *Integrity Hazard* can be defined as **any situation or event or condition** (e.g., coating damage) **able to initiate or grow an integrity treat** (e.g., corrosion).

Whereas, threat is defined by Webster's dictionary as both "*an expression of intention to inflict … damage*" and "*an indication of something impending*". Thus, an Integrity Threat can be defined as an *abnormal state affecting a pipeline that may have the* **capacity to cause a failure**. Abnormal states or threats (e.g., Stress Corrosion Cracking) are typically created, made active or dormant, or grown by one or more hazards (e.g., tape coating damage, stress levels, pressure changes).

One example of applicability is the integrity threat assessment process, which requires the discovery of all hazards and their potential to initiate (i.e., susceptibility) or grow (i.e., identification) integrity threats and assessing their capacity to cause failure.

1.2.4 Type of Pipeline Systems and Their Integrity Challenges

Pipeline systems are divided into three (3) types, which are described herein with their integrity challenges:

Gathering Pipelines transport untreated fluids from onshore or offshore wells to battery/treatment, processing or refining facilities for extracting natural gas, natural gas liquids (e.g., propane, butane, and ethane) or crude oil. Gathering pipelines are also known as feeders.

Untreated fluids contain water, sour fluids and impurities that can cause internal corrosion and if not timely treated, releases. Pipe joints and fittings in gathering pipelines are the places with the highest frequency of releases. Gathering pipelines are mainly located in remote areas with low population density, but increased exploration is seeing gathering systems getting closer to consequence areas.

Gathering pipelines are mostly designed for low pressures typically less than 25% to 30% of the Specified Minimum Yield Strength (SMYS) of the pipe to transport volumes for relatively short distances commonly above ground. They are designed with thicker pipe walls to resist the effects of product corrosivity for longer duration; however, the pipeline life is usually short requiring continuous replacements, product chemical treatment, internal corrosion resistant materials (e.g., reinforced fiberglass pipe) or internal coating.

Transmission Pipelines transport treated fluids from processing, production or refining facilities to distribution centers or terminals. Transmission pipelines are also known as trunk lines. Facility laterals or interconnecting pipelines can be also included within the transmission pipeline category, if the fluids have been treated.

Transmission pipelines may cross populated, developed and/or environmentally sensitive areas to reach their destination and are usually more prone to high exposure to excavation or drilling damage. Transmission pipelines are typically designed for medium to high pressures (i.e., 30% to 72–80% SMYS) transporting larger volumes for longer distances with thinner pipe underground. They are typically cathodically protected going through multiple geographies (e.g., elevation, water bodies) and infrastructure (e.g., road and rail crossing).

Distribution Pipelines transport treated fluids from transmission pipelines, distribution centers or terminals to the end-user or customer. Distribution pipelines are also known as city or town networks. Distribution pipelines cross populated and environmentally sensitive areas to reach houses, factories or dwellings with high exposure to excavation or drilling damage. Distribution pipelines are typically designed for low pressures (i.e., <30% SMYS) transporting smaller volumes for shorter distances with thinner pipe walls underground.

1.2.5 Pipeline Integrity Management Program (IMP)

Harvard Leadership Weblog provides some definitions of Management as "*the guidance and control of action required to execute a program*" or "*the process of planning, leading, organizing and controlling people within a group in order to achieve a goal*". Oxford Dictionary defines Program as "*planned series of future events or performances*" [1].

These definitions substantiate the linkage between Management and Program to create a leadership-driven process to achieve goals through a series of planned and controlled actions. Hence, an **Pipeline Integrity Management Program (IMP)** can be defined as a *engineering-sound-process to manage the integrity of a pipeline* *through the identification, susceptibility, assessment, prevention, mitigation and monitoring of risks to protect people and environment providing a reliable service to shippers and customers.*

1.2.6 Pipeline Integrity Management System (PIMS)

A Management System (MS) is defined as a *"set of interrelated or interacting elements to establish policy and objectives and to achieve those objectives"* [2]. A MS interlinked with an IMP would become a Pipeline Integrity Management System (PIMS) enabling the organization to achieve its goals, objectives and targets obtaining an state-of-art adequacy, timely implementation and measured effectiveness.

PIMS provides the integration between engineering and management processes such that stakeholder requirements are met with respect to pipeline system integrity at all times. That is, all participants of PIMS (e.g., operations, maintenance, integrity, auditors, management, human resources) interact continuously to promote its effectiveness throughout the life of the pipeline system.

As illustrated in Figure 1.1, the operational principle of a Management System (MS) PLAN-DO-CHECK-ACT can provide the platform of an Integrity Management Program (IMP) for pipelines enabling functionality, efficiency and effectiveness to accomplish the goals, objectives and targets of an organization with the senior management leadership.

The following *MS core elements* capture the high level management system processes mainly originated in Europe and Japan. Those MS core elements assist in providing leadership, guidance, support, verification and evaluation of programs (e.g., IMP):

1. *policy and commitment*
2. *planning (plan)*
3. *implementation (do): performance assessment and KPIs*
4. *conformance and compliance verification and action Plans (Check), and*
5. *management review (Act)*

In the other hand, IMP core elements have been diversely named, listed and improved by some industry standards and regulations starting in Canada and USA, and later on improved around the world. In a nutshell, the intent of these engineering and management core elements essentially focuses on hazard/threat, consequence, and risk processes such as:

1. *susceptibility,*
2. *identification,*
3. *assessment,*
4. *mitigation,*
5. *prevention, and*
6. *monitoring*

Chapter 3 describes a more comprehensive list of PIMS elements and their linkage, which are depicted in Figure 3.1 PIMS a Management System Approach (MS + IMP). In summary, a **Pipeline Integrity Management System (PIMS)** can be defined as a *leadership driven-process to direct, plan, implement, verify, measure and continuously improve the integrity of a pipeline(s) to protect people and environment providing a reliable service to shippers and customers.* **PIMS** *is supported at the core by an engineering-sound Integrity Management program (IMP).*

1.2.7 Pipeline Life Cycle

The pipeline is comprised of six (6) stages. Their definitions are complemented with an integrity perspective via examples, as follows:

1. Design
2. Manufacturing
3. Construction
4. Operation
5. Deactivation or Reactivation
6. Abandonment

1.2.7.1 Pipeline Design with an Integrity Perspective The pipeline design stage defines the engineering characteristics and financial feasibility as well as the environmental, human and socio-economic impacts and associated mitigation and monitoring plans. Community consultation and regulatory approvals are required before manufacturing and construction stages.

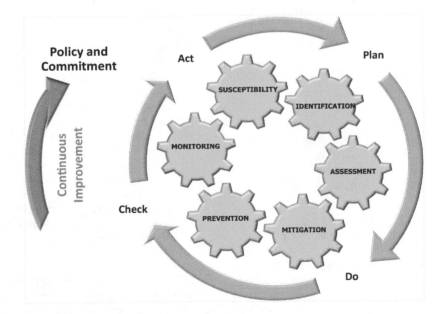

FIG. 1.1 PIMS CONCEPTUAL: CONNECTING MANAGEMENT SYSTEMS AND INTEGRITY MANAGEMENT PROGRAMS

Examples of pipeline design practices to prevent or minimize integrity threats:

- Safety margins and tolerances for resistance to corrosion
- Pipe fracture toughness for resisting cracking and geometry damage
- Seam weld type and quality for minimizing cracking initiation and growth
- Non-shielding coating and non-telescopic wall thickness for minimizing Stress Corrosion Cracking
- Depth of cover for reducing the chance for external interference damage and river crossing scour
- Strain and girth weld capacity for resisting pipeline buckling/wrinkling and outside forces from hydro-geotechnical loadings such as landslides, thaw settlement, subsidence and upheaval
- Automated start-up and shutdown devices including Variable Frequency Drives (VFDs) for smoothing up the pressure-induced crack growth due to pressure spikes or cycling

Examples of pipeline design practices to reduce potential consequences:

- Automatic or remotely-operated valves with an optimal distance separation for reducing drain-down volume in the event of a release
- Leak detection systems with optimal performance to timely detect and locate small releases in different flow scenarios with a minimum number of false alarms
- Field monitoring systems for early warning of events such as hydro geotechnical (e.g., water crossings, landslides, subsidence), corrosion (e.g., corrosion growth) and external interference damage

1.2.7.2 Pipeline Manufacturing with an Integrity Perspective The pipeline manufacturing stage provides the fabrication of pipe, valves, and appurtenances required to be installed in the pipeline system. Lessons learned from incidents associated with seam weld quality control, lack of pipe thermal treatment inducing stress corrosion cracking on the pipe body, and hydrostatic testing levels not matching the expected operating condition can help to prevent repeating incident history in the pipe manufacturing stage.

Examples of pipeline manufacturing practices to prevent or minimize integrity threats:

- Pipe and seam weld material specifications resisting fatigue and minimizing residual stresses
- Quality assurance and control of pipe seam weld reducing imperfections that could become defects during operation
- Records of material testing including composition, stress, and test duration for in-service assessments of pipe anomalies

1.2.7.3 Pipeline Construction with an Integrity Perspective The pipeline construction stage provides the pipeline and facilities including the associated environmental remediation as well as building the long-term community relationships for the upcoming operation stage. This stage includes *pre-commissioning* and *commissioning* activities such as hydrostatic testing, drying and corrosion prevention, cleaning pigging, baseline in-line inspection, final pipeline surveying and filling up the pipeline with the designed fluid that ensures the pipeline system is ready prior to start-up of the operation.

Examples of pipeline construction practices to prevent or minimize integrity threats:

- Proper placement of pipe (e.g., clock position) during transportation to avoid fatigue on seam welds
- Pipe protection avoiding damage handling, in-ditch laying and backfilling
- Field inspection of girth weld anomalies detecting any connections to seam weld imperfections
- Pipe surface preparation for proper adhesion of coating avoiding cathodic protection shielding
- Compatibility and sealing overlap of factory-applied pipe coating and field-applied girth weld coating
- Pipe bending ensuring wrinkles and ripples are not exceeding maximum operating stress (i.e., PRCI criteria)

1.2.7.4 Pipeline Operation with an Integrity Perspective Pipeline operation stage provides the fluid transportation service (e.g., gathering, transmission, distribution) ensuring that people and environment are protected while meeting shippers and customers' commitments.

Example of pipeline operation practices to prevent or minimize integrity risks:

- Representative integrity threat assessments (i.e., Susceptibility and identification)
- Continuous estimation of consequence severity and extent and optimizing valve location and closure (e.g., remotely operated or auto shut off), if needed
- Validated Risk Assessments for prioritizing actions in a timely matter
- Appropriate, accurate and timely in-line inspections for applicable integrity threats
- Mitigation via operational measures (e.g., overpressure controls)
- Mitigation via maintenance practices (e.g., inhibition, cleaning)
- Mitigation via integrity (e.g., repairs, pressure reduction)
- Mitigation via design of system modification (e.g., trap installation, variable frequency drive pumps)
- Effective cathodic protection and coating repairs minimizing corrosion initiation and growth
- Smooth system operation minimizing both controller human factor effects, and material fatigue and upsets causing crack initiation and growth
- Periodic monitoring of integrity hazards (i.e., threat susceptibility) and existing threats establishing growth and new emerging threats
- Effective damage prevention and public awareness programs to prevent external interference damage (i.e., 1st, 2nd, and 3rd party)

1.2.7.5 Pipeline Deactivation/Reactivation with an Integrity Perspective The pipeline deactivation stage provides the operator with the opportunity to deactivate the pipeline with the option to reactivate it in the future.

Examples of pipeline deactivation practices to prevent or minimize integrity threats:

- Effective cathodic protection during deactivation establishing a barrier to external corrosion
- Removal of stagnant water and residue that could potentially induce active internal corrosion
- Adequate signage and patrolling to avoid potential external interference damage or to early identify hydro-geotechnical hazards (e.g., water crossing scour, landslides)

Prior to pipeline reactivation integrity threats should be prevented or mitigated by determining an engineering assessment of the integrity threats and consequences expected at the time of the reactivation and implementing mitigative, preventative, and monitoring measures.

Examples of pipeline reactivation practices recommended for the engineering assessment:

- Identifying the condition of a pipeline prior to reactivation: deterioration is expected to have occurred during the deactivation
- Quantifying the location, extent and size of corrosion, dents, and cracking via in-line inspections: information gleaned to be used for mitigation and prevention programs before and after the reactivation
- Reviewing the controller competency, workplace and procedures: increase confidence in a sound operation not affecting integrity of the pipe
- Monitoring plan: determining changes in remaining life and re-inspection interval of the active integrity threats

1.2.7.6 Pipeline Abandonment with an Integrity Perspective The pipeline abandonment stage provides a pipeline with the integrity that would not affect population, environment or property under non-operating conditions. Typically, abandonment requires pipelines to be emptied free of corrosive, toxic and pollution products and filled with an inert material (e.g., grout, concrete, inert gas). In some jurisdictions, abandonment requires pipelines to be pulled out of the ground and the adjacent lands to be remediated.

1.3 FIRST HYDROCARBON TRANSMISSION PIPELINES, RESERVOIRS AND LEGISLATION

The first pipelines were used for gathering and distribution of water or gas for lighting and made out of materials such as cooper, bamboo, rock, wood or baked clay used by ancient civilizations of China, Egypt, Mesopotamia, Persia and Rome. However, hydrocarbons (i.e., crude oil, natural gas, and refined products) were initially transported in barrels by teamsters driving horses and transferred into steamboats to be taken on rivers up to refineries for processing and later on to nearby consumers.

In 1792, William Murdock, a Scottish engineer, produced gas from heating coal with a closed iron retort that was taken through a hollow pipe producing a steady flame. Later on in 1794, he piped the house by lighting a series of burners. In 1807, the Pall Mall Street in the City of Westminster, London had the first gas lamps lit by coal gas.

From the late 1850s, the European need for energy motivated the exploration in Canada, USA, Russia, Iran and Mexico as Europe had not developed those resources yet. In 19th century, the exploration went farther to South America, the Middle East, and Australia requiring pipelines to transport crude oil and gas to refineries, processing plants and then customers.

Transmission pipelines demonstrated to be more economical and safer transportation than teamsters and railcars by reducing cost and loss of product on land and water. Refineries were initially built near the hydrocarbons production areas and refined products were transported by multiple modes of transportation to dispersed customers and markets. However, over time pipelines were built from the gathering sites directly to large markets (i.e., cities) to be closer to the customers, extending the pipeline lengths even farther to deliver refined products.

The construction of transmission pipelines in some countries was done several years after the reservoir discovery making the pipeline timeline different than the hydrocarbons exploitation timeline due to the use of other means of transportation used first (e.g., teamsters, bulk-boat and railcars). Figure 1.2 depicts a hydrocarbons pipeline timeline capturing the chronological order of the first transmission pipelines (not comprehensive) in those countries and regions between 1862 and 1980, based on the publicly and readably available information gathered at the time of the book write-up. The timeline shows the following countries: *Canada, USA, Azerbaijan— formerly South of Russia region, Mexico, Iran, Colombia, Saudi Arabia, Argentina, Brazil, Europe, China, Australia and Peru.*

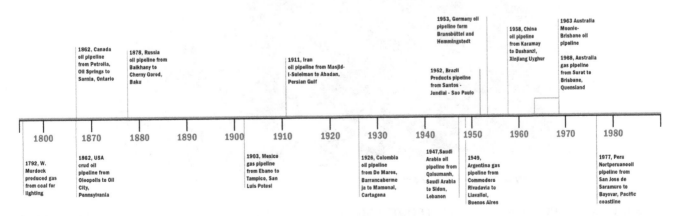

FIG. 1.2 TIMELINE OF THE FIRST HYDROCARBONS TRANSMISSION PIPELINES: 1862 TO 1977 (NOT COMPREHENSIVE)

1.3.1 Canada

In 1857, the first oil well in North America was commercially drilled in Oil Springs, Lambton County, Ontario by James M. Williams. Figure 1.3 depicts a photograph of the Canadian oil field of Oil Springs. In 1860, James M. Williams reincorporated his firm as *The Canadian Oil Company* establishing an integrated oil company to produce, refine, and market hydrocarbons [3].

In 1862, the first Canadian crude oil pipeline went from Petrolia near Oil Springs to Sarnia, Ontario [4]. In 1869, the Union Pipeline company was established merging Parker, Thompson & Co., and Karns & Parker pipelines. By 1870, Canada was exporting to Europe supported by about 100 refineries in operation. In 1911, Union Gas amalgamated three (3) gas pipeline companies expanding throughout Ontario and later in 1942 connected to an USA pipeline by a pipeline from Windsor, ON to Detroit, MI. Figure 1.4 depicts a photograph of some of the workers contributing to the construction of the Union Gas Pipeline System in Southern Ontario.

Subsequently, Canadian Western Natural Gas built a 275 km pipeline from Bow Island to Calgary, AB [5, 6]. In 1947, a major discovery of crude oil reserve was made in Leduc, Alberta. The reserve was owned by Imperial Oil. In order to refine the crude in the USA, it was required transporting it to Regina, SK and then into Superior, WI, USA. In 1948, the company made a proposal to the Federal government, which approved an interprovincial pipeline of 720 km from Edmonton, AB to Regina, SK. However, federal legislation did not exist for allowing the construction and operation of pipelines traversing Canadian provinces or into the USA. On 30 April 1949, the Canadian Parliament approved the first Canadian Federal Pipe Line Act. This Act facilitated the incorporation of the following five (5) pipeline companies:

1. Interprovincial Pipe Line Company or IPL (Edmonton, AB—Superior, WI pipeline)
2. Westcoast Transmission Company (Taylor, BC to USA pipeline)
3. Trans-Northern Pipe Line Company (Ontario to Quebec including Ottawa lateral)

FIG. 1.3 CANADIAN OIL FIELDS OF OIL SPRINGS, LAMBTON COUNTY, ONTARIO (*Source:* Courtesy of Petrolia, Canada's Victoria Oil Town at www.petroliaheritage.com)

Early Construction Workers on Union Gas Pipeline System in Southern Ontario

FIG. 1.4 EARLY CONSTRUCTION WORKERS IN SOUTHERN ONTARIO (*Source:* Courtesy of Pipe Line Contractors Association of Canada http://www.pipeline.ca/history.html)

4. Western Pipe Lines Ltd, which later merged with TransCanada Pipelines [7] (Alberta-Saskatchewan Border to Toronto and Montreal pipelines)
5. British American Pipe Line Company

In the 1950s, the global oil supply situation promoted the construction of Trans Mountain Pipe Line (Edmonton, AB to Vancouver, BC pipeline).

Officers from the International Petroleum Company working in Colombia and Peru [8] were brought to North America due to their experience on pipelines and assigned to support the new pipeline company IPL. The pipeline objective was to transport the Leduc, Alberta crude oil to Regina Refinery and then via Gretna, MB to the Superior, WI Refinery in the USA and farther to Great Lakes. This pipeline was completed in 1950 [9, 10].

In 1954, Premier Ernest C. Manning ordered TransCanada and Western Pipe Lines [11] to merge making an all-within-Canada natural gas transmission pipeline from Alberta going to as far east as Montreal Refinery, QC [12]. In 1959, the federal government established the National Energy Board which has the authority to regulate Canadian pipelines crossing provincial or international borders. The NEB Act was designed to ensure that pipeline projects will be economically feasible; tolls would be just and reasonable; the exported energy would be surplus of the Canadian needs; and all projects were to be for the public convenience and necessity. Confirming the focus, the Board used to report to the Minister of Trade and Commerce.

1.3.2 United States of America (USA)

In 1859, Edwin I. Drake drilled the first oil and natural gas well in the USA at Titusville, Pennsylvania. In those days, bulk-boats and line of tank cars such as the "Star Line" were used first to transport crude oil from Pithole to Oil City [13]. Steamboats navigated on Allegheny River transporting oil from Oil City to Pittsburgh [14].

In 1862, the first USA pipeline (6 inch) was built from Oleopolis to Oil City followed by the construction of a 4 mile pipeline from West Pithole to Pithole named as "The Star Pipe Line". As illustrated in Figure 1.5, Samuel Van Syckel started, against teamster ridicule and opposition, a 2-inch wrought iron pipeline from Pithole to Miller farm railroad station in 1865.

The transportation monopoly dominated by railroad with their own pipelines did not allow for other pipelines to cross the rail tracks [15].

In 1872, the public sentiment drove Pennsylvania and Ohio legislatures granting "common carrier pipelines" the privilege to acquire right-of-way over others, which promoted the development of transmission pipelines. Pennsylvania legislature also passed the first anti-pollution bill preventing running of tar and distillery refuse into certain creeks. By 1876, pipeline transportation companies had used pumps in pipelines, which were typically arriving at ports for steamboat transportation onto major rivers.

In 1879, the first long pipeline with a length 177 km and a diameter of 6 inches named *Tidewater* was completed for transporting oil from Coryville to Williamsport, Pennsylvania driving the railroad business to reduce their "monopolized" rate. By 1881, transmission pipelines in the USA had expanded to a total length of 966 km. Figure 1.6 depicts the Standard Oil's Trunk Pipeline System in 1885.

In 1938, the USA Federal Government started regulating the natural gas industry focusing on the tolls from interstate natural gas transmission companies.

During World War II (WWII), USA built the War Emergency Pipelines (WEP) in response to the German U-boat submarine attacks on tankers transporting hydrocarbons from the Gulf of Mexico to ports along the Atlantic east coast. In 1943, the Big Inch (NPS24)

and Little Big Inch (NPS 20) WEP were built within the year to start transporting crude oil and refined products from Longview (Texas) across the Mississippi River to Southern Illinois and then east to Phoenixville, Pennsylvania, to New York City and Philadelphia [16].

In 1967, the USA President Lyndon Johnson signed into law the Public Law 89-670 establishing the Department of Transportation (DOT), the regulator. In 1968, DOT Office of Pipeline Safety (OPS, now PHMSA) enforced the first Natural Gas Safety Act. Late in 1979, the Act included liquid pipelines having updates in 1988, 1992 and 1996.

1.3.3 Russia, South of (Today's Azerbaijan)

Since 1821, the Baku oil fields were formally assigned for exploitation using a lease system on a rotating basis. Figure 1.7 shows the oil gushers of Baku and some of the early pipelines. However, the first Russian well, drilled with a method was at Bibi-Eybat, Baky did not achieve success. Later on in 1864 at Anapa or Анапы, this well turned out successful and their method expanded across Russia [3].

In 1878, the engineer *Vladimir Grigoryevich Shukhov* [17] designed and built the first oil pipeline, with heating and pumps, between Balkhany and Cherny Gorod with a diameter of 88.9 mm (3 inch) and a length of 12 km in the Baku area. By 1883, Baku had exceeded 94 km of pipelines.

In 1894, rules for oil fields on the lands of the *Kuban* and *Tersk Cossack* hosts were imposed. The design of the first long pipeline, the 835 km Trans-Caucasian kerosene pipeline between Baku (today Azerbaijan) and Batumi (today Georgia) concluded in 1895 and its construction finished in 1901.

1.3.4 Mexico

In 1862, *Antonio Del Castillo* drilled the first oil well near the Tepeyac Hill, Mexico City sprouting a mixture of water and oil in significant amounts, which ended up being used as lighting oil. A year later, 1863, *Manuel Gil y Saenz*, priest and historian, spotted oil sprouting from the ground while walking between San Fernando and San Carlos, Macuspana Municipality, State of Tabasco [18].

By 1903, *Ezequiel Ordoñez*, a Mexican geologist, illustrated in Figure 1.8, recommended drilling the Cerro de La Pez, El Ébano, State of San Luis Potosí to an American petroleum company from California. This successful discovery of a reservoir led to building the first oil pipeline from Doheny's wells in the Ébano area to Tampico using threaded pipe [19].

In 1916, Cerro Azul in the state of Veracruz, another discovery by Ezequiel Ordoñez reached a significant production level, which drove the construction of the first gas transmission pipeline from Cerro Azul, South of Tampico to Mara Redonda, Tampico [20].

In 1917, the Mexican Constitution Article 27 defined the absolute property of the subsoil legalizing its ownership to the Mexicans; however, the Article 27 was not promulgated until 1925. The oil companies responded by creating the Association of Petroleum Producers.

In 1938, the Mexican President *Lázaro Cárdenas Del Río* nationalized the hydrocarbons resources after an unsuccessful labor dispute between the workers and the foreign oil producing companies. The labor dispute was about a request for a workday of 8 hours and an increase in salaries. Thus, *Petroleos Mexicanos* (PEMEX) was created for representing the national interest by directly managing Mexico's hydrocarbon resources.

Later on in 2008, the Mexican government created the *Comisión Nacional de Hidrocarburos* (CNH) to regulate and supervise the exploration and extraction of hydrocarbons including the process, transportation and storage. In addition, the *Comisión Reguladora*

Van Syckel, very determined to lay his pipeline to Pithole in 1865, was personally the subject of much ridicule by his fellow oilmen, and his pipeline was attacked by teamsters. Their tune soon changed when the pipeline showed that it could deliver nearly 2000 BOPD to the Miller Farm terminal. Many teamsters were suddenly out of work, and the sceptical oilmen turned sheepish but some took up the pipeline business. All benefited in the end. Van Syckel showed the way.

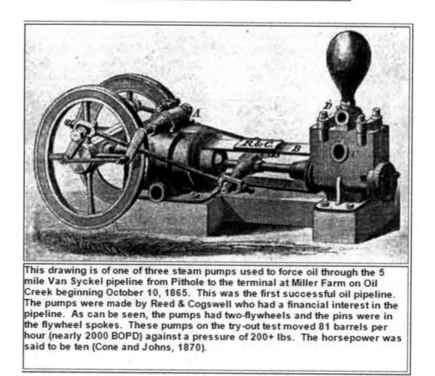

This drawing is of one of three steam pumps used to force oil through the 5 mile Van Syckel pipeline from Pithole to the terminal at Miller Farm on Oil Creek beginning October 10, 1865. This was the first successful oil pipeline. The pumps were made by Reed & Cogswell who had a financial interest in the pipeline. As can be seen, the pumps had two-flywheels and the pins were in the flywheel spokes. These pumps on the try-out test moved 81 barrels per hour (nearly 2000 BOPD) against a pressure of 200+ lbs. The horsepower was said to be ten (Cone and Johns, 1870).

FIG. 1.5 VAN SYCKEL (TOP) AND STEAM PUMP FROM THE 5 MILE VAN SYCKEL PIPELINE, 1865 (*Source:* Samuel T. Pees, Petroleum History Institute http://www.petroleumhistory.org/OilHistory/pages/Pipelines/van_syckel.html) [15]

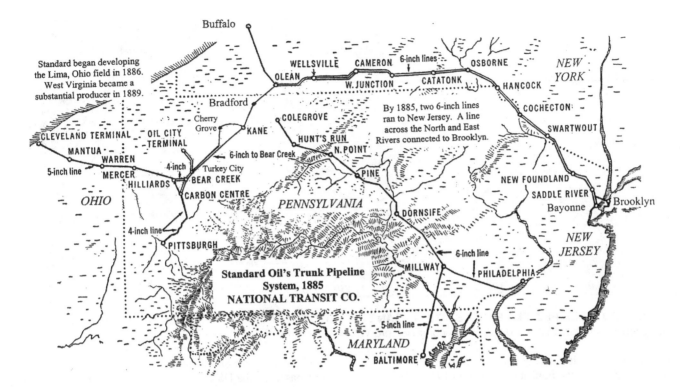

FIG. 1.6 STANDARD OIL'S TRUNK PIPELINE SYSTEM, 1885 NATIONAL TRANSIT CO'S, USA (*Source:* McElwee, N., 2007, The National Transit Co., Standard Oil's Great Pipeline Company, Oil Creek Press, USA)

FIG. 1.7 RUSSIA'S EARLY OIL PIPELINE BALAKHANI—BLACK CITY (*Source:* Wikimedia Commons http://museum.rosneft.ru/past/chrono/year/1879)

de Energía (CRE) was created to regulate gas, refined products, and products derived from hydrocarbons and electricity [21].

1.3.5 Iran

In 1908, the first commercial oil well sprouted 15 m high in *Masjid-i-Suleiman*. Subsequently, the first pipeline was built to transport this oil to Abadan near the Persian Gulf coastline in 1911.

In 1911, the Iranian Parliament granted to a Russian/later English-established company the construction of the second Iranian

oil pipeline transporting the oil for a length of 13.5 km, which was received by ship at the small port of Bijar Khaleh, Caspian Sea and taken to Rasht, the market [22]. More pipelines continued to be built from 1921 as documented by the Iranian Oil Pipeline and Telecommunication Company [23].

In 1951, Iranian parliament nationalized the reservoirs and infra-structure of the oil company; and later in 1954, the assets were shared with international oil companies to allow Iranian oil to be brought back to the international market. The Iran Ministry of

FIG. 1.8 CERRO DE LA PEZ, EL ÉBANO, STATE OF SAN LUIS POTOSÍ (LEFT) AND PIPELINES GOING TO TAMPICO (RIGHT) (*Source:* Courtesy of Manuel Aguirre, http://mexicomaxico.org/Voto/pemex.htm)

Energy founded in 1975 is in charge of the regulation of energy and roles have changed after the Iranian revolution in 1979 [24].

1.3.6 Colombia

On 6 March 1926, the Andean National Corporation Limited in Colombia, completed the construction of the first crude oil pipeline from De Mares Concession wells in Barrancabermeja to the Mamonal Port in Cartagena on the Atlantic Coastline. These assets were consolidated into the International Petroleum Company (IPC) from the ownership of Andean (pipeline) and Tropical Oil's assets including Barrancabermeja and El Centro, Santander [25].

In 1926, the South American Gulf Company (SAGOC) owned by the Mellon family from Pennsylvania purchased the Colombian Petroleum Company (COLPET) becoming 75.3% SAGOC, 23.7% the Carib Syndicate and 1% Barco's. In 1927, the Colombian government voided the concession purchase due to the non-fulfillment of the existing contract and hydrocarbons law by the previous COLPET's owners. However, the SAGOC and the COLPET

obtained the rights on the concession for 50 years from 1931 till 1981 based on the US National City Bank loan conditional upon the Colombian government recognizing the right of SAGOC + COLPET company over the Barco Concession. In 1935, SAGOC sold COLPET's shares to Texaco and Mobil and kept its participation in the consortium [26–28].

In 1938–1939, SAGOC built the NPS 12 pipeline from the Barco Concession crude oil wells in Petrolea (illustrated in Figure 1.9) near to Tibú in the Bari Tribe territory, Norte Santander to Coveñas Terminal, Sucre on Atlantic Coast. Years later, the pipeline Right-Of-Way (ROW) became the road between Tibú-Orú-Convención, Norte Santander to Ayacucho, Cesar.

In 1951, the De Mares concession reverted from the International Petroleum Company (i.e., Tropical Oil) to the Colombian government, which created the *Empresa Colombiana de Petroleos (ECOPETROL)* in 1948 to represent the national interests on hydrocarbons. Shell and Texaco-Mobil continued the exploration and production in the country.

FIG. 1.9 323.85 MM (NPS 12) PETROLEA—COVEÑAS SAGOC PIPELINE (LEFT) AND PETROLEA CAMP AND PUMP STATION IN 1938 (RIGHT) (*Source:* Época de la SAGOC, La Historia de Coveñas, Gabriel Moré Sierra)

In 1994, the Colombian energy regulator, *Comisión de Regulación de Energía y Gas (CREG)*, was created to regulate electricity, natural gas and liquefied petroleum gas with a strong focus on the engineering and technical aspects. In 2003, the *Agencia Nacional de Hidrocarburos (ANH)* is created to strategically manage the hydrocarbon resources associated to the national supply and to ensure 50–60% royalties out of the foreign exploitation [29].

1.3.7 Saudi Arabia

In March 1938, the first oil well was discovered in Dammam Well No. 7, illustrated in Figure 1.10, starting to produce 1,585 bpd at nearly 1.5 km of depth, in the Eastern Province of Saudi Arabia. As illustrated in Figure 1.10, the first major pipeline was the Trans-Arabian Pipeline (TAP) with 1,212 kilometer in length and 760 mm (30 inch) in diameter that was built between 1947 and 1950 from Qaisumah in Saudi Arabia to Sidon in Lebanon. The pipeline transported the crude oil to the Mediterranean Sea avoiding the Persian Gulf and Suez Canal. A gathering system fed the TAP pipeline from Abqaid in the southeast of the country coastline with the Persian Gulf [30, 31].

In 1981, the two (2) East-West Pipelines were built for transporting natural gas liquids and crude oil linking Eastern Province fields at Abqaiq (same starting point for the TAP) near the Dhahran metropolitan area with Yanbu on the west coast.

1.3.8 Argentina

In 1887, the *Compañía Mendocina de Petróleo* drilled 20 wells of which four became productive near Mendoza, and got abandoned 10 years later. In 1907, a major oil discovery (illustrated in Figure 1.11) was found nearby the Comodoro Rivadavia town, which was initially exploited by the government company *Yacimientos Petrolíferos Fiscales* (YPF) [32, 33].

In 1929, YPF announced the government decision to control the fuels prices, which were established by international production companies alleged to be operating under the *Achnacarry Agreement* for setting production quotas outside of the USA. In 1946, Argentina's government created its gas company (*Dirección Nacional de Gas del Estado*) following YPF advice to create an independent government company for transporting and commercializing gas across Argentina.

In 1949, *Gas del Estado* built the first major gas transmission pipeline running for 1605 kilometers from Comodoro Rivadavia to Llavallol, Buenos Aires [34]. In 1992, *Gas del Estado* was privatized and a regulator was created, "*Ente Nacional Regulador del Gas (ENARGAS)*", to ensure adequate pipeline maintenance and prices for the end consumers [35].

1.3.9 Brazil

In 1897, the farmer Eugenio Ferreira de Camargo drilled the first oil well in Brazil, but the production was very little. Brazil imported all hydrocarbons until the 1930s, but a visionary, Monteiro Lobato, started the national oil and gas exploration with the *Companhia Petróleos do Brasil* leading the first gasoil or LPG production in 1936 at *Riacho Doce* [36, 37].

In 1939, the federal government represented by the Agriculture Ministry announced the first oil reservoir at the Recôncavo Baiano. Then, Brazil started nationalizing subsoil resources creating the *Conselho Nacional do Petróleo (CNP)* and the hydrocarbons law for directly managing the exploitation, production, transportation, distribution, imports, and exports of hydrocarbons and refined products.

In 1951–1952, *CNP* built the first major transmission pipelines between Santos-Jundiaí-Sao Paulo for transporting gasoline, diesel, kerosene and fuel oil. In 1953, *Petróleo Brasileiro S.A.* (Petrobras) was created introducing the government management of all hydrocarbon activities, which would be modified in 1997 to open opportunities for private investments. This opening to industry was followed by the

FIG. 1.10 DAMMAM NO. 7 OIL WELL (LEFT) AND TRANS-ARABIAN PIPELINE (RIGHT) (*Source:* Wikimedia Commons)

FIG. 1.11   COMODORO RIVADAVIA (CR) OIL WELLS ON THE ATLANTIC COASTLINE (*Source:* Wikimedia Commons)

creation of a regulator named as the *Agência Nacional do Petróleo (ANP)* for overseeing hydrocarbons, gas and biofuels [38, 39].

1.3.10   Europe

Even though the oil-bearing sands (i.e., oil sands) were discovered in 1856 by farmer Peter Reimers, the first European oil well was drilled at Wietze, Germany in 1859, illustrated in Figure 1.12 [40, 41].

In 1939, the first UK commercial oilfield was located at Dukes Wood and Eakring, UK. Eakring oil well No. 1 first sprouted and the oil that was transported by rail to Edinburg, Scotland. Eakring personnel, drilling innovations (e.g., slant, water injection, jet powered-drill using lubricant mud) and training were used for the following major discoveries in the North Sea. The same year at the beginning of World War II (WWII), the UK established the *Government Pipeline and Storage System (GPSS)* to provide a secure hydrocarbon distribution network [42, 43].

During WWII, the American military engineers built a pipeline from the Port of Cherbourg, France, where ships offloaded hydrocarbon products at a floating dock, to Rhine River, Germany. This pipeline provided gasoline daily to General Patton's Third Army mainly following the Red Ball highway, but avoiding some cities and major ammo depots. This pipeline experienced multiple sabotages from Nazi sympathizers causing gasoline distribution problems delaying the war's end.

In 1953, the first European oil transmission pipeline was built from between Brunsbüttel and Hemmingstedt in the North of Germany. In 1959, the second European transmission pipeline, the North-West Oil Pipeline (NWO), was built to receive the imported oil at Wilhelmshaven Port to be transported either to Wesseling near Cologne, or to Hamburg, Germany. During the following years, the Rotterdam-Rhine Pipeline (RRP) was built and extended to the Frankfurt area. In 1953, the first commercial European products pipeline was completed from Havre to Paris, France [44, 45].

FIG. 1.12   WIETZE OIL WELLS (*Source:* Grantville Gazette, Volume 23, 1 May 2009 by Jeff Corwith https://grantvillegazette.com/wp/article/publish-303/)

In 1959, natural gas was discovered at the Groningen field in the Netherlands, followed by the first discoveries in the UK sector of the North Sea (e.g., Forties oil field). In the 1970s, substantial reservoirs of gas were discovered in Norway, which were mainly developed for exporting to Europe [46].

In 1974, the UK Health and Safety Executive (HSE) was created to encourage, regulate and enforce workplace health, safety and welfare, and for research into occupational risks in England, Wales, and Scotland.

1.3.11   China

In 1958, the first large oil field was discovered in Daqing, Heilongjiang Province by Li Siguang and drilled by Wang Jinxi with the 1205 drilling team [47]. China's first oil transmission pipeline from Karamay to Dushanzi was built in 1958.

The train (Sinotrans) was not able to move out all oil from the Daqing field, which affected the production and the national economy. Then, the State Council and the Military, the Ministry of Chemical Industry, Shenyang fuel CMC in conjunction with the three (3) northeastern provinces organized the construction of the Northeast pipeline construction called *"eight three" (83 battle) pipeline*. This pipeline was the first long distance oil

pipeline constructed from 1970 to 1975 with a length of 667 km [48, 49].

1.3.12   Australia

In 1953, the first high grade crude oil well was discovered in Rough Range, Western Australia near Perth. However, the first hydrocarbon transmission pipeline did not get built until 1963 from Moonie oil field to Brisbane, illustrated in Figure 1.13 [50].

In the 1960s, the first commercial gas reservoirs were found in the Surat Basin in Queensland followed by four (4) discoveries: the Gidgealpa-2 well in the Cooper Basin in South Australia, Perth Basin in Western Australia, Amadeus Basin in Northern Territory and Gippsland Basin in Victoria [51].

In 1969, a pipeline transported natural gas from the Surat Basin to Brisbane in Queensland on the eastern coastline. The same year, pipelines were built from Gippsland Basin to Melbourne and from Cooper Basin to Adelaide. Longer pipelines were built from Perth Basin to Perth and from Cooper Basin to Sydney in 1971 and 1976, respectively [52].

In the early days, pipeline transportation was mainly owned and operated by the state and federal governments. In 1993, Professor Fred Hilmer, Dean of the Australian Graduate School of

FIG. 1.13   THE 1963 MOONIE-BRISBANE PIPELINE, APRIL 2005 (*Source:* Barry Wood and Roger Woodman, Great Southern Press)

Management, University of New South Wales proposed a national competition policy removing restrictive practices.

In 1995, the Australia government created the Australian Competition and Consumer Commission (ACCC) as independent authority of to protect consumer's rights, business' rights and obligations, to develop industry regulation and price monitoring, and to prevent illegal anti-competitive behavior. In 2005, the government created the Australian Energy Regulator (AER) under ACCC to regulate the transmission network of electricity and gas.

1.3.13 Peru

In November 1863, Peru drilled the first crude oil well at a 24-m depth in Zorritos, Tumbes at the North coastline. In 1969, the Peruvian law and hydrocarbons strategy changed making the government, via PetroPerú, the exclusive owner and operator of all hydrocarbon production wells, refineries, pipelines and commercialization excepting exploration and exploitation activities. PetroPerú took over the reservoirs and refineries owned by International Petroleum Company (IPC) [53, 54].

During 1975–1977, PetroPerú built the first crude oil pipeline system named NorPeruano from Estación 1 at San Jose de Saramuro to the Bayovar Terminal along the Pacific coastline. In 1978, PetroPerú built a North branch from the Andoas station to connect to the Estación 5 of the NorPeruano pipeline system [55, 56].

In 1993, the government enacted the privatization of PetroPeru's assets and activities creating PeruPetro to represent the state in contracting matters. However, PetroPeru did not disappear as the purpose of privatization did not get achieved due to the lack of competition in keeping the fuel prices self-regulated [56].

In 1996, the Peruvian regulator *Organismo Supervisor de la Inversión en Energía* (OSINERG) was created to supervise the Electricity and Hydrocarbon industry ensuring that a permanent, reliable, safe and quality service. Later on in 2007, OSINERG became OSINERGMIN by increasing its mandate to regulate the mining industry [57].

1.4 PIPELINE INTEGRITY MANAGEMENT ERAS: HISTORY AND EVOLUTION

This section walks you through the history and evolution of Integrity Management (IM) starting with pipeline maintenance developing into IMP and then expanding to become PIMS. These eras were mostly started by either the occurrence of serious incidents causing fatalities and/or damage to environment or regulatory and industry initiatives to prevent incidents. The pipeline incidents typically produced national or international news influencing stakeholders to request regulators and companies to improve pipeline integrity performance. They were most of the time followed by regulatory changes, advocate groups and industry initiatives for improvement.

However, pipelines are a safer form of transportation [58–62], and transport most of the oil and gas around the world today, Figure 1.14 [63].

Some of the existing pipelines around the world are over 40 years old, and can be expected to continue to operate for many more years, some reservoir regions still have over 50 years from today (2016) of proven, recoverable reserves of oil and gas. This creates the need for using an effective integrity management of the existing pipelines and its eventual replacement ultimately driven by societal risk. This raises questions about the "design life" of a pipeline system. Pipeline design standards rarely specify a design life for a pipeline, as this life is dependent more on operation rather than design. ASME B31.4 and B31.8 both define the design life as:

'*a period of time used in design calculations, selected for the purpose of verifying that a replaceable or permanent component is suitable for the anticipated period of service. Design life [does/may] not pertain to the life of the pipeline system because a properly maintained and protected pipeline system can provide … service indefinitely.*'

As pipelines age, they are faced with many threats such as corrosion and fatigue. These threats can lead to failures. Figure 1.15 summarizes these causes in European gas pipelines. Similar causes are reported in the USA [64].

Some of the regulatory responses to pipeline companies experiencing failures were in the form of severe orders or corrective actions including operational restrictions, administrative monetary penalties and sometimes prison sentences. Societal responses impacted the legislative and regulatory framework of the pipeline life cycle (e.g., design, manufacturing, construction, operation, deactivation/reactivation and abandonment) increasing approval and integrity requirements as wells as easier public access and participation to pipeline company information and applications for new and modified pipeline systems. Other societal or public responses were the birth and growth of pipeline safety advocacy groups or associations of incident-affected people, higher risk aversion to pipeline incidents, and more research and development focus on integrity technologies, processes and methodologies.

These events induced a change in the approach to pipeline integrity going from probability-driven to consequence-driven, and then to risk-driven (a 3-part approach: probability, consequence and risk). Later on, integrity management programs evolved from being led by technical management to an all inclusive management system (MS). MS emphasizes the need for senior management

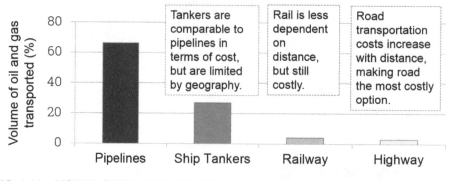

FIG. 1.14 MODES OF TRANSPORTATION FOR OIL AND GAS (*Source:* Phil Hopkins)

PIPELINE INTEGRITY MANAGEMENT SYSTEMS – A PRACTICAL APPROACH • 15

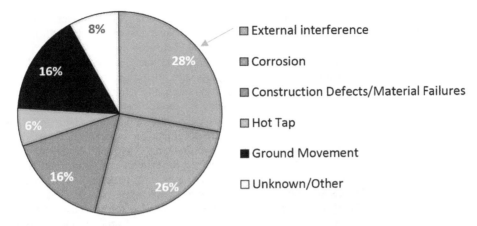

FIG. 1.15 CAUSES OF GAS PIPELINE FAILURES IN EUROPE (*Source:* Phil Hopkins)

commitment and responsibility with multiple departmental supports (e.g., integrity, operation and maintenance, supply, human resources and legal) to the integrity management program.

Three (3) distinctive eras can be identified that have started in different times around the globe, as follows:

1. Pipeline Maintenance Era: *Focus on Criticality*
2. Pipeline Integrity Management Program Era: *Focus on Consequence and Risk*
3. Pipeline Integrity Management System Era: *Focus on Senior Commitment and Effectiveness*

1.4.1 Pipeline Maintenance Era: *Focus on Criticality*

From the pipeline perspective, the maintenance era focused on pipeline anomaly criticality to retain or restore the pipeline transportation function driven by business continuity. The term *Safety* focused on employees within the working areas. Integrity of the pipeline accounted for public safety and environmental consequence and risk prioritization was at infancy stages. The main focus was criticality mitigating identified integrity threats.

Consequence of pipelines was considered for emergency response purposes, but not as a driver for developing pipeline mitigation, prevention and monitoring programs. In the late 1990s deaths caused by pipeline spill fires and explosions (e.g., Bellingham, Rapid City) and repeated pipeline failures with no clear understanding of the failure mechanism (e.g., Stress Corrosion Cracking) caught the public, media and regulatory attention creating a call for action and demand for a change. These incidents questioned the need for pipeline maintenance to focus on consequence.

This maintenance era developed methodologies for corrective, preventative and predictive maintenance establishing performance indicators such as maintainability, availability and cost efficiency for tracking and improvement. Advanced repair methods for pipelines such as stopples and packers were developed within this era minimizing business interruption. In-line inspection (ILI) technologies for corrosion, geometry and cracking were under development and improved as they were used. Improvement of sensor resolution and data interpretation was the focus for mainly bigger diameter pipelines.

1.4.1.1 Changes Driven by Stakeholder Agreement In 1994, some Canadian regulators, pipeline industry associations and Canadian Standard Association members formed the *Pipeline Risk*

Assessment Steering Committee (PRASC) to guide the development of a process to determine and manage the risk of pipeline operations to the general public. The PRASC membership was comprised of the National Energy Board (NEB), the Alberta Energy and Utilities Board-EUB- (named changed to Alberta Energy Regulator), the Transportation Safety Board of Canada (TSB), Canadian Standards Association (CSA) committee members, the Canadian Gas Association (CGA), the Canadian Association of Petroleum Producers (CAPP), and the Canadian Energy Pipeline Association (CEPA).

In 1996, PRASC achieved the development of the Risk Assessment guidelines for incorporation into CSA-Z662 as an Appendix. This event marked the beginning of a formalized and publically available risk-based approach for pipelines.

1.4.1.2 Changes Driven by Pipeline Failures In 1944, one major turning point for the operation and maintenance legislation of gas pipelines in United States came with the interstate pipeline rupture near a densely populated apartment community in Edison, NJ (23 March 1994). The failure was due to cracking within a gouged dent exposed to fluctuating operating stresses. The explosion and fire destroyed eight (8) apartment buildings and caused the evacuation of approximately 1,500 residents; one person died of a heart attack. The gas in the pipeline continued fueling the fire for over 2 hours due to difficulties in isolating the pipeline [65–68].

NTSB recommended the following areas for the pipeline industry and DOT OPS to address in maintaining the integrity of USA's pipeline network:

- Steel pipeline toughness properties
- Pipeline marking
- Surveillance procedures
- Damage prevention programs
- Rapid detection and shutdown
- Internal inspection and
- Land use management

For hazardous liquids, multiple pipeline failures causing environmental damage occurred between 1994 and 1997 such as St Helena Parish, LA (1997); Lodge Grass, MT; Athens, Georgia (1997); Mount Morris, IL (1997); Hammond, IN (1996); Donaldson, MN (1996); Reedy River, SC (1996); Gramercy, LA (1996); Plover, WI (1994) and San Jacinto River, TX (1994) [69].

In October 1996, USA President Bill Clinton signed the Accountable Pipeline Safety and Partnership Act, which adopted performance-based standards for gas pipelines.

In 1997, the Title 49 Code of Federal Regulations (CFR) Part 195—Transportation of Hazardous Liquids by Pipeline described Operation and Maintenance requirements. Those maintenance-based requirements covered elements such as Procedural Manual for Operation, Maintenance and Emergencies; Training, Maps and Records; Maximum Operating Pressure; Inspection of Right-Of-Way and crossings under navigable waters, cathodic protection; external and internal corrosion control; valve maintenance; repairs; pipe movement; overpressure safety devices; public education and damage prevention programs. However, regulations and industry standards did not contain an specific Integrity Management section accounting for consequences and risk.

1.4.1.3 Changes Driven by Geotechnical Failures In 1987, the end of the construction of a NPS 18/20/24 pipeline in Northern Colombia from the Caño Limón reservoir going through the Eastern branch (i.e., Arauca and Catatumbo basins) of the Andes Mountains up to the Coveñas Caribbean Coastline started a comprehensive maintenance approach era. This era was characterized by the business driver to transport a high volume of crude oil (i.e., 240,000 Bbl/day) requiring the maintenance of the Right-Of-Way (ROW) with high potential of interrupting the pipeline operation due to geotechnical and social unrest.

The root-cause analysis of the significant geohazard impacts on the pipeline identified that the design and construction state-of-the-art of the time did not have an understanding of geotechnical response after construction and the need for integrated practices between pipeline construction and operation for managing the ROW. These factors contributed to and increased the geohazard magnitude and severity affecting the pipeline and environment. However, increased financial support and operations awareness stimulated advances in engineering design and techniques for maintaining the significant ROW erosion, landslides, water crossing scour and deforestation over the following decade.

Colombian pipeline industry drove the creation of publications focused on geotechnical maintenance for pipelines (i.e., Garcia-Lopez, Manuel; Amortegui, Jose V.) influencing other South American countries capturing and sharing their geotechnical knowledge and practices.

1.4.2 Pipeline Integrity Management Program (IMP) Era: *Focus on Consequence and Risk*

The Integrity Management Program (IMP) era was characterized by a heightened focus on consequence triggering regulations, regulatory enforcement and monetary or service penalties. Requirements in the USA were segmented based on consequence (i.e., High Consequence Areas) defined in terms of population (i.e., density and concentration) and specific environment affectation (i.e., drinking water, commercially navigable ways).

This era promoted the development and improvement of methodologies and technologies for assessing the integrity of pipelines: In-Line Inspection (ILI), Hydrostatic Testing and Direct Assessment (DA). For piggable pipelines, in-line inspection technology improved on detecting, characterizing and sizing threats with high frequency incidents such as dents (e.g., third party damage), corrosion and cracking (i.e., SCC and seam weld).

For non-piggable pipelines, hydrostatic testing became an alternative to verify the integrity of the pipeline through strength and leak testing for the short-term. However, a special hydrostatic test at 100–105% SMYS test pressure for ½ hour, named the "Spike Test" by Canadians, extended the life of pipelines with cracking by delaying their growth (e.g., crack tip "plastification" or blunting).

For other non-piggable pipelines that were not amenable to hydrostatic testing due to the introduction of incompatible liquid test medium, the Direct Assessment methodology was developed using susceptibility factors and protection results (e.g., cathodic protection, stagnant water, coating and soil type). DA methods were developed for external and internal corrosion, and stress corrosion cracking. The methodology required a field validation for adjusting the criteria as well as iterative investigation to develop a final mitigation plan and DA reassessment interval.

Consequence assessment methods were developed in both simplified and detailed forms to estimate the extent of vapor clouds, thermal radiation (if ignited) and overpressure (i.e., blast) for gas as well as liquids on land and water bodies accounting for diverse types of weather, toxicity, flammability and volatility. Consequence assessments contributed to emergency response planning, selecting valve type and location for minimizing consequences and identification of areas for earlier mitigation schedule, public awareness (i.e. prevention measures) and surveillance (i.e. monitoring measures).

The IMP era formalized the use (and requirement) of risk assessment in the USA for prioritizing integrity assessments, mitigative and preventative measures. The beginning of risk assessment for pipelines started with a qualitative approach based on indexed attributes and scores highly dependent on the historical knowledge of operation and maintenance personnel and Subject Matter Experts (SME).

The utilization of in-line inspection data and methodologies for probability calculation as well as the introduction of data management platforms such as Geographical Information Systems (GIS) provided a discrete and integrated data for better quantification of criticality. These changes allowed for semi-quantitative risk assessments that better represent risk and support more defendable and assertive decisions.

1.4.2.1 North America
1.4.2.1.1 Canada. As illustrated in Figures 1.16 through 1.18, Canada's mid to late 1990s, pipeline operating companies experienced challenges from one common threat named Stress Corrosion Cracking (SCC) causing pipeline failures such as Rainbow, AB (1993); Virginia Hills, AB (1994); Rapid City, MB (1995); Vermillion Bay, ON (1995); Summit Lake, BC (1995) and Glenavon, SK (1996) triggering a regulatory and industry framework about Stress Corrosion Cracking (SCC) integrity management [70].

In August 1995, the National Energy Board (NEB) of Canada initiated a public inquiry about SCC due to the multiple failures; revealing the need for better understanding of how to manage SCC in its early stages. As a result of the findings of the 1996 Inquiry on Stress Corrosion Cracking on Canadian Oil and Gas Pipelines (the Inquiry Report on SCC), the NEB took the initiative to require companies to proactively identify, mitigate and prevent the SCC hazard using a programmatic approach.

The inquiry identified the following areas that needed to be addressed:

- Implementation of an SCC management program by each pipeline company;
- Changes to the design of pipelines;
- Continued research;
- Establishment of an SCC database;
- Improved emergency response practices; and
- Continued information sharing.

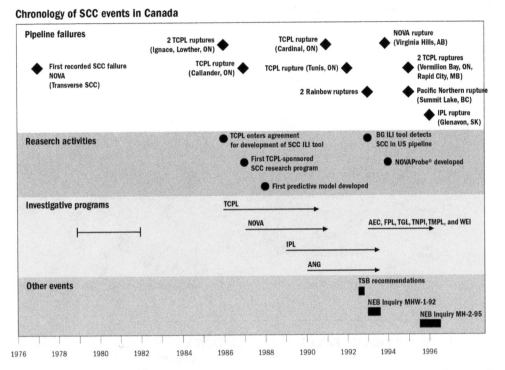

FIG. 1.16 CHRONOLOGY OF SCC EVENTS IN CANADA FROM 1976 TO 1996 (*Source:* National Energy Board, 1996, Public inquiry concerning stress corrosion cracking on Canadian oil and gas pipelines: MH-2-95)

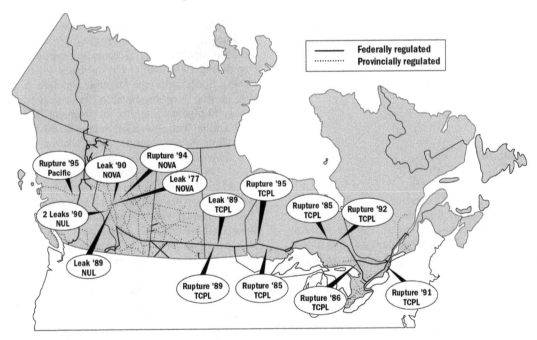

FIG. 1.17 CANADIAN NATURAL GAS PIPELINE SCC FAILURES BY LOCATION BEFORE 1996 (*Source:* National Energy Board, 1996, Public inquiry concerning stress corrosion cracking on Canadian oil and gas pipelines: MH-2-95)

Recommendation 6-1 of the Inquiry Report on SCC stated that "… *the Board requires each pipeline company to develop and implement an SCC management program by 30 June 1997*". Recommendations 6-2 to 6-7 provided further recommendations related to accountability, scope, consideration of consequence and probability of failure when prioritizing activities for identification, mitigation and prevention of SCC. In addition, the NEB was tasked with auditing documentation for SCC management programs. The above recommendations contain all of the essential elements of a management program or system (Plan, Do, Check, Act) that have

Location of SCC failures on natural gas pipelines

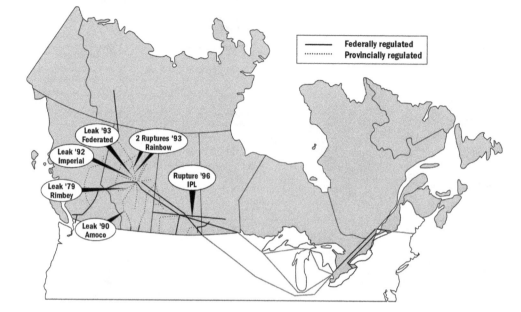

FIG. 1.18   CANADIAN LIQUIDS PIPELINE SCC FAILURES BY LOCATION BEFORE 1996 (*Source:* National Energy Board, 1996, Public inquiry concerning stress corrosion cracking on Canadian oil and gas pipelines: MH-2-95)

been used to achieve the goal of reducing the probability of failures attributed to SCC.

In May 1997, the Canadian Energy Pipeline Association (CEPA) published the first edition of Stress Corrosion Cracking Recommended Practices (CEPA SCC RP) in response to the 1996 NEB continued information sharing requirement. The CEPA SCC RP compiled guidelines to develop and implement SCC integrity management programs including:

- SCC Integrity Management Program
- Field program development
- Inspection
- Data collection
- Integrity assessment
- Prevention and Mitigation
- Risk assessment

Between 1996 and 1997, PRASC contributed to the issuance of the *Major Industrial Accidents Council of Canada* (MIACC) guidelines for risk acceptance criteria based on land use. Prior to the dissolution of MIACC in 1999, the document was transferred to and later on updated by the Process Safety Management (PSM) division of the Chemical Institute of Canada/Canadian Society for Chemical Engineering (CSChE) becoming "Risk Assessment—Recommended Practices for Municipalities and Industry" [71]. In 2008, the PSM reviewed the *Allowable Land Uses* based on experience and developments in land use planning recommending individual risk acceptance criteria per land use described in Figure 1.19.

In 2005, the Canadian Standard Association (CSA) introduced the Guidelines for Pipeline Integrity Management Programs in Annex N (non-mandatory), which would be adopted by the following provincial regulators in the subsequent years:

- British Columbia Oil and Gas Commission (BC OGC), Directive 2011-01;

- Alberta Energy Regulator (formerly ERCB), Directive 77-2006;
- Technical Standards and Safety Authority (TSSA)—Code Adoption Amendment for Ontario.

1.4.2.1.2 United States of America. For hazardous liquids, the Integrity Management Program era was highly influenced by multiple pipeline failures causing environmental damage occurring during 1999 such as Crystal Falls, MI (1999); Knoxville, TN (1999). Specifically, the Olympic Pipeline failure in Bellingham, WA causing 3 deaths became the critical point for a significant change in pipeline integrity management. The change started with the National Transportation Safety Board findings from the pipeline failure investigation led by its chairman Jim Hall.

Bellingham, Washington Olympic (Products) Pipeline Rupture [72].

At 3:28 pm Pacific Daylight Time (PDT) on 10 June 1999, a 406.40 mm (NPS 16) Olympic products pipeline ruptured within the Ferndale-Allen section, and 1 ½ hours later, ignited and burned 2.4 kilometers (1 ½ miles) along the pipeline sending a fireball down Whatcom Creek, illustrated in Figure 1.20. One 18-year-old and two 10-year-old people were killed by the explosion. Eight (8) additional injuries were documented and a single-family residence and the City of Bellingham's water treatment plan were severely damaged. The pipeline leaked approximate 900,000 liters (237,000 gal) of gasoline into the Hanna and Whatcom Creeks within the City of Bellingham, Washington state (WA). The failure is illustrated in Figure 1.20.

On 18 June 1999, US DOT Office of Pipeline Safety (OPS) issued a Corrective Action Order (CAO) restricting Olympic Pipeline operations of the Ferndale–Allen section and specifying conditions to be met before returning to normal operations. The CAO required reducing the maximum operating pressure to 80% of the normal pressure or the surge pressure at the failure point, whichever was lower. The CAO also required reviewing, fixing and monitoring the SCADA as well as procedures for abnormal conditions and testing of valves within HCA, installing a check valve; adding remotely-operated valves for consequence reduction,

Annual Individual Risk
Chance of fatality per year

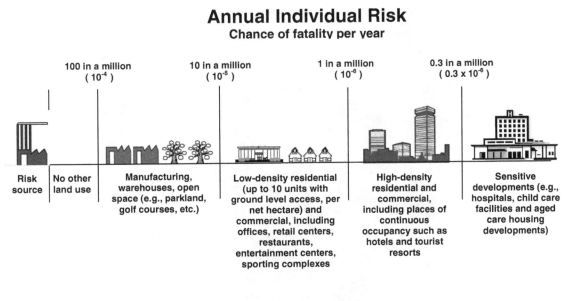

Allowable Land Uses

FIG. 1.19 RISK-BASED LAND USE PLANNING GUIDELINES—RECOMMENDATION (*Source:* Chemical Institute of Canada/ Canadian Society for Chemical Engineering—Process Safety Management)

cathodic protection review and scheduling for the remediation, pressure testing, ILI and repairs.

On 27 October 1999, Jim Hall testified about the progress on the Bellingham explosion before the House of Representatives. Mr. Hall expressed the following in regards to the Regulatory framework: *"Ultimately, RSPA and OPS are answerable for the regulatory context in which this pipeline company operates and in which this accident occurred. NTSB has, for many years, argued that periodic verification of pipeline integrity must be a requirement of service. Internal inspections done in the Bellingham pipe identified anomalies in the area that ultimately failed, but that inspection data produced no change and the regulatory processes did not require a correction. Office of Pipeline Safety (OPS)'s response to our (NTSB) 1987 recommendation is in an "unacceptable" status. Mandatory internal detection programs are needed to protect the public and industry alike and federal action is long overdue"* [73].

The NTSB report identified the following probable causes of the pipeline rupture:

1. Pipe damage made (gouged deformation) by a General Construction company (excavator: high-chromium bucket) during the 1994 Dakin-Yew water treatment plan modification project and Olympic pipeline company's inadequate supervision during the project construction
2. Inaccurate evaluation of 1996/1997 in-line inspection results (i.e., 23% wall loss—possible mill/mechanical defect and wrinkle bend) and data integration (i.e., new foreign installations) leading the company to not excavating the pipe damage
3. Failure to test, under operating conditions (e.g., upstream control valve closure), all safety devices (i.e., relief valve for appropriate pressure) associated with the Bayview product facilities before activating them
4. Failure to investigate and correct repeated unintended valve closure events including not developing procedures for preparing controllers to the operational challenges

5. Slow or non-response of the SCADA system did not allow for a timely sequential shutdown during the pipeline overpressure caused by the valve closure and upstream active pumps

On 1 December 2000, the Department of Transportation (DOT) Research and Special Program Administration (RSPA) issued the first Integrity Management Program (e.g., Title 49 CFR 195.452) regulation marking a historical milestone for pipelines with 500 or more miles of pipeline focusing on High Consequence Areas (HCA).

Thus, the integrity management program era for hazardous liquid pipelines started with a systematic and formalized process for identifying, assessing, mitigating, monitoring and preventing risks focusing on areas of high consequence to the public and environment.

Carlsbad, New Mexico Gas Transmission Pipeline Rupture [74].

At 5:26 am Mountain Daylight Time (MDT) on 19 August 2000, a 762-mm (NPS 30) diameter gas transmission pipeline (line 1103) ruptured and subsequently exploded. The failure resulted in killing 12 people camping under a concrete-decked steel bridge on the east side the Pecos River crossing, destroying three (3) vehicles and affecting the gas supply to California. The failure is illustrated in Figure 1.21.

On 20 June 2001, US DOT RSPA issued a Notice of Probable Violation (NPV), Proposed Civil Penalty and Proposed Compliance Order that defined requirements that can be summarized at a high level in two (2) groups, as follows [75]:

Personnel Qualifications:

- Qualification plan related to Covered Tasks (i.e., Internal Corrosion for operations and maintenance)
- Completion of qualification of individuals who performed those Covered Tasks
- An evaluation method that does not rely solely on work performance history

FIG. 1.20 THE 406.40-MM (NPS 16) PRODUCTS PIPELINE RUPTURE AND FIRE IN BELLINGHAM, WA (*Source:* National Transportation Safety Board (NTSB), USA Report PAR-02/02)

Accuracy of and Access to Drawings:

- Review of pipeline elevation profile drawings identifying low points (i.e., internal corrosion susceptibility)
- Review of practices for drawing updates
- Process to make updated drawings available to corrosion personnel

On 6 August 2002, RSPA/OPS finalized the definition of High Consequence Area for gas transmission pipelines based on Class Location designations 3 and 4 as well as "*identified sites*" within the *Potential Impact Circle* (PIC) defined as any facility housing persons of limited mobility (e.g. nursery homes, hospitals), concentration of people in outdoor areas (e.g. campgrounds, recreational areas), and buildings (e.g. schools, churches) within a specified frequency (i.e., days per week and/or year) [76].

On 17 December 2002, the Pipeline Safety Improvement Act became law after being passed by Congress as a response to the number of significant incidents occurred and setting forth requirements to prevent pipeline incidents with large consequence. This law requires "RSPA/OPS to issue regulations, not later than December 17, 2003, prescribing standards for an operator's conduct of risk analysis and adoption and implementation of an integrity management program and for defining direct assessment". Furthermore, the Act defined minimum IMP requirements for gas transmission pipelines in HCAs as well as permitting a State interstate agent access to an operator's IMP.

On 11 February 2003, the NTSB identified the probable cause of the pipeline rupture as a significant reduction of the pipe wall thickness due to a large internal corrosion pitting associated with microbial influenced corrosion (MIC) at the bottom of pipe on a bend. The NTSB pipeline accident report identified as a finding that the absence of facilities and programs to run cleaning pigs for removing liquids and solids would likely have contributed to the severity of internal corrosion. Furthermore, the partial clogging of the "drip" upstream of the failed pipe contributed to some liquids bypassing the drip and accumulating at the ruptured site. On the contrary to what is expected, recent aerial and ground patrols prior to the incident did not identify evidence of leaking.

The National Transportation Safety Board (NTSB) investigation also identified major safety issues in the adequacy of the federal safety regulations and oversight. The regulations did not provide guidance to gas pipeline operators in mitigating internal corrosion. The Office of Pipeline Safety (OPS) oversight did not conduct accurate pre-accident assessment of the company's internal corrosion control programs nor did identify the deficiencies in a timely manner.

Photo of ruptured El Paso Natural Gas Company pipeline in crater at Carlsbad, New Mexico

Photo #7 - camping area on east side of Pecos River

FIG. 1.21 THE 762 MM (NPS 30) GAS TRANSMISSION PIPELINE RUPTURE AND EXPLOSION IN CARLSBAD, NM (*Source:* National Transportation Safety Board (NTSB), USA Report PAR-03/01 Docket # 15291 NTSB Accident ID# DCA00MP009)

Carlsbad's accident influenced the USA Pipeline Integrity Management rule for gas transmission pipelines capturing the people concentration areas such as beaches, playgrounds, recreational facilities, campgrounds, outdoor theaters, stadiums into the "*identified site*" definition.

On 15 December 2003, the Department of Transportation (DOT) Research and Special Program Administration (RSPA) issued the title 49 CFR 192 Subpart O regulations marking the beginning of the Pipeline Integrity Management Program for gas transmission pipelines focusing on HCAs. The rule required gas transmission pipeline operators to perform ongoing (i.e. baseline and continual) assessment of pipeline integrity (e.g. ILI, Hydrotest, Direct Assessment), data gathering and integration, mitigative and preventative actions, and monitoring of IMP effectiveness to modify it for improvement. This rule required consensus standards to use threat-by-threat analysis for ascertaining adequate protection and was expected to provide coordinated risk control measures needed to improve pipeline safety (integrity). Gas gathering and distribution did not get included at this time, but the gas pipeline distribution

IMP rule would happen years later using a similar approach applicable to these types of pipelines.

1.4.2.2 South America In 1997, the 830 km (518 miles) NPS 16/30/36/42 Oleoducto Central S.A. (OCENSA) pipeline owned by Ecopetrol, BP, Total, Petrominerales and Triton started operation transporting crude oil from Cusiana to Morrosquillo Gulf on the Atlantic coastline. Enbridge, a former 24.7% owner of Ocensa, started operating the pipeline via a 100% owned entity called CITCol earning a fee for the transportation service. Enbridge's Canadian and USA experience on IMP, technology development and application on pipeline integrity introduced advances in Colombian practices for in-line inspection, defect and risk assessment beyond the cathodic protection/corrosion approach to pipelines. In the early 2000s, even though a regulation did not get created, an annual Pipeline Conference (i.e., Jornada Andina de Ductos) led by pipeline operators started capturing and sharing pipeline integrity practices associated with geohazards [77].

On 11 December 1998, Argentina experienced a tragic pipeline explosion in Mesitas near Salta City, Salta Province causing nine (9) deaths and one (1) maintenance worker's injury during a pipeline repair. ENARGAS, the regulator, issued multiple orders focusing on reducing the pipeline operating pressure, looking for alternatives for gas transportation to critical customers, and conducting a failure investigation to determine the origin of the ignition source for the explosion using independent and internationally-recognized parties. The failure investigation identified the root-cause to be external Microbial Influenced Corrosion (MIC) with a high growth rate initiated by a longitudinal coating defect and low level of cathodic protection in the area. A key contribution factor was the increase of the pipeline pressure increasing the propagation of the fracture causing the rupture [78].

In 2006, the Argentina Secretary of Energy approved the regulation 1460 *Reglamento Técnico de Transporte de Hidrocarburos Líquidos por Cañería* providing the direction for integrity management programs of pipelines excluding the production or concession areas transporting liquid hydrocarbons.

In the period of December 2004 to April 2007, Peru experienced six (6) failures in the Camisea Natural Gas Liquids (NGL) pipeline releasing products into the environment, but without injuries or deaths. The root-causes of the pipeline failure were time-delayed hydrogen-induced cracking at a girth weld grown by hydrostatic and in-service pressure cycles, unstable geotechnical conditions, landslide overload on the pipeline, and construction mechanical damage (i.e., dent) overloaded by flash floods that exposed the pipe due to riverbed scour [79].

In 2010, the Mexican Secretary of Energy issued the regulation *Administración de la Integridad de Ductos de Recolección y Transporte de Hidrocarburos* to develop and implement an Integrity Management Program. This regulation provides the guidance to determine the current and ongoing integrity condition of pipelines enabling timely detection, mitigation and prevention of the risk. This regulation is expected to increase safety and environmental protection by preventing incidents via both probability and consequence reduction [80].

1.4.2.3 Europe A United Kingdom (UK) Health and Safety Executive (HSE) origin dates back to 1833 with the creation of the first factory inspectors as the HM Factory Inspectorate [81]. In 1985, the explosion of a domestic gas pipe caused eight (8) fatalities triggering the recommendation for the British Gas Corporation to review its priorities for replacing cast iron gas pipelines.

In 1988, the Piper Alpha offshore platform experienced a series of catastrophic explosions causing 167 fatalities. This incident produced the Lord Cullen Inquiry recommending the need for an offshore installation regulation and appointing the HSE to enforce occupational health and safety in the offshore oil and gas industry. In 1991, the new regulation introduced a shift from prescriptive regulations to goal-setting regulations with a Safety Case System to demonstrate the control of major hazards and the use of a suitable management system.

1.4.3 Pipeline Integrity Management System (PIMS) Era: *Focus on Senior Commitment and Effectiveness*

The Integrity Management System era mainly came about from the realization of regulators in the UK and Canada of the need for a safety management system. This system required a clear and visible senior management commitment, leadership and accountability in conjunction with an organization with processes and procedures that were essential in preventing incidents. Management System operational principles described by ISO (i.e., plan, do, check and act) are life cycle processes that can be integrated with the IMP for achieving systemic, systematic and effective results.

1.4.3.1 North America In Canada, based on the success of the management of SCC (measured against the SCC management program audit results and a significant reduction in SCC failures following the 1996 inquiry), the NEB proceeded to revise its regulations. The revised regulations required companies to use a programmatic approach for all potential integrity threat (of which SCC was just one potential threat) on liquid or gas pipeline systems. After extensive consultation, the NEB revised its *Onshore Pipeline Regulations* (OPR) and in 1999 issued a new version (OPR-99) that required companies, in Section 40, to "… *develop a pipeline integrity management program*". The goal of the requirement is to "…*ensure that pipelines are suitable for continued safe, reliable and environmentally responsible service.*" The NEB also issued the companion Guidance Notes to the OPR-99 that explained how to achieve the goal of Section 40 employing a program and system approach.

On 26 May 1999, the Canadian Governor General in Council approved the first regulation made by the National Energy Board requiring pipeline companies to develop a pipeline integrity management program (IMP) via the Onshore Pipeline Regulations (NEB OPR-99) using a "goal-oriented" approach. The OPR-99 Guidance Notes described the four (4) essential elements of IMP, as follows [82]:

1. Management System
2. Records Management System
3. Condition Monitoring
4. Mitigation Program

Even though Management Systems were included in the regulations, the understanding of their benefits and implementation process would take almost a decade to be realized. This drove the regulators and pipeline industry to formalize Management Systems via CSA-Z662-mandatory clauses applicable for the entire pipeline cycle.

In 2012, the NEB and BC OGC regulators [83] published a paper at ASME IPC 2012 showing and explaining the trends of integrity management system audits that revealed that the planning and verification processes of the PIMS had improved from the 2001–2015 period to the 2006–2010 period.

In the USA, as of 2013, PHMSA regulations do not require a pipeline safety management system; however, the IMP regulations had introduced some management system elements such as Management of Change, Communication, Performance Plans, Records, and Personnel Qualifications. Furthermore, the USA Senate Committee on Commerce, Science and Transportation [84] has led congressional hearings bringing on as witnesses, executives of companies such as PG&E (incident: San Bruno, CA) and El Paso Energy Pipeline Group (incident: Carlsbad, NM) emphasizing the importance of senior management leadership and commitment, a key management system element for achieving effectiveness.

1.4.3.2 South America In 2007, Peru established the law "*Decreto Supremo # 081-2007-EM*" [85] defining the regulatory framework for managing the integrity of pipelines as well as improving the requirements for authorizing the operation of pipelines, landowner protection in ROW agreements and safety requirements for pipelines. This regulation introduced both IMP and Management System elements capturing the experience from the USA, European and Canadian regulatory frameworks from previous years.

On 23 December 2011, a Colombian products pipeline within a landslide exploded in Dosquebradas, Risaralda causing 33 death and 80+ injured people [86]. In 2012, the Colombian Institute of Technical Standards (ICONTEC) issued the standard Norma Técnica Colombiana (NTC) for *Pipeline Integrity Management System (PIMS)* for hazardous liquids resulting from the experience in Canada, USA and Colombia. As further explained in Section 1.5.4.8 of this book, the standard combined the Canadian approach toward Management System for pipelines and USA approach towards IMP in conjunction with the Colombian knowledge and experience on pipeline geohazards to create a PIMS. ASME Pipeline System Division created the International Pipeline Geotechnical Conference to worldwide compile and share geotechnical knowledge for pipelines in 2013 [87].

1.4.3.3 Europe In 1996, HSE issued the Pipelines Safety Regulation (PSR-96) [88] providing safety requirements for design, construction, operation, maintenance, and decommissioning. PSR-96 requires pipeline operators to have a safety management system to ensure the risk of a major accident is As Low As is Reasonably Practicable (ALARP). The same year, Marcogaz, the Technical Association of European Natural Gas Industry, developed a high level document about *"Pipeline Integrity Management System"* describing key elements such as technical safety, pipeline integrity, safety management system, emergency planning, land use planning, public consultation and the role of competent authorities.

1.4.3.4 Russia In 2013, the amendments to the Federal Law No. 116-FZ introduced the requirement for implementing safety management systems and ensure adequate functioning of industrial safety of hazardous production sites including pipelines. The 116-FZ amendment highlights the identification, analysis and prediction of risks as well as the planning, implementation and timely adjustment of measures to reduce risk of accidents. This is a significant advance into management systems seeking for adequacy, implementation and effectiveness.

1.5 PIPELINE INTEGRITY MANAGEMENT FRAMEWORK

This section describes the pipeline integrity societal framework and their participants, regulatory oversight approaches and jurisdiction types including some case studies, pipeline industry standards and associations.

1.5.1 Societal Framework and Their Participants

As illustrated in Figure 1.22, the pipeline integrity societal framework of pipeline integrity management is comprised of a legislative, regulatory and industry standards hierarchy influenced by multiple government agencies, outreach associations and nongovernmental organizations.

Legislative institutions (e.g., congress, parliament) create Acts and Laws stating government's policy, expectations and oversight mechanisms to protect people and the environment while ensuring transportation of fluids needed by the economy. Some legislative institutions establish Hearings with the participation of regulators, industry and associations to gain an understanding of the issues or challenges affecting stakeholders and requiring law-making strategy.

Acts and Laws (e.g., PIPES) need to be codified into *regulations* (e.g., CFR 192, 195) specifying requirements, timelines and processes to comply with. Regulators may adopt industry standards and recommended practices via reference, text imported or modified onto regulations (e.g. Code Adoption Document with amendments). Regulators also use consultation processes to assess the impact or benefits of a proposed rulemaking. This process minimizes non-desired results decreasing the effectiveness of the rulemaking.

Industry standards and recommended practices are typically developed by industry, academia and government representatives (e.g., CSA-Z662). Industry standards are produced through multi-level development and scrutiny processes. Industry standard vetting requiring representative member participation for decision making as well as multi-level technical review and consensus go through public scrutiny before the standard can be published. Recommended Practices (RP) have a lesser vetting process not requiring public scrutiny, which can be typical of some specific or evolving topics.

Official failure investigations are conducted by specialized and independent government agencies (e.g., TSB, NTSB) to identify root-causes and contributing factors of an incident. Agencies focus

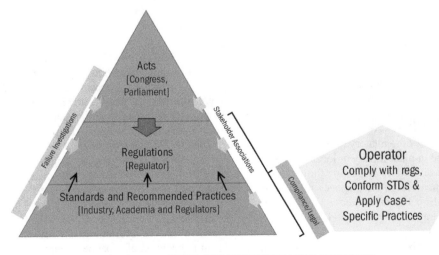

FIG. 1.22 PIPELINE INTEGRITY FRAMEWORK

on investigating incidents with loss of life, significant environmental or property damage.

These agencies should be independent from Executive (e.g., government) or Judicial (e.g., Courts) institutions reporting exclusively to Legislative institutions (e.g., Congress, Parliament) to avoid bias in the determination of recommendations.

Incident investigation agencies may provide recommendations as a call for action aimed directly to regulators, executive government or other agencies for preventing reoccurrence of the investigated incident. However, they may not be able to issue orders, directions, penalties or fines to any party involved in the incident.

Stakeholder Associations advocating for focused areas (e.g., Pipeline Safety Trust), industry (e.g., CEPA, INGAA, and AOPL), environment (e.g., Green Peace), landowners (e.g., CAEPLA, OLA) or think-tankers (e.g., Pembina) can influence acts, laws, regulations and industry standards. Stakeholder associations commonly participate in public hearings, industry panels and events to voice their concerns and proposed changes.

Compliance and Legal offices of pipeline operating companies identify, assess and interpret the law, regulations and industry standards required to be followed. In the integrity context, the difference between Compliance and Legal offices is that *Compliance* acts as a "*internal regulator*" providing guidance on regulatory requirements and overseeing their implementation; whereas, *Legal* act as a "*internal advisor*" providing legal interpretation of the law associated with the requirements and preventing the company from legal exposure risks.

In summary, pipeline operating companies are required to comply with the Law and Regulations as well as the adopted industry standards and consider the context of the societal framework.

1.5.2   Regulatory Oversight Approaches

There are four (4) regulatory approaches that can be easily distinguished as to being used by regulators to oversee compliance with regulations around the world. The following are the regulatory oversight approaches defined in terms of the integrity performance focus (refer to sections 3.8 and 14.3.2.3):

1. Prescriptive (focus on input, process and output stages) Provides oversight on adequacy and implementation, but effectiveness assessed if safety and environmental protection outcomes did not reach objectives
2. Goal Setting or Performance-based (focus on outcome stages) Provides oversight on effectiveness
3. Goal-Oriented (focus on process, output and outcome stages) Provides oversight on adequacy, implementation and effectiveness
4. Management-Based (focus on process stage) Provides oversight on plan adequacy

Those regulatory oversight approaches are explained in detail in Chapter 11.

1.5.3   Regulatory Jurisdictions

There are typically three (3) types of regulatory jurisdictions: federal/national, provincial/state and in-plant/facility or local/municipality. They can be consolidated or divided by delegating further down into regions or departmental offices. Regulatory agency mandate is typically focused on protecting the public in the areas of safety, environment, transportation fees, energy infrastructure and market.

1.5.3.1 Federal or National Jurisdiction   Federal or National jurisdictions would cover all hydrocarbons pipelines within the

country. Federal regulators may delegate some powers onto provincial or state regulators provided that qualifications, competency and reporting processes and verification are developed by the delegate.

The Pipeline and Hazardous Materials Safety Administration (PHMSA) is responsible for oversight of national pipelines in the USA; however, State regulators annually certified by PHMSA can inspect and enforce pipelines provided that the state follows the Federal regulations and enforcement actions substantially the same as defined by the pipeline safety statutes.

In Canada, the *National Energy Board (NEB)* regulates the pipelines crossing either a provincial or international border. They are mostly transmission pipelines extending for long distances transporting large volume of Hydrocarbons fluids. Pipelines within Provinces are regulated by Provincial regulators following Acts issued by the Province. The Federal regulators as well as all Canadian provincial regulators continue adopting the latest version of Canadian industry standard for oil and gas pipeline system (i.e., CSA-Z662).

In the UK, the *Health and Safety Executive (HSE)* is responsibility for the oversight of all pipelines in the UK under the Pipeline Safety Regulation. Pipelines transporting hazardous fluids are categorized within the group of Major Accident Hazard Pipeline (MAHP) for which the regulations apply additional requirements in protecting the health and safety of employees and the public.

1.5.3.2 Provincial or State Jurisdiction   Provincial or State jurisdictions cover all hydrocarbons pipelines within the province or state either as a direct mandate or delegation from the federal regulator. When these regulators are mandated by the province or the state, they are independent from the other regulators; however, Memorandums of Understanding (MOU) are signed between regulators to share information about the common regulated companies, lessons learned and best regulatory practices.

In Alberta, a Canadian Province in the west of the country, the Alberta Energy Regulator (AER) provides regulatory oversight of full-cycle of pipelines from application to abandonment and reclamation following its mandate of conserving water resources, managing public lands, and protecting the environment while providing economic benefits for all Albertans [89].

In Texas and Minnesota, USA states in the south and North central of the country, both state regulators have something in common and one difference. The Texas Railroad Commission (TRC) and Minnesota Office of Pipeline Safety (MNOPS) inspect intrastate pipelines within their states. However, TRC can only enforce Federal violations directly; whereas, MNOPS reports Federal violations for PHMSA to take an enforcement action. MNOPS can issue State violations [90].

1.5.3.3 In-Plant/Facility or Local/Municipality Jurisdiction
In-plant/facility or local/municipality regulators are focused on safety providing approval for construction and installation as well as operational oversight. In-Plant/facility jurisdictions cover all piping, equipment including pressure vessels and instrumentation within a facility (e.g., station, plant, batteries, communications); however, some jurisdictions are exclusively assigned the pressure vessels; whereas, piping and other installations are assigned to either the provincial/state or federal/national regulators. They may be independent from each other; however, MOUs are written to work together in achieving safety practices by the common regulated pipeline companies.

In Ontario, a Canadian province in the east of the country, Technical Standards and Safety Authority (TSSA) promotes and enforces Safety of, in the context of pipeline systems, boilers and pressure vessels, and fuels. TSSA, as a provincial regulator, has been empowered for overseeing pipeline systems including facilities; whereas, other provinces have separated the in-plant from the pipeline oversight [91].

1.5.4 Industry Standards

Pipeline integrity industry standards from CSA, ASME, API, NACE, BS, IGEM and ICONTEC are explained as to focus and coverage related to pipeline integrity.

1.5.4.1 CSA Z662 Oil and Gas Pipeline Systems

The CSA-Z662 standard covers the pipeline cycle describing requirements for design, construction, operation and deactivation/reactivation of pipelines transporting oil and gas fluids.

In 1999, the CSA Z662 industry standard included Clause 10.11.1 requiring for the first time that companies "… shall establish effective procedures (see Clause 10.2) for managing the integrity of pipeline systems so that they are suitable for continued service, including procedures to monitor for conditions that may lead to failures and to eliminate or mitigate such conditions." The 2003 version of the CSA Standard was revised such that it replaced the term "procedures" with "management program" to add the programmatic nature of pipeline integrity. A 2005 supplement to the 2003 version of CSA Z662 included Annex N Guidelines for pipeline integrity management programs, which provided a framework that may be used when developing an IMP. In the 2007 version of CSA the addition of Clause 10.2 *Safety and Loss management system* requirements was introduced for the first time along with the referenced Annex A named with the same title.

In the 2011 version of CSA Z662 the *Pipeline System Integrity Management Program* section was moved from the Operating, Maintenance and Upgrading section (Clause 10) to the beginning of the standard (Clause 3, *Safety and loss management, integrity management programs, engineering assessments*) to remove any ambiguity that the IMP requirements should be applied to the overall pipeline life cycle (design, construction, operation) within the safety and loss MS requirements. All Canadian energy regulators have adopted the mandatory components of CSA-Z662 such as all chapters and mandatory annexes (i.e., H on pipeline failure records).

The version 2015 of the CSA-Z662 consensus standard required the implementation of a risk management process that identifies, assesses, and manages the hazards and associated risks for the full life cycle of their pipeline system. The clause 3.4 risk management lists the following elements:

(a) risk acceptance criteria;
(b) risk assessment, including hazard identification; risk analysis, and risk evaluation;
(c) risk control;
(d) risk monitoring and review;
(e) communication; and
(f) documentation.

Canadian energy Provincial and Federal regulators such as Alberta Energy Regulator (AER), British Columbia Oil and Gas Commission (BC OGC), Technical Standard and Safety Authority (TSSA) from Ontario have also adopted the informative Annex N making it mandatory in those provinces.

1.5.4.2 API 1160 Managing the Integrity of Hazardous Liquid Pipelines

In 2001, API released the first edition of its integrity standard. API and some of its members participated in the consultation of and contributed to the CFR 195.452 Hazardous Liquid Integrity Management regulation. API 1160 is referenced in some sections of the Rule as a more detailed guidance for developing and measuring performance of IMP. This version of the standard provided structured guidance for data gathering and integration, HCA designation, risk assessment, integrity assessment methods, mitigation and IMP evaluation including a chapter for facility integrity management.

In 2013, API changed the 1160 from Standard to Recommended Practice, a developing process that requires less external scrutiny prior to its release. The 2013 version resulted from updating the 2001 version via a contract with Kiefner & Associates and API internal review. The 2013 version introduced changes such as introducing the concept of critical locations and adding remaining life, re-inspection frequencies and preventative measures. HCA was referenced to the USA regulatory framework.

1.5.4.3 ASME B31.8S Managing the Integrity of Gas Pipelines

In 2001, ASME released a supplement to the B31.8 gas pipeline standard describing the gas pipeline integrity management as the B31.8S. This 2001 version was used as an input for the regulation USA DOT CFR 192 subpart O issued in 2003. The B31.8S provided an overall process that could be applied to any pipeline regardless of their transported fluid provided that their specific conditions were considered.

As illustrated in Figures 1.23 and 1.24, ASME B31.8S outlined IMP elements as well as some Management System (MS) elements such as plans for communication, quality, integrity and performance. The IMP focused on the management of integrity threats including their identification, assessment, mitigation, and prevention as well as their prioritization considering consequence.

USA DOT CFR 192 regulation adopted ASME B31.8S to provide more details in the implementation of the requirements; however, in case of conflict between the CFR 192 Subpart O and ASME B31.8S, then regulation controls.

1.5.4.4 BS PD 8010 Part 4-Risk-Based Integrity Management of Steel Pipelines

In 2012, the British Standard Institute released a Published Document (PD) as the Part 4 of the Code of practice for pipelines providing guidance for integrity management of land and subsea pipelines. The PD covered the pipeline life cycle from design to decommissioning aimed at management and engineering practitioners.

The PD included IMP elements such as risk, process and condition monitoring, in-service inspection as well as MS elements such as information management, audits and review. Furthermore, the PD included maintenance (i.e., maintainability, spares) and emergency response.

1.5.4.5 UK IGEM TD/1

The UK Institute of Gas Engineers and Managers has created standards for dry gas transmission and distribution steel pipelines providing guidance for their design, construction, inspection, testing, operation, and maintenance.

IGME TD/1 defines the integrity management requirements and process including:

- Identification of threats and risks
- Identification of risk control measures for each threat
- Preparation of a Pipeline Integrity Management Plan defining the how-to for the implementation and assessment of risk control measures
- Implementation of surveillance, monitoring and inspection, maintenance and repair
- Reporting
- Evaluation ensuring all threats were managed

Integrity management is required to be integrated with and supported by a safety management system. The safety management system provides the policy, organization, planning and implementation, performance measure, audits and ongoing reviews to ensure the effectiveness of the integrity management. UK HSE indicates on its website that IGEM standards of practice including TD/1 and TD/3 should be used for natural gas major hazard pipelines [92].

1.5.4.6 BS PAS-55 Asset Management British Standard Institution (BSI) incorporated the *Publicly Available Specification* (PAS) about Asset Management. The Institute of Asset Management initiated the PAS-55 following the rigor of consultation, but not obliged to reach consensus as it is required by the BSI standards.

As illustrated in Figure 1.25, PAS 55 follows the ISO operating principle of Plan-Do-Check-Act for managing physical assets in an optimal and sustainable manner within the entire life cycle to achieve the organizational strategic plan. PAS 55 focuses on physical assets linked to the other categories of assets such as human, financial, information and intangible. As illustrated in Figure 3.5, the strategic organizational plan derives into the policy, which defines asset management strategy, objective, and plan and their contingency. With asset management enablers and controls such as leadership, communication, information management, risk management, and management of change would create the conditions for the implementation of the asset management plans. Plans would focus on life cycle activities of the assets:

1. Creation, acquisition and enhancement
2. Utilization of assets
3. Maintenance
4. Decommissioning and/or disposal

Pipelines are assets that can be managed using the PAS 55 approach provided that the process includes specific areas such as prevention of integrity threats and the emphasis on formal integrity risk assessments for prioritization and decision making.

ISO 55000 has taken the core elements and processes of PAS 55 expanding the scope to all type of assets, redefining asset management and providing a higher level approach. ISO 55000 may still rely on PAS 55 and other practices as it requires detailed processes for specific assets [93].

1.5.4.7 ICONTEC NTC 5747 Management Integrity of Gas Pipelines In 2009, the Colombian Institute of Technical Standards (ICONTEC) issued the standard for managing the integrity of gas

pipelines following and customizing ASME B31.8S with the due authorization.

1.5.4.8 ICONTEC NTC 5901 Integrity Management System for Hazardous Liquid Pipelines In 2012, a new standard combined the Canadian approach toward Management System for pipelines and USA approach towards IMP in conjunction with the Colombian knowledge on pipeline geohazards to create a PIMS.

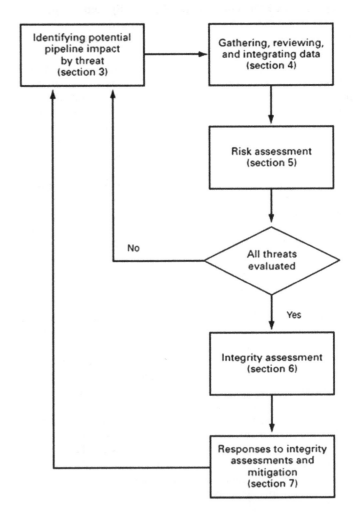

FIG. 1.24   ASME B31.8S-2014 INTEGRITY MANAGEMENT PLAN PROCESS FLOW DIAGRAM

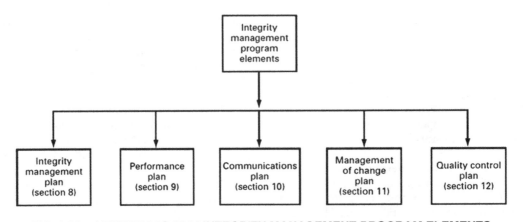

FIG. 1.23   ASME B31.8S-2014 INTEGRITY MANAGEMENT PROGRAM ELEMENTS

This standard formalized the approach for integrity threat assessments as well as consequence assessments beyond high consequence areas. Performance indicators included the performance cycle associated with the response to the pipeline system to the integrity actions monitoring the effectiveness of the actions taken.

1.5.5   Pipeline Industry Associations

1.5.5.1 Canadian Energy Pipeline Association (CEPA) CEPA [94] started in 1993 with a couple of major Canadian transmission pipeline companies and 20 years later represents more than 95% of the Canadian transmission pipelines transporting natural gas, dense gas, crude oil, and refined products. Some CEPA member companies also operate in USA and Mexico. CEPA's mission is "*to enhance the operating excellence, business environment and recognized responsibility of the Canadian energy transmission pipeline industry through leadership and credible engagement between member companies, governments, the public and stakeholders.*"

In 2007, CEPA launched the *Integrity First* initiative to develop a common set of principles and commitments, confirm minimum operating standards, and performance reporting metrics; and demonstrate commitment to continuous improvement and transparency through annual public reporting.

In 2014, CEPA *Integrity First* evolved becoming a management system to strengthen pipeline performance, communication and engagement with stakeholders. The intent of this strategy is to addresses integrity, reputational, regulatory and business risks

in a unified approach. The initiative priorities include damage prevention, emergency response, reclamation and education.

CEPA reports to the public the member pipeline integrity performance providing the frequency of failure and significant incidents, release volume for liquids and gas, and fatalities.

1.5.5.2 Canadian Association of Petroleum Producers (CAPP) CAPP [95] dates its origins back to 1927 resulting from Canadian petroleum groups associations such as the Canadian Petroleum Association (CPA), Alberta Oil Operators' Association and the Independent Petroleum Association of Canada (IPAC). In 2014, CAPP represents the Canadian upstream oil, oil sands and natural gas industry. CAPP's mission is to "*enhance the economic sustainability of the Canadian upstream petroleum industry in a safe and environmentally and socially responsible manner, through constructive engagement and communication with governments, the public and stakeholders in the communities in which we operate.*"

CAPP has developed multiple industry recommended practices focused on the mechanical integrity of upstream pipelines such as practices for mitigating internal corrosion for sour and sweet gas gathering systems, oilfield water and oil effluent pipelines as well as mitigating external corrosion of buried pipelines.

CAPP has also contributed with their member professional participation in the development of industry standards such CSA-Z662 Oil & Gas Pipeline Systems. Furthermore, CAPP contributed to the upcoming CSA-Z247 Damage Prevention, in an effort to change the increasing trend of external interference incidents since 2010.

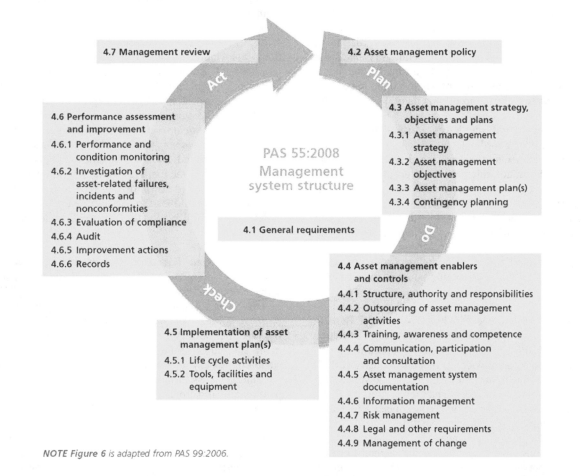

NOTE *Figure 6 is adapted from PAS 99:2006.*

FIG. 1.25   STRUCTURE OF BSI PAS 55-1:2008 (*Source:* British Standard Institute and Institute of Asset Management, BSI Group)

In 2014, CAPP initiated two (2) task groups looking into High-Impact Areas and Pipeline Performance Reporting.

1.5.5.3 Interstate Natural Gas American Association (INGAA) INGAA [96] represents interstate natural gas transmission companies in the USA; however, some member-companies are headquartered in Canada. INGAA is a trade association monitoring and advocating regulatory and legislative positions on behalf of its member companies. INGAA made contributions during to the CFR192 Subpart O gas transmission rulemaking process as well as the PHMSA and NTSB hearings triggered by incidents with significant consequences.

INGAA has provided guidance documents about interacting integrity threats, integrity assessment methods and tools, and proposed safety improvement initiatives such as:

- Extending Integrity Management Beyond HCA to Cover 100% of People Living Near Pipelines
- Pre-Regulation Pipelines: Assessing Records and Managing the Integrity
- Fitness for Service: Defined and Applied to Pre-Regulation Pipelines
- Integrity Management Principles: Prevention, Assessment and Mitigation Practices
- INGAA Responds to NTSB Recommendation: Development of Advanced In-line Inspection Platforms
- Resident Manufacturing and Construction Threats

1.5.5.4 Association of Oil Pipe Lines (AOPL) AOPL [97] was established in 1947 and represents the interests of owners and liquid hydrocarbon pipeline operators in USA; furthermore, AOPL has members owning and operating pipelines in Canada and Mexico. AOPL members transport crude oil, gasoline, diesel, home heating oil, kerosene, propane, and biofuels to communities and refineries as wells carbon dioxide to oil and natural gas fields for increase production.

AOPL monitors legislative, regulatory and judicial developments providing member coordination and leadership on common issues, public comments to rulemaking as well as testifying to Congress. AOPL has made rulemaking proposals introducing pipeline repair criteria to continuously improve pipeline integrity practices. AOPL has supported pipeline industry applied research organizations such as PRCI for the development of knowledge, technology and application in the pipeline integrity area.

AOPL has grouped the liquid pipeline industry to adopt safety and environmental initiatives for promoting safer and more reliable pipelines.

1.5.5.5 Pipeline Performance Tracking System (PPTS) In 1999, the American Petroleum Institute (API) and Association of Oil Pipe Lines (AOPL) created a voluntary reporting of performance indicators of industry-wide oil pipeline industry (PPTS) [98] focused on knowledge to preventing incidents and spills. API/AOPL PPTS captures release incidents and their conditions determining causes and causative factors and consequences to public safety and the environment as well as remediation.

PPTS released Advisories to the members, regulators and the public providing lessons learned and knowledge gained from the release incidents to prevent incidents. The following are some of the key topics covered by the advisories as of 2014:

- New Findings on Releases from Facilities Piping
- The Role of One-Call Partners
- The Role of Landowner/Tenant on 3rd party damage
- Operator Excavation Damage Incidents
- Excavation Damage Overview Advisory
- Operator Error or other incorrect operation
- Crude Oil Releases
- Refined Product Releases Requirement of New Strategies for Continued Improvement

1.5.5.6 Regional Association of Oil, Gas and Biofuels (ARPEL) Since 1965, ARPEL [99] groups hydrocarbon, technology and goods and services companies to promote industry integration and growth as well as seeking ways to maximize its contribution to sustainable energy development in South America and the Caribbean. Since 1976 ARPEL holds Special Consultative Status with the United Nations Economic and Social Council (ECOSOC). In 2006, the association declared its adherence to the Ten principles of the United Nations Global Compact.

In 2012, ARPEL held the first workshop about "Geotechnics in Oil & Gas Pipeline Right of Ways" sharing lessons learned and new knowledge in response to geotechnical pipeline incidents affecting people and environment in the region. ARPEL developed a training program to accredit professionals as Assessors in Pipeline Integrity Management through a series of theorical and practical courses in external and internal corrosion, natural forces, operational errors and third party damage, and mechanical integrity evaluation and risk assessment. The course is guided by a Pipeline Integrity Management Manual, which is updated regularly by the member companies.

1.5.5.7 Marcogaz Marcogaz [100], as a non-profit international association registered in Belgium was created in 1968 to represent the European natural gas industry in monitoring and influencing regulation, standardization and certification in the safety, environment, health, and integrity of natural gas systems.

In 2005, Marcogaz developed "*Guidelines for the definition of Performance indicators for Safety Management Systems*". The guidelines included steps for developing quantitative and qualitative performance indicators through the use of event, fault and bowtie tree methods. Tree methods would provide the identification of barriers to be tracked down towards monitoring and preventing the conditions that could lead to a release. Barriers to integrity threats can be identified following these methods to develop performance indicators.

1.5.5.8 United Kingdom Operator's Pipeline Association (UKOPA) UKOPA [101] was created in 1996 and "*exists to provide the recognised and authoritative view of UK Pipeline Operators on strategic issues relating to safety management, operations and integrity management of pipelines*". UKOPA represents the pipeline industry in monitoring and influencing the development and implementation of pipeline regulations and industry standards related to its mission. UKOPA key areas are land use planning and risk criteria, safe working practices, fitness for purpose and revalidation, and safety issues.

UKOPA periodically collects pipeline and product loss incident data from onshore Major Accident Hazard Pipelines (MAHP) as defined by the UK Pipeline Safety Regulations, 1996 including only product loss onto the public areas. UKOPA conducts analysis to determine incident frequencies per year for product loss and pipeline fault data or features (e.g., external interference, corrosion, construction damage) confirmed by field investigation. UKOPA provides multi-year frequency trends that can help in identifying areas for improvement for association member discussion and path forward.

1.5.5.9 Australian Pipeline and Gas Association (APGA)
APGA [102] was initially named Australian Pipeline Industry Association (APIA) established in 1968 to initially represent pipeline industry contractor interests. In 2015, APIA name was changed to APGA to represent the interests of pipeline owners, operators, contractors and engineering companies as well as to continue contributing to gas policy development. APGA is the advocate for the Australian pipeline industry assisting in maintaining the leadership in the pipeline management of environment, health and safety.

1.5.6 Pipeline Industry Research

1.5.6.1 Ministry of National Resources Canada (NRCan)
NRCan [103] is the Canadian Federal Ministry that manages the *Pipeline Research Program*. One of the program objectives is to *"meet the federal government's needs for science and technology information on the regulation and maintenance of aging pipelines and the regulation and construction of new pipelines, in order to help federal decision-makers fulfill their regulatory responsibilities and reduce environmental impacts."*

NRCan has also worked with the Canadian pipeline industry in developmental research to be incorporated in the Canadian Standard for design and reliability of pipelines aimed to improving their safety, environmental and service performance.

1.5.6.2 Pipeline Research Council International (PRCI)
PRCI [104] started in 1952 as the Pipeline Research Committee (PRC) of the American Gas Association motivated by the integrity challenge of the long-running brittle fractures in natural gas transmission pipelines. In 1954, the PRC substantially resolved this challenge demonstrating the benefits of industry collaboration research.

In 2014, PRCI membership is comprised of pipeline operating companies, service providers, manufacturers and consultants to advance pipeline knowledge, methodologies and technologies for improving pipeline design, operation, maintenance and integrity efficiency, effectiveness and performance. PRCI defines itself as the *"preeminent global collaborative research development organization of, by, and for the energy pipeline industry."*

PRCI's focus is strategically directed by an *Executive Committee* comprised of senior management of PRCI members, who are drawn from the PRCI Board of Directors. As a result of the Executive committee strategy, *Technical Committees* represented by PRCI technical representatives of the member companies plan and develop research projects as well as identify research avoiding duplication. PRCI staff facilitates the development of the strategy, the planning of the tactical actions, and their implementation via contracting of knowledge-based companies to achieve the PRCI research objectives.

Technical Toolboxes markets PRCI's reports and software to the public.

1.5.6.3 European Pipeline Research Group (EPRG)
EPRG [105] was formed in 1972. EPRG focuses on addressing integrity challenges of gas transmission pipeline in design, manufacturing, construction, operation and maintenance. EPRG strives for ensuring a pipeline safety and reliability performance record in Europe. EPRG develops research programs developing recommendations and guidelines, which provide knowledge, methodologies and best practices for the gas pipeline industry.

1.5.6.4 APGA Research and Standards Committee (APGA-RSC)
APGA conducts research and development through the Research and Standards Committee (RSC), which is a separately funded by the APGA committee. The RSC [106] was formed in the early 1980s and formalized in 1996. APGA-RSC focuses on improving the safety and reliability of pipelines and reducing the costs of pipeline design, construction and operations. RSC research is promoted to the Standards Australia for their application within the AS 2885 *Australian Standard, Pipelines—Gas and Liquid Petroleum.*

1.6 REFERENCES

1. Harvard University, 2010, Leadership Weblog, http://blogs.law.harvard.edu/leadership/management-definition/.

2. *ISO*, Anon., 2015, *Quality Management Systems*, ISO 9001, International Organization of Standardization, Geneva, Switzerland.

3. Government of Alberta, *Baku, Azerbaijan*, Glenbow Archives, NA-302-9, Canada http://history.alberta.ca/energyheritage/oil/pre-modern-global-history/early-human-pre-industrial-history/baku-azerbaijan.aspx#page-1.

4. Petrolia Heritage, *Oil Springs*, Canada's Victorian Oil Town, Ontario, Canada http://www.petroliaheritage.com/oilSprings.htm.

5. Pipe Line Contractors Association of Canada, *History and Purpose*, Ontario, Canada http://www.pipeline.ca/.

6. CEPA, *History of Pipelines*, Canadian Energy Pipeline Association, Canada http://www.cepa.com/about-pipelines/history-of-pipelines.

7. Funding Universe, TransCanada Pipe Lines Ltd History, http://www.fundinguniverse.com/company-histories/transcanada-pipelines-limited-history/.

8. Reference for Business, *History of Enbridge*, http://www.reference-forbusiness.com/history2/0/Enbridge-Inc.html.

9. Wikipedia, *History of the Petroleum Industry in Canada—Pipeline Networks*, http://en.wikipedia.org/wiki/History_of_the_petroleum_industry_in_Canada#Pipeline_networks

10. Alberta Government, *Conventional Oil Timeline*, http://history.alberta.ca/energyheritage/oil/default.aspx.

11. Breen, D. H., 1993, *Alberta's Petroleum Industry and the Conservation Board*, University of Alberta Press, Canada.

12. Bradley, Jr. R. L., 2011, *Edison to Enron: Energy Markets and Political Strategies*, John Wiley & Sons, USA.

13. Giddens, P. H., 1941, *Pittsburgh and the Beginnings of the Petroleum Industry to 1866*, Western Pennsylvania Historical Magazine, Volume 24, September 1941, Number 3, http://ojs.libraries.psu.edu/index.php/wph/article/viewFile/2197/2030.

14. McElwee, Neil, 2007, *The National Transit Co., Standard Oil's Great Pipeline Company*, Oil Creek Press, USA.

15. *Pees, S. T.*, 1865, *The Van Syckel Pipeline*, Oil History, Petroleum History Institute, USA http://www.petroleumhistory.org/OilHistory/pages/Pipelines/van_syckel.html.

16. America Oil & Gas Historical Society, *Big Inch Pipelines of WW II*, http://aoghs.org/oil-and-natural-gas-transportation-history/pipelines-of-texas/).

17. Wikipedia: http://en.wikipedia.org/wiki/Vladimir_Shukhov and NLMK Group Corporate Magazine #3, 2011, http://nlmk.com/docs/tree/3_2011.pdf.

18. Beltrán, J. E., 1988, *Petróleo y Desarrollo de la Política Petrolera en Tabasco*, Gobierno del Estado de Tabasco, Villahermosa, México.

19. Fernández, L., *Historia del Transporte de Hidrocarburos por Ducto en México* http://www.ref.pemex.com/octanaje/o55/o.htm.

20. Mexico Maxico, 2010, *Pemex, Cronología, Expropiación y Estadísticas* http://www.mexicomaxico.org/Voto/pemex.htm.

21. México, *Comisión Nacional de Hidrocarburos* and *Comisión Nacional de Energía* http://www.mexicomaxico.org/Voto/pemex.htm.

22. Hosseini, M., *Nobel Brothers Rasht Pipelines*, http://www.fouman.com/Y/Get_Iranian_History_Today.php?artid=1341.

23. Iranian Oil Pipeline and Telecommunication Company, *The History of the Company—The Iranian Pipeline Chronicle* http://www.ioptc.ir/english/Thehistoryofthecompany.aspx.

24. Wikipedia, *Anglo-Persian Oil Pipeline*, http://en.wikipedia.org/wiki/Anglo-Persian_Oil_Company.

25. ECOPETROL, *Oil in Colombia, 1900–1950—Positioning of Jersey Standard in Colombia*, http://www.ecopetrol.com.co/especiales/Libro60anios/eng/cap1-5.htm.

26. Bucheli, M., 2006, *Confronting the Octopus: United Fruit, Standard Oil, and the Colombian State in the Twentieth-Century*, University of Illinois, International Economic History Conference, Helsinki, Finland http://www.helsinki.fi/iehc2006/papers3/Bucheli94.pdf.

27. Koelling, G. W., 1971, *Mineral Industry of Colombia*, Minerals Yearbook Area Reports, Volume 3, Bureau of Mines, International, http://digicoll.library.wisc.edu/cgi-bin/EcoNatRes/EcoNatRes-idx?type=turn&id=EcoNatRes.MinYB1971v3&entity=EcoNatRes.MinYB1971v3.p0252&isize=text.

28. Colombian Petroleum Company, 1967, *COLPET al Día*, Edition No. 43 and 44, Bogotá, June 1962 Page 2, Colombia https://www.facebook.com/media/set/?set=a.34408509805.41204.5634859805&type=3.

29. Agencia Nacional de Hidrocarburos, *Historia*, Colombia http://www.anh.gov.co/la-anh/Paginas/historia.aspx.

30. Saudi Aramco, *History Milestones*, Saudi Arabia http://www.saudiaramco.com/en/home/about/history/milestones.html.

31. Wikipedia, *Trans-Arabian Pipeline*, http://en.wikipedia.org/wiki/Trans-Arabian_Pipeline.

32. Solberg, Carl E., 1979, *Oil and Nationalism in Argentina*, Stanford University Press, USA.

33. Alonso, M. E., Vasquez, E. C., 2005, *Historia: la Argentina contemporánea (1852–1999)*, 3rd edición, Aique, pp. 112–113. ISBN 950-701-622-8.

34. Mewett, L., 2009, *South America Snapshot*, Pipelines International Magazine, September, http://pipelinesinternational.com/news/south_america_snapshot/008026.

35. Wikipedia, *Gas del Estado—Argentina*, http://es.wikipedia.org/wiki/Gas_del_Estado.

36. Fernando, L., *Petróleo no Brasil*, Ebah, http://www.ebah.com.br/content/ABAAABkrAAK/petroleo-no-brasil.

37. Wikipedia, *Companhia Petróleos do Brasil*, http://pt.wikipedia.org/wiki/Companhia_Petr%C3%B3leos_do_Brasil.

38. Cubatao,1973, *O oleoduto da Serra do E.F.S.J.*, http://www.novomilenio.inf.br/cubatao/ch028b.htm.

39. Petrobras, *70 anos da exploração de petróleo na Bahia*, http://fatosedados.blogspetrobras.com.br/2011/12/14/70-anos-da-exploracao-de-petroleo-na-bahia/ http://www.youtube.com/watch?feature=player_embedded&v=Wg_2FWQc62Q.

40. Raffinerie Heide, *Mineral Oil: A presence at Dithmarschen since 1856*, http://www.heiderefinery.com/en/the-company/history.html.

41. Corwith, J., 2009, The Oil Mines at Wietze and Pechelbronn, Grantville Gazette, https://grantvillegazette.com/wp/article/publish-303/.

42. Dukes Wood Museum, *Eakring Personnel*, http://www.dukeswood-oilmuseum.co.uk/eakring.htm.

43. UK Ministry of Defense, *Legislation to enable sale of the Government Pipeline and Storage System (GPSS)* https://www.gov.uk/government/uploads/system/uploads/attachment_data/file/35893/gpss_booklet.pdf.

44. Nord-West Oelleitung, *Company—Securely transporting crude oil*, https://www.nwowhv.de/c/index.php/en/.

45. Trapil, *The Le Havre-Paris Pipeline Network* (TRAPIL), http://www.driee.ile-de-france.developpement-durable.gouv.fr/IMG/pdf/PlaquetteLHPBAT_cle2c33b1.pdf.

46. Stern, J., *Natural Gas in Europe—The Importance of Russia*, Oxford Institute for Energy Studies, Centrex, Vienna, Austria http://www.centrex.at/en/files/study_stern_e.pdf

47. CNPC, 2015, Daqing Major Events, http://www.cnpc.com.cn/en/xhtml/flash/daqing_major_events.swf.

48. CPPB, 2014, CPP historical heritage, China Petroleum Pipeline Bureau http://cpp.cnpc.com.cn/gdj/lscc/qywh_index.shtml.

49. Li Xinmin/Langfang, 2010, *The backbone of China's oil and Gas Pipeline Network basically Completed*, Economic Information Daily, Xinhua News Agency Network Center, China http://jjckb.xinhuanet.com/gnyw/2010-08/09/content_247114.htm.

50. Wood, B., Woodman, R., 2005, The Moonie Oil Pipeline—1963, The Australian Pipeliner, Australia http://pipeliner.com.au/news/the_moonie_oil_pipeline_-1963/43267.

51. Ellis, G. K., Jonasson, K. E., 2002, *The Rough Range Oil Field, Carnarvon Basin*, The Sedimentary Basins Of WA 3 P707-718, Petroleum Exploration Society of Australia (PESA), Australia.

52. Kimber, Max J., 1996, *The Changing Face of the Australian Pipeline Industry*, AGA PRC 9th Symposium on Line Pipe Research, Houston, TX, USA.

53. Touzett, P., 2007, *Marcos Regulatorios y el Rol de las Empresas Estatales de Hidrocarburos, Estudio de Caso: Perú*, CIDA, OLADE; University of Calgary, AB, Canada.

54. Glenbow Museum, *Exploration and Production—South America, Alaska, United States and Middle East: 1915–1940s*, Series IP-3, Calgary, AB, Canada https://www.glenbow.org/collections/search/findingAids/archhtm/iolphotos.cfm.

55. PerúPetro, 2010, *Actividades de Exploración y Explotación de Hidrocarburos en el Perú*, Perú.

56. PetroPerú, *Oleoducto Norperuano*, http://www.petroperu.com.pe/portalweb/Main.asp?seccion=76.

57. OSINERGMIN, 2013, *Qué es Osinergmin?*, Organismo Supervisor de la Inversión en Energía y Minería, http://www.osinergmin.gob.pe/newweb/pages/Publico/589.htm?5465.

58. Anon., US Office of Pipeline Safety Incident Database https://hip.phmsa.dot.gov/analyticsSOAP/saw.dll?Portalpages.

59. API, AOPL, 811, Pipeline 101, http://www.pipeline101.com/Are-Pipelines-Safe/What-Is-The-Safety-Record.

60. Anon., 'Gas Pipeline Incidents'. 9th Report of the European Gas Pipeline Incident Data Group. Doc. Number EGIG 14.R.0403. February 2015. www.egig.nl.

61. P M Davis et al., 'Performance of European cross-country oil pipelines. Statistical summary of reported spillages in 2013 and since 1971', CONCAWE (conservation of clean air and water in Europe). Report 4/15. May, 2015. www.concawe.be.

62. Anon., 'Pocket Guide to Transportation. 2014', US Department of Transportation. RITA Bureau of Transportation Statistics http://www.ntsb.gov/investigations/data/Pages/Data_Stats.aspx.

63. Anon., 2013, *Onshore Pipelines: The Road to Success,* International Pipe Line & Offshore Contractors Association, 3rd edition, USA http://wiki.iploca.com/pages/viewpage.action?pageId=1803629.

64. USA, DOT PHMSA, Significant Incidents, https://hip.phmsa.dot.gov/analyticsSOAP/saw.dll?Portalpages.

65. INGAA Foundation, 1997, *Natural Gas Pipeline Safety 1994–1997: The Change to a New Safety Paradigm*, http://www.ingaa.org/INGAAFoundation/Studies/FoundationReports/567.aspx.

66. Jseaton, *Edison Natural Gas Explosion—Durham Woods*, USA http://www.youtube.com/watch?v=NyMbaZ9FVjA. Jseaton/Jean-Luc Ponty.

67. NTSB, 1995, *Natural Gas Pipeline Explosion and Fire, Edison, New Jersey, 23 March 1994*, PAR-95-01, PB95-916501, National Transportation Safety Board (NTSB), Washington, D.C., USA http://www.ntsb.gov/investigations/AccidentReports/Pages/PAR9501.aspx.

68. Beitler, S., *Edison, NJ Gas Pipeline Explosion, Mar 1994*, GenDisasters, http://www3.gendisasters.com/new-jersey/19145/edison-nj-gas-pipeline-explosion-mar-1994.

69. Wikipedia, *List of Pipeline Accidents in the United States from 1995 to 1999* http://en.wikipedia.org/wiki/List_of_pipeline_accidents_in_the_United_States_1975_to_1999.

70. NEB, 1996, *Stress Corrosion Cracking on Canadian Oil and Gas Pipelines*, Report of the Inquiry MH-2-95, National Energy Board (NEB), Canada.

71. CIC/CSChE, 2008, *Risk-based Land Use Planning Guidelines*, Process Safety Management (PSM) division of the Chemical Institute of Canada/Canadian Society for Chemical Engineering, http://www.cheminst.ca/sites/default/files/pdfs/Connect/PMS/the%20accompanying%20cover%20note.pdf.

72. NTSB, 2002, *Pipeline Rupture and Subsequent Fire in Bellingham, Washington* June 10, 1999, Pipeline Accident Report, NTSB/PAR-02/02, PB2002-916502, National Transportation Safety Board (NTSB), USA http://www.ntsb.gov/investigations/AccidentReports/Reports/PAR0202.pdf.

73. Hall, J., 1999, *Testimony before the House of Representatives*, Chairman of the National Transportation Safety Board, USA.

74. NTSB, 2003, *Natural Gas Pipeline Rupture and Fire Near Carlsbad, New Mexico, August 19, 2000*, NTSB/PAR-03/01, PB2003-916501, National Transportation Safety Board, Washington, D.C., USA http://www.ntsb.gov/investigations/AccidentReports/Reports/PAR0301.pdf.

75. US DOT RSPA, 2001, *Notice of Probable Violation, Proposed Civil Penalty and Proposed Compliance Order to El Paso Energy Pipeline Grou*p, Research and Special Programs Administration, Southwest Region, Office of Pipeline Safety, Houston, TX, USA http://primis.phmsa.dot.gov/comm/reports/enforce/documents/420011004/420011004_Notice%20Letter_06202001.pdf.

76. GPO e-CFR, 2015, *Title 49 Transportation Part 192 Subpart O Gas Transmission Pipeline Integrity Management*, U.S. Government Publishing Office, USA http://www.ecfr.gov/cgi-bin/text-idx?SID=89782ac3510ac2d1303a4a3d19a5b37b&node=49:3.1.1.1.8.15&rgn=div6.

77. Enbridge, Inc., 2008, *OCENSA/ CITCOL/ CLH, Annual Report 2008*, http://ar.enbridge.com/ar2008/management-discussion-analysis/international/ocensa-cit-col-clh/.

78. ENARGAS, 1999, Resolución a Transportadora de Gas del Norte S.A. sobre Reventón 11 Diciembre 1998, Buenos Aires, Argentina http://www.enargas.gov.ar/MarcoLegal/Resoluciones/Data/R99_1262.htm.

79. Exponent, 2007, *Pipeline Integrity Analysis of the Camisea Transportation System*, Failure Analysis Associates, California, USA http://camisea.org/Pipeline_Camisea_BID_2007.pdf.

80. Gonzalez, J. L., 2013, *Implementation of Pipeline Integrity Management in Mexico*, Mexico, D.F., a_i Mexico http://www.ai.org.mx/ai/archivos/ingresos/jorge_gonzalez/presentacion.pdf.

81. UK HSE, *The History of HSE*, UK Health and Safety Executive, UK http://www.hse.gov.uk/aboutus/timeline/.

82. NEB, 2013, *Guidance Notes for the Onshore Pipeline Regulations*, National Energy Board (NEB), Canada https://www.neb-one.gc.ca/bts/ctrg/gnnb/nshrppln/index-eng.html.

83. Mora, R. G., Paviglianiti, J., Slocomb, R., Bourassa-Mota, A., Zaidi, M., 2012, *Trends on Integrity Management Programs and Management Systems Audit and Incident Findings*, IPC2012-90046, ASME International Pipeline Conference, Calgary, AB, Canada.

84. USA GPO, San Bruno, CA—PG&E *Senate Committee on Commerce, Science and Transportation—Congressional Hearings*, USA http://www.gpo.gov/fdsys/pkg/CHRG-112shrg74986/pdf/CHRG-112shrg74986.pdf.

85. OSINERGMIN, *Reglamento de Transporte de Hidrocarburos por Ductos*, Decreto Supremo Nº 081-2007-EM, Perú http://www2.osinerg.gob.pe/MarcoLegal/DS-081-2007-EM-CONCORDADO.doc.

86. Ministerio de Salud y Protección Social, 2012, Resolución Evento Catastrófico Dosquebradas, Colombia.

87. ASME IPTI, ASME International Pipeline Geotechnical Conference – IPG 2013, July 24–26, 2013, Bogota, Colombia http://www.asme-conferences.org/IPG2013/.

88. UK HSE, 1996, *A Guide to the Pipelines Safety Regulations 1996*, Guidance on Regulations, Health and Safety Executive, United Kingdom http://www.hse.gov.uk/Pubns/priced/l82.pdf.

89. Alberta Energy Regulator, *AER Who we are*, website http://www.aer.ca/about-aer/who-we-are.

90. Minnesota Office of Pipeline Safety, website https://dps.mn.gov/divisions/ops/about/Pages/history.aspx.

91. Ontario Pipeline Regulator, *Technical Standards and Safety Authority (TSSA)*, website http://www.tssa.org/regulated/about/profile.aspx.

92. UK HSE, *Use of Pipeline Standards and Good Practice Guidance*, Health and Safety Executive (HSE) website, United Kingdom www.hse.gov.uk/pipelines/resources/pipelinestandards.htm.

93. IAM, *What is PAS55?*, The Institute of Asset Management (UAM), website www.theiam.org/products-and-services/pas-55/what-pas55.

94. Canadian Energy Pipeline Association website www.cepa.com.

95. Canadian Association of Petroleum Producers website http://www.capp.ca.

96. Intrastate Natural Gas Pipeline Association website www.ingaa.org.

97. Association of Oil Pipe Lines website www.aopl.org.

98. American Petroleum Institute, Pipeline Performance Tracking System (PPTS) website www.api.org/oil-and-natural-gas-overview/transporting-oil-and-natural-gas/pipeline-performance-ppts/ppts-overview.

99. Regional Association of Oil, Gas and Biofuels Sector Companies in Latin American and the Caribbean website www.arpel.org.

100. Marcogaz—Technical Association of the European Natural Gas Industry website http://www.marcogaz.org/.

101. UKOPA United Kingdom Onshore Pipeline Operators' Association website http://www.ukopa.co.uk/.

102. Australian Pipelines & Gas Association (APGA) website http://www.apga.org.au/.

103. Natural Resources Canada website www.nrcan.gc.ca.

104. Pipeline Research Council International website www.prci.org.

105. European Pipeline Research Group website www.eprg.net.

106. APGA Research and Standards Committee http://www.apga.org.au/issues/apia-committees/research-and-standards-committee/.

ENGINEERING CONCEPTS FOR PIPELINE INTEGRITY

Pipeline integrity management starts with a good pipeline design. Accordingly, this Chapter introduces the reader to the basics of pipeline design and the basics of pipeline engineering that contribute to the safe operation of a pipeline.

A pipeline is designed to reside in a right of way, at a distance from the general public. Its diameter is selected to give the required throughput, and its wall thickness is calculated to ensure it can withstand all loads imposed on the pipeline.

Pipelines are routed away from the general public and environmentally-sensitive areas, but inevitably pipelines can sometimes pass near people. Pipeline standards impose location classifications, setbacks to limit the chances of the general public damaging the pipeline, and hence creating a threat.

2.1   INTRODUCTION

Pipeline operators want their pipeline to operate safely[1] for its entire 'design life'. This 'design life' is dependent on how the operator cares for his/her pipeline, as both ASME B31.4 [1] and B31.8-2014 [2] define the 'design life' as: "… *a period of time used in design calculations, selected for the purpose of verifying that a replaceable or permanent component is suitable for the anticipated period of service. Design life [does/may] not pertain to the life of the pipeline system because a properly maintained and protected pipeline system can provide … service indefinitely.*"

Pipeline integrity management is aimed at ensuring a pipeline system maintains its structural integrity (it does not leak or rupture), by performing many maintenance and protection activities to ensure releases do not occur throughout the pipeline's design life.

Design and integrity management are inter-related: a badly designed pipeline can be very difficult to manage, and may not operate safely throughout its design life. This inter-relation means pipeline integrity engineers need an understanding of pipeline design.

Pipeline integrity standards and pipeline design standards both have the same intent: safe pipeline operation, Add this intent to the fact that an engineer's prime responsibility is safety, and the general public expect a safe pipeline, gives the 'take home' message of "*safety is an engineer's prime concern in pipeline design, construction, and operation.*"

This Chapter introduces basic design criteria and parameters of specific interest to pipeline integrity engineers. The Chapter is given a structure by aligning it with the pipeline design process,

and selecting key elements either used in, or related to, pipeline integrity.

2.2   DESIGN OF PIPELINES

Pipelines need to be designed to operate safely; therefore, published standards are used for all design, construction, and operation. There are many pipeline design standards around the world (international [3], European [4, 5], and national standards [e.g., 6 to 10]), and many recognized practices (e.g., [11]). But, what is 'design'?

2.2.1   'Design'

The word 'design' is usually associated with other words such as… 'creativity', 'flair', 'style', etc. In 'engineering' design the important words are… 'safety', 'standard compliance', 'regulatory compliance', 'fit-for-purpose', 'within cost and schedule', etc. This pragmatic view of design is supported by pipeline design standards:

- DNV OS F101 [10] considers 'design': "*All related engineering to design the pipeline including structural as well as material and corrosion.*"
- ASME B31.4 [1] states that "*… engineering design [is] detailed design developed from operating requirements and conforming to Code requirements, including all necessary drawings and specifications, governing a piping installation.*"

Pipeline design standards have the objective of routing, designing, and constructing a pipeline that will:

- operate safely;
- have minimal impact on the environment; and,
- be cost effective (both in terms of capital expenditure and operational expenditure).

2.2.2   Line Pipe and Routing of Onshore and Subsea Pipelines

This Chapter will now cover pipeline design, but it briefly covers line pipe (the pipe used to construct the pipeline) and pipeline routing first. It will focus on onshore pipelines as most pipelines are onshore. Subsea pipelines are built using a 'lay barge' (a ship that performs most of the pipeline construction activities), and do not have the scale and variety of obstacles (for example, railway crossings, mountains, sensitive environmental areas, golf courses, etc.)

---

[1] Safety means protecting people and the environment.

and surrounding population that need to be addressed when routing an onshore pipeline.

Subsea pipelines do have routing issues, such as variable seabed topography, currents, anchorage areas, shipping lanes; therefore, the essential principle of routing (of avoiding obstacles, constraints, and hazards) applies to both subsea and onshore pipelines. ISO 13623 [3] gives a summary of both onshore and subsea routing issues.

2.3   LINE PIPE

Pipelines are constructed from 'line pipe', specially made in 'pipe mills' for use in transmission pipelines. Line pipe has been manufactured since the 19th century and has developed and improved every decade [12–14]. Line pipe is made from steel. Steel is made by melting iron ore and coke (carbon) in a furnace. The molten steel is then poured into moulds, where it cools into ingots. The ingots are then rolled into coils or slabs or billets.

The steel is low in carbon (approximately 0.1% by weight, alloyed with iron). This type of steel gives a good combination of strength, ductility, toughness, and 'weldability' (ease of welding (joining) to other metals). Other iron-based materials [e.g., 'wrought iron' (almost pure iron) and 'cast' iron (iron with a high carbon content)] are either too low strength, or too brittle, to function well as structural materials. Stainless steels are used for special pipeline applications (e.g., in highly corrosive environments), but they are not economically viable in the quantities needed for large structures such as long distance pipelines.

Before discussing the line pipe used in pipelines, it is necessary to briefly mention key mechanical properties possessed by the line pipe.

2.3.1   Mechanical Properties

Line pipe is ordered (usually in lengths of 12 m (40 feet)) according to its mechanical properties such as:

- strength (yield and ultimate);
- ductility (the ability to deform);
- toughness (the ability to resist the presence of defects such as cracks).

This Section will introduce the reader to the key mechanical properties.

2.3.1.1 Strength   The main tensile properties of line pipe are measured from a stress-strain curve: the yield strength; and, the ultimate tensile strength (UTS), Figure 2.1.

It is not practical for the pipe supplier to test every 12 m (40 feet) length of pipe provided. The supplier tests a representative sample, and gives 'specified minimum' values: 'specified minimum yield strength' ('SMYS'), and 'specified minimum tensile strength' ('SMTS'). The actual line pipe ordered is expected to exceed these values. The yield strength of line pipe is also known as 'grade'; for example, if line pipe is quoted as 'grade X52', the SMYS is 52,000 lbf/in^2 (359 N/mm^2). These specified minimum strength values are usually below the actual strength values, Figure 2.1.

2.3.1.2 Ductility   Materials can behave in different ways when under load; for example, when chalk is bent it quickly breaks. There is no deformation, so it is a 'brittle' material. When a silver spoon is bent, it does not break, and there is extensive deformation, so it is a 'ductile' material.

Ductility is the ability to deform, Figure 2.2. In a tensile test a specimen can deform extensively before failure: this is ductile behavior. It can also show little deformation (brittle behavior), or, it can show a mixture of brittle and ductile behavior.

Line pipe needs to be able to bend (line pipe is bent during onshore construction to create bends), deform (pipelines can be dented during operation), and stretch (pipelines can be subjected to strains higher than predicted in design; for example, due to ground movement).

Ductility is also a description of how a material fails and how the resulting fracture faces appear:

- 'ductile' means extensive deformation before failure, and a fracture surface that shows ductile features; and,
- 'brittle' means limited deformation before failure, and a fracture surface that shows brittle features.

The amount of ductility on a fracture surface is usually called 'shear area'. The higher the percentage of shear area, the larger the amount of ductility: 100% shear area is fully ductile, whereas 0% shear area is zero ductility (fully brittle).

2.3.1.3 Toughness   Small defects in a structure can fail at very low stresses; for example, ceramic tiles (high strength) cannot tolerate a small scratch: tap the tile, and it fractures along the scratch.

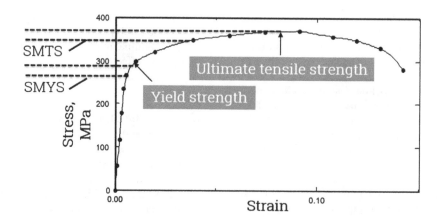

FIG. 2.1   DETERMINING YIELD AND ULTIMATE TENSILE STRENGTHS

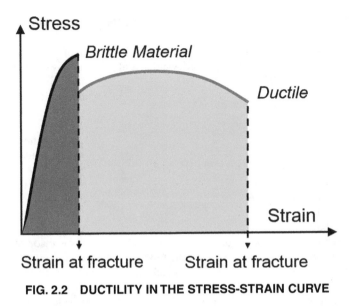

FIG. 2.2 DUCTILITY IN THE STRESS-STRAIN CURVE

Plastics (low strength) can contain scratches, but do not fracture at the scratches when they are tapped.

Some materials can withstand the presence of a defect. These materials are called 'tough'. When a defect is in a material, it is the material's 'toughness' that resists its presence. Toughness is the ability of a material to withstand the presence of a defect such as a crack. Low toughness material (such as glass) cannot tolerate cracks, and can fail in a 'brittle' manner. Materials such as aluminum, can withstand large cracks and are high toughness. Line pipe needs to be tough, as it may need to withstand defects during the pipeline's service.

Line pipe toughness is measured using a small test called a 'Charpy' test. This is a small 'impact' (it is tested by hitting it with a pendulum) specimen. A full size specimen is 55 mm × 10 mm × 10 mm (2.17 in. × 0.39 in. × 0.39 in.). The toughness is measured in Joules (or ft lbs).

Toughness is sensitive to temperature, Figure 2.3, and the toughness transitions from high toughness to low toughness. At high temperatures line pipe has high toughness, but at low temperatures it can have low toughness. The ductility on the fracture faces of the

toughness specimens also varies with temperature: the lower the test temperature, the lower the ductility, Figure 2.3.

2.3.1.4 Hardness 'Hardness' generally means the resistance to local indentation or deformation. Hardness is not a material property, as it depends on other material properties (e.g., strength and ductility), but the greater the hardness of the metal, the greater resistance it has to local deformation.

Hardness is normally measured by pressing a steel ball or cone into the metal, and measuring the depth of indentation for a specified load. The indentation depth is them related to the material hardness. There are various methods for this measurement (Brinell test, Rockwell test, Vickers test). The Vickers hardness test uses a square-based pyramidal diamond indenter. The applied load may range from 1 to 120 kgf. 'HV10' indicates a Vickers Hardness ('HV') obtained using a 10 kgf load ('10').

2.3.1.5 Weldability Line pipe steels need to be weldable, as each section will be welded together in the field to create the pipeline. 'Weldability' is a measure of how easy it is to make a weld in a particular steel:

- without cracks;
- with adequate mechanical properties for service; and,
- with resistance to service degradation.

An index called 'carbon equivalent' helps determine if a steel is able to be welded. For modern pipeline steels, the lower the value of carbon equivalent, the more weldable the steel. The carbon equivalent (CE) is a value calculated from the chemical composition of the steel. The calculation incorporates the elements carbon, manganese, chromium, molybdenum, vanadium, copper, and nickel. Here is an example of a CE formula (from the International Institute of Welding):

$$CE = \%C + \%Mn/6 + \%(Cr + Mo + V)/5 + \%(Ni + Cu)/15 \qquad (1)$$

In pipelines a CE < 0.4 is usually needed. At values of CE > 0.4 there may be welding problems. Note that CE formulae are only guides: on major projects weldability testing (trials) will be needed

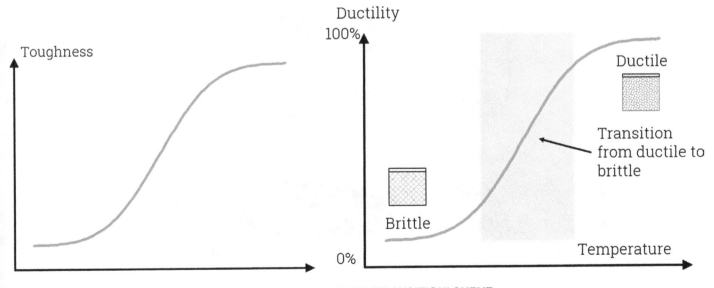

FIG. 2.3 TOUGHNESS 'TRANSITION' CURVE

2.3.2 Types of Line Pipe

The line pipe is usually supplied in lengths of 12 m (40 feet), with the requested diameter, wall thickness, and material properties. This line pipe is often classified according to how it was formed; for example (Figure 2.4):

- longitudinally welded;
- spirally welded; or,
- seamless (no weld).

Longitudinally welded pipe needs to be shaped before welding: it is shaped into a 'U' and then an 'O'. Most large diameter line pipes are 'cold' expanded ('E') diametrically in the pipe mill (but it is unusual to cold expand spiral welded line pipe). 'Cold' means no heating is used; therefore, the expansion is at ambient temperature. The line pipe is strained to at least 0.3%, and usually 0.5–1.5%, to give: an increased yield strength; and, the correct diameter and roundness.

Welded line pipe is made using one of two processes (Figure 2.4):

- A process that adds weld metal to join the ends of the plate (for example, double submerged arc welded, DSAW, line pipe). Line pipe made using a double submerged arc weld has two longitudinal welds, Figure 2.4, and has been made since the 1940s.
- A process (autogenous) that welds the edges of the plate together without deposited weld metal. 'Electric resistance welded' (ERW) line pipe (modern version is 'high frequency induction', HFI) is an example of an autogenous weld. The first ERW line pipe was produced in the 1920s.

Seamless line pipe was first produced in larger diameters in the late 1930s, and is made from a solid 'billet' (a large solid cylinder) of steel. A history of line pipe types is given in Table 2.1.

The processes used to produce line pipe have evolved as the steel-making processes evolved [15]. The ends of the plate are joined together in different ways. This joining started with simple butt welds along the longitudinal axis, made in furnaces or lap and hammer welded. Up to 1930 most line pipe was made by 'furnace lap-welding' in which the tapered edges of a cylinder of steel (created by heating) were mechanically bonded with heat and pressure.

Lap-welded pipe had a weak longitudinal seam weld, which was reflected in design standards that limited its design stresses. Modern seamless pipe and ERW pipe became alternatives to

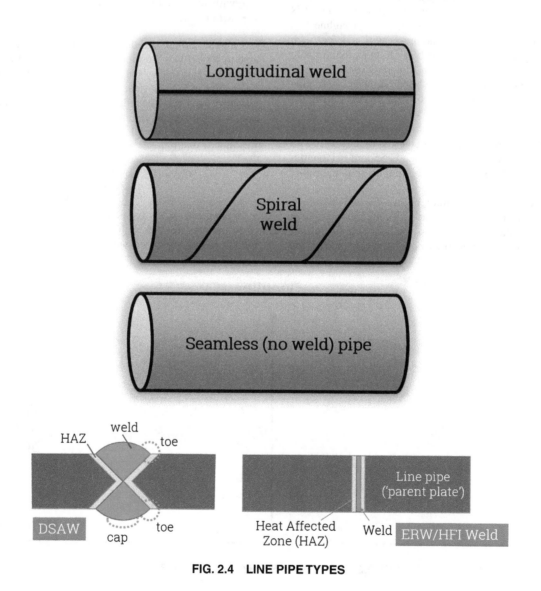

FIG. 2.4 LINE PIPE TYPES

TABLE 2.1 EVOLUTION OF LINE PIPE USED IN PIPELINES[1] [13]

Process	Process dates Start	End	Common diameters (in.)	Max. length (feet)	Unique identifying characteristic(s)
Furnace butt welded	1832	1954	0.125 to 3	20	No visible weld relatively short joint length
Continuous butt welded	1923	Current	0.125 to 4½	40	Uniform wall thickness with no visible weld
Lap weld	1887	1962	1.25 to 30	22 to 26	Waffle-like pattern over the weld seam
Hammer weld	1917 to 1921	1942 (or later)	20 to 96	30	
Electric resistance weld	1928	Current	1.25 to 24	80	Occasional 'trim tool marks' near the weld zone
Flash weld	1930	1972	8.625 to 36	40	Square weld bead on the inside (ID) and outside (OD) diameter
Single sided arc weld (SAW)	1925	1952 (or later)	to 96	40	Elliptical weld bead on the OD
Double submerged arc weld (DSAW)	1946	Current	16 to 48	40	Elliptical weld bead on the ID and OD
Seamless	1890 1899 1938	Current	to 6 to 16 to 26	40	Surface roughness, and helical variation in wall thickness
Spiral weld	1948	Current	to 56	40	Helical weld

[1] The first six processes in this Table 2.1 do not use weld 'filler' metal. The ends are forged together.

lap-welded pipe in the late 1920s, and by 1950 lap-welded pipe manufacturing disappeared, as ERW and seamless pipe proved to be superior [15]. It should be noted that early ERW pipe (pre-1970), welded using a low frequency welding process, was prone to weld problems [e.g., poor fusion in the weld line, and heat affected zone (HAZ[2])] (Figure 2.4) and this caused failures [16–18]. This resulted in ERW pipe being used only for low pressure operation. Modern HFI/ERW line pipe, from a reputable line pipe supplier, using good quality steel, does not have these problems, but the pipeline industry still has legacy issues from this older ERW line pipe [16–18].

The 'submerged arc process' ('SAW') followed the production of seamless pipe, and became the most common means of producing large diameter (≥24" (610 mm)) line pipe, but by the 1940s the more reliable double submerged arc welded line pipe superseded the single SAW.

2.3.3   Steel-Making

Pipelines have gradually increased in both diameter and operating pressures, Table 2.2.

These increases are due to improvement in line pipe manufacturing processes, better welding, and better steel. Early (pre-1960s) line pipe steel was made from 'semi-killed' steel. 'Semi-killed' steel is partially deoxidized: oxygen in combination with other elements forms non-metallic inclusions (e.g., hydrogen sulfide) which are considered impurities. These steels had relatively poor toughness and low strength [13].

'Fully-killed' steels (i.e., fully deoxidised) were then introduced to give better toughness, but high strength could only be achieved by alloying (adding other elements to the steel mix), which caused poor weldability. Much line pipe contains a

TABLE 2.2 SIZE INCREASES IN PIPELINES OVER THE PAST CENTURY [14]

Year	Pressure, barg (psig)	Diameter, mm (in.)
1910	2 (29)	400 (15.5)
1930	20 (290)	500 (19.7)
1965	66 (957)	900 (35.4)
1980	80 (1160)	1420 (55.9)
2000	120 (1740)	1620 (63.8)

TABLE 2.3 INCREASE IN LINE PIPE STRENGTHS BY DECADE [14]

Time	Highest yield strength
1950	X52
1960	X60
1970	X70
1980	X80
2000	X100

longitudinal or spiral weld, and the line pipe is welded together to make a pipeline; therefore, the line pipe must be easily welded together.

'Controlled' rolled [or Thermo Mechanical Control Process (TMCP)] steels were then developed that gave high toughness and high strength. Controlled rolling gives a finer grain size within the steel which leads to high strengths and toughnesses [13]. Tables 2.3 and 2.4 summarize the increase in mechanical properties by decade: these increases in strength and toughness allow pipelines to operate at higher stresses and tolerate bigger defects.

[2] The heat affected zone is the material around a weld that is affected (material properties change) by the welding.

TABLE 2.4   INCREASE IN TOUGHNESS BY DECADE [19]

Decade	1950	1960	1970	1980	1990
Grade	X42/52	X52/60	X60/65	X65/70	X75
Typical CVN, J (ft lb)	27 (20)	41 (30)	54 (40)	88 (65)	109 (80)

2.3.4   Line Pipe Coatings

2.3.4.1 Corrosion   All steel structures, such as pipelines, can corrode. Corrosion is an electro-chemical process; therefore, all corrosion cells must have:

- An anode (+). The anode is the point at which metal dissolution occurs and electrons are created. This is an electrode in an electrochemical cell where oxidation (electrons are lost) occurs. Electrons flow away from the anode in the external circuit. These electrons flow to the cathode.
- A cathode (–). Oxygen reduction occurs at the cathode and the electrons are consumed. This is an electrode in an electrochemical cell where reduction (gaining electrons) is the principal reaction.
- A metallic path connecting the anode and cathode (e.g., the pipe).
- An electrolyte (e.g., solutions such as water).

If all these element are present, corrosion usually occurs and metal ions enter the solution at the anode, Figure 2.5. If one of these is not present then the corrosion process will stop, and this is how most corrosion protection systems work; for example, by creating a barrier (the coating) and eliminating the electrolyte.

2.3.4.2 Line Pipe Coating Types   Pipelines will be buried or underwater. Steel will corrode if exposed to the surrounding soil or sea; therefore, the line pipe must have an anti-corrosion coating on its outer surface. The coatings used today are:

- asphalt (from crude oil), coal tar (from coal);
- powder systems (based on epoxy resins);
- polyethylene, polypropylene systems; and,
- liquid systems.

Most modern line pipe uses fusion bonded epoxy (FBE), 'two-layer' polyethylene or 'three-layer' polyethylene or polypropylene, Figure 2.6.

Figure 2.6 gives examples of coatings, and also a pipe with concrete 'weight coating'. Some line pipe used under water has this concrete coating to prevent the pipeline floating back to the surface. This is called a 'weight coat'. Coal tar (anti-corrosion) coatings are often used below the concrete coating.

Some pipelines also have internal coatings, as the pipeline may be needed to be protected against aggressive products, but on most pipelines the internal coatings are used to improve flow. Accordingly, internal coatings are often called 'flow' coats. Coatings can increase flow by several percent [21].

2.3.4.3 'Cut-Back' Areas   The line pipe is welded 'in the field'. The area around the girth field weld will need coating; consequently, the ends of each section of line pipe do not contain coating between 75 mm (3") and 150 mm (6"), as illustrated in Figure 2.7. These areas are the 'cut-back' areas, and a variety of coatings are used on this cut-back area (e.g., 'shrink sleeves', or the girth weld area has a coating to match the line pipe coating).

It should be noted that all actions conducted outside the controlled conditions in a factory must be carefully supervised. Any type of coating in the field will be exposed to both the construction environment and weather. This can lead to coatings on the cut-back areas being inferior to that on the line pipe. Operators often report a disproportionate amount of external corrosion on their pipelines around the girth weld, which can be due to poor field application of coatings around the girth weld.

2.3.4.4 Field Coatings versus Factory Coatings   Line pipe can be coated in a factory (controlled, clean conditions), or in the field (less controlled, difficult conditions). Most modern line pipe has a factory-applied coating, but many older pipelines were field-coated, and these field coatings may not be as good as factory-applied.

The earliest pipelines were buried without any external coatings, but by the 1920s pipeline operators were aware of soil causing corrosion, and started to coat their pipelines as it was being laid, 'in the ditch' [12]. Usually, coal tar or wraps were used, but these older coatings are susceptible to disbonding, and other faults (called 'holidays').

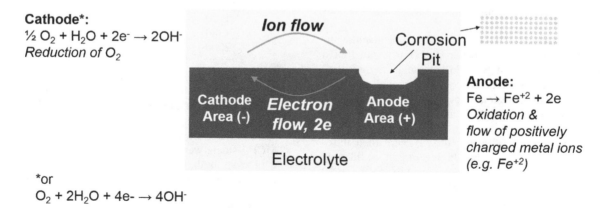

Cathode*:
$\frac{1}{2} O_2 + H_2O + 2e^- \rightarrow 2OH^-$
Reduction of O_2

*or
$O_2 + 2H_2O + 4e^- \rightarrow 4OH^-$

Ion flow

Corrosion Pit

Cathode Area (-)   *Electron flow, 2e*   Anode Area (+)

Electrolyte

Anode:
$Fe \rightarrow Fe^{+2} + 2e$
Oxidation & flow of positively charged metal ions (e.g. Fe^{+2})

FIG. 2.5   CORROSION CELL [20]

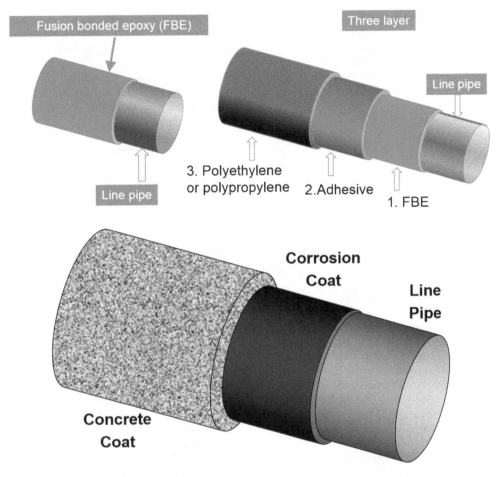

FIG. 2.6   EXAMPLES OF PIPELINE EXTERNAL COATINGS

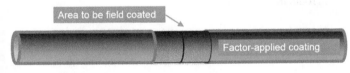

FIG. 2.7   "CUT-BACK" AREAS ON LINE PIPE TO BE COATED IN THE FIELD

Pipeline operators determined that coatings were the primary defense against corrosion, but could not provide complete corrosion protection [12].

2.3.5   Cathodic Protection

Coatings on early pipelines were good, but the engineers noted that where coating was damaged, missing, or disbonded, rapid corrosion could still occur [12]. Operators started to fit 'cathodic protection' (CP) to their pipelines, as a secondary corrosion protection method.

CP is a technique to control the corrosion of a metal surface by making that surface the cathode of an electrochemical cell, Figure 2.5. Current (electrons) is supplied to the pipeline by connecting the pipeline to a 'sacrificial anode' (e.g., a magnesium block), or to a commercial electricity supply from an 'anode bed', Figure 2.8. By the late 1940s pipelines were built with both an external coating and CP [12].

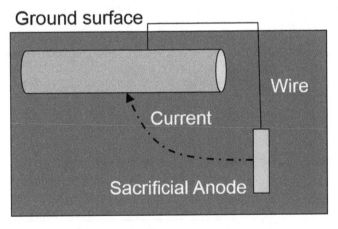

FIG. 2.8   EXAMPLE OF A CATHODIC PROTECTION SYSTEM

2.4   PIPELINE ROUTING

2.4.1   Introduction

Pipelines need a route to transport their product safely (for both people and environment) from the product's source, to the product's destination. Standards and laws usually do not give detailed guidance on routing: they generally limit guidance to matters of public safety. They can have provision for rejecting applications for 'superfluous' pipelines, and may specify minimum proximities of pipelines from buildings, landfill sites. They can also have provisions for the safeguarding of the environmental and water supplies, and obligations for the reinstatement of agricultural land, but the actual detail of pipeline routing is left to the routing engineers.

2.4.2 Environmental Considerations

Pipelines must pass two tests to be shown to be beneficial to society:

- environmental impacts; and,
- consumer benefits (is this a 'least cost' option?).

A pipeline project must first satisfy all environmental requirements before consumer benefits are considered. Therefore, the start of planning a pipeline route will be the start of the assessment of possible environmental impacts, as the construction of a pipeline can affect the environment.

All environmental considerations are included in an environmental impact assessment (EIA). Federal, state, and local regulators want to ensure the pipeline has minimal impact on the environment, and will require an EIA.

This EIA will identify potential environmental effects caused by the construction and operation of the pipeline. It comprises a series of studies, surveys, and consultations that enables the pipeline to be routed to minimize its effects, and identifies measures to ensure successful reinstatement after construction. Table 2.5 gives a typical outline of an environmental assessment.

2.4.3 Pipeline 'Line Pipe', 'Components', and 'Facilities'

The pipeline route has to accommodate the pipeline, but the pipeline is a 'system' that includes many types of equipment and plant. The location of all this equipment and plant must be planned in the routing.

The pipeline to be routed is made up of lengths of line pipe, welded together. This line pipe is specially manufactured for use in pipelines, and it is strong, ductile, tough, and weldable. The line pipe is usually ordered with an anti-corrosion coating, as it will be buried below ground or under water.

TABLE 2.5 TYPICAL OUTLINE OF AN ENVIRONMENTAL ASSESSMENT [22]

Considerations	
Air quality	Planning/future land use
Geological/Ecological/ Biological/Flora/Fauna environment	Emissions
Cultural environment	Archaeology
Cumulative impacts (considering other local projects)	Geology, topography, soils, seismicity
Grow-inducing impacts	Socio-economic
Hydrology and water impact	Visual/scenic resources
Land use/landscape/soils	Transportation/traffic
Noise	Socio-economic environment
Public health and safety	Public service and utilities

The line pipe is described by its outside diameter, wall thickness, and strength. The supplier of the pipe will produce the pipe within specified tolerances; therefore, the diameter and wall thickness will have plus and negative tolerances around the specified dimensions.

The pipeline will also need 'components' ('fittings'). These are any items that are part of the pipeline, other than a straight line pipe or line pipe bent in the field to allow direction changes. Components are valves, flanges, tees.

Pipelines also contain 'facilities' ('installations') such as compressor stations (to drive gas along a pipeline), pumping stations (to drive liquid along a pipeline), metering stations (to measure the input and output of the pipeline). Facilities include all plant and equipment for the extraction, production, chemical treatment, measurement, control, storage, or offtake of the transported fluid [9].

2.4.4 The Routing Process

Routing is a key element of pipeline design. The process of routing a pipeline is:

- a pipeline is needed;
- an economic study and basic engineering and route planning is conducted;
- an approximate route (or routes) is selected;
- permission to build is sought from Government Agencies;
- the pipeline company must notify and consult with many authorities and satisfy many regulatory requirements and laws (the route must not damage the environment and must be acceptable to local communities);
- landowners, etc., must be consulted;
- a more detailed route is planned;
- the land is acquired;
- the pipeline is built.

The route must allow:

- access for all construction traffic up to commissioning;
- access during operation for maintenance; and,
- be a distance from the general public during operation.

Hence, the operator needs to consider certain boundaries around the pipeline (Figure 2.9):

- a 'right of way' (ROW) or 'permanent easement' along the route;
- a 'working width' during the construction;
- a 'zone' ('setback', or 'location classification' primarily for hazardous gas lines) beyond the right of way, to avoid or limit adjacent buildings, where the location, size, and number of buildings allowed in these zones are specified in pipeline design standards.
- 'high consequence areas' are areas around a pipeline where a failure during operation would have high consequences, and regulations now require pipeline operators to conduct a 'pipeline integrity validation' (through inspection, testing, and analysis) of pipelines that run through/near these (high consequence areas (HCAs)).

2.4.4.1 Right of Way The pipeline operator will need access to their pipeline for operational and maintenance reasons; therefore, they will need to purchase or lease the land the pipeline will reside

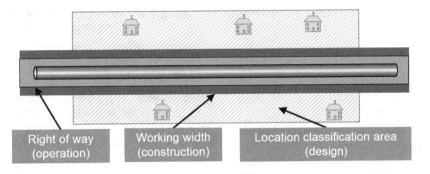

FIG. 2.9   ROUTING BOUNDARIES

in. This right of way is the narrow corridor of land within which a pipeline operator has the right to conduct activities in accordance with an agreement with the landowner [9].

The ROW is sometimes called a 'permanent easement': an 'easement' is a legally binding agreement between the landowner and the owner of the pipeline, either in perpetuity or for a defined period, which details the rights and obligations of both parties. Ownership of the land remains with the landowner, and the landowner has only given up defined rights on the portion of land used for the right of way [9].

The width of the ROW needs to be sufficient to allow the operator to perform all routine operations on the pipeline, including its excavation, Figure 2.10.

2.4.4.2 Working Width   The pipeline will be constructed using heavy engineering equipment, and a large workforce. This will require a width of land, for the period of time it takes to construct the pipeline, and this width will necessarily be wider than the ROW.

This width of land needed for the construction is called the 'working width', Figure 2.11, and this width must be able to accommodate the moving traffic supplying materials and equipment, stationary traffic excavating the trench, and laying the pipe, and also storage for the 'top soil' (the rich organic soil used by landowners for crops and livestock), and the 'subsoil' from the pipeline's trench.

The working width is not needed by the operator on a permanent basis, and hence it is often called a 'temporary easement'. This means the pipeline operator and the construction company have temporary rights to use the land, throughout the pipeline construction and commissioning.

2.4.4.3 Setbacks and Location Classifications   Pipeline standards limit the location and density of the population around a new pipeline, if it carries a hazardous substance such as natural gas. These limits are not imposed on less hazardous products, such as crude oil: ASME B31.8 (for gas pipelines) contains these restrictions ('location classifications'); but, ASME B31.4 (liquid pipelines) does not contain these restrictions.

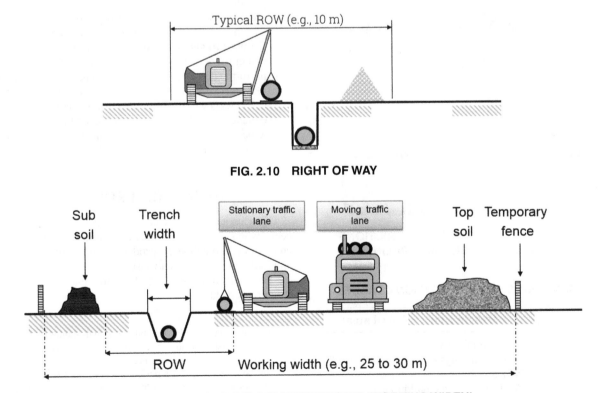

FIG. 2.10   RIGHT OF WAY

FIG. 2.11   PIPELINE CONSTRUCTION'S 'WORKING WIDTH'

Standards limit population around a new pipeline by one or two methods (Figure 2.12):

- ensuring a low density of buildings, or people, along the pipeline route; and,
- specifying a minimum distance (or 'proximity' or 'setback') for buildings from the pipeline.

Most countries impose these limitations by specifying 'location classes' for their natural gas pipelines. ASME B31.8 classifies gas pipelines in four ranges: 'Class 1' to 'Class 4'. The classification depends on the number of buildings, traffic density, around a gas pipeline, within a 1 mile by 440 yd area, Figure 2.9. Canada's (CSA Z662) standard has a boundary of 1600 m × 200 m (either side).

This classification is important as it limits the 'design factor' of the pipeline: Table 2.6 gives the classification in ASME B31.8. Design factor is covered later in this Chapter, but the effect of a reduced design factor is usually thicker-walled pipe needing to be used in more densely populated area [23]. This will mean higher line pipe costs and higher construction costs, and so needs to be avoided.

2.4.4.4 High Consequence Areas Regulations can require pipeline operators to conduct a 'pipeline integrity validation' (through inspection, testing, and analysis) of pipelines that run through/near high consequence areas (HCAs). These HCAs are where a pipeline failure would have high consequences. It is common-sense to route pipelines away from such areas.

The consequences of natural gas and hazardous liquid pipeline releases differ [24], Table 2.7:

- HCAs for natural gas transmission pipelines focus on populated areas, as environmental and ecological consequences are usually minimal; whereas,

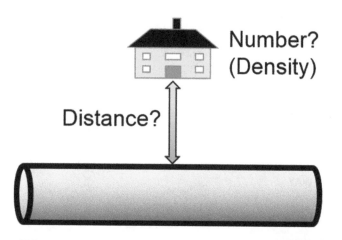

FIG. 2.12 LOCATION AND DISTANCE RESTRICTIONS IN PIPELINE STANDARDS FOR GAS PIPELINES

TABLE 2.6 ASME B31.8'S FOUR LOCATION CLASSES

Class	Area (in 1 mile × 440 yd area)
Class 1	0–10 buildings (rural)
Class 2	11–45 buildings (areas around towns)
Class 3	46+ dwellings (e.g., suburban)
Class 4	Multi-storey-type buildings

TABLE 2.7 HIGH CONSEQUENCE AREAS IN THE USA AS DEFINED BY PHMSA IN 2011

HCAs for liquid lines in the USA include:	HCAs for gas lines in the USA include:
High population areas (e.g., 10,000 people/sq. mile). 'Busy' commercial navigable waterways.	ASME B31.8 Class Location 3 or 4. Class Location 1 or 2 with 'potential impact radius (PIR)' > 660 ft. (200 m) containing ≥ 20 occupied buildings.
Unusually sensitive areas (as defined in USA 49 CFR 195.6 as drinking water or ecological resource area).	The area within the PIR circle with an 'identified site' such as a school or camping ground.

- HCAs for hazardous liquid pipelines focus on populated areas, drinking water sources, and unusually sensitive ecological resources.

2.4.5 Typical Routing Sequence

A pipeline is routed using an iterative process: a wide 'corridor of interest' (several kilometres) is first selected, and as more detailed data (e.g., soil types) are obtained, this corridor is narrowed down to a preferred 'route' [e.g., 500 m (1640 feet) wide]. This wide route allows changes in the final position of the pipeline, as even more detailed data is received. Finally, the working width and ROW are selected, and maps produced for the pipeline construction company.

The typical stages followed in routing a pipeline are:

- identify a search area based on the start and end points of the pipeline;
- identify potential corridors [for a short pipeline this will be ~1 km (3281 feet) wide] within which the pipeline will eventually be routed, using desk-based information;
- gather more data and select a preferred corridor;
- perform site visits and conduct studies (soils, ecology, archaeological) to allow a preliminary pipeline route [e.g., 500 m (1640 feet) wide] within the preferred corridor;
- conduct more detailed surveys and studies to allow a final route to be mapped, clearly indicating the boundaries of the working width and the right of way, and the pipeline centerline.

2.5 THE DESIGN PROCESS

This Chapter will now cover pipeline design. Design is a 'process', which starts with simple and limited data (e.g., the start and end point of a pipeline, its product, its intended flow rate, and inlet and outlet pressures[3]), but must end with a detailed design with all the necessary drawings, specifications, maps, materials, and construction schedule.

---

[3] Pressures in the pipeline business and standards are quoted in 'gauge' pressure. Gauge pressure, p_g, is usually measured as the pressure above atmospheric pressure, p_{atmos}. 'Absolute' pressure, p_a, is pressure in excess of a perfect vacuum; therefore, $p_a = p_g + p_{atmos}$. Gauge pressure is written as 'psig' or 'barg', and absolute pressure as 'psia' or 'bara'.

TABLE 2.8 THE BUSINESS PROCESS

Player	Role
Producers	Who provide oil and gas to...
Shippers	Who buy the oil or gas from the producers to sell to their...
Customers	Who receive the oil and gas from the...
Pipeline companies	Who transport the oil and gas through pipelines, on behalf of the shippers, to the shippers' customers... and finally...
Regulators	Who are government bodies who protect the public interest, and ensure that tolls charged for transportation services are just and reasonable, and there is fair access to pipeline.

2.5.1 The Business Process

The business process of pipelines has five main 'players', Table 2.8, but it is the pipeline company that designs the pipeline, to satisfy the needs and requirements of the other players.

2.5.2 The Legal Process

The design, construction, and operation of transmission pipelines are usually controlled by national regulations or laws.[4] The selection of a design standard, or design calculations, is often limited by these regulations/laws.

Before a pipeline can be built, the pipeline company must obtain:

- a permit or certificate from a government agency; and,
- various local, county, and state permits.

The agencies represent 'public interests'. The pipeline cannot be built or operated without these permits. The permit application needs to provide detailed information about the pipeline and facility designs (basic technical data), route, and environmental impacts.

The government agencies begin a review process of the permit application by notifying affected property owners. In addition, the operator will send all affected landowners and other stakeholders information about the project.

This can include:

- a project map;
- responses to questions frequently asked by landowners; and,
- other project information.

2.5.3 The Social Impact and Environmental Process

The environmental impact assessment (EIA) of the pipeline route, construction, and operation is undertaken at the earliest opportunity. The construction of the pipeline can impact on the environment, and this impact must be assessed.

---

[4] Laws ('statutes') are created by Governments (e.g., the USA Congress). Regulations are 'rules' based on an interpretation of these laws, usually written by Federal Departments. They are guidance to implement, interpret, or make specific the law enforced or administered. Regulations have the same effect as laws: both are enforceable. Failure to comply with either the laws or regulations could result in legal proceedings.

All pipelines will require a full environmental assessment during their design;

Example: an onshore pipeline can include [25]:

- impact (social and economic) on owners and occupiers of the land around the pipeline;
- the disturbance to local communities (noise, increased numbers of construction vehicles);
- the accidental transmission of plant and animal pests;
- damage (e.g., erosion) to river and stream banks or adverse downstream effects of fisheries;
- scarring of landscape patterns by hedgerow or woodland removal;
- accelerated erosion of fragile habitats such as moorlands or heathland ecosystems.

2.5.4 Onshore and Subsea Pipelines

The design process starts with a simple question... is the pipeline onshore (or 'on land'), or subsea (or 'offshore'). This is because differing standards and regulations apply to onshore and subsea pipelines.

BSI PD 8010 [9] states an onshore pipeline is a: "...*pipeline laid on or in land, including those sections laid under rivers, lakes and inland watercourses.*" ISO 13623 [3] states an offshore pipeline is a: "... *pipeline laid in maritime waters and estuaries seaward of the ordinary high water mark.*"

2.5.5 Liquid and Gas Pipelines

Many pipeline standards and regulations and laws differentiate between liquids (e.g., crude oil) and gases (e.g., natural gas). For example, in the USA, one pipeline standard (ASME B31.4) is used for liquid (incompressible fluid) pipelines; whereas another (ASME B31.8) is used for gas (compressible fluid) pipelines. Hence, the pipeline product needs to be confirmed. When this is confirmed, and the location (onshore or subsea) is known, a formal 'design process' is followed.

2.5.6 The Design Process

Pipeline design is a process. It starts with feasibility and conceptual studies, Figure 2.13. These lead into the 'detailed design', where all the design must be specified in great detail. The feasibility and conceptual phases are often called 'front end engineering design' (FEED).

A 'feasibility study' attempts to determine the practicality of a project. The simplest type of feasibility study answers the 'yes-or-no' question: '*should we undertake the project?*'

The conceptual studies include:

- consultations with 'statutory' (government) bodies on constraints and mitigation;
- several corridors for the pipeline, with a preferred option;
- more detailed engineering information, maps, and land acquisition needs;
- cost estimates (+15%); and,
- scope for of the detailed design phase, and procurement and construction strategies.

Detailed design is the start of moving from the 'definition' phase of design, to the 'execution' phase. The detailed design provides:

- detailed design calculations and surveys;
- route maps and drawings for civil, mechanical, electrical, instrumentation, and communication disciplines;

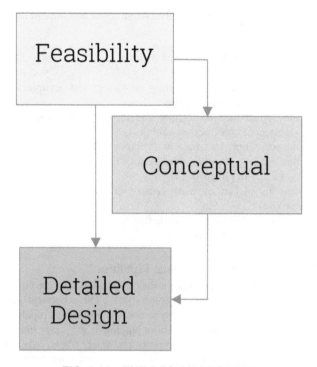

FIG. 2.13 THE DESIGN PROCESS

- full materials' schedules; and,
- commissioning and operational procedures, and the scope of work for construction.

2.6 PIPELINE DESIGN

2.6.1 Design Standards: Good Practice but Minimum Requirements

Pipelines need to be designed to operate safely; therefore, published standards are used for all design, construction, and operation. There are many pipeline standards around the world. Usually, there will be a recognized (by both the pipeline owner and the regulatory authority) standard already in use in the country where the pipeline is to operate.

A standard is [26] "*a set of technical definitions and guidelines that function as instructions for designers, manufacturers, operators or users of equipment to provide consistent and comparable results.*" It is [27] "*an agreed, repeatable way of doing something.*"

It should be emphasized that a standard is not a law, it is merely 'good practice' [28, 29]:

"*... a code is not a law... it is... written by engineers, operators and managers... as a result of their experience and their knowledge of the engineering and scientific principles involved, state what they agree is good practice from the standpoint of public safety... a code is merely a statement of what is generally considered good practice...*"

Accordingly, standards are considered voluntary because they serve as guidelines, but do not have the force of law. Their use becomes mandatory when they have been incorporated into a business contract or incorporated into regulations. When a standard is adopted by one or more governmental bodies, or has been incorporated into a business contract, it is called a 'code' [30].

Standards ensure essential and minimum safety requirements. CSA Z662 [6] states: "*This Standard is intended to establish essential requirements and minimum standards for the design, construction, operation, and maintenance of oil and gas industry pipeline systems.*"

The design, construction and operation of a pipeline will be expected to meet more than minimum requirements. Indeed, the UK's regulator states [31]: "*... depending on the level of risk and complexity of the situation, it is possible that meeting good practice [such as designing to standards] alone may not be sufficient to comply with the law.*"

It should be emphasized that standards are not 'specifications'. The pipeline industry uses specifications (for example, API 5L [32]), but a specification gives requirements on material properties, tolerances, acceptance levels, and is usually written for a specific item/job/contract. It is very detailed and 'specific'. Standards are not as detailed, and can give the designer choices and options, rather that requirements, and can allow the designer to use different procedures for unusual conditions [29].

This section of the Chapter should be noted by staff involved in integrity management of pipeline, as these staff also work to standards (for example, References [33] to [38]): these standards are minimum requirements, and companies would be expected to do more than detailed in these standards.

2.6.2 Design Standards Are Safety Standards

The most popular pipeline design standards are from the American Society of Mechanical Engineers (ASME):

- ASME B31.8 for pipelines carrying gases, such as methane;
- ASME B31.4 for pipelines carrying liquids, such as crude oil.

They were first published in their modern form in 1955 [39] when the main industry/standards concerns were:

- maintaining the safety of the pipeline system while economically transporting product; and,
- establishing requirements for safe design, construction, inspection, testing, operation, and maintenance of pipeline systems.

These standards focus on the safety aspects of a pipeline's design and operation, and provide generic guidance: they cannot cover unusual conditions or special designs.

As pipelines were built in other regions, other standards were developed. These new standards often used the ASME standards as a basis, but many countries needed to adapt the ASME standards to allow for differing environments; e.g., mountains, deserts, jungles, and hence produced their own standards.

Pipeline standards usually cover the design, fabrication, installation, inspection, and testing of pipeline facilities used for the transportation of a fluid. They also cover safety aspects of the operation and maintenance of those facilities. Their intent is to ensure that pipeline systems are adequate under normal operating conditions, and they are concerned with the safety of the general public, and employee safety.

2.6.3 Design Standards and Competency

Designing a pipeline requires competent designers. CSA Z662 [6] states: "*This Standard is not a design handbook, and competent engineering judgment should be employed with its use.*"

'Competence' is a mix of practical and thinking skills, experience, and knowledge (gained through experience or study [40]), and also depends on the individual's values. Developing and maintaining a competency involves education/training, mentoring, and experience.

An individual (or organization) is competent when he/she has: sufficient knowledge of the tasks to be undertaken and the risks involved; and, the experience and ability to carry out their duties, and recognize their limitations [41].

2.6.4 Substance Classification in Standards

Pipelines transport different substances, such as natural gas or diesel. Standards have differing requirements for differing substances; therefore the substance in the pipeline has to be 'classified' before starting design.

This substance classification is important, as pipelines carrying hazardous gases have differing, and more conservative design requirements compared to pipelines carrying liquids. For example, highly stressed pipelines carrying natural gas are not permitted in heavily populated areas, but pipelines carrying gasoline are permitted.

Substance classification can be simple [for example, dividing all substances into compressible fluids (gases) or incompressible fluids (liquids)], or a more detail assessment based on the differing hazard potential of fluids.

Some countries (e.g., USA) have different standards for different fluids: ASME B31.8 is for 'gases', and ASME B31.4 for 'liquids'. This demarcation (between natural gas and hydrocarbon liquids) is a simple way to account for hazard potential: natural gas would be viewed as high hazard, as its major consequences would be casualties; whereas liquid lines would have major consequences on the environment. Other countries (e.g., Canada and Australia) have standards covering both liquids and gases.

Standards usually classify substances in terms of their 'hazard potential'. Gases will usually have higher hazards than liquids. Table 2.9 [9] gives examples of this hazard potential.

Pipeline substances are classified using these hazard potentials in pipeline standards. ISO 13623 [3] gives the classifications in Table 2.10. Each substance classification then follows a specific design route in the standard, with the more hazardous substances attracting more stringent design criteria.

TABLE 2.9 HAZARD POTENTIAL OF SOME PIPELINE SUBSTANCES

Substance	Hazard potential
Liquids	Such as crude oil and refined petroleum products are flammable (and radiate a high heat on ignition) and will be released as a liquid which can infiltrate water courses.
Liquid petroleum gas (LPG)	Is flammable and although conveyed by gas or liquid in a pipeline, will be released as a heavier-than-air gas (propane and butanes).
Methane	Is flammable, lighter than air, radiates heat on ignition, and can form a vapor cloud.
Ethylene	Is flammable, radiates high heat on ignition, and is slightly lighter than air.

TABLE 2.10 CLASSIFICATION OF PIPELINE SUBSTANCES IN ISO 13623 [3]

Classification	Description
A	Non-flammable water-based fluids.
B	Flammable and/or toxic substances which are liquids at ambient temperature and atmospheric pressure conditions; e.g., oil, petroleum products, methanol.
C	Non-flammable substances which are non-toxic gases at ambient temperature and atmospheric pressure conditions; e.g., nitrogen, air, argon, carbon dioxide.
D	Non-toxic, single phase natural gas.
E	Flammable and/or toxic substances which are gases at ambient temperature and atmospheric pressure conditions and are conveyed as gases and/or liquids; e.g., hydrogen, methane (other than 'D'), LPG, natural gas liquids, ammonia, chlorine.

Gases and liquids mixtures should be classified in relation to their composition and hazard potential. If the classification is not clear, the more hazardous classification is used.

2.6.5 'Sizing' the Pipeline

Pipeline design calculations start with the determination of the 'size' (diameter) of the pipeline to satisfy the demands of the 'shippers'; i.e., deliver the required flow rate.

The shippers are the companies who will use the pipeline to transport their products, to their customers, at a specified flow rate. 'Sizing' a pipeline means determining the required internal diameter (D_i) to satisfy this flow rate, using a 'hydraulic analysis'.

The flow rate (Q) will depend of the fluid velocity, v, [which will be a function of the input pressure (p_i), output pressure (p_o)], internal pipe diameter (or pipe cross-sectional area, A), and the fluid and pipe properties, c (e.g., surface roughness and gas gravity).

Simplistically... $v = Q/A$.... therefore, $Q = vA = v\pi D_i^2/4$... so... $D_i = (4Q/v\pi)^{0.5}$... but, it is not that simple, and is missing key parameters. Table 2.11 summarizes the factors affecting flow rate.

The equations describing flow in a pipeline are called 'equations of state', and have the general form:

$$Q = cD_i^{2.5}[(p_i^2 - p_o^2)/L]^{0.5}, \text{ or,} \qquad (2)$$

$$D_i = [Q/c]^2/[(p_i^2 - p_o^2)/L]^{0.4} \qquad (3)$$

The actual equations used are more complicated; for example, an equation used on gas pipelines is the 'Weymouth' equation:

$$Q = 18.06[T_o/p_r]\cdot[(p_i^2 - p_o^2)D_i^{16/3}/(GTLZ_a)]^{0.5} \qquad (4)$$

where:

Q Volumetric discharge (ft³/hour), measured at a reference pressure (p_r, psia) and reference temperature, T_o.

L pipeline length

D_i pipeline internal diameter

G gas gravity (air = 1)

TABLE 2.11   KEY PARAMETERS AFFECTING FLOW RATE

Parameter	Effect on flow rate
Size (diameter) of the pipe	To be calculated.
Inlet and outlet pressures	The higher the differential, the higher the flow rate.
Velocity of the fluid	Depends on the pressures (inlet/outlet) driving the fluid.
Friction of the fluid in contact with the pipe	The smoother, cleaner, and larger a pipe is, the less effect pipe friction has on the overall fluid flow rate.
Viscosity of the fluid	'Viscosity' is the measure of the internal friction in a liquid, or the resistance to a flow: the higher the viscosity, the more difficult it is to flow. Low viscosity fluids (e.g., water, alcohol) flow easily, whereas high viscosity fluids (e.g., cold honey) flow slowly. Generally, the higher a fluid's viscosity, the lower the flow rate.
Density of the fluid	Density = mass/volume. More dense fluids require more inlet pressure to maintain a desired flow rate.

Z_a	average compressibility factor (typically 1)	p_r	reference pressure (e.g., atmospheric pressure) in psia
p_i	inlet or upstream pressure (psia),	T_o	reference temperature (e.g., ambient) in deg R (= deg F +460)
p_o	outlet pressure (psia),	T	Temperature of gas in deg R

The pressure the fluid will enter the pipeline ('inlet pressure') could be the pressure from the oil/gas reservoir, or be provided by a pump (for a liquid) or a compressor (for a gas). This pressure will usually be the pipeline's 'design pressure'. 'Design pressure' is the maximum pressure a pipeline can normally operate at, and is calculated using the line pipe's yield strength, wall thickness, outside diameter, and appropriate 'joint' and 'design' factors. This calculation and these factors are covered later in the Chapter.

Pipeline pressure will decrease as the product flows along a pipeline due to:

- friction between product and the inside wall of the pipeline;
- elevation changes;
- friction in the product (viscosity);
- components (e.g., valves);
- offtakes.

This means the pipeline may need additional pumping/compressor stations along its route.

Having established the required diameter, the required wall thickness to withstand the imposed stresses on the pipeline needs to be calculated.

2.6.6   Pipeline Wall Thickness Calculation

Pipeline design standards aim to calculate the wall thickness that is able to withstand the internal pressure, and all other loadings; for example:

- ground movement;
- external pressure;
- fatigue;
- damage by external interference.

The main stress on a pipeline is usually the hoop stress (σ_h) created by internal pressure, Figure 2.14. This stress must not exceed the pipeline material's yield strength; therefore, the maximum pressure the pipeline can withstand needs to be calculated—its 'design pressure', p_d. The design pressure is the pressure needed to satisfy all the design requirements of the standard, and the required flow/delivery in the pipeline [42].

2.6.6.1 The Barlow Equation The wall thickness (t) of the pipeline has to withstand all stresses on it: stresses from internal and external pressure, stresses from external bending, tensile, and compressive loads, etc. Usually, the main stress in the pipeline is caused by internal pressure, and this main stress is the hoop stress, Figure 2.14.

This hoop stress is calculated using the design pressure (p_d) in the 'Barlow' formula. The Barlow formula is for 'thin wall' (high D/t ratio) pipe. At low D/t (≤ 20) the formula becomes inaccurate:

$$\text{Hoop stress } \sigma_h = p_d D_o / 2t \qquad (5)$$

$$t = p_d D_o / 2\sigma_h \qquad (6)$$

The diameter (D_o) used in this equation is the pipeline's outside diameter. North American standards use the 'nominal' ('specified') wall thickness, t_{nom} in this equation, but other standards (e.g., [3 to 5]) use the 'minimum' wall thickness (t_{min}), Figure 2.14.

The pipeline cannot operate at very high hoop stresses, as this would lead to the pipeline yielding and possibly failing. Standards control and limit stresses in pipelines by using 'design factors'. The design factors are specified in the standards as a percentage of the specified minimum yield strength, SMYS (e.g., 72% SMYS) or a ratio of pipeline stress to SMYS (e.g., 0.72).

2.6.6.2 Design Factor Design factors account for uncertainties in the pipeline's design [43, 44]. These uncertainties include:

- variability in materials;
- variability in construction practices;
- uncertainties in loading conditions;
- uncertainties in in-service conditions.

The design factors in pipeline standards do not exceed '1', and hence the pipeline stresses are maintained below the SMYS. Note the 'design factor' is not the same as a 'safety factor' (the ratio of operating pressure to failure pressure), as it is not a measure of how close to failure a pipeline is: it is a measure of how close it is to SMYS, Figure 2.15. See Section 2.6.10.

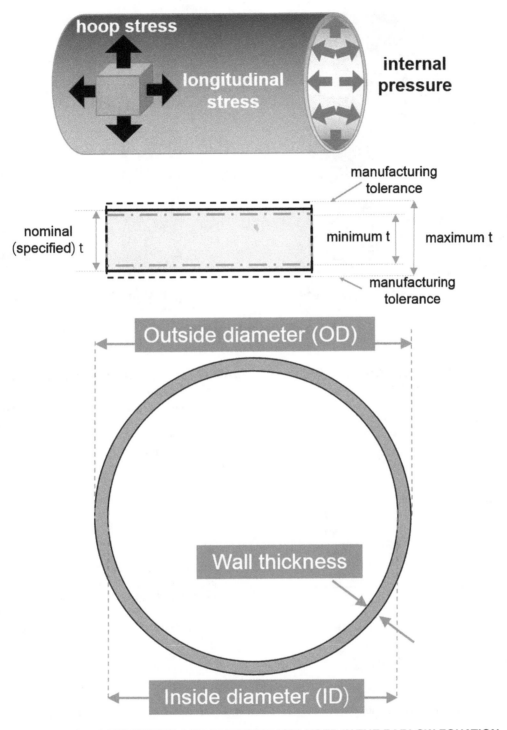

FIG. 2.14   WALL THICKNESS AND DIAMETER (OD) USED IN THE BARLOW EQUATION

Standards incorporate these design factors in the wall thickness calculation. For example, the hoop stress is limited by the hoop stress design factor:

$$\text{Design factor, } \phi = \text{Stress/SMYS} \tag{7}$$

$$\text{Hoop stress design factor} = \text{Hoop stress/SMYS} \tag{8}$$

$$\text{Hoop stress design factor} = \phi = p_d \cdot D_o / 2.t \cdot \text{SMYS} \tag{9}$$

$$t = p_d \cdot D_o / \{2(\phi \cdot \text{SMYS})\} \tag{10}$$

Table 2.12 gives examples of maximum hoop stress design factors quoted in standards around the world.

2.6.6.3 Joint and Temperature Factors   Pipeline standards include other parameters in the wall thickness calculation; for example, ASME B31.8 gives the design pressure, p_d, as:

$$p_d = (2.\text{SMYS} \cdot t/D_o) \cdot \phi \cdot \text{E.T} \tag{11}$$

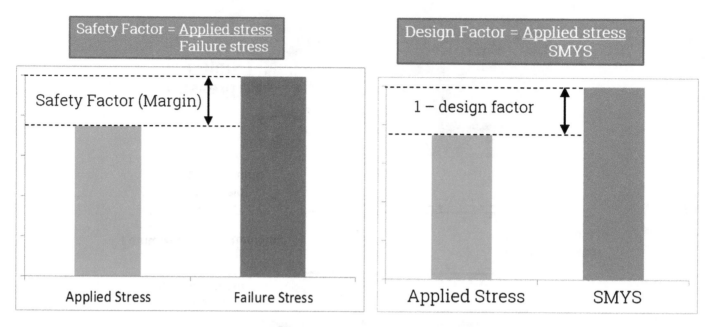

FIG. 2.15 DESIGN FACTOR

TABLE 2.12 HOOP STRESS DESIGN FACTORS IN STANDARDS

Standard	Hoop stress (σ_h) equation	Maximum hoop stress design factor
ASME B31.4 [1]	$\sigma_h = p_d D/2t_{nom}$	0.72 (0.80 for slurries)
ASME B31.8 [2]	$\sigma_h = p_d D/2t_{nom}$	0.80
BS 8010-1 [9]	$\sigma_h = p_d D/2t_{min}$	0.72
CSA Z662 [6]	$\sigma_h = p_d D/2t_{nom}$	0.80
AS 2885.1 [7]	$\sigma_h = p_d D/2t_{nom}$	0.80
ISO 13623 [3]	$\sigma_h = p_d (D - t)/2t_{min}$	0.83
EN 1594 [4]	$\sigma_h = p_d D/2t_{min}$	0.72

Therefore, wall thickness is:

$$t = p_d \cdot D_o / \{2(\phi \cdot \text{E.T. SMYS})\} \qquad (12)$$

where:

E = weld joint or joint efficiency factor, and T = temperature derating factor.

The joint efficiency factor, E, is to allow for the quality of the longitudinal or spiral seam weld in the line pipe. It is based on a history of line pipe's weld quality (strength). Most hydrotest failures in the 1950s and 1960s were caused by poor quality line pipe welds. This lead to some types of line pipe being considered inefficient to contain the full pipeline's pressure. In ASME and CSA Z662, E varies from 0.6 to 1, depending on the line pipe's weld type.

The temperature derating factor, T, accounts for the fact that line pipe's strength (and modulus of elasticity) decreases with increasing temperature. In ASME pipeline standards T varies from 0.867 to 1, depending on the design temperature.

2.6.7 Design Pressure and Maximum Operating Pressure

The design pressure is the maximum pressure permitted by a standard. The operating pressure of the pipeline must be less than or equal to the design pressure.

Equation 11 shows that the design pressure is the pressure calculated using:

- the line pipe's yield strength;
- wall thickness;
- outside diameter; and,
- appropriate 'joint' and 'design' factors, and 'derating' factors.

The design pressure is not necessarily the maximum operating pressure. CSA Z662 [6] states: *"The designer selects the design pressure for each segment of the pipeline system, and such design pressures are required to be not less than the intended maximum operating pressure. The maximum operating pressure is ultimately established by pressure testing...."*

This is important: the pipeline's operating pressure is not 'set' by the design pressure, it is set by pressure testing to defined pressure levels. The design pressure sets a 'ceiling' for this operating pressure, Figure 2.16.

ASME B31.8 [2] gives the following definitions:

- *"design pressure or internal design pressure: the maximum pressure permitted by this Code, as determined by the design procedures applicable to the materials and locations involved. It is used in calculations or analysis for pressure design of a piping component."*
- *"maximum allowable operating pressure (MAOP): the maximum pressure at which a pipeline system may be operated in accordance with the provisions of this Code."*
- *"maximum operating pressure (MOP): sometimes referred to as maximum actual operating pressure, the highest pressure at which a piping system is operated during a normal operating cycle."*

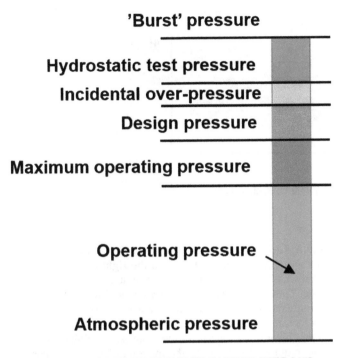

FIG. 2.16   OVER-PRESSURES IN PIPELINES [45]

2.6.8   Over-Pressures

Sometimes there are sudden increases in pressure in pipelines—a 'surge' or 'incidental pressure' or 'over-pressure'. These surges are usually in liquid pipelines, and can be caused by a sudden change of pressure; for example, a valve quickly closing. These over-pressures can send the pipeline pressure over its design pressure, or maximum operating pressure.

Most standards allow over-pressures of about 10% of the design or maximum operating pressure, Figure 2.16, and Table 2.13. This allowance for over-pressures allows operators to 'set' their pressure

TABLE 2.13   OVER-PRESSURE ALLOWANCES IN STANDARDS

Standard	Hoop stress factor (Using thickness specified in the standard/code (t_{code}))	Maximum over-pressures (% of design or maximum operating pressure[1])
ASME B31.4 [1]	0.72 (0.8 for slurry pipelines)	10%
ASME B31.8 [2]	0.80	10% (for design factors ≤0.72) 4% (>0.72)
BS 8010-1 [9]	0.72	10%
CSA Z662 [6]	0.80	10%
AS 2885.1 [7]	0.80	10%
ISO 13623 [3]	0.83	10%
EN 1594 [4]	0.72	15%

[1] Check individual standards for specific pressure.

relief devices at a pressure above their design or maximum operating pressure.

This allowance is also important for engineers who are responsible for the integrity of the pipeline: the maximum pressure a pipeline can experience is not the design pressure or the maximum operating pressure, but these pressures plus the allowance for over-pressures, or the set point of the pipeline's protective devices.

2.6.9   Location Classification and Proximity to Buildings/People

Pipeline standards limit the location and density of the population around a new pipeline, if it carries a hazardous substance such as natural gas.

These limits are not usually imposed for less hazardous products, such as crude oil:

- ASME B31.8 (for gas pipelines) contains these restrictions;
- ASME B31.4 (liquid pipelines) does not contain these restrictions.

Subsea pipeline standards do not usually have 'location classifications' as these pipelines are not near populated buildings. See DNV-OS-F101 [10] for 'safety classes' if a subsea pipeline poses a safety risk.

Standards limit population around a new pipeline by (Figure 2.12):

- ensuring a low density of buildings, or people, along the pipeline route; and,
- specifying a minimum distance (or 'proximity' or 'setback') for buildings from the pipeline.

2.6.9.1 Location Classification   Most standards impose limitations on the density/proximity of people around a pipeline by specifying location classes, usually for hazardous gas pipelines. This is simply a classification based on the number of buildings around the pipeline, within a specified area, Figure 2.17.

ASME B31.8 classifies its gas pipelines in four ranges: 'Class 1' to 'Class 4' [23]. The classification depends on the number of buildings, traffic density, around a gas pipeline, within a 1 mile by 440 yards area, Table 2.14.

ISO 13623 [3] has similar classifications (albeit spread over five Classes), and other standards/guidelines have simpler classifications based on people density rather than buildings; for example [11]:

- Type 'R'—rural areas with a population density not exceeding 2.5 persons per hectare (2.47 acres);
- Type 'S'—areas intermediate in character between Types R and T in which the population density exceeds 2.5 persons per hectare and which may be extensively developed with residential properties, schools, shops;
- Type 'T'—central areas of town or cities, with a high population density, many multi-storey buildings, dense traffic and numerous underground services.

The location classification affects the design factor that the pipeline can be operated at, Table 2.14.

The main purpose of location classifications is to limit excavation activities around a pipeline, as excavation has caused many serious pipeline failures in onshore pipelines. ISO 13623 states [3]:

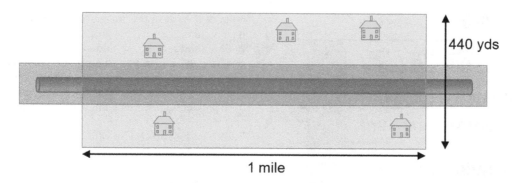

FIG. 2.17   AREA FOR CLASSIFICATION IN ASME B31.8

TABLE 2.14   LOCATION CLASSIFICATIONS IN ASME B31.8

Class	Area (in 1 mile × 440 yd area)	Design factor
Class 1	0–10 buildings (rural)	0.72–0.80 (depends on pre-service pressure test level)
Class 2	11–45 buildings (areas around towns)	0.60
Class 3	46+ dwellings (e.g., suburban), etc.	0.50
Class 4	Multi-storey-type buildings	0.40

"A significant factor contributing to the failure of pipelines is line damage caused by third-party activities... Determining location classes based on human activity provides a method of assessing the degree of exposure of the line to damage and consequent effect on public safety."

Using location classes should reduce damage to pipelines, by reducing the number of people (and hence activities) near the pipeline. Obviously, location classification can limit any consequence of failure, by ensuring limited population around the pipeline, but that is not its main purpose: its main purpose is protecting the pipeline from damage and this is a proactive way to design a pipeline. Location classification protects the pipeline from the people, and the people from the pipeline.

2.6.9.2   Decreasing Design Factors in Populated Area   Operators have historically been aware of increased excavation activities associated with high population areas. This led to increased wall thicknesses [43]. Also, operators had historically thickened the wall at road and rail crossings. *"Based on good engineering practice and a relatively safe record dating back to early last century, pipeline designs required thicker wall pipe in locations with higher population densities"* [43]. Therefore, design factor is reduced in higher location classifications by using thicker wall pipe, Equation 12.

Reducing design factor by increasing wall thickness can also:

• ensure more resistance to external interference;
• reduce the stress levels, which gives the pipe an increased ability to withstand pipeline damage from excavations without rupturing; and,
• provide additional safety if corrosion occurs in the higher population areas.

2.6.9.3   Changes during Operation   The pipeline operator has no control over land outside the right of way, and consequently has no direct power to prevent building around the pipeline. This means that more buildings within a location classification can change the pipeline's classification, and hence its design factor.

A class location change resulting from new buildings around a pipeline requires a revision to the maximum allowable operating pressure, or increased wall thickness, or an engineering assessment to show a change is not needed. Regulations currently require that pipelines with higher local population density operate at lower stresses, to provide extra safety in those areas [46].

ASME B31.8 notes *"the number of buildings intended for human occupancy is not an exact or absolute means of determining damage-causing activities,"* so judgment can be used to determine changes to operating stress levels, when additional buildings are constructed. Indeed, a table in ASME B31.8 gives concessions on location classifications when extra buildings are built around a pipeline during operation.

Land use around a pipeline is usually controlled by local authorities, and this should prevent location classification changing, as these authorities should be aware of the pipeline and its location classification. The local 'planning departments' will generally want to accommodate developments, but consultation is needed between the pipeline operator, property developers, and the local authorities, before any development.

Sometimes local authorities use 'consultation zones' (zones around a pipeline where developments require consultation between all parties), or 'setbacks' (a minimum distance from a building to a pipeline).

2.6.9.4 Proximity Distances ('Setbacks')   When a gas from a pipeline ignites it can cause damage; for example, a large diameter pipeline at a pressure of 69 bar (1000 psi) would cause burn damage up to a distance of ~200 m (~650 feet) either side of its centre line [47].

Locating a building a 'proximity' distance or 'setback' distance from a pipeline can reduce damage incidents, and their consequences. Some standards and regulations require these proximity distances; for example, BSI PD 8010-1 [9] states: *"the initial route and population density assessment should take account of the proximity to occupied buildings."*

These proximity distances are given in Reference 11 for natural gas pipelines, Figure 2.18. The distances are based on a fire model using a radiation level of 32 kW/m². This is high thermal radiation at the proximity distance (strong sunlight has a thermal radiation of

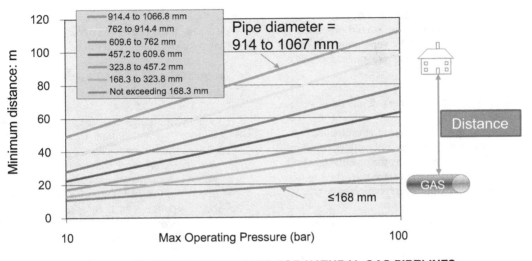

FIG. 2.18 PROXIMITY DISTANCES FOR NATURAL GAS PIPELINES

1 kW/m²). This level was not chosen as a 'safe' level but to reflect [48]:

- the low frequencies of pipeline failure;
- the possibility of escape to take cover from direct radiation;
- the fact that the most of the population is indoors most of the time; and,
- the need to route pipelines in a densely populated country.

Building setbacks are also used by local governments to provide separation between the community and potential risks from a pipeline [49], by avoiding encroachment on the pipeline right-of-way, thereby reducing the likelihood of excavation damage to the pipeline [50].

2.6.10 Pipeline Protection Design

Onshore and subsea pipelines can be damaged during construction and operation, Table 2.15 [51], and therefore require protection. This protection starts with a good design (e.g., locating pipelines in remote regions), but earth moving equipment, drilling machines, ships' anchors, working on, or near, the pipeline can damage it during operation. These impacts ('external interference') can cause 'mechanical damage', such as gouges and dents, and is a major cause of pipeline failures.

Design standards do not give detailed guidance on pipeline protection other than marking the pipeline ROW, or generic depths of cover; consequently, pipelines sometimes need extra defenses against interference.

TABLE 2.15   PROTECTING PIPELINES AGAINST DAMAGE

Phase	Protection
Protection during transportation.	Separation pads.
Protection during handling and storage.	Protection pads, sand berms, wood pads.
Protection during installation (lowering in, backfilling).	Sand padding, concrete coatings, nonwoven geo-textiles.
Protection during pipeline operation.	Above-ground pipeline markers, coatings, concrete slabs.

2.6.10.1 Awareness Methods   Good communication ('awareness' methods, Table 2.16 [52]) between excavators, fishing fleets, and the pipeline owners can reduce damage, by making parties who work near pipelines aware of the pipelines.

2.6.10.2 Surveillance   Pipelines can be regularly surveyed by air or land, to check that no excavations are being conducted along the pipeline route, or no shipping activity is occurring over the subsea pipeline.

One purpose of surveillance is the periodic examination of the pipeline right-of-way to detect unauthorized excavation, or recently completed excavation, by observing earth disturbance. Pipeline company personnel, or contractors, perform the surveillance, which involves periodic ground-based or aerial patrols.

Aerial surveillance is typically conducted every two weeks, but its effectiveness depends heavily on patrol frequency [53]: surveillance may be ineffective if the interval between patrols exceeds the time required for an excavation contractor to mobilize on site and commence digging. It has been estimated that a surveillance frequency of less than one patrol per month can detect less than five percent of unreported excavations. The effectiveness of a periodic patrol does not improve significantly unless the patrol is performed more frequently than weekly [36, 53].

TABLE 2.16   EXAMPLES OF AWARENESS METHODS

Awareness methods	
Printed materials (posters, calendars, diaries).	Trade shows.
Electronic communications methods.	Informational or educational items.
Mass media communications (e.g., radio).	Pipeline marker signs.
Specialty advertising materials.	'One-Call' centre outreach.
Door-to-door, face-to-face meetings with landowners, excavators.	Town hall meetings.
Operator websites.	Education and reminders, to all who excavate.

2.6.10.3 Physical ('Separation') Barriers Physical barriers (e.g., concrete slabs) can be placed over and around pipelines to protect them. The protection is achieved by separating the pipeline from the excavator or impactor. There are three separation methods [7]:

- separation by burial where the (deeper) depth or cover is designed to eliminate damage;
- separation by exclusion by restricting access to the pipeline (e.g., by fencing), but this method is usually only effective where access to pipeline facilities is controlled by the pipeline licensee;
- separation by barriers such as concrete slabs over the top of a pipeline, or crash barriers on bridges carrying pipelines.

Examples of separation by barriers are:

- markers (buried electronic markers are passive markers placed above pipelines, which can later help pinpoint the facility with a locator unit), and buried tape is made of stretchable material and placed 150–300 mm (6–12 in.) below ground level, along the pipeline, to warn of the presence of the pipelines [54];
- depth of cover (standards require a minimum depth of cover for onshore pipelines [e.g., 0.9 m (2ft. 11") to 1.1 m (3ft. 7")], but research work reduced by a factor of 10 as the depth of cover is increased from 1.1 m (3 ft. 7") to 2.2 m (7 ft. 2");
- wall thickness (increased pipe thickness offers protection against damage; for example, very few (about 5%) of excavating machinery used in suburban areas can penetrate pipe wall thicknesses of 11.9 mm (0.469") [56];
- concrete sections, channels, enclosure, pipe sleeves, rock dumping and mattresses for subsea pipelines;
- protective steel meshes or steel or plastic slabs covering the pipe and/or surrounds (concrete slabs can reduce damage frequency by 0.16, and concrete slabs, plus visible warning, can reduce the frequency by 0.05 [36]).

Further guidance on protective measures is given in Reference [36].

2.6.10.4 Terrorism and Theft Pipeline product is increasingly being stolen from pipelines. Liquid pipelines carrying anything from crude oil to diesel are being targeted. The theft is simple: an 'illegal tap' (a drilled hole regulated by a valve) is attached to the pipeline, and the fluid is extracted. Similarly, oil and gas pipelines have a long history of being attacked by terrorists.

2.6.10.4.1 Terrorism/Sabotage. The oil and gas industry is the target attacked most frequently by terrorists [57, 58]. Terrorists preferred targets are [57, 58]:

- government, diplomatic and security forces;
- transportation;
- property;
- infrastructure, utilities and manufacture;
- retail;
- hospitality, leisure, and entertainment.

Oil and gas production facilities are 'high value' targets, and often attacked, but these facilities are relatively easy to secure/

protect. Pipelines are difficult to protect, and very easy to damage. A typical attack may involve two or three terrorists, three shovels, less than 5 kg of explosive, a roll of wire, and a battery [57].

The risk of terrorist attacks on pipelines can be reduced by (for example):

- identify key facilities, their vulnerabilities, and produce a security plan and counter-measures for them;
- security checks on pipeline personnel;
- security exercise programs with rapid restoration plans;
- 24 hour security patrols; and,
- employing local communities, or even local militia, to protect a pipeline, can reduce local violence, increase local employment, and, redistribute the wealth (this actually satisfies the main aim of local militia and people) [60].

Above ground pipelines can be buried to protect them from sabotage, and other methods such as CCTV, isolation valves, pipeline-wide computerized sensor systems (to detect pressure changes), have been used to improve pipeline security, but with limited success [60].

2.6.10.4.2 Product Theft. A major threat to pipelines is product theft. Historically, African countries had major theft problems, but this is now an international problem, costing the industry billions of dollars every year [61, 62], and causing many fatalities [63]. Theft is rapidly increasing all over the world, particularly in Europe, and is now endemic in some developed and under developed countries.

Various factors contribute to theft/vandalism [60]:

- poverty;
- lack of basic services;
- corruption amongst government officials;

Theft in poor regions can take several forms:

- Small scale, local opportunist theft: this is usually local theft, for local consumption, by 'amateurs'. This usually has the highest failure consequences.
- Small scale, local compensation claims, driven by compensation gains (food, farming) following a deliberate leak.
- Larger scale, product pipeline theft, by local organized crime, allowing them to fill road tankers.
- Large scale crude oil theft, by organized criminals. These can be international criminals, using sophisticated equipment and facilities (including storage), allowing them to fill ocean tankers.

Theft in the developing world is organized crime.

Reducing theft is difficult, as pipelines are spread over large, remote, geographical areas, and are easily accessible. Some methods of reducing theft are:

- education of local people into the dangers of these illegal taps;
- internal inspection of the pipeline using 'smart' pigs to detect small, through-wall holes;
- dismantle organized crime (many of the thefts are for a criminal with an organization selling to customers, including illegal refineries);

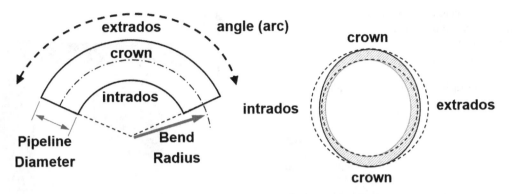

FIG. 2.19 PIPELINE BEND TERMINOLOGY

- review internal staffing (often criminals are working with pipeline staff); and,
- patrols.

2.6.11 Pipeline Bends

A pipeline route will not be straight: as it has many obstacles to cross or avoid; therefore, it will need bends. 'Field' bends are formed in the field from 12-m lengths of line pipe, using special bending machines and sometimes internal mandrels to prevent buckling.

Field bends are used for small angle bends, and 'manufactured' bends are used for sharp angle bends. These manufactured bends (or 'elbows') are specially made for pipeline use.

Bends are defined by (Figure 2.19):

- their radius (in units of pipe diameter D[5]; for example 3D and 5D); and,
- their angle (or 'arc').

The outside of the bend is called the 'extrados', and the inside of the bend is called the 'intrados'.

Older pipelines may have problematic bends: 'hot wrinkle bends' were produced many years ago by heating the intrados of the bend to allow softening of the line pipe and easing bending, and deliberate wrinkling: these type of bends have poor resistance to compressive stresses, and are not allowed in modern pipelines.

Similarly, older pipelines can have 'mitre' bends: these bends were popular before manufactured bends were available. They were made in workshops by cutting and welding the ends of two consecutive pipe joints at an angle. Their use is now usually limited to lowly stressed pipelines [1, 2], or not permitted at all [3]. These mitre bends can pose a variety of problems, including restricting the passage of 'smart' pigs.

2.6.12 Pipeline Crossings

Pipelines will need to cross many 'barriers' such as rivers, railways, highways, services, and utilities. The crossing can be achieved by: cutting through (trenching) a barrier (an 'open cut');

drilling/tunneling beneath the barrier ('trenchless'); or, going over the barrier ('aerial'). It is less disruptive to bore or tunnel ('trenchless' crossing) under a barrier by using a boring machine (an engine with an 'auger') to drill horizontally; or, a ramming machine.

Preventing damage to the pipeline at the crossing is the key consideration in the design and operation of crossings [1]. Standards give protection requirements, such as those detailed in Section 2.6.10.

2.6.13 Pipeline Valves

A valve is a mechanical device for controlling the flow of fluids, and pipelines contain many valves. There are many types in pipelines, e.g., 'gate', or 'ball'. Valves have 'actuators': these are devices that utilize a source of power to operate (move or control) a valve. Actuators can be manual (hand turns, with gears), or motorized (electrical, hydraulic, or pneumatic).

Design standards give little guidance on the types of valves to be used, but they do have requirements for the location and spacing of 'block valves'.

A 'block' ('mainline', 'isolation', 'line break') valve is a major installation in a pipeline. It is a mechanical device (within a valve assembly) that can be closed to block the flow of oil or gas through the line. It allows the pipeline to be divided into smaller sections to allow isolation and minimize the effects of a rupture, can divert products; and segregate plant (e.g., pigging facilities).

Block valve positions vary depending on standards, but ease of access and site location is important [2]. It is common sense to place isolating valves at the beginning and end of a pipeline, and they are usually required on both sides of pump stations and major waterways [e.g., 100 ft. wide (30 m) or greater].

Block valves are important in liquid lines as liquid lines can be in industrial, commercial, and residential areas, where construction activities pose a particular risk of external interference to the pipeline [1]. Standards require appropriate spacing and location of block valves consistent with the type of fluids being transported, to minimize the consequences of any leak [1].

2.6.13.1 Location The location of block valves should be determined for each pipeline, considering the following factors [7, 9]:

- topography;
- ease of access;
- the pressure and the fluid, and the security of supply required;
- the ability to detect events which might require isolation;
- the response time to events;

---

[5] Bends are often specified by their 'nominal pipeline size' (NPS). NPS is a dimensionless designator for pipe in the USA. It indicates a standard pipe size by the appropriate number. Do not confuse NPS with 'specified outside diameter'; for example, NPS 12 has a specified outside diameter (D_o) of 12.750", NPS 8 has a D_o = 8.625", but NPS 24 has D_o = 24".

- the volume and consequences of fluid release;
- operating and maintenance procedures;
- protection from vandalism; and,
- proximity to occupied buildings.

2.6.13.2 Spacings Historically, valves are placed along a pipeline, at varying distances;

Example: [1, 2]:

- every 20 miles (32 km) or less for natural gas pipelines in rural lines;
- every 5 miles (8 km) or less for natural gas pipelines in suburban locations; and,
- every 7.5 miles (12 km) or less for liquid pipelines.

Standards today retain these fixed spacing, but also allow spacing based on an engineering assessment taking account of the amount of fluid released, time to release, duration of failure, the impact in the area of the release, continuity of service, and security, operating and maintenance flexibility of the system.

The origin of these fixed spacing is: pressure from governments to install valves in public areas; perception of improved safety; good 'public relations'; and, a positive effect in the event of liability case, following failure [64].

There are reasons for small spacing in oil and natural gas lines as small spacing can minimize loss of product in case of failure. This is particularly important in liquid lines where the liquid can flow into rivers, drains, unless quickly stopped, although some studies show little correlation between spill size and block valve spacing [34].

Use of more block valves in gas pipelines for safety reasons is also contentious. Some gas pipeline failures ignite after a short delay: a study [65] of 55 international pipeline failures that ignited, found a delay in ignition in 29 (53%) of the failures with an average delay of 3 minutes. This means that emergency response (valve closure, fire-fighting mobilization) would not have an effect on casualties.

It might be thought that the use of rapid-closure valves would reduce casualties following a failure. A modern remotely controlled valve is very reliable and can be closed within 2 to 3 minutes, but a safe condition is only reached when the gas between the valves is released which 'almost always' is more than 10 minutes [66, 67]. Most damage occurs within 10 minutes of a gas pipeline failure, and virtually all fatalities and injuries in a gas pipeline failure occur at, or very near (within 3 minutes), the time of initial rupture [68, 69]. It is difficult to find pipeline failures where casualties would have been reduced by remotely controlled valves [68, 69].

It is emphasized that faster incident response time could reduce the amount of property damage from secondary fires (after an initial pipeline rupture) and faster incident response time could result in lower costs for environmental remediation efforts and less product lost [70].

2.6.14 Safety Factor versus Design Factor

All engineering structures have loads imposed. These loads must not cause the structure to fail; i.e., the operational load must not exceed the failure load of the structure. There is always uncertainty in both the calculations of operational loads, and the structure's failure load; therefore, a 'factor of safety' is imposed between the operational load, and the failure load.

Designers use 'safety factors' (Figure 2.20) in their calculations of design stresses, to account for uncertainties in the design, construction, and operation of engineering structures. For example,

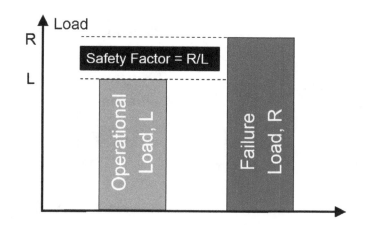

FIG. 2.20 SAFETY FACTOR

in the first ASME code (adopted in 1914) that covered boilers, a 'design factor' of 5 was applied to the tensile strength to establish the allowable tensile stress for internal pressure design. The same factor was later used in the 1920s on pressure vessels [43].

The safety factor is made large enough to easily compensate for uncertainties in the values of both the load (L) on the structure and the resistance (R) to this load (for example, the failure load of the structure), Figure 2.20.

$$\text{The safety factor is } S = R/L \tag{13}$$

$$\text{The safety margin} = (R/L) - 1 \tag{14}$$

Failure occurs if the safety factor is less than one, or if the safety margin becomes negative.

The load on the structure and the resistance of the structure to this load are not a single load or resistance, but a range of loads and resistances, and sometimes the strength and resistance ranges may overlap: this leads to uncertainty. There are many considerations when deciding on a safety factor for any engineering structure; for example, confidence in the materials and loads, types of maintenance and inspection [44]. Additionally, safety factors need to consider the consequence of failure; e.g., possible casualties may require increased safety factors.

Another consideration is the ability to 'prove' the condition of a new structure: if a structure can be proven to withstand a load (e.g., a pipeline surviving a pre-service pressure test) then a low safety factor can be used [44]. Therefore, the pre-service pressure test allows pipelines to operate at lower safety factors.

Pipeline standards limit stresses by specifying design factors below the value of 1; hence, the design factor is a simple engineering way of ensuring that working stresses in a pipeline are well below the yield and ultimate tensile strength of the pipeline material, Figure 2.15.

Design factor allows for uncertainties in design calculations. The factor is a percentage of SMYS. The SMYS of line pipe is not a failure load for defect-free line pipe; therefore, the design factor is giving a margin of safety on yielding (SMYS), not failure. Design factor is often considered a 'safety factor', but it is not a true 'safety factor', as safety factor is the ratio of failure load to design load.

2.6.15 Good Practice versus Best Practice

Integrity management starts with good pipeline design and construction, and satisfying all legal requirements. Pipeline design standards are 'good practice' [28, 29], but designers and operators are expected to do more than this good practice [31].

The same expectations apply to pipeline integrity management programs, as historically pipeline operators have controlled all threats by adopting good practices.

2.6.15.1 Good Practice
'Good practice' is the generic term for standards which control risk which have been judged and recognized by a Regulatory Body as satisfying the law when applied to a particular relevant case in an appropriate manner [71].

Examples of 'good practice' are:

- guidance produced by government departments;
- standards produced by standards-making organizations (e.g., ISO);
- guidance agreed by a body (e.g., trade federation, professional institution) representing an industrial/occupational sector.

Satisfying 'good practice' can be viewed as meeting a minimum requirement. Depending on the level of risk and complexity involved, it is possible the adoption of good practice alone may not be sufficient to comply with the law. For example, in high hazard situations, where the circumstances are not fully within the scope of the good practice, additional measures may be required to reduce risks [72, 73].

2.6.15.2 Best Practice
Integrity management programs carry with them inherent legal issues and exposures as they are specified in regulations, and failure to comply with regulations can mean a civil liability is certain, as this is viewed as negligent: liability is assumed if a law or regulation is broken [74].

Negligence is conduct *"which falls below the standard established by law for the protection of others against unreasonable risk of harm"* [74]. The real standard to achieve is 'best practice'.

Good practice satisfies a law/requirement [71], and can be viewed as meeting a minimum requirement, but 'best practice' goes beyond this requirement [72, 73]. 'Best practice' usually means a standard of risk control above the legal minimum [71, 75]. A consensus on current best practice will be standards developed with a period of public enquiry and full consultation, incorporating the views and expertise of a very wide range of interests from consumers, academia, special interest groups, government, business and industry [71].

Best practice will include 'risk-based' management. The USA's regulator says [76]: *"Good engineering is only part of the solution... a strong risk-based approach [ensures] the safety and reliability of ... pipeline infrastructure."*

2.7 PIPELINE CONSTRUCTION

Pipelines have been constructed since the 1860s. The early days used manpower and horse power, but today, pipelines are constructed by modern equipment and technologies which have developed over time, Figure 2.21. This Section will first briefly cover subsea pipeline construction, then give an overview of onshore pipeline construction.

2.7.1 Subsea Pipeline
Lay barge construction is used for constructing most pipelines offshore (and swamps and marshes). Lay barge methods were developed in the Gulf of Mexico from the 1940s. These ships contain all the equipment needed to weld together line pipe and form a pipeline. The ship will have a number of welding 'stations', an inspection station, and coating stations. Lay barges can lay several kilometers of pipeline/day.

Most lay barges use the 'S-lay' method where the pipe is welded on a ramp at several stations as the barge moves forward. Tensioners apply a force at the end of the ramp as the pipe leaves the barge. The welded pipeline goes into an overbend, supported by a 'stinger': a structure at the end of the barge to support the pipeline as it leaves the barge. A stinger can be large and long [e.g., 100 m (328 feet)]. The pipe continues downwards through a long unsupported span, creating a sagbend. The combination of the overbend and sagbend create an 'S' shape.

The 'J-lay method' allows for a more vertical drop (and hence lower tension) and is, therefore, the method of choice for laying pipe in deeper waters. J-lay uses a steep ramp (the shape of the pipeline during laying is a vertically-elongated J); hence, there is no overbend, and no, or only a short, stinger.

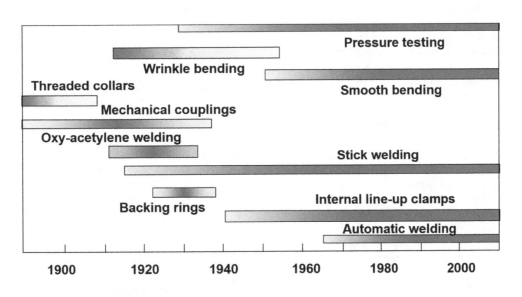

FIG. 2.21 TECHNOLOGY DEVELOPMENTS IN PIPELINE CONSTRUCTION [13]

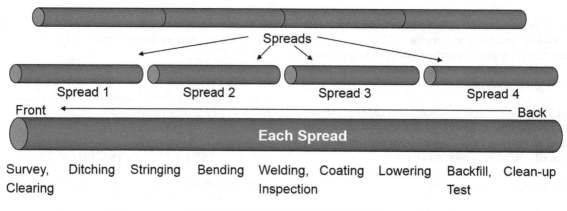

FIG. 2.22 THE 'SPREAD' METHOD

J-lay cannot be used in shallow water, as the water depth does not allow enough length of pipe to curve around from the ramp angle to the horizontal seabed without overstressing.

2.7.2 Onshore Pipelines

A complete set of equipment—for preparing the right of way, excavating the trench, welding the line pipe together, lowering into the trench, and backfilling the trench—is called a 'spread'. Most onshore pipeline construction use this spread method to build a pipeline: the spread method is a production line, but the pipeline is static, and the workforce moves along the line.

Multiple spreads may be needed on long pipelines (Figure 2.22), or pipelines that are being built quickly.

The construction advances a step at a time, and allows high speed construction. A construction spread may be 20 to 60 km (12 to 37 miles) in length, with the front of the spread clearing the right-of-way and the back restoring the right-of-way (Figure 2.22).

There may be several sets of construction equipment operating along the working width at any given time. The distance along the pipeline over which this equipment is deployed is typically less than a mile (1.5 km). Special construction crews will be used for road, railroad, and waterway bored crossings, and block valve installation.

The construction process for a rural pipeline construction will be:

- delivery and storage of equipment, and line pipe;
- formation of workers' camps;
- 'set out' (survey and mark) the working width;
- erect temporary fencing if livestock are around the working width;
- prepare the working width (clear bushes, debris);
- 'grade' (level) the working width;
- where required, the nutritious topsoil is removed and stored separately from other excavated and backfill materials on the working width (Figure 2.11);
- lay out ('string') the line pipe along the working width;
- bend pipe sections as required;
- (girth) weld the pipe sections together;
- inspect these welds, then coat these welds;
- excavate trench, placing the subsoil next to the trench;
- install pipe in trench;
- complete crossings, special sections, and 'tie-ins';
- complete backfilling and re-instatement of land around pipeline;
- complete facilities and pipeline control;
- clean and 'gauge' (check for blockages) inside of the pipeline;

- pressure test (using a gas or liquid);
- dry line where necessary;
- fill line with product.

The two key considerations throughout this process are safety of staff and the general public, and protecting the environment.

2.8 REFERENCES

1. Anon., 'Pipeline Transportation Systems for Liquids and Slurries', ASME B31.4-2012, American Society of Mechanical Engineers. New York USA. 2012.
2. Anon., 'Gas Transmission and Distribution Piping Systems'. ASME B31.8-2014, American Society of Mechanical Engineers. New York USA. 2014.
3. Anon., 'Petroleum and natural gas industries—Pipeline transportation systems', ISO 13623:2009, International Organization for Standardization, 2009.
4. Anon., 'Gas infrastructure. Pipelines for maximum operating pressure over 16 bar. Functional requirements', European Harmonised Standard: BS EN 1594:2013. British Standards Institution. 2013.
5. Anon., 'Petroleum and natural gas industries - Pipeline transportation systems', European Harmonised Standard: BS EN 14161. British Standards Institution 2015.
6. Anon., 'Oil and gas pipeline systems', CSA Z662-15, Canadian Standards Association. 2015.
7. Anon., 'Pipelines—Gas and liquid petroleum—design and construction', AS 2885.1-2012. Standards Australia, 2012.
8. Anon., 'Code of Practice for Pipelines-Part 2: subsea pipelines', BSI PD 8010-2. British Standards Institution Published Document, UK. 2004.
9. Anon., 'Code of Practice for Pipelines-Part 1: steel pipelines on land', BSI PD 8010-2. British Standards Institution Published Document, UK. 2004.
10. Anon., 'Submarine Pipeline Systems', Offshore Standard. Det Norske Veritas. DNV-OS-F101. Norway. October 2013.
11. Anon., 'Steel pipelines and associated installations for high pressure gas transmission IGEM/TD/1. Edition 5. Communication 1735. Institution of Gas Engineers and Managers. UK.
12. C J Trench, J F Kiefner, 'Oil Pipeline Characteristics and Risk Factors: Illustrations from the Decade of Construction'. American Petroleum Institute. December 2001.
13. E Clark, B Leis, R J Eiber, 'Integrity Characteristics of Vintage Pipelines', The INGAA Foundation Inc., Report E-2002-50435. Interstate Natural Gas Association of America 2005.
14. J Capelle, G Pluvinage, 'Evaluation of failure risk due to use of high strength steels in pipelines', Journal of Pipeline Engineering. 1st Quarter. 2013. p. 51.

15. J F Kiefner, 'Fatigue Strength of Seamless Line Pipe and Modern ERW Line Pipe', Pipeline Research Council International Catalog No. L51847. No. PR-201-9707. September, 2001.
16. J F Kiefner, E B Clark, 'History of Line Pipe Manufacture in North America', ASME Research Report. CRTD Vol 43. ASME, USA. 1996.
17. J F Kiefner, 'Dealing with Low-frequency-welded ERW Pipe and Flash-Welded Pipe with respect to HCA-related Integrity Assessments', ETCE 2002. ASME Engineering Technology Conference on Energy. February 4–6, 2002. Houston, Texas. Paper No. Etce2002/pipe-29029.
18. http://primis.phmsa.dot.gov/gasimp/docs/TTO05_LowFrequency ERW_FinalReport_Rev3_April2004.pdf.
19. B N Leis, T Thomas, 'Linepipe Property Issues in Pipeline Design and Re-establishing MAOP', Paper ARC-17, PEMEX International Congress on Pipelines, Merida, Mexico, November 2001.
20. Anon., 'Standard Practice—Pipeline External Corrosion Direct Assessment Methodology'. NACE SP0502-2010. Standard by National Association of Corrosion Engineers, UUSA. June, 2010.
21. G de Vries. 'The Other Class of Coating', World Pipelines. October 2005. p.29.
22. Anon., 'Pipeline Route Selection for Rural and Cross-country Pipelines', ASCE Manuals and Reports on Engineering Practice No. 46. American Society of Civil Engineers. USA. 1998.
23. T Shires, M Harrison, 'Development of the B31.8 Code and Federal Pipeline Safety Regulations: Implications for Today's Natural Gas Pipeline System', Gas Research Institute Report GRI-98/0367.1. December 1998.
24. https://primis.phmsa.dot.gov/comm/FactSheets/FSHCA.htm.
25. J Barnett, M Jordin, 'Pipelines: A worm's eye view', Issue 2, March 1998. Transco, UK.
26. W Berger, 'Codes and Standards', ASME Europe Info. ASME. Issue 10. July 2008. p.1.
27. http://www.hse.gov.uk/work-equipment-machinery/standard.htm.
28. F A Hough, 'The Gas Industry has Approved its New Safety Code', Gas Magazine, November 1954.
29. F A Hough, 'The New Gas Transmission and Distribution Piping Code (ASA B31 Section 8)', Gas Magazine, Series in 8 Parts, January through September 1955.
30. Anon., 'Introduction to ASME Codes and Standards'. American Society of Mechanical Engineers, New York. USA.
31. http://www.hse.gov.uk/pipelines/resources/designcodes.htm.
32. Anon., 'Specification for Line Pipe', American Petroleum Institute. API 5L-2012. Edition 45. 2012.
33. Anon., 'Managing System Integrity of Gas Pipelines', ASME B31.8S-2014. American Society of Mechanical Engineers. New York, USA. 2014.
34. Anon., 'Managing System Integrity for Hazardous Liquid Pipelines', American Petroleum Institute. API Recommended Practice 1160. Second Edition. September, 2013.
35. Anon., 'Pipeline systems Part 4: Steel pipelines on land and subsea pipelines - Code of practice for integrity management' BSI PD 8010-4:2012. British Standards Institution. UK. 2012.
36. Anon., 'Pipeline systems Part 3: Steel pipelines on land. Guide to the application of pipeline risk assessment to proposed developments in the vicinity of major hazard pipelines containing flammables—Supplement to PD 8010-1:2004', BSI PD 8010-3: 2009. British Standards Institution. UK. 2009.
37. Anon., 'Integrity Management of Submarine Pipeline Systems'. Recommended Practice. DNV-RP-F116. Det Norske Veritas. Norway. October 2009.
38. Anon., 'Petroleum and natural gas industries - Pipeline transportation systems - Recommended practice for pipeline life extension', Technical Specification. ISO/TS/12747. First edition. International Organization for Standardization. 2011.
39. M J Rosenfeld, R W Gailing, 'Pressure testing and recordkeeping: Reconciling historic pipeline practices with new requirements', Journal of Pipeline Engineering. 1st Quarter 2013. p. 15.
40. https://primis.phmsa.dot.gov/oq/glossary.htm.
41. http://www.hse.gov.uk/construction/cdm/faq/competence.htm.
42. Anon., 'Recommended Practice for Assessment and Management of Cracking in Pipelines', API RP 1176. American Petroleum institute. 2015.
43. E Michalopoulos, S Babka, 'Evaluation of Pipeline Design Factors', Gas Research Institute, 1999/2000.
44. B Leis, T Thomas, 'Linepipe Property Issues in Pipeline Design and Re-establishing MAOP', International Congress on Pipelines, PEMEX, Merida, Mexico, Paper ARC-17, November 2001.
45. Anon., 'Design, Construction, Operation, and Maintenance of Offshore Hydrocarbon Pipelines (Limit State Design)'. Pipeline Segment. API Recommended Practice 1111. Third Edition. American Petroleum Institute. July 1999.
46. Federal Register / Vol. 69, No. 124 / Tuesday, June 29, 2004 / Notices. http://primis.phmsa.dot.gov/classloc/docs/ClassLocationChangeWaiver CriteriaNotice062904.pdf.
47. E Golub et al., 'Safe Separation Distances Natural Gas Transmission Pipeline Incidents', International Pipeline Conference. ASME. Calgary, Canada. Vol 1, ASME 1998, p. 37.
48. P Hopkins, H Hopkins, I Corder, 'The Design and Location of Gas Transmission Pipelines using Risk Analysis Techniques', Risk and Reliability Conference, Aberdeen, UK. May 1996.
49. Anon., 'Partnering to Further Enhance Pipeline Safety In Communities Through Risk-Informed Land Use Planning Final Report of Recommended Practices', Pipelines and Informed Planning Alliance (sponsored by the United States Department of Transportation, Pipeline and Hazardous Materials Safety Administration, Office of Pipeline Safety). November 2010.
50. Anon., 'Model Setback Ordinance for Transmission Pipelines', Municipal Research and Services Center of Washington. 2007. mrsc.org/getmedia/6F3E7C3F-E358-4ECC-8678-.../modsettrans.aspx.
51. http://wiki.iploca.com/display/rtswiki/11.4+Mechanical+Protection +Selection+Guide.
52. Anon., 'Public Awareness Programs for Pipeline Operators', American Petroleum institute. Recommended Practice. API 1162. December, 2003.
53. Anon., 'Mechanical Damage Final Report. Michael Baker Jnr Inc., Department of Energy'. PHMSA OPS. USA. April 2009.
54. A Muradali et al., 'Effectiveness of New Prevention Technologies for Mechanical Damage', Gas Research Institute Report No. 8410. Project L027. June 2004.
55. E Jager et al., 'The influence of land use and depth of cover on the failure rate of gas transmission pipelines'. Proc. of the International Pipeline Conference. IPC2002. IPC02-27158. Calgary Canada. 2002.
56. P Hopkins, I Corder, P Corbin, 'The Resistance of Gas Transmission Pipelines to Mechanical Damage', International Conference on Pipeline Reliability, Calgary, June 1992.
57. M Stone, 'The Issue of Terrorism', World Pipelines. May 2005.
58. R Brown, 'Iraq's Pipeline War', World Pipelines, May 2005.
59. C Fleming, 'Security Challenge', World Pipelines, April 2009.
60. H-C Imig et al., 'Mitigating Risks', World Pipelines, Vol 10, No. 3. March 2010. p. 109. The Independent (online), UK press. 29 March 2012.
61. The Independent (online), UK press. 29 March 2012.
62. A Harrup, D Luhnow, 'Mexican Crime Gangs Expand Fuel Thefts'. The Wall Street Journal, 18th June 2011 (online).
63. E Castillo. Associated Press. 19 Dec 2010, on MSNBC.
64. R Eiber et al., 'Valve Spacing Basis for Gas Transmission Pipelines', Pipeline Research Council international. PRCI Report L51818. PR-249-9728. January, 2000.
65. J Haswell, 'Emergency Planning for High Pressure Gas Pipelines'. UK Onshore Pipeline Operators Forum, UKOPA/01/0006. www.ukopa.co.uk.
66. Anon., 'Pipeline Safety: Rapid Isolation of Ruptured Sections of Gas Transmission Pipelines'. Pipeline and Hazardous Materials Safety Administration (PHMSA), USA DOT. Docket No. RSPA-97-2879.

67. Anon., 'Pipeline Integrity Management in High Consequence Areas (Gas Transmission Pipelines); Proposed Rule'. USA DOT RSPA 49 CFR Part 2. Pipeline Safety: Docket RSPA-00-7666. Jan. 28, 2003.

68. Anon., 'Remotely Controlled Valves on Interstate Natural Gas Pipelines', US Department of Transportation, Sept 1999.

69. C. R. Sparks et al., 'Remote and Automatic Main Line Valve Technology Assessment', Final Report to Gas Research Institute, Report No. GRI-95/0101, July 1995.

70. Anon., 'Pipeline Safety: Better Data and Guidance Needed to Improve Pipeline Operator Incident Response', United States Government Accountability Office. Report GAO-13-168. January 2013.

71. Anon., 'Assessing compliance with the law in individual cases and the use of good practice', Health and Safety Executive, UK. http://www.hse.gov.uk/risk/theory/alarp2.htm. May 2003.

72. Anon., 'Guidance on conveying carbon dioxide in pipelines in connection with carbon capture and storage projects', Health and Safety Executive, UK. http://www.hse.gov.uk/pipelines/co2conveying.htm.

73. S Chatfield, 'Decision-making on assessment of high pressure gas transmission pipelines', Health and Safety Executive, UK Publication. http://www.hse.gov.uk/gas/supply/gsmrassess/symposium.pdf.

74. C A Paul, 'Legal issues in pipeline integrity programmes', The Journal of Pipeline Engineering, 1st Quarter, 2007. p. 49.

75. http://www.bsigroup.com/en/Standards-and-Publications/About-standards/Differences-between-Consensus-and-Commissioned-standards/.

76. S Gerard, Pipeline and Hazardous Materials Safety Administration, Written Statement to Subcommittee on Oversight and Investigation Committee on Energy and Commerce. United States House of Representatives, May 16, 2007.

ELEMENTS OF A PIPELINE INTEGRITY MANAGEMENT SYSTEM (PIMS)

Chapter 3 describes the purpose, elements, and interrelationships of a *Pipeline Integrity Management System* (PIMS). PIMS has the ability to explain a company's pipeline integrity management approach driven by the organizational and engineering integrity processes to achieve the integrity goals, objectives, and targets ensuring the safety of employee and the public, the protection of the environment, and a reliable service.

The PIMS elements detailed along the book chapters follows the ISO 14000 structure MS operational principle. As illustrated in Figure 3.1, the PLAN-DO-CHECK-ACT principles and elements led senior management's policy and commitment and their integration with the IMP engineering and management elements create a PIMS, as follows:

1. PIMS Policy and Commitment
2. PIMS Planning (Plan)
3. PIMS Implementation (Do)
4. PIMS Conformance and Compliance Verification and Action Plans (Check)
5. PIMS Management Review (Act)

A Pipeline Integrity Management System (PIMS) can be defined as a *leadership driven-approach to direct, plan, implement, verify measure and continuously improve the integrity of a pipeline(s) to protect people and environment providing a reliable service to shippers and customers providing a reliable service to shippers and customers. PIMS is supported at the core by an engineering-sound Integrity Management Program (IMP).*

3.1 MANAGEMENT SYSTEMS (MS) AND PIPELINE INTEGRITY

3.1.1 Brief History of Management Systems

A Management System (MS) is defined as a "*set of interrelated or interacting elements to establish policy and objectives and to achieve those objectives.*" Harvard Leadership Weblog provides the definitions of Management as "*the process of planning, leading, organizing and controlling people within a group in order to achieve goal*" [1].

The origins of the MS can be linked back to the scientific method cycle (i.e., observation/question, hypothesis and predictions, experimentation/testing, and analysis) as the basis for Shewhart's, Deming's and Japan's post World War II manufacturing developments.

In 1939, Walter Shewhart defined a statistical quality control method aligned with the scientific method by following three steps illustrated in Figure 3.2: *specification, production, and inspection.*

The linear application of this method evolved over time becoming a cyclical process to create knowledge and feedback; now it is known as Shewhart's cycle [2].

In 1950, W. Edwards Deming introduced his first cycle or Deming's Wheel at the Japanese Union of Scientists and Engineers (JUSE) seminar by applying Shewhart's principles to manufacturing as four (4) steps: *product design; production and associated testing; sales; and in-service testing and market research.* In 1951, Japanese executives correlated Deming's Wheel of four (4) steps to the Plan-Do-Check-Act (PDCA) cycle, which became the Quality Control and Improvement foundation of Japan's industry, as illustrated in Figure 3.3.

In 1971, the British Standard Institute (BSI) published the first standard about quality assurance (BS 9000) for the newly emerging electronics industry at the time. In 1974, BSI published the BS 5179 *Guidelines for Quality Assurance* focused on manufacturing, which had started during World War II when bombs were going off at the factories due to lack of consistent procedures and inspection. This era was characterized by understanding quality as conformance (i.e., consistently getting the same output as prescribed) by following approved procedures and standards; instead of focusing more on product improvement.

In 1980s, Deming reintroduced the Shewhart cycle *Plan-Do-Study-Act* as a quality control program for management, as illustrated in Figure 3.4. The intent of this Deming's cycle was to provide a cycle of learning and improvement of a product or a process.

In 1987, ISO developed the first management system standard, ISO 9000 Quality Management System. ISO 9000:1987 had three "models" for quality management systems:

- ISO 9001 creation of new products;
- ISO 9002 without the creation of new products; and
- ISO 9003 final inspection of finished product.

This first version used the same structure of the BS 5750 based on the British industry experience and standards (i.e., BS 5179) as well as the USA Ministry of Defense (MIL-Q-9858) terminology with emphasis on conformance with procedures and inspection. However, this first ISO 9000 version did not capture Deming's and others Japanese's companies understanding of quality as to including both conformance control and product/service continual improvement [3].

The year 2000 version of ISO 9000 introduced the concept of process management, senior management responsibility and commitment, and performance metrics for measuring and improving

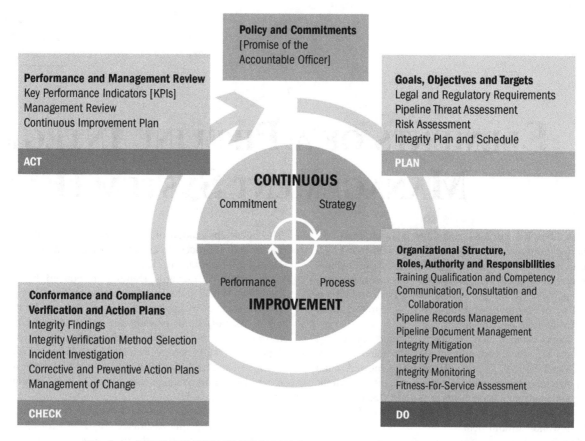

Policy and Commitments
[Promise of the Accountable Officer]

Performance and Management Review
Key Performance Indicators [KPIs]
Management Review
Continuous Improvement Plan

ACT

Goals, Objectives and Targets
Legal and Regulatory Requirements
Pipeline Threat Assessment
Risk Assessment
Integrity Plan and Schedule

PLAN

CONTINUOUS
Commitment   Strategy

Performance   Process
IMPROVEMENT

Organizational Structure, Roles, Authority and Responsibilities
Training Qualification and Competency
Communication, Consultation and
    Collaboration
Pipeline Records Management
Pipeline Document Management
Integrity Mitigation
Integrity Prevention
Integrity Monitoring
Fitness-For-Service Assessment

DO

Conformance and Compliance Verification and Action Plans
Integrity Findings
Integrity Verification Method Selection
Incident Investigation
Corrective and Preventive Action Plans
Management of Change

CHECK

FIG. 3.1   PIPELINE INTEGRITY MANAGEMENT SYSTEM (PIMS) ELEMENTS

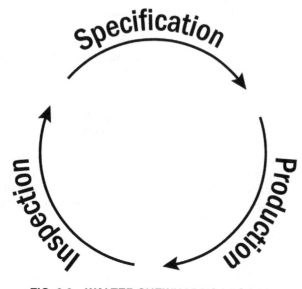

Specification
Inspection
Production

FIG. 3.2   WALTER SHEWHART'S PROCESS

Act   Plan

Check   Do

FIG. 3.3   THE JAPANESE PDCA CYCLE EVOLVED FROM DEMING WHEEL

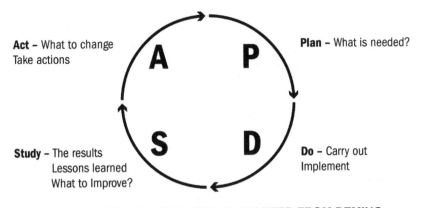

FIG. 3.4   THE 1993 PDSA CYCLE ADAPTED FROM DEMING

effectiveness. These are essential elements for continual process improvement to meet and satisfy customers' requirements and expectations. The ISO 9000: 2000 process included the following elements [4]:

1. Management Responsibility
2. Resource Management
3. Product and Service Realization
4. Measurement, Analysis and Improvement

Using the British Standard BS 7750 as a template, ISO 14000 *Environmental Management System* standard was published in 1996. This standard focused on process management with systemic and systematic approaches to minimize company's effects onto the environment (i.e., air, water, or land) ensuring compliance with environmentally-oriented laws, regulations, and standards using a continual improvement cycle. The ISO 14000 process included the following elements [5]:

1. Environmental Policy
2. Planning
3. Implementation and Operation
4. Checking and Corrective Action
5. Management Review

As illustrated in Figure 3.5, British Standard Institution (BSI) Publicly Available Specification (PAS) 55: 2008 (now evolved into ISO 55000) lists some principles and attributes such as holistic, systematic, systemic, risk-based, optimal, sustainable and integrated of an asset (e.g., pipeline) management system that can be applied in the development of a PIMS [6].

Other ISO standards such as *ISO 26000:2010 Guidance on social responsibility* and *ISO 31000:2009 Risk management—Principles and guidelines* can be used for complementing the integrity management approach in developing and implementing a holistic PIMS.

3.1.2   Management Systems (MS) for Pipeline Integrity

The application of the International Organization for Standardization (ISO) operational principle "Plan-Do-Check-Act" driven by senior management leadership has proven beneficial in achieving the integrity goals, objectives and targets linked with the rest of

the organization (e.g., pipeline production and operations, maintenance, projects, finance and accounting, human resources).

> *Example:* Linkage Between Pipeline Integrity and Rest of the Organization
> The decreasing integrity status of pipelines may create decreasing production and operations conditions (e.g., pressure/flow reduction). Consequently, the reduction of the overall company's income would reduce the ability to improve both operational flows and integrity remediation creating a downward spiral effect.

Figure 3.1 depicts an endless-rotating water wheel propelled by a healthy continuous improvement driving a management cycle. The MS elements could be described for pipeline integrity, as follows:

- The policy and commitment of senior management towards focused goals and overarching company processes
- Effective planning (e.g., need focused, state-of-the-art, integration) of integrity processes
- Efficient implementation (e.g., time, quality) in the field of the integrity plans
- Transparent, truthful and proactive verification (i.e., Corrective and Preventive Action Plans) capturing lessons learned from integrity inspections, audits, and incidents;
- Focused senior management review of integrity performance developing strategies for integrity risk reduction, process improvement and motivation for teams

MS elements also identify the accountability, responsibilities, and sources to comply with the pipeline integrity laws and regulations (i.e., external) and the company and industry standards requirements (i.e., internal) for reaching both pipeline integrity societal and corporate goals.

Furthermore, senior management *phronesis* or practical wisdom approach helps in managing the integrity of pipelines by accounting for the context (e.g., societal risk and responsibility) toward continuous improvement [7]. However, a MS needs the integrity engineering core to support the risk reduction at the program level; thus, an integrity management program is essential to it.

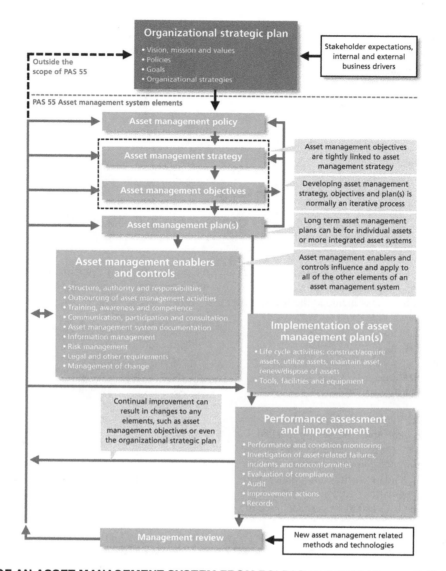

FIG. 3.5   ELEMENTS OF AN ASSET MANAGEMENT SYSTEM FROM BSI PAS 55-1:2008 (*Source:* British Standard Institute and Institute of Asset Management, BSI Group)

3.2   INTEGRITY MANAGEMENT PROGRAM (IMP)

Oxford Dictionary defines Program as "*planned series of future events or performances.*" Hence, an Integrity Management Program (IMP) can be defined as an engineering-sound process to manage *the integrity of a pipeline through the identification, susceptibility, assessment, prevention, mitigation and monitoring of risks to protect people and environment providing a reliable service to shippers and customers.* As illustrated in Figure 3.1, the inner core of the water wheel is "geared" by the IMP specific elements providing the engineering integrity process to adequately (what), timely, (when) and effectively (results-oriented)

1. Identify,
2. Assess,
3. Mitigate,
4. Prevent, and
5. Monitor

Risks resulting from pipeline integrity threats causing consequences such as safety, environmental, security of supply and business. However, an IMP needs the Management System (MS) platform to become effective across the organization instead of becoming an isolated "technical" process.

3.3   PIMS: MANAGEMENT SYSTEM AND INTEGRITY MANAGEMENT PROGRAM LINKAGE

Even though the pipeline MS and IMP elements have originated from different continents (i.e., North America and Europe, respectively) in response to different circumstances (i.e., pipeline failures and quality product realization), their historical development have both pursued common goals.

One of the functions of a management system (MS) is to systemically enable senior management to provide direction, guidance, support and evaluation to sustain continuous improvement; whereas, the Integrity Management Program (IMP) would provide the engineering integrity core for substantiating the risk reduction continuously. Thus, a MS interlinked with an IMP would become a Pipeline Integrity Management System (PIMS) enabling the organization to achieve state-of-art adequacy, timely implementation, and measured effectiveness of the integrity goals, objectives, and targets toward continuous improvement.

3.4 PIMS POLICY AND COMMITMENT [MS]

Corporate values create the foundation of integrity management policy and commitment aligned with the vision (i.e., future expectation), mission (i.e., reason for existence), and corporate goals (i.e., organizational expectations in context).

An Integrity Management (IM) policy and commitment statement should include following characteristics:

- All inclusive policy promoting a safety culture among all company employees and contractors with senior management leadership: *What to focus on and Who is involved?*
- Acknowledgement of current and future societal risk tolerance and expectations: *What are the stakeholders' expectations?*
- Declaration towards safety, environment and reliable pipeline service ensuring their continual improvement: *Leader's commitment*
- Declaration of commitment to comply with the law and regulations: *Company's and leader's responsibility*
- Expression of values and principles at company's foundation used for pipeline integrity: *Core basis directing what to and not to do*
- Assurance of company's proactive, timely and responsible actions within the nature, scale, and potential risks: *Company's willingness to properly act*

IM policy and commitment statement should be communicated from the top of the organization to all employees and demonstrate it with clear and consistent actions. The policy needs to be periodically reviewed to check whether continues meeting stakeholders' needs (external and/or internal).

Example: Pipeline Integrity Management Policy and Commitment Statement

Pipeline Company's senior management is committed to providing our customers with a reliable service ensuring the safety of our employees and the public, protecting the environment of the areas and communities in which we operate and conducting business with social responsibility.

Pipeline Company's values are founded on honesty, transparency, respect, people's growth and teamwork making a contribution to humanity through all company's activities including working with our communities. Pipeline Company abides to the law and regulations performing all actions in a responsible and diligent manner.

Pipeline Company understands and shares the increased stakeholders' expectations (e.g., society, regulators and communities) for which the Pipeline Company strives for applying State-of-the art planning, efficient implementation, and transparent internal verification using a continual improvement approach to achieve zero releases.

Pipeline Company is committed to diligently anticipate and identify potential integrity risks and manage them timely and effectively working with all employees and contractors to continue improving our safety culture. President (signed)

3.5 PIMS PLANNING: PURPOSE AND HIGH LEVEL PROCESSES [MS]

3.5.1 Goals, Strategy, Objectives and Targets [MS]

The organization identifies the *integrity goals* to be achieved in the short, middle, and long term. Integrity goals are typically described at a high level in the policy representing the aims towards to all company activities (e.g., anticipate risks, safety culture development) are ultimately conducted for.

Integrity goals conversely reflect company's challenges that can only be achieved with an ongoing, consistent, and restless effort over a period of time (i.e., not overnight). Those challenges are driven by multiple and complex risk factors (e.g., weather, human error, unauthorized activities) governed by both internal (i.e., company's own) and external forces (e.g., nature, environment, society).

Example: Pipeline Integrity Goals

- Zero pipeline failures (e.g., from landslides, seam weld cracking or any other threat).
- One standardized/consistent integrity management process across all company affiliates.
- A safety culture instilled within the entire company's employees.

Integrity strategies are major or overall aims identifying the how-to at a high level. They may need to account for the current knowledge of the integrity health condition and the future degradation of the pipeline systems as well as the company's integrity management processes. The integrity strategy should be designed to be applied across the organization in an integrated and harmonious manner to achieve the integrity goals.

Integrity strategies are developed by understanding how the integrity threats (e.g., seam weld cracking, incorrect operation), and their hazards/weaknesses (e.g., pipeline aging, pressure cycling, and controller's fatigue) can be overcome by the strengths (e.g., company's focus, support, and global knowledge) and opportunities (e.g., ongoing technology and methodology developments).

Example: Pipeline Integrity Strategies

- Implement monitoring and mitigating technologies for timely identifying and mitigating the effects of geohazards on the pipelines.
- Utilize the latest and appropriate in-line inspection technologies, state-of-the-art integrity methodologies, and operational improvements for timely identification, assessment, and mitigation of the effects of seam weld cracking on pre-1970s vintage pipelines.
- Exchange experiences and lessons learned with global affiliates and industry to increase integrity management effectiveness.

Integrity objectives can be described as to what to achieve and by when to operationalize the integrity strategy in the mid and long-term. They define the actions to be taken within a given timeframe that can be measured and achieved demonstrating their completion. They should also express progress in mitigating, preventing, and monitoring the condition criticality of the pipelines as well as the management processes. Integrity objectives should account for the expected effectiveness from the methods, technologies, and tools used.

Example: Pipeline Integrity Objectives

- Identify all pipeline crossings that can be affected by river bed scouring due to a potential 100 year-water flood return by the end of the year.
- Conduct the crack in-line inspection baseline on all seam weld cracking susceptible pipelines within the next 3 years.
- Improve the overpressure protection response time and severity profiles of hydraulic and control systems under unintended valve closure scenarios within the next two (2) years.

Integrity targets can be described as detailed what to achieve and when towards to an objective in the short term. They describe the expected end-results and deadlines towards to the integrity objectives to be achieved within a timeline. Multiple or staged integrity targets with shorter deadlines (e.g., <1 year) are typically needed to achieve one overarching integrity objective and its deadline. Objectives and targets are interconnected in defining the expected performance and results.

Example: Pipeline Integrity Targets for the objective *"Conduct the crack in-line inspection baseline on all seam weld cracking susceptible pipelines within the next 3 years"*

- Identify the pipeline segments susceptible to seam weld cracking by 2nd quarter of the year.
- Develop the in-line inspection program (e.g., technologies, potential vendors, pipe modifications, process and procedures) by 3rd quarter of the year.
- Formalize the in-line inspection plan identifying awarded vendors, field personnel, supply schedule and reporting deadlines by the end of the first semester

3.5.2 Legal and Regulatory Requirements [MS]

Even though pipeline integrity regulators require and verify compliance with the laws and regulations, the focus on the *"right thing to do"* toward the society (i.e., safety and environmental protection) allows for a holistic understanding of the *"spirit of the law."* This approach has proven to provide better outcome in the short and long-term.

Compliance with legal and regulatory requirements associated to integrity management require processes and competent personnel to clearly identify what legal framework is applicable and how it should be implemented into the existing practices to demonstrate compliance.

The creation and ongoing updates of laws, regulations, and adopted industry standards (i.e., referenced by the law/regulation) should be monitored for anticipating their impact on time and resources as well as planning their timely implementation.

A proactive, formal, and systematic process to manage legal and regulatory changes will prevent potential non-compliances resulting from one or more of the following:

- Lack of knowledge (e.g., did not know),
- Incorrect interpretation (e.g., thought it was not applicable), and
- Delayed implementation (e.g., not in place when law became effective).

A formal and documented root-cause investigation should be conducted when non-compliance is identified. Communication of those lessons learned to all personnel would benefit in preventing non-compliance reoccurrence.

The root-cause investigation should be followed by a Corrective and Preventive Action Plans (CPAP) identifying both the action items to correct and prevent the root-causes as well as the personnel responsible to carry out their completion. Management of Change (MoC) process is typically used to effectively introduce changes in the processes identified in CPAPs. Root-cause investigation, Corrective and Preventive Action Plans (CPAP) and Management of Change (MoC) are discussed in more detail in Chapter 13.

A Continuous Assessment occurs to update the integrity condition of the pipeline established by a previous assessment (e.g., Baseline). However, the last assessment may have determined different re-inspection intervals for each integrity threat hindering the ability to conduct all the next integrity assessments (e.g., ILIs, hydrotests, Direct Assessments) at the same time. This difference in integrity assessment timelines does not preclude the pipeline operator from continuously updating the integrity condition of the pipeline by estimating growth (e.g., corrosion, cracking) and/or project the trends (e.g., unauthorized third party activities), as applicable.

Hence, the Continuous Assessment is recommended to be conducted for all integrity threats at the time as an integrity threat is required for re-inspection. The benefit of this approach is to integrate all integrity threats at the same time for evaluating criticality.

3.5.3 Pipeline Integrity Threat Assessment [IMP]

The purpose of a pipeline integrity threat assessment is to identify, classify, analyze, and evaluate the susceptibility to and the capacity of integrity threats (i.e., likelihood) to cause a failure in a pipeline. Table 3.1 provides a correlation of the integrity threats mentioned in the ASME B31.8S, CSA-Z662 (Canada), API RP 1160 and UKOPA (United Kingdom) that may help the reader with following the chapter within their own jurisdiction. A pipeline threat assessment is an IMP process, which contribute to determining the risk and subsequent prioritization of mitigation, prevention, and monitoring measures.

The frequency of integrity threat assessments depends on the changes in the conditions (e.g., operational, environmental, population) leading to hazards (e.g., corrosivity, flooding, increased right-of-way activity) that could create or increase the severity of integrity threats; and consequently, a higher likelihood of pipeline failure. By default, minimum fixed periods of time can also be setup for conducting pipeline integrity threat assessments (e.g., at 1–2 years).

Example: Information recommended for review to conduct a pipeline integrity threat assessment by a company, consultant, or regulator:

a. Information from before the pipeline was placed into service such as design and material characteristics, manufacturing and construction records, hydrostatic testing parameters, methods and results, commissioning;
b. Historical operation of the pipeline such as leaks and rupture history (e.g., root-cause) as well as future/expected operation including any operational changes (e.g., pressure, temperature, flow, cycling);
c. Integrity assessments (e.g., ILI, "in-service" hydrostatic testing, direct assessment);
d. Mitigation (e.g., digs, repairs), monitoring (e.g., cathodic protection and leak detection effectiveness), prevention (e.g., damage prevention, surveillance) and Fitness-For-Service assessments conducted;

TABLE 3.1   CORRELATION OF PIPELINE INTEGRITY THREAT NAMES FROM INDUSTRY STANDARDS

ASME B31.8S Gas transmission pipelines	CSA-Z662 (Canada) Gas, liquid hydrocarbons, oilfield water and steam, liquid or dense phase carbon dioxide pipeline systems	API 1160 Hazardous liquid pipelines	UKOPA (United Kingdom) Major accident hazard pipelines
External corrosion	Metal loss	External corrosion	External corrosion
Internal corrosion	Metal loss	Internal corrosion	Internal corrosion
Stress corrosion cracking—SCC (i.e., environmentally-assisted cracking)	Cracking	Crack and crack-like	Other
Manufacturing-related (e.g., seam weld and pipe body)	Material or manufacturing	Design and material	Seam weld defect
Welding/fabrication related (e.g., girth weld, wrinkle bend, threads)	Construction	Construction	Girth weld and pipe defect, construction/material
Mechanical damage (e.g., immediate/delayed failure, vandalism/theft)	External interference	Third party damage	External interference
Incorrect operations (e.g., procedure)		Operation errors	Operator error
Weather-related and outside force (e.g., hydro-geotechnical, cold weather, lightning, heavy rains or floods, Earth movements)	Weather Geotechnical failures	Ground movement	Ground movement
Pipeline equipment (e.g., valves, seals)	Ancillary equipment	System operations	N/A

e.   Current regulatory and industry standard requirements and near future changes
f.   State-of-the-art knowledge including the lessons learned about initiation and growth of integrity hazards and threats

Figure 3.6 describes a recommended pipeline integrity threat assessment process using a management system approach PLAN-DO-CHECK-ACT led by an integrity policy and commitment from senior management.

Figure 3.7 describes a pipeline integrity hazard and threat management cycle [8]:

i.   Identification
ii. Susceptibility
iii. Assessment
iv. Mitigation
v.   Prevention
vi. Monitoring

Please refer to Chapter 4 for details about the pipeline integrity threat assessment.

3.5.4   Pipeline Consequence Assessment [IMP]

The purpose of a pipeline consequence assessment is to determine the extent and severity of the consequence resulting from pipeline release scenarios. Pipeline consequence assessments contribute to IMP by identifying the direct areas (i.e., along the pipeline right-of-way) and indirect areas (i.e., further from ROW, but reached through water crossings, valleys or canyons) that could be affected, if a released occurred.

Pipeline consequence assessments should use both leak and rupture scenarios in determining the extent of the consequence. Even though leaks have lower discharge rates than rupture, the detection time could be longer creating a larger consequence area or reaching further than a rupture release. The severity of both scenarios also needs to be assessed to determine the worst consequence cases from multiple views: volume as well as reached length, area and affected receptors.

As illustrated in Figure 3.8, the following is a recommended high level process for conducting a pipeline consequence assessment PLAN-DO-CHECK-ACT led by a safety and environmental policy and commitment from senior management:

a.   *Consequence Assessment Planning (Plan)*: goals, targets, objectives, data gathering and schedule
b.   *Consequence Analysis (Do)*: determine the extent of the consequence, the receptors reached and their grade of affectation using data gathering, field reconnaissance, mathematical formulation, computational algorithms and desktop verifications;
c.   *Consequence Classification along the Pipeline*: (Do) evaluates the results of the consequence analysis to determine the classification either qualitatively or quantitatively of the consequence along the pipeline based on predefined consequence criteria;
d.   *Results Validation (Check)*: field confirmation of the consequence analysis (e.g., discharge, dispersion) and classification (e.g., elevated consequence)
e.   *Management Review* on Assessment and Implementation (Act)

Please refer to the Chapter 5 for details about pipeline consequence assessment.

3.5.5   Risk Assessment [IMP]

The purpose of a pipeline risk assessment is to quantify the risk levels of pipelines and their associated acceptability. Risk assessment is attainable by estimating the probability, consequence and combined risk and ranking of relative or absolute values. Risk assessment aggregates and amalgamates pipeline characteristics,

Pipeline Integrity Threat Assesment Process

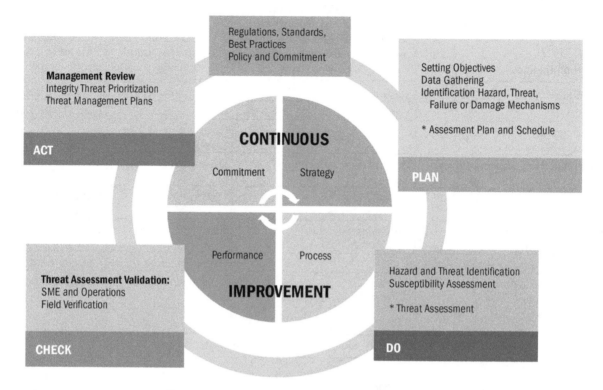

FIG. 3.6   PIPELINE INTEGRITY THREAT ASSESSMENT PROCESS USING A MS APPROACH

FIG. 3.7   PIPELINE INTEGRITY HAZARD AND THREAT MANAGEMENT CYCLE

conditions, and attributes reflected as risk factors to establish a risk magnitude (relative or absolute) facilitating the prioritization of pipeline segments and identification of pipeline segments of interest. This definition allocates pipeline sections into categories (i.e., acceptable, tolerable, and intolerable/unacceptable) for risk management purposes.

The use of risk assessments can be broadened to include the following:

- Optimization of pipeline maintenance, inspections and replacement
- Evaluation of change of service (e.g., type of product being transported)
- Changes of operating condition (e.g., pressure, temperature)
- Upgrading class location or consequence (i.e., population, environment)
- Assessment of safety of the reliability-based pipeline design with the aim of demonstrating the structural adequacy
- Increasing the availability and reliability of a pipeline

The selection of the risk assessment methodology (e.g., qualitative, quantitative, and semi-quantitative) is influenced by the expectations in terms of:

- End-use of the risk assessment results (e.g., ranking for ILI, cost-effective risk reduction)
- Level of expected effort in gathering and processing integrity data (e.g., limited data at hand)
- Time to conduct the assessment (e.g., months, years)

Figure 3.9 describes a recommended pipeline integrity risk management process using a management system approach PLAN-DO-CHECK-ACT led by a risk policy and commitment from senior management:

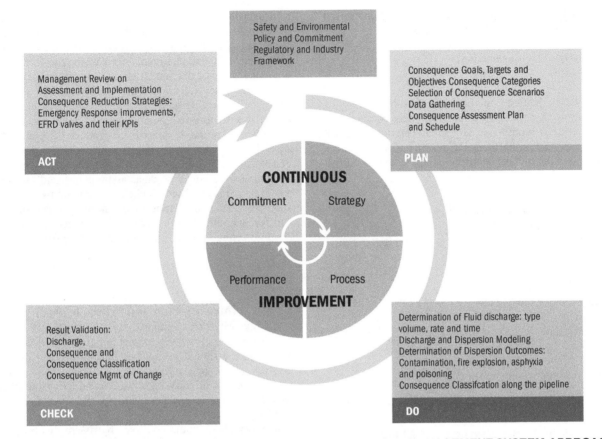

Safety and Environmental Policy and Commitment Regulatory and Industry Framework

Management Review on Assessment and Implementation Consequence Reduction Strategies: Emergency Response improvements, EFRD valves and their KPIs

ACT

Consequence Goals, Targets and Objectives Consequence Categories Selection of Consequence Scenarios Data Gathering Consequence Assessment Plan and Schedule

PLAN

CONTINUOUS

Commitment   Strategy

Performance   Process

IMPROVEMENT

Result Validation: Discharge, Consequence and Consequence Classification Consequence Mgmt of Change

CHECK

Determination of Fluid discharge: type volume, rate and time Discharge and Dispersion Modeling Determination of Dispersion Outcomes: Contamination, fire explosion, asphyxia and poisoning Consequence Classifcation along the pipeline

DO

FIG. 3.8   PIPELINE CONSEQUENCE ASSESSMENT PROCESS USING A MANAGEMENT SYSTEM APPROACH

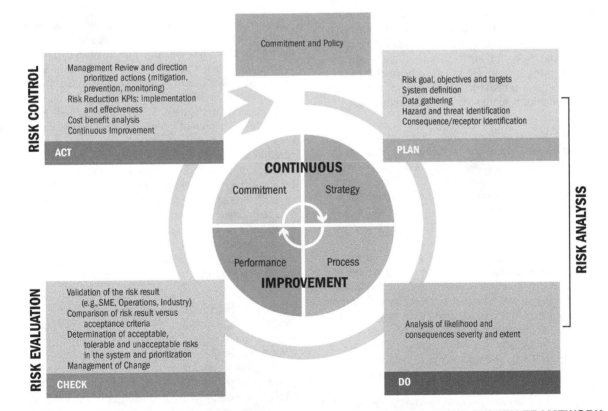

Commitment and Policy

RISK CONTROL

Management Review and direction prioritized actions (mitigation, prevention, monitoring) Risk Reduction KPIs: implementation and effeciveness Cost benefit analysis Continuous Improvement

ACT

Risk goal, objectives and targets System definition Data gathering Hazard and threat identification Consequence/receptor identification

PLAN

CONTINUOUS

Commitment   Strategy

Performance   Process

IMPROVEMENT

RISK ANALYSIS

RISK EVALUATION

Validation of the risk result (e.g., SME, Operations, Industry) Comparison of risk result versus acceptance criteria Determination of acceptable, tolerable and unacceptable risks in the system and prioritization Management of Change

CHECK

Analysis of likelihood and consequences severity and extent

DO

FIG. 3.9   PIPELINE RISK MANAGEMENT PROCESS WITHIN THE MANAGEMENT SYSTEM FRAMEWORK

The pipeline risk assessment process is conceptually comprised of the following two (2) high level sub-processes:

1. Risk analysis and risk estimation
2. Risk evaluation

The risk analysis is the first step of the process that assesses the probability of failure (e.g., integrity threats) and their consequence (e.g., leak, large leak, or rupture) per segment to determine the risk per pipeline segment and profile along the pipeline. Risk drivers are the risk factors with the highest contribution to the estimated risk for a given pipeline segment.

> *Example:* SCC risk factors such as coating type (e.g., tape disbondment), cathodic protection (e.g., ineffective), stress level (e.g., >40% SMYS) may be the factors with the highest contribution to the estimated risk for a given segment, and would therefore become the risk drivers.

The validation of a risk model involves the following reviews:

a. Completeness of the risk factors applied to each threat or consequence
b. Adequacy of the formulation
c. Accuracy and uncertainty of the data input
d. Make-sense of the results
e. Representativeness of the estimated risks versus reality

The validation should be conducted by personnel working in the pipelines being assessed (e.g., integrity, operations, leak detection, right-of-way inspection and surveillance) and the Subject Matter Experts (SME) from multiple disciplines such as risk, integrity, safety, environment. This validation may involve running multiple iterations adjusting the risk model and their results; particularly, cases where the risk values and ranking are not representative of the actual reality of phenomena portraying the risk to be unrealistically high or low, and a ranking inverse. However, sometimes the data-driven results provide "new or unexpected perspective" that should be further investigated with potential for risk to be mitigated.

> *Example:* A pipeline risk assessment reported section A-B with the highest probability of failure due to manufacturing issues. The validation team disregarded this result based on no leak history in the segment. If this pipeline section failed years later, would you consider as a failure contributing factor "risk assessment validation reliance only on historical data instead of susceptibility"?

The risk evaluation process identifies the pipeline segments exceeding the risk acceptance criteria and defines alternatives for risk reduction by using risk sensitivity analysis. The evaluation process uses the ranking of the pipeline segments by risk, probability and consequence to identify the ones exceeding the risk criteria drilling down to discover the risk drivers of those highly ranked segments. The sensitivity analysis reveal the effect of the risk drivers on the overall risk of a selected pipeline providing an understanding of what risk factors/drivers are required to effectively reduce the risk.

Regulatory, legal and company framework should be considered in selecting the most appropriate preventive and mitigation actions. The level of risk in the risk evaluation process may lead to a decision to perform further analysis or to manage the risk by maintaining controls. This decision will be influenced by the organization's risk goals and objectives. A selection of mitigation plan is the next step that involves balancing the cost and efforts of implementation of risk reduction plan against the benefits derived.

The frequency for conducting risk assessment can be defined on a fixed interval (e.g., every 5 years) and/or driven by a change in the risk factors or conditions (e.g., new consequence areas, higher/lower pressure, service or product change, and pipeline acquisitions).

The results of the risk assessment should document the risk methodology, sources of data, assumptions and challenges, and results. The results should include the ranking by risk, probability, and consequence indicating their risk drivers, and options identified for risk reduction through the sensitivity analysis. Decisions about selecting options to reduce risk will be part of the risk management process.

Please refer to the Chapter 6 for details about pipeline risk assessment.

3.5.6   Integrity Plan and Schedule [MS]

The integrity management strategy (e.g., identifying the extent of a given threat) and objectives (e.g., complete ILIs within the next 3 years) should be applied during the development of the integrity plans. Identifying integrity threats, consequences, and risks become a primary task prior to initiating the planning; an engineering assessment would provide guidance on what and how to manage the risks.

Integrity plans should be driven by risk results (i.e., acceptance criteria), prioritization (i.e., ranking) and drivers (e.g., seam weld cracking, high consequence) starting with the risk pipeline segments not meeting the acceptance criteria and/or top ranked pipeline segments (i.e., what and where).

The plan to get-to-know the integrity condition of every pipeline segment should follow the timing and/or prioritization identified by the risk assessment. A pipeline has completed its Baseline Assessment when a pipeline has been integrity assessed (i.e., ILI, Hydrotest and/or Direct Assessment) for all identified (i.e., already found) and susceptible (i.e., probable to be found) integrity threats, as illustrated in Figure 3.10. Some pipelines get their baseline assessment a couple of years after their commissioning; others get their Baseline Assessment after many years into to service triggered by new regulations or company's own initiative.

A Baseline Assessment typically contains both the execution of the integrity assessment method and the validation of these results as well the data integration with other non-integrity assessment data (e.g., encroachment survey) capturing all the integrity threats identified as a risk driver. The Baseline Assessment should provide an understanding of the integrity condition at the time of the evaluation provided that the limitations of the applied integrity assessment method(s) are offset with complementary integrity data.

> *Example:* Hydrostatic testing may be selected as an integrity method in conjunction with cathodic protection surveys (e.g., CP potential, Close Interval Survey—CIS-, Direct Current Voltage Gradient—DCVG) to baseline risk drivers such as seam weld cracking and corrosion; however, if mechanical damage/external interference is identified as a risk driver for this pipeline, further surveys (e.g., ILI Caliper) and data integration (e.g., unauthorized activities/damage prevention issues, public awareness) may be needed.

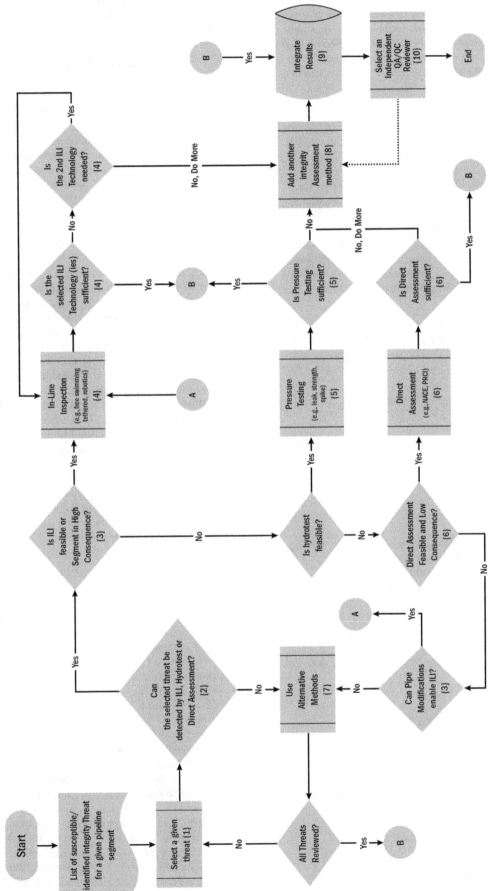

FIG. 3.10 INTEGRITY ASSESSMENT METHOD SELECTION

For cases when a Baseline Assessment has been conducted in the past, the recent discovery, suspicion or susceptibility of a new integrity threat may trigger the baseline assessment to be supplemented with this particular threat; this means that selected pipeline segments within an entire pipeline system may be subject to a "Integrity Threat-specific Baseline Assessment" and the data integration with other threats to be also re-evaluated, if applicable.

Example: The occurrence of a pipeline failure (e.g., Low-Frequency Electrical Resistance Welding—LF ERW—seam weld rupture) revealing new knowledge about risk drivers to failure (e.g., presence of defects from a manufacturing type and era) may initiate specialized in-line inspections (e.g., dual/multiple cracking in-line inspection technologies), associated repairs and operational pressure change monitoring to estimate next re-inspections for the new integrity threat(s). Furthermore, corrosion, mechanical damage, and incorrect operations may need to be revisited as their interaction with the new integrity threat may increase the integrity criticality of the combined anomalies (e.g., dent or corrosion on seam weld cracking).

Every activity in the plan should contain the following components or attributes, as a minimum, to allow for grouping and characterization (e.g., external/internal corrosion mitigation, monitoring) of the plan:

- Pipeline system (e.g., Martigues, Sarnia, Pattaya)
- Pipeline segment (e.g., Martigues—Marseille)
- Activity description (e.g., crack in-line inspection)
- Type of activity (e.g., ILIs, cathodic protection, engineering)
- Integrity threat (e.g., corrosion, cracking, mechanical damage/external interference)
- Consequence (e.g., highly populated area, environmentally sensitive area, product supply)
- Risk categorization and ranking (e.g., safety, environmental, and business)
- Integrity process (e.g., identification, mitigation, monitoring)
- Responsible (e.g., Project Engineer and Manager)
- Year and total (i.e., multi-year) budgets
- Actual expenditure (i.e., as of the cut-off date or carry-over)
- Forecast expenditure (i.e., estimated value to the end of the activity at a given time)
- Approval status (e.g., being prepared, submitted, approved, rejected)
- Activity status (e.g., on hold, execution, completed, deferred, cancelled)

3.6 PIMS IMPLEMENTATION [MS]

The process of implementation would need, as a minimum, the following elements to carry out the integrity planning:

- Organizational Structure [MS]: Roles, Authority, and Responsibilities
- Training, Qualification, and Competency [MS]
- Communication, Consultation, and Collaboration [MS]
- Records Management [MS]
- Document Management [MS]
- Integrity Mitigation [IMP]

- Integrity Prevention [IMP]
- Integrity Monitoring [IMP]
- Fitness-for-Service Assessment [IMP]

3.6.1 Organizational Structure: Roles, Authority and Responsibilities [MS]

Starting in the 21st century, oil and gas companies with pipeline assets started progressively re-structuring their pipeline maintenance groups to be more effective, as during this time, the pipeline industry had been heavily criticized due to their significant pipeline failures causing fatalities, environmental damage, and suspension of critical utilities (e.g., gas for heat and electricity, furnace oil). The public did not distinguish between one company and another. All pipeline companies were under the public scrutiny regardless their individual safety, environmental and public (e.g., services, transportation) performance.

Pipeline maintenance typically reported either to Operations or Engineering/Technical Services in the midst of other services such as SCADA, leak detection and projects. Senior management found themselves facing ongoing and increasing internal (e.g., owners, CEOs, presidents) and external (e.g., society, regulators, media) expectations related to integrity of their pipelines.

The organizational structure changed empowering pipeline integrity teams with resources (e.g., personnel, budget) for improving integrity remediation, mitigation and prevention, and access to senior management for influencing the operation (e.g., pressure reduction, shutdowns, system modifications). Needs that were previously filtered by multiple layers of the organization driven by non-holistic approaches (e.g., looking only inside) causing delays and sometimes no action resulting in pipeline failures; something that senior management and company owners were no longer willing to tolerate, but sought for changes.

Simultaneously, regulations introduced requirements going further than technical, but managerial such as the formalization of senior management commitment including the public definition of an accountable officer with the organizational authority, leadership and responsibility to prevent failures.

3.6.2 Training, Qualification, and Competency [MS]

Training, qualification, and competency are three (3) stages in the development and utilization of employee's skills to adequately and effectively do an integrity job or any other for that matter. In some countries, the three (3) terms have been merged into two (2) by making training and qualification one as to be "Training" and naming competency as "Qualification," but their overall concept and purpose were the same.

Training is the stage in which an employee receives the required information and knowledge to do the job; theoretical and practical exercises contribute to their learning process. Written and oral tests may be used for verifying the learning. This stage typically comes with a certificate of training for the personnel that have successfully completed the course. Training does not enable an employee to resolve all necessary situations that a job can be faced with; however, they may be able to understand them.

Qualification is the stage in which a trained employee reaches the level of skills needed for executing the job. Working on-the-job with initial supervision, guidance, and periodic evaluations would enhance the training to become qualified. This stage may come with a certificate of qualification or a ticket in a given job (e.g., welder, controller) that may be referenced to a region, country, or jurisdiction. If the qualification process is conducted within the same conditions (e.g., specific operational conditions, pipeline type, product, jurisdiction) as where the employee would normally work, the employee has become competent to this specific job.

Competency is the stage in which a qualified employee applies their skills to a *specific job* with its own characteristics, conditions and challenges that may vary from the region, country or jurisdiction where their qualifications were achieved; meaning that competency has an unique contextual requirement.

Example: a qualified and competent employee in a given country may not be competent in another where the industry standards and regulations are different; however, specific training and work under regional supervision and/or examination may satisfy this need for a contextual exercise of the skills required to become competent.

3.6.3 Communication, Consultation, and Collaboration [MS]

Communication, consultation and collaboration are enablers to increase efficiency and effectiveness of pipeline integrity management plans. All three (3) enablers can be connected depending on how much involvement from the parties is needed to improve (e.g., efficiency) or achieve (e.g., effective) the results.

Communication is a two-way enabler where clear understanding of the message enables effective and efficient actions focusing what exactly is needed, thereby reducing activity recycle and duplication. Communication requires not just emitting the message, but also confirming its receipt which may need verbal interaction not just emails. The communication can be top-down and/or bottom-up.

Example: planning an excavation of a pipeline requires an effective communication with the control center. The communication shall ensure that the pipeline controllers receive and confirm the message understanding the work being conducted and the potential unintended events (e.g., release) that may require them *to immediately act upon.*

Consultation is a two-way enabler requiring the emitter (who requests) to be proactive in submitting material (e.g., opinion, document, survey) for review and to be open mind in understanding the receiver's (who is consulted) responses. However, while consultation captures people's views, it does not create a commitment towards to end results. Consultation may be employed when leading into collaboration (the next step).

Example: requesting industry's feedback on a proposed new standard's section requires a coordination team's open mind to understand different needs and views than their own as well as their ability to consistently and systemically integrate feedback for everyone's benefit. Ignoring feedback due to the coordination team's own specific interests instead of everyone's interest will cause a loss of credibility of the coordination team and lack of support during implementation.

Collaboration is a team enabler requiring all members, not just the coordinator, to move from "my perspective" to a "team's perspective" by focusing on the end-user needs. An indicator of collaboration, as to whether teamwork is occurring, is finding a common solution, not perfect, but agreeable by all.

Example: building an integrated process by organizationally diverse teams requires that all members of the new team should speak up respectfully communicating freely their concerns and needs; and for all to listen, while understanding the integrity objectives (e.g., no releases) and how to mold a participative process with shared objectives. The new process would promote teamwork with collective responsibilities and accountabilities towards a continuous improvement culture.

3.6.4 Pipeline Records Management [MS]

Pipeline records management (PRM) is a process to define, catalog, store, and retrieve, access, maintain, retain, and dispose of pipeline records. PRM process shall identify roles, responsibilities and authorities as well as access security and backup/restore systems for managing pipeline records. PRM provides the corporate memory for legal, operations and maintenance, pipeline integrity, risk management and decision making. However, design, manufacturing, and construction records area also managed.

Pipeline retention of records is typically dictated by regulations in each jurisdiction.

Example: Key pipeline records and minimum retention

- Construction procedures, mill certificates, equipment name-plate data: *pipeline service*
- Location of buried facilities, crossings: *pipeline service*
- SCADA data including discharge and suction pressure, temperature and flow: *1 year*
- Maintenance Welding Installation: *pipeline service*
- Surveillance and monitoring reports: *pipeline service*
- Defect information and incidents: *pipeline service*
- Training Conducted: *1 year*
- Inspection and Audits: *5 years or two consecutive audits*
- Abandoned facilities: *2 year after abandonment*

In the context of pipeline integrity, records management provides the basis for conducting the following processes:

a. Hazard identification, Threat Assessment and Susceptibility, and Probability of Failure;
b. Consequence Assessment;
c. Data Integration and Risk Assessments;
d. Fitness for Service Assessments;
e. Mitigation, Prevention, and Monitoring;
f. Verifications: Inspections and Audits;
g. Incident Investigation;
h. Management of Change;
i. Risk Management;

3.6.5 Pipeline Document Management [MS]

Pipeline document management is a process to create, review, update (e.g., version, control), approve, communicate, and dispose of strategic plans, programs, processes, procedures, and activity plans.

Example: Pipeline Document Management and Management of Change

1. Pipeline Document Management process to review and update documents providing the *How-to adequately change the document* such as roles and responsibilities of the initiator, reviewers, endorsers, and approvers.
2. Management of Change process to assess and mitigate the implications of a change providing a *How-to safely and effectively transition into the newly changed process* such as the need for additional personnel training/qualifications, equipment, redesign of approval and budgeting processes, time for completion.

3.6.6   Integrity Prevention [IMP]

Prevention is defined by the Oxford Dictionary as the *"The action of stopping something from happening or arising."* Prevention implies an anticipatory counteraction requiring advancing planning or action. Prevention is focused on hazards (e.g., coating damage) creating a barrier (e.g., coating selection) that would prevent from having conditions leading to an integrity threat (e.g., external corrosion). Prevention of integrity threats can be applied in all stages of the pipeline life; however, their highest benefit/cost is achieved in the operational long-term.

In early 2000s, several pipelines started utilizing extensive prevention measures prior to the start-up (e.g., pipe wall thickness and fracture toughness, mill quality, coating type and application, depth of cover) creating a high value savings in "avoided costs" (e.g., failures, license-to-operate/production losses, affected reputation and societal/regulatory lack of confidence for new projects); whereas, other ageing pipelines continue inevitably paying higher in-service mitigation costs for integrity threats such as corrosion, manufacturing and hydro-geotechnical as adequate prevention was not employed.

Example: Quality control and quality assurance during manufacturing and construction of pipelines can reduce the likelihood of defects being introduced into the system at the beginning of life. It is easier to prevent something from occurring than to mitigate once it has occurred.

Please refer to the Chapter 11 for details about integrity prevention measures for pipelines.

3.6.7   Integrity Mitigation [IMP]

Mitigation is a reactive process for reducing risk via minimizing the effect of integrity threats (e.g., corrosion) and/or their consequence (e.g., leak on water). If threat severity is not detected early, mitigation would be focused on restoring the function (e.g., pipeline pressure/flow) under special operating conditions (e.g., pressure reduction, shutdown). The difference between mitigation and prevention resides in the status of the integrity threat at the time of action.

Example: Mitigation versus prevention (a) the integrity threat is already present (i.e., corrosion in the pipeline) requiring mitigation or (b) the integrity threat is probable (i.e., pipeline susceptible to corrosion) to occur due to the conditions or hazards (e.g., coating damage, ineffective cathodic protection) requiring prevention.

Mitigation should be focused on reducing both the identified integrity threat (i.e., reactive approach, but necessary) and the hazards creating the conditions leading to an integrity threat (i.e., proactive approach).

The duration and frequency of mitigation measures are linked to integrity threat monitoring. Levels of integrity threat acceptability (e.g., corrosion growth >0.1 mmy) should be defined as to the starting point for mitigation. The assessment of mitigation results over a period of time would indicate the level of effectiveness of the mitigation showing whether the objective (i.e., integrity health status) has been accomplished or changes need to be introduced.

Mitigation of integrity threats can be applied to all stages of the pipeline life (i.e., design, manufacturing, construction, in-service, deactivation/reactivation and abandonment). However, the increase of mitigation can be typically seen in early years of operation (e.g., geotechnical, incorrect operations, and mechanical damage) and as the pipeline ages over time (e.g., corrosion, cracking, weather-related).

Example: When a crack detection ILI survey is performed, the tool can report defects such as SCC. SCC features found during excavation can be mitigated using a number of techniques. SCC can be mitigated by grinding out the feature, placing a pressure containing sleeve over the feature (i.e., Type B sleeve), or performing a pressure reduction.

Please refer to the Chapter 11 for details about integrity mitigation measures for pipelines.

3.6.8   Integrity Monitoring [IMP]

Integrity monitoring is a process focused on determining either conditions or processes that may capture initiation and/or growth of integrity threats.

- Monitoring Integrity Conditions for
    - New hazards (e.g., unauthorized RoW activity) and/or threats (e.g., Theft)
    - Changes in hazards (e.g., pressure cycling) and/or threats (e.g., crack growth)
- Monitoring Integrity Processes for
    - Completeness (e.g., no cracking ILI on pre-1970 vintage pipeline)
    - Effectiveness (e.g., corrosion growth after inhibition)

Integrity monitoring may trigger additional measures such as mitigation and prevention as well as focused inspections for dimensioning the extent of the findings.

Example: Most atmospheric paint systems have a three layer system. This system includes: a primer coat, an intermediate coat, and a top coat. If an asset is monitored for paint system performance, it is significantly cheaper to repaint when the top layer first starts to fail than when the asset shows rust bleed or metal loss.

Please refer to the Chapter 11 for details about integrity monitoring measures for pipelines.

Figure 3.11 describes a recommended pipeline integrity mitigation, prevention and monitoring process using a management system approach PLAN-DO-CHECK-ACT led by an integrity policy and commitment from senior management.

Please refer to the Chapter 11 for details about mitigation, prevention, and monitoring.

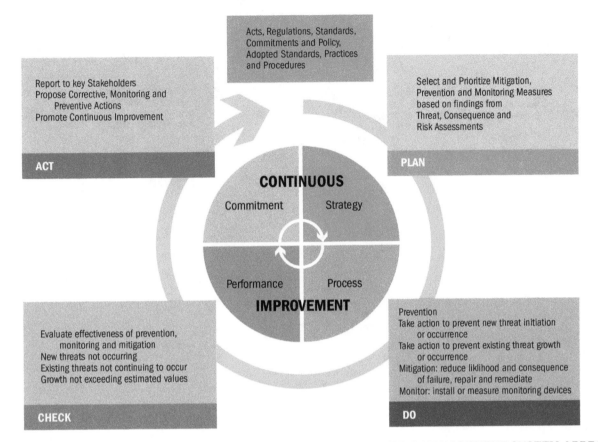

FIG. 3.11   MITIGATION, PREVENTION, AND MONITORING PROCESS USING A MANAGEMENT SYSTEM APPROACH

3.6.9   Fitness-for-Service Assessment for Pipelines [IMP]

Fitness-for-Service Assessment (FFS) for pipelines is an engineering assessment evaluation of the integrity of a pipeline under current and/or future conditions providing a path forward required to safely and reliably operate for a defined period.

The FFS evaluates the integrity threats identified or deemed susceptible accounting for their likelihood, consequence, and risk. The assessment accounts for the regulatory, industry and company framework as well as the integrity assessment (e.g., ILI, Hydrotest and Direct Assessment) applicability and limitations (e.g., anomaly detection, sizing) and the representativeness of their validation programs. Integrity data integration is used providing a more complete understanding of severity (e.g., coincidental anomalies), potential consequences (e.g., leak or rupture) and barriers in place or to be added (e.g., pressure reduction, localized cathodic protection groundbeds).

When selecting the industry standards (e.g., construction, in-service) to be used for the FFS, they should be scrutinized in light of the objective of the FFS to ensure those standards provide a safe and reliable operation instead of "paper compliance."

A Fitness-for-Service assessment provides the short and long-term integrity mitigation, prevention and monitoring measures as well as potential system modifications for ensuring a safe, environmentally-responsible and reliable service during a defined period until the next pipeline integrity assessment. FFS provides a "Fit-for-service" timeframe for which the integrity condition of a pipeline is acceptable provided the FFS proposed measures are found to be effective over time.

Figure 3.12 describes a recommended Fitness-For-Service assessment process using a management system approach PLAN-DO-CHECK-ACT led by an integrity policy and commitment from senior management.

Please refer to the Chapter 12 for details about mitigation, prevention, and monitoring.

3.7   CONFORMANCE AND COMPLIANCE VERIFICATION AND ACTION PLANS [MS]

Checking is a process to verify the adequacy (i.e., completeness), the implementation (i.e., walk the talk) and effectiveness (i.e., meet the objective) of the integrity management activities including the management system (e.g., training) and program (e.g., risk assessment). The end-result of any verification should be pointing out the strengths and the areas for improvement instead of search for fault.

The integrity management checking process is aimed to identify the areas that:

1. Meet requirements and expectations or *"No Finding."*
   a. If exceeded, to be recognized as a *"Best Practice"* for other business units' benefit.
2. Meet with recommendation for continuous improvement: optimize efficiency or effectiveness (non-mandatory change).
3. Partially or did not meet adequacy, implementation, and effectiveness becoming a *"Finding"* (mandatory change).

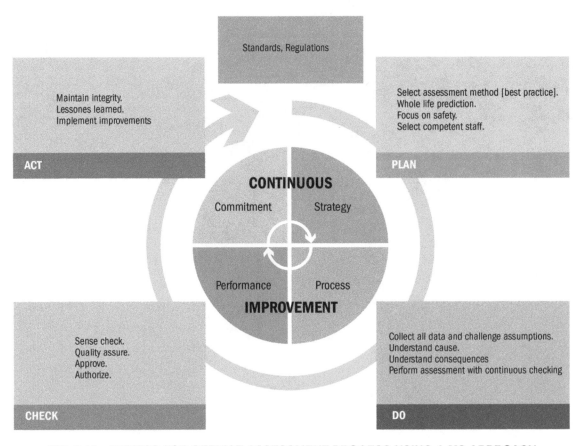

FIG. 3.12 FITNESS-FOR-SERVICE ASSESSMENT PROCESS USING A MS APPROACH

Figure 3.13 describes a recommended Fitness-For-Service assessment process using a management system approach PLAN-DO-CHECK-ACT led by an integrity policy and commitment from senior management.

Please refer to the Chapter 13 for details about Compliance and Conformance Verification and Action Plans.

3.7.1 Integrity Findings: Non-Conformance, Non-Compliance [MS]

Findings can be classified as Non-Conformance or Non-Compliance.

Non-conformances are findings during the verification process that did not meet the expectation required by a company's own or industry standard requirements not adopted by the law or regulations.

Non-compliances are findings during the verification process that did not meet the expectation required by the law, regulations, or industry standards incorporated by reference or adopted as of a mandatory requirement.

For instance, in 2006, the USA Department of Transportation, Pipeline and Hazardous Materials Safety Administration incorporated by reference all or part of industry standards and specifications from the American Petroleum Institute (API), American Gas Association (AGA), American Society of Mechanical Engineers (ASME), American Society for Testing and Materials (ASNT), Manufacturers Standardization Society (MSS), National Fire Protection Association (NFPA) and Pipeline Research Council International (PRCI). These standards became mandatory.

Findings should be written using a plain language identifying what (e.g., process, result), action (e.g., expectation), and the impact or effect (e.g., lacking of).

> *Example: "The defect assessment analysis did not conduct data integration to determine the criticality."* Thus, this finding is indicating the process of *"defect assessment analysis,"* an action (*not conducting data integration*) and the lack of "criticality."

The findings and actions to be taken should be also written by the verifier in a plain language. The following are some recommended verbs to be used (e.g., ISO, CSA-Z662) such as:

- *Shall or Must:* indicates that the requirement needs to be met to conform or comply with
- *Should:* indicates that it is recommended or advised, but not required.
  Some regulations use "should" expecting that the regulated user provides a justification to whether it is not applicable or to show other methods to meet the intent of the regulation.
- *May:* indicates option(s) for the user to consider, but not required.

3.7.2 Integrity Verification Method Selection [MS]

The integrity checking process requires the definition of *where* (i.e., selection) and *how* (i.e., tool) the verification would take

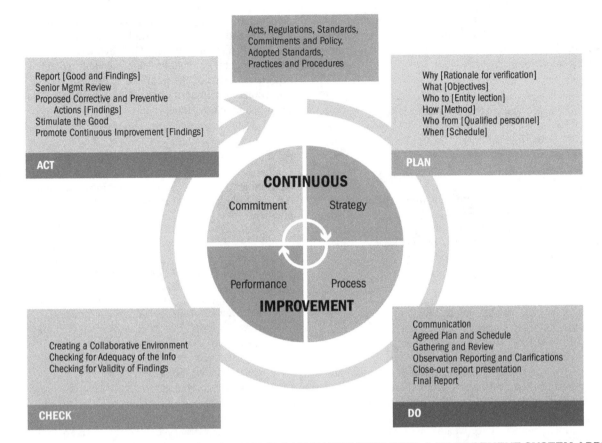

FIG. 3.13 COMPLIANCE AND CONFORMANCE VERIFICATION PROCESS WITH A MANAGEMENT SYSTEM APPROACH

place. In the integrity context, the selection can random or focused. The following are some examples for "*where*" to check:

- Integrity incidents (e.g., leaks)
- Integrity performance (e.g., lower than expected)
- findings reoccurrence (e.g., non-compliances)
- Integrity process (e.g., crack management) of all or specific segment
- Risk ranking of pipeline segments (e.g., high probability, high consequence, combined)

The objective of the checking including type (e.g., adequacy, implementation, and effectiveness) and expected in-depth would determine the type of tool to be used in verifying conformance and/or compliance.

Example: Tools used for integrity verification

- Integrity Information Meeting
- Integrity Inspections
- Integrity Audits

Information Meetings' objective is to exchange integrity information between two (2) entities (e.g., departments, companies, company and association, and company and regulator) in both directions: *verifier* and *verified*. The *verifier* provides background and rationale for the meeting indicating the information needed at that time. The *verifier* may identify areas for additional information or verification based on the answers received from the *verified*. The

verified also may ask for information about the verification framework and interpretations as well as recommended practices for continuous improvement. Typically, the *verified* has the opportunity to provide feedback to the *verifier* as to the findings draft as well as proposed improvements for future verifications.

Integrity Inspections' objective is to verify the implementation of a given requirement. Integrity inspections may be conducted in the office and/or the field following a "check-list" approach. Inspectors are typically empowered to halt or suspend activities until the referenced requirements are met. Inspection may typically be scheduled informing the *verified* in advance; however, inspections may also be conducted any given time without advance notice to the *verified*.

Integrity Audits' objective is to verify the adequacy, implementation, and effectiveness of an integrity management system, program, procedures, activities, and records. Integrity audits are conducted in the office with some field verification following a "protocol" approach that is provided to the *verified* with advance notice. Auditors are not typically empowered to halt or suspend activities requiring an inspector to exercise a non-compliance order. Audits review the documentation, interview personnel, verify execution, and analyze trends requiring multidisciplinary teams, and sometimes legal counsel. One of the greatest values of the integrity audits is their ability to identify systemic findings, that when corrected, provide an increase of the integrity system effectiveness.

3.7.3 Incident Investigation: Root-Causes and Contributing Factors [MS]

Pipeline companies and regulators develop incident investigation processes to determine the root-causes and the contributing factors in the event of a pipeline incident as well as providing the basis for preventing their reoccurrence (i.e., Corrective and Preventive Action Plans). Pipeline incident definition changes within jurisdictions and industry standards.

Example: Defining pipeline incidents

- People injury or fatality resulting from a pipeline incident
- Unintended or uncontrolled release of the fluid transported by the pipeline system
- Unintended pipeline fire or explosion
- Pipeline operation exceeding pressure limits dictated by the law, regulations or standards
- Natural forces (e.g., landslide, subsidence, flooding) that could cause a pipeline damage and/or release
- Security situations that could put in danger people, property or environment related to pipeline systems

The following are some of the stages involving an incident investigation, not including the emergency response, clean-up, and remediation [9, 10]:

- Notification of the incident to affected parties (e.g., community) and authorities (e.g., emergency responders, regulators)
- Securing and making the site safe for the investigation
- Site data gathering (e.g., scaled photographs, videos, soil and wreckage samples, witness info, weather and environmental conditions before, during and after incident, referenced incident map)
- Field preparation of the failed pipe for laboratory analysis (e.g., examination protocol)
- Office data gathering (e.g., operations, maintenance and integrity historical records; technical and management interviews)
- Incident data integration (e.g., site, office, laboratory)
- Reconstruction and simulation of events
- Determination of root-causes and contributing factors
- Incident Investigation Report and Documentation
- Lessons Learned, and Corrective and Preventive Action Plans (CPAP)

3.7.4 Corrective and Preventive Action Plans [MS]

Corrective and Preventive Action Plans (CPAP) are developed to address non-conformances, non-compliances, and incident root-causes and contributing factors. They typically focus on reducing likelihood of reoccurrence as it is more feasible than consequence reduction. CPAP should also identify the deadlines for completion as well as both the accountable management personnel and the personnel responsible for the execution.

Example: Information typically provided in a CPAP

- Finding (e.g., non-conformance, non-compliance, incident root-cause)
- Action to be taken (proposed/approved)
- Accountable (e.g., senior management)
- Responsible (e.g., doer, implementer)
- Estimated and actual completion date
- Remarks (e.g., observations, challenges, updates)

Accountable management personnel should, in a timely manner, monitor and review the implementation and effectiveness of the CPAP measures taken. The review should identify whether improvements or changes are needed to achieve the CPAP objectives. Effectiveness of the CPAP should be realized in a reduction of the integrity risks (e.g., leak, rupture, threat growth) of the pipeline in the future.

3.7.5 Management of Change—MoC [MS]

Changes in operations, fluid, consequence (e.g., population, environment), processes and technology may introduce variation (i.e., increase/decrease) in the integrity risk exposure to/from the pipeline.

Example: Changes that may affect integrity

- Faster aging due to nature of growth (e.g., deep cracks) or hazard conditions (e.g., coating damage)
- New integrity threats acting on old stable-threats (e.g., Corrosion or SCC on construction Dents)
- Higher than normal operational effects (e.g., cycling, spikes) on the pipeline
- New transported fluids causing more corrosivity (e.g., water content) or cycling (e.g., lighter products)
- Manufacturing anomalies activated (i.e., now growing) by abnormal operational conditions (i.e., overpressure)
- Weather (e.g., rain) causing higher hydro-geotechnical effects (e.g., flooding, landslide) on pipeline crossings
- New urban development increasing in population density (i.e., third party damage)
- New environmentally sensitive areas increasing protection (e.g., new high consequence)
- Improvements in inspection technology requiring new procedures and acceptance criteria
- Reduction or elimination of corrosion chemical programs in the past

The following are some recommended elements and questions in developing pipeline integrity Management of Change (MoC) processes:

- Description: *What is the change?*
- Rationale: *Why do we need to change?*
- Risk Assessment of the Change
  - Risk Estimation (increase/decrease): before, during, and after change
  - Risk Evaluation (Pass/Fail)
- Proposed risk mitigative measures (e.g., training, pressure reduction, new responsibilities)
- Proposed response protocol for abnormal conditions during the change (e.g., right-of-way patrol, integrity review, management approval to re-start)
- Communication: *Who needs to be informed?*
- Technical Review: *Who has to endorse the change?*
- Management Review: *Who needs to approve the change to proceed?*

A MoC may introduce new or modifications in the organizational structure, roles and responsibilities, policy, process and procedures as well as the technology to be used in managing the risk during and after the change. A MoC closure may require assessing the risk back to normal or new future condition to ensure a continued safe operation.

3.8 MANAGEMENT REVIEW: INTEGRITY PERFORMANCE AND KPIs [MS]

As illustrated in Figure 3.14, the pipeline integrity management review monitors and closes the management system cycle initiated by the setting of stakeholder-driven integrity goals, objectives, and targets supported by the integrity policy and commitment. Management has the right to follow-up and request status reporting granted by their responsibility taken before the stakeholders (e.g., internal and external).

As illustrated in Figure 3.15, the integrity performance cycle is explained in four (4) stages:

- Integrity Feed (Input)
- Integrity Process
- Integrity Output (Direct or Response)
- Integrity Outcome (End-results)

Thus, KPI attributes are comprised of:

- Approach (i.e., IMP or MS)
- Stage (i.e., Feed, Process, Output, and Outcome)

- Type (i.e., leading, lagging, qualitative, and signal)
- Measurement Unit (e.g., magnitude, frequency, percentage, cumulative)

The Chapter 14 describes in more detail the integrity performance cycle and associated Key Performance Indicator (KPI) attributes with strategies for their development and implementation.

This ongoing management review cycle requires introducing motivation and incentives to the people performing the process being reviewed. Introducing changes during the review with the purpose of increasing performance should be made with the participation of the doers/implementers to gain their ownership and motivation.

Example: Initiating questions to trigger a thought process towards to review

- Is anything wrong or not working right? Is anything missing? [Adequacy]
- Are we getting what we need? Are we achieving the goals with this PIMS process? [Effectiveness]
- Are there systemic issues emerging? [tsunamis or undercurrents]
    - Are minor incidents continuing or increasing? [Emerging]
- Are integrity findings (defects, indications) consistent with expectations? [Reality Check]

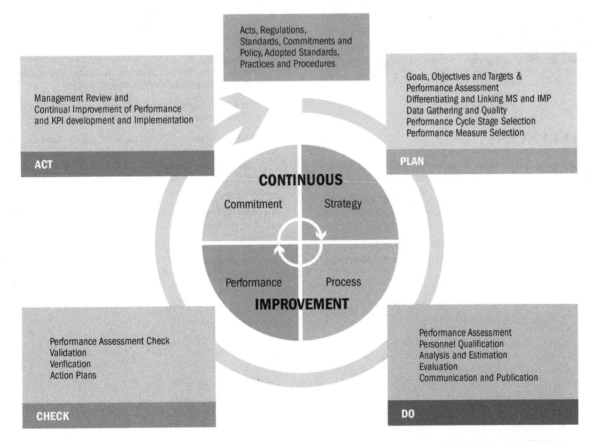

FIG. 3.14   INTEGRITY MANAGEMENT REVIEW WITH A MANAGEMENT SYSTEM APPROACH

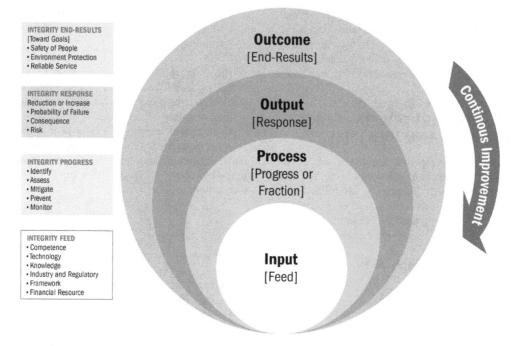

INTEGRITY END-RESULTS
[Toward Goals]
• Safety of People
• Environment Protection
• Reliable Service

INTEGRITY RESPONSE
Reduction or Increase
• Probability of Failure
• Consequence
• Risk

INTEGRITY PROGRESS
• Identify
• Assess
• Mitigate
• Prevent
• Monitor

INTEGRITY FEED
• Competence
• Technology
• Knowledge
• Industry and Regulatory
• Framework
• Financial Resource

FIG. 3.15   INTEGRITY MANAGEMENT PERFORMANCE CYCLE [MS]

- Do Regulator's, Landowner's, or Public's expectations appear to be increasing? [External Effects]
- Is the Corporate Goal and Commitment being followed? [Implementation]
- Is there over-delegation to inexperienced or untrained staff? [Competency]
- Are courses linked to corporate needs and requirements? [Training]

3.9   REFERENCES

1. Harvard Leadership Blog, http://blogs.law.harvard.edu/leadership/management-definition.

2. Moen, R., Norm, C., 2010, *Circling Back—Clearing up myths about the Deming cycle and seeing how it keeps evolving*, Quality Progress, http://www.apiweb.org/circling-back.pdf.

3. Vanguard, 2015, *Brief history of ISO 9000, Vanguard Consulting Ltd*, http://vanguard-method.net/library/iso-9000/a-brief-history-of-iso-9000-2/.

4. Anon., 2005, *ISO 9000: Quality Management Systems*, International Organization of Standardization, https://www.iso.org/obp/ui/#iso:std:iso:9000:ed-3:v1:en.

5. Anon., 2004, ISO 14001: *Environmental Management Systems*, International Organization of Standardization, https://www.iso.org/obp/ui/#iso:std:iso:14001:ed-2:v1:en.

6. Anon., 2008, *BS Publicly Available Specification (PAS) 55 Asset Management*, British Standard Institute, http://shop.bsigroup.com/ProductDetail/?pid=000000000030171836.

7. Nonaka, I., Takeuchi, 2011, *The Big Idea: The Wise Leader*, Harvard Business Review, https://hbr.org/2011/05/the-big-idea-the-wise-leader.

8. Chad, B., Cameron, G., Mora, R. G., 2010, *Guidelines to Conducting Threat Susceptibility and Identification Assessments of Pipelines Prior to Reactivation*, IPC2010-31585, American Society of Mechanical Engineers, Proc. of the 2010 International Pipeline Conference, Calgary, Alberta, Canada.

9. National Transportation Safety Board (NTSB), 2015, *The Investigative Process*, http://www.ntsb.gov/investigations/process/Pages/default.aspx.

10. Transportation Safety Board (TSB) of Canada, 2015, *Investigation Process*, http://www.tsb.gc.ca/eng/enquetes-investigations/index.asp.

INTEGRITY HAZARD AND THREAT SUSCEPTIBILITY AND ASSESSMENT

4.1 INTRODUCTION

Despite the fact that pipelines are the safest mode of transportation, hazards, and threats occur during a pipeline's lifecycle. These hazards and threats may result in an undesirable events or loss of containment and affect waterways or populations of people. If a pipeline release occurs, it can result in fire, explosion, and contamination of the environment that affects communities. Pipeline operators should establish processes to identify and analyze all hazards and threats with their corresponding root causes for evaluating and managing the associated risk (i.e., likelihood and consequence).

Hazard and Threat Assessments are an essential element of the process of risk analysis. Its primary purpose is to identify all types of hazards and threats leading to degradation of the pipeline condition. Hazard types in a pipeline include:

- Active (i.e., SCC defect growing into time),
- Inactive (i.e., dormant ERW imperfections),
- Potential (i.e., soil movement)

Hazards that cause failure modes (i.e., Corrosion threat) require estimation of the Probability of Failure (POF).

The Hazard and Threat Assessment process includes Hazard Identification. Hazard Identification identifies potential sources of failure, hazardous events, circumstances, and situations that might affect the integrity of a pipeline and its consequences. Impact to the pipeline integrity is defined as any damage or failure mechanism, which may lead to a loss of containment. That is the failure or inability of the pipeline to transport any products safely and reliably on the pipeline system. There are several techniques for performing hazard identification. This chapter outlines examples of technique to identify hazards and threats.

This chapter describes the meaning of hazard and threat: identification, susceptibility, and assessment. The chapter then steps the user through the steps they may take to implement a robust program for identifying hazards, as illustrated in Figure 4.1. The hazard identification process has several levels of robustness. This begins with a recognized industry process that meets regulatory requirements and follows an industry standard (e.g., ASME B31.8S, API 1160 and CSA Z662) to a well-developed company process that leads and drives a company culture of continual improvement. The process then leads to a well-developed company process that drives a company culture of continual improvement. The level of hazard identifying and analyzing should be selected wisely. The objective is to reach a better engineering decision process related to pipeline integrity. An insufficient analysis may lead to poor decisions and excessive analysis wastes resources.

4.2 THREAT MANAGEMENT CYCLE

Threat assessment is designed to identify, classify, analyze, and evaluate hazards and threat likelihood, susceptibility and severity. This process should take place throughout a pipeline's lifecycle (i.e., prior to operation, in-service, and through end of life). If a pipeline is considered susceptible to threat and the evaluation of the likelihood is estimated as a high, pipeline operators may implement prevention, mitigation, or monitoring measures (Chapter 11).

Hazard and threat mitigation measures and programs are intended to reduce the probability of an undesired pipeline event and check effectively the monitoring program. Figure 4.2 depicts the cycle and elements proposed for performing threat assessment and threat reduction and monitoring program. The sequence of steps for managing threats to pipeline integrity includes company's commitment and responsibilities, quantifiable objectives and methods for identifying hazards and threat susceptibility (Initiation and Growth), assessing current and potential hazards and threats, implementing integrity threat mitigation and prevention actions, and performing threat monitoring program.

Threat assessment contributes to the overall prioritization of preventive, mitigative, and monitoring measures to increase people safety and reduce consequence to environment. Additionally, threat assessment might address specific issues within in a context of engineering assessment. Threats may arise from the following:

- Increase population resulting from a class location change
- Change of pipeline service (i.e., pipeline transporting sour gas changing to liquid pipeline to transport refined products)
- Aging pipeline
- Change of flow direction of the pipeline
- Revised MOP from increasing the operating conditions.

4.2.1 Pipeline Integrity Hazard and Threat Management Roadmap

Figure 4.3 describes a recommended steps and elements for conducting hazard and threat management. The threat management road map comprises mainly six processes:

- Performance of hazard identification process,
- Determination of threat exposure and susceptibility analysis,
- Confirmation of the threat existence,
- Estimation of the likelihood of failure process,
- Ranking and treatment of the threat, and
- Implementation of mitigative and preventive measures, and monitoring programs.

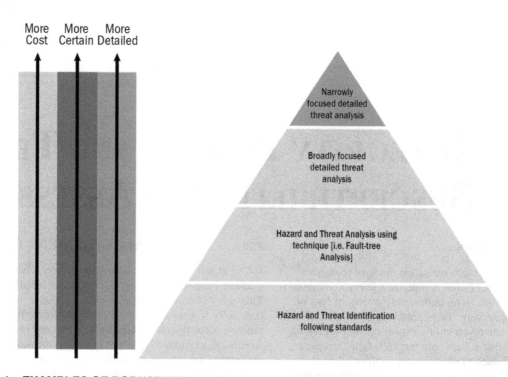

FIG. 4.1 EXAMPLES OF ROBUSTNESS LEVELS OF HAZARD IDENTIFICATION AND EVALUATION PROCESS

FIG. 4.2 EXAMPLES OF THREAT MANAGEMENT CYCLE PROCESS

The first framework step in understanding the potential integrity threats and assessment of any consequence area along the pipeline system is the planning process. This involves setting objectives, and assembling and reviewing information about potential risk factors. It leads into the threat identification step and then on to assessment. In this element, the pipeline operator conducts the initial collection, review, and integration of data that is needed to understand the condition of the pipe and identify the location-specific threats. To support the pipeline threat assessment information collected should include:

- Operational data
- Maintenance records

- Surveillance practices
- Pipeline design and material specification and characteristics
- Pipeline construction procedures, methods, and failures (i.e., welding procedures and non-destructive testing)
- Pipe manufacturing process and records
- Operating history
- Previous failure modes and failure mechanisms
- Anomalies, imperfections, and its repairs
- Concerns that are unique for each system and segments

Subsequently, a confirmation of the existence of threat may be conducted through performing integrity assessment techniques

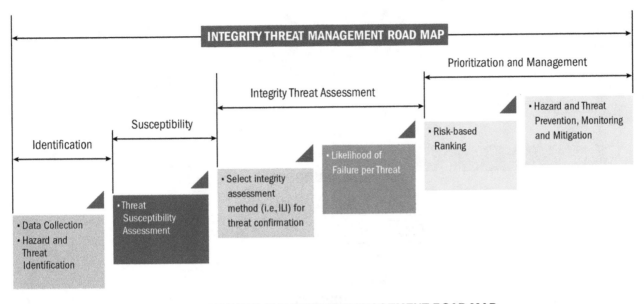

FIG. 4.3   HAZARD AND THREAT MANAGEMENT ROAD MAP

(i.e., ILI) and susceptibility analysis following of an evaluation of the probability of failure for each encountered threat. The ranking and treatment of threats are the final process of the hazard and threat management. The threat assessment is outlined in this chapter.

4.3   PIPELINE INTEGRITY THREAT ASSESSMENT PROCESS

Threat Assessment helps significantly to understand and evaluate the condition of the pipeline integrity. Pipeline companies are required to perform threat assessment for determining threat exposure, reviewing the attributes for all potential threats, assessing the relevance and severity of each exposed threat, validating the threat assessment, and performing management review. The threat assessment should include assessment of individual threats, and interacting threats that are also key components in managing pipeline safety. Determination of hazards associated to threat and threat exposure can be conducted through threat identification techniques, threat analysis process or susceptibility analysis. Relevance and severity of threat assessment is typically the primary goal of a threat assessment for failure likelihood estimation based on the availability, quality, and completeness of the data for each threat.

Threat assessment validation and management review are described in further chapters. For instances, threat management plan includes prevention, mitigation, and monitoring for pipelines is depicted in Chapter 11. As illustrated in Figure 4.4, the threat assessment process is structured using a PLAN-DO-CHECK-PLAN (PDCA) integrated management system approach.

The objective of the threat assessment is to identify and assess the site-specific hazards exposure to a given threat, or new integrity threat, and evaluate if existing threat has vanished or became inactive to establish the most effective preventive and mitigation actions and address these relevant hazards in the most effective manner. In doing so, the threat assessment categorizes additionally the safety level inherent in the options of mitigation action. Pipeline operators typically used levels of threat assessment to evaluate conformances and compliances and it goes from a simple pipeline data review to a performance of detailed qualitative/quantitative analysis to estimate the probability of likelihood.

4.4   PIPELINE INTEGRITY HAZARDS AND THREATS

Hazards and threats are terms in the pipeline integrity management that are used interchangeably to denote the same meaning. However, pipeline operators also acknowledge that Hazard and Threat have different meanings and are independent to emphasize vulnerability or danger to pipeline integrity. This chapter differentiates the two terms. Differentiation between hazard and threat allows integrity personnel to be confident that a thorough integrity hazard and threat assessment process is available. This allows an operator to identify all hazards subsequently assess threat susceptibility of a pipeline.

Anomalies or defects causing modes of failure might be introduced into a pipe at any point during the lifecycle of the pipeline. This includes manufacturing of the pipe, pipeline construction, and development during normal operation. For instances, high pressure cycling stress growing a longitudinal seam weld anomalies, coating disbondment developing corrosion, or stress concentrators initiating cracking during service. These threats may worsen over time under specific operational or environmental conditions. These include failure mechanisms such as fatigue-induced or environmentally assisted cracking and ultimately cause in-service pipeline failures.

The purpose of a pipeline integrity hazard and threat assessment is to identify hazards and threats through structured techniques (i.e., event tree analysis), performance of threat susceptibility analysis, and evaluate the susceptibility to the capacity of integrity threats (i.e., likelihood) to cause a failure in a pipeline. Additionally, a pipeline threat assessment is an Integrity Management Program (IMP) process contributing to determine the risk and subsequent prioritization of mitigation, prevention and monitoring measures.

4.4.1   Failure or Damage Mechanism

A phenomenon that induces harmful micro and/or macro changes in the material conditions that are harmful to the material condition or mechanical properties. Damage mechanisms are usually incremental, cumulative, and unrecoverable. Common damage mechanisms are associated with chemical attack, creep, erosion, fatigue, fracture, embrittlement, and thermal ageing [1].

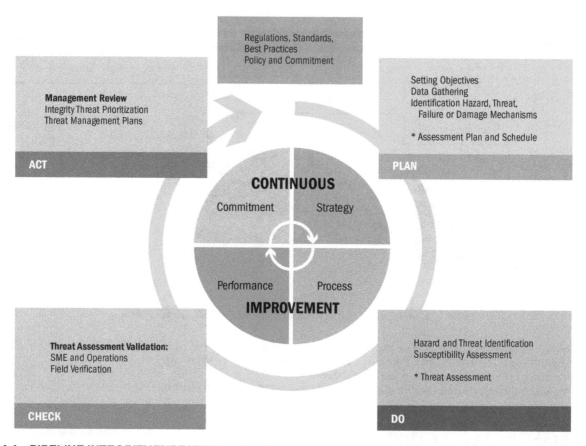

FIG. 4.4   PIPELINE INTEGRITY THREAT MANAGEMENT PROCESS WITHIN MANAGEMENT SYSTEM APPROACH

4.4.2   Pipeline Integrity Hazard

Hazard is defined by Webster's dictionary as "*a source of danger*" and danger is explained as the *exposure or liability to harm or loss*. Hazard means anything that can cause harm (e.g., soil type, coating type, lack of fusion on the long seam weld, low Charpy-V-notch value, etc.) to the pipeline integrity. Hazards within the pipeline integrity context could be defined as a source, event, circumstance, and situation of harm or loss that might affect or impact the integrity of a pipeline. Thus, an *Integrity Hazard* can be defined as *any detrimental situation, event, or condition* (e.g., coating damage) *able to initiate or grow an integrity treat (e.g., corrosion)*.

The UK's HSE defines broadly a "hazard" as: "... the potential for harm arising from an intrinsic property or disposition of something to cause detriment..." A hazard is something (e.g., an object, a property of a substance, a phenomenon, or an activity) that can cause adverse effects.

4.4.3   Pipeline Integrity Threats

An Integrity Threat is defined as an *abnormal state affecting a pipeline that may have the capacity to cause a failure*. Abnormal states or threats (e.g., Stress Corrosion Cracking) are typically created, made active, or grown by one or more hazards (e.g., tape coating damage, stress levels, pipe manufacturing thermal treatment) [2].

DNV RP F116 defines "threat" as: "An indication of an impending danger or harm to the system, which may have an adverse influence on the integrity of the system." Thus, an "*Integrity Threat*" within the context of pipeline integrity can be defined as an indication of the pipeline degradation due to "*Integrity Hazard*" that may causes pipeline damage or develop a pipeline failure.

ASME B31.8S considers that each of the 22 causes representing a "*Threat*" to pipeline integrity and those threats have been grouped into nine categories. ASME B31.8S, CSA-Z662 and API 1160 have integrity threat types that are described in Table 4.1.

Hazards, contributing factors, threat attributes, and risk factors are commonly names identified in the pipeline industry for determining the basic causes of the pipeline failures or answering the question of why reasons pipeline are failing.

The failure mechanisms are typical physical degradation of components due to operational conditions combined with components features like design, materials, and surface treatment, and commonly are divided into the following categories with examples of integrity threats within parenthesis related to metal structure components:

a.   Mechanical and Metallurgical Failure (Cracking, Fatigue, Fracture)
b.   Uniform or Localized Loss of Thickness (i.e., corrosion)
c.   High Temperature Corrosion (i.e., Embrittlement)
d.   Environment Assisted Cracking (i.e., SCC, SSC, HIC)
e.   Deformation
f.   Friction (i.e., abrasion)
g.   Wear

There are failure mechanisms for metal structure components (i.e., pipeline), polymer (i.e., Extrusion, Bad resilience, etc.), and electric components (i.e., dielectric breakdown, induced current, voltage drop, etc.).

Failure mode definition is what the equipment of pipeline component failed from or means the way or mode in which pipeline might fail (i.e., corrosion fatigue). Root cause is what caused the failure mode to occur and what can be changed to prevent re-occurrence

TABLE 4.1   CORRELATION OF PIPELINE INTEGRITY THREAT NAMES FROM INDUSTRY STANDARDS

ASME B31.8S Gas Transmission Pipelines	CSA-Z662 (Canada) Gas, Liquid Hydrocarbons, Oilfield Water and Steam, Liquid or Dense Phase Carbon Dioxide Pipeline Systems	API 1160 Hazardous Liquid Pipelines	UKOPA (United Kingdom) Major Accident Hazard Pipelines
External corrosion	Metal loss	External corrosion	External corrosion
Internal corrosion	Metal loss	Internal corrosion	Internal corrosion
Stress corrosion cracking—SCC (i.e., environmentally-assisted cracking)	Cracking	Crack and crack-like	Other
Manufacturing-related (e.g., seam weld and pipe body)	Material or manufacturing	Design and material	Seam weld defect
Welding/fabrication related (e.g., girth weld, wrinkle bend, threads)	Construction	Construction	Girth weld and pipe defect, construction/material
Mechanical damage (e.g., immediate/delayed failure, vandalism/theft)	External interference	Third party damage	External interference
Incorrect operations (e.g., procedure)		Operation errors	Operator error
Weather-related and outside force (e.g., hydro-geotechnical, cold weather, lightning, heavy rains or floods, Earth movements)	Weather	Ground movement	Ground movement
	Geotechnical failures		
Pipeline equipment (e.g., valves, seals)	Ancillary equipment	System operations	N/A

or is a hazardous factor that caused non-conformance and should be permanently eliminated/mitigated through process improvement (i.e., over-pressurization and temperature cycles, pressure cycles, poor condition of coating, etc.). Following Table 4.2 depicts examples of hazards associated with identified threats.

4.5   PIPELINE INTEGRITY HAZARD IDENTIFICATION

Hazard identification in the risk analysis process (see Figure 6.2) generally focuses on identifying all hazards, threats, and failure

Example: Damage/Failure Mechanisms, Threats, and Hazards Source from Industry Standards

TABLE 4.2   EXAMPLE OF FAILURE MECHANISMS, INTEGRITY THREAT, AND ITS ASSOCIATED INTEGRITY HAZARDS

Damage/failure mechanism	Integrity threats/Failure mode (leak or rupture)	Integrity hazards (Contributing hazards)
Wall thickness reduction due to, for example, but not exclusively, to corrosion or erosion [3]	External metal loss	• Coating type and condition • Disbondment of coating • Cathodic protection faulty or shielding • Number of mainline CP readings not within specs • Pipe aging—pipe is old • Soil type • Microbiological induced corrosion • Type of cathodic protection • Wall thickness • Casing condition and type • External metal loss anomaly • River crossing • Wetland areas • MFL In-line inspection run • Pipeline piggable • CP installation age • Remaining life of external metal loss anomaly • Seam weld type • High external failure rate • High external corrosion growth rate

(Continued)

TABLE 4.2   EXAMPLE OF FAILURE MECHANISMS, INTEGRITY THREAT, AND ITS ASSOCIATED INTEGRITY HAZARDS (Continued)

Damage/failure mechanism	Integrity threats/Failure mode (leak or rupture)	Integrity hazards (Contributing hazards)
	Internal metal loss	• Microbiological induced corrosion • Product service or liquid on the pipeline system • Internal coating • Product velocity • Water • Remaining life of internal metal loss anomaly • Pipeline piggable • Internal metal loss anomaly • Wall thickness • Pipe aging • MFL in-line inspection run • Level of chemical in the fluid • Level of hydrogen sulfide • Seam weld type • Low point of pipeline • Mill loss corrosion rate • Internal failure rate • No Inhibitor treatment and usages
	Microbiologically induced corrosion (MIC)	• Aqueous environments • Coating condition • Hydrogen sulfide presence • Operating temperature
	Selective ERW Seam Corrosion	• Coating deterioration • Metal loss anomaly on LSW
	Soil Corrosion	• Soil-to-air interference • Dissimilar soils • MIC • Operating temperature • Soil resistivity and characteristics • Soil type (water drainage) • Coating type, coating age, coating condition • Cathodic protection • Stray current drainage • Material of construction • Presence of water/moisture
	External Interference (AC/DC)	• Alternate current interference • Direct current interference • Induced AC voltages on pipelines • Electromagnetic field interference (EMI) • Soil resistivity • High voltage power line crossing • AC/DC sources near pipeline
	Gouging/Grooving	• Gouging anomalies • First, second, or third party
Cracking—Mechanically driven or environmentally assisted cracking of pipe or component	Stress corrosion cracking (SCC)	• Pipeline operating tensile stress levels • Pressure cycling range (rainflow analysis) • Corrosive medium • Operating temperature • Pipe aging • Corrosion coating system • Pipe material properties • Terrain condition

TABLE 4.2 EXAMPLE OF FAILURE MECHANISMS, INTEGRITY THREAT, AND ITS ASSOCIATED INTEGRITY HAZARDS (Continued)

Damage/failure mechanism	Integrity threats/Failure mode (leak or rupture)	Integrity hazards (Contributing hazards)
		• Longitudinal seam weld type • Anomaly type (i.e., corrosion, slivers, etc.) • Residual stress (bending stress, denting) • Distance downstream from station
	Sulfide stress cracking (SSC)	• Pipeline operating tensile stress levels • Pressure cycling range (rainflow analysis) • Corrosive medium (i.e., sour gas) • H_2S concentration • Water • Operating temperature • Presence of water • Post-weld heat treatment • High hardness • High-strength material
	Hydrogen-induced cracking—HF	• Steel hardness levels • H_2S concentration • Presence of water • Operating temperature • pH • Residual tensile stress • Pipe material properties (high strength low alloy steel, hard microstructure, etc.) • Aqueous HF acid environments • Coating condition • High level of applied tensile stress • Material of construction • Sour products
	Mechanical damage delayed cracking	• Pipe denting • Operating stress levels • Cyclical stress or pressure fluctuation • Pipe material properties (strength, hardness, toughness Charpy-V-Notch, microstructure)
	Corrosion fatigue cracking	• pH of soil • Longitudinal seam weld type • Preferential corrosion on LSW • Pressure fluctuation • Corrosive environment • Stress raises and concentrators (pits, notches, surface defects, etc.)
		• Operating temperature (thermal stress) • Operating stress level • Pipe material properties
	Mechanical fatigue cracking	• Cyclical stress or pressure fluctuation • Thermal cycling range • Operating stress level • Pipe material properties (strength, hardness, toughness Charpy-V-Notch, microstructure) • Stress raisers (mechanical notches) • Stress concentrators (grinding marks, tool markings, groves, corrosion, etc.)

(Continued)

TABLE 4.2   EXAMPLE OF FAILURE MECHANISMS, INTEGRITY THREAT, AND ITS ASSOCIATED INTEGRITY HAZARDS (Continued)

Damage/failure mechanism	Integrity threats/Failure mode (leak or rupture)	Integrity hazards (Contributing hazards)
	Other cracking (delaying cracking)	• Cyclical stress or pressure fluctuation • Operating temperature • Pipe aging • Pipe material properties (strength, hardness, toughness Charpy-V-Notch, microstructure) • Stress raisers (mechanical notches) • Stress concentrators (grinding marks, tool markings, groves, corrosion, etc.) • Denting (mechanical damages) • Land uses
	Defective long seam weld (i.e., corrosion fatigue on LSW, lack of fusion, cracking, etc.)	• Pipeline operating tensile stress levels • Pressure cycling range (rainflow analysis) • Corrosive medium • Operating temperature
		• Pipe aging • Coating system • Coating type • Pipe material properties • Longitudinal seam weld type • Anomaly type (i.e., corrosion, slivers, etc.) • Residual stress (bending stress, denting)
Geotechnical and hydro technical	Landslides/mass movement threat	• Climatic factors • Topography and morphology • Geology • Soil type • Hydrology • Land uses • Avalanches • Slow ground movement • Heavy rains/washout in slopes • Landslides caused by third party (i.e., overloading, underground mining) • Specialized geohazards: man-made, thaw off, residual and sensitive soils, desert mechanisms including dune migration, volcanic overburden, geochemical karst, and acid rock drainage
	Inland flood	• Precipitation • Snow melt • Ice-jams
	Scouring	• River flow volume • River flow velocity • Stream shape (stream characteristic factor) • Sediment size and type • Annual flooding frequency • Depth of cover
		• Mean water depth • Maximum span allowable (pipe perpendicular to flow) • Scouring events previously • Flood areas
	Washout/erosion	• River scour • Bank erosion

TABLE 4.2   EXAMPLE OF FAILURE MECHANISMS, INTEGRITY THREAT, AND ITS ASSOCIATED INTEGRITY HAZARDS (Continued)

Damage/failure mechanism	Integrity threats/Failure mode (leak or rupture)	Integrity hazards (Contributing hazards)
	Undermining subsidence	• Mining activities
	Seismic	• High seismic zone
	Freeze-thaw	• Ground movement by cyclic freezing • Thawing of the supporting soil • Frost heave susceptibility
	Hydro-geotechnical	• River scour • Bank erosion • Channel migration
Mechanical damage/Deformation	External interference (External activities causing damage to the pipe or component)	• Company employee • Company contractor • Third party • Vandalism • Others • One call activity • Excavation damage actions • Shallow sections identified/not mitigated • Outside force actions
	Denting	• Dent anomaly • Improper construction procedure • Improper maintenance procedure • High agricultural activity • Pipe material properties • Inadequate depth of cover • Excavation activity
Material or manufacturing failure	Weld cracking	• Incomplete fusion • Toe cracks • Off seam weld • Hook cracks • Cold welds • Weld metal crack • Burnt pipe edges • ILI detection difficulty of anomaly
	Defective pipe body	• Anomaly type (i.e., laminations, sliver, blistering, scabs, hard spots, etc.) • Stress risers
	Defective long seam weld	• Defective long seam weld (i.e., impurities) • Defective helical weld • Pipe material properties • Decade installed • Preferential corrosion on LSW • Type of long seam weld method (i.e., lap weld, LF ERW, Flashed weld, DSAW) • Treatment type and process • ILI detection difficulty of anomaly
Construction failure causing defect, damage, or deficiency in the pipe or component	Wrinkle, buckle, and ripple	• Geotechnical hazards • Pipe material properties • Inadequate construction practices • Soil movement • Settlement

(Continued)

TABLE 4.2   EXAMPLE OF FAILURE MECHANISMS, INTEGRITY THREAT, AND ITS ASSOCIATED INTEGRITY HAZARDS (Continued)

Damage/failure mechanism	Integrity threats/Failure mode (leak or rupture)	Integrity hazards (Contributing hazards)
	Defective girth weld	• Anomalies due to welding procedures • ILI detection difficulty of anomaly • Span or unsupported pipe • Pipe material properties • Soil movement • Strain • Slag inclusions • Porosity • Lack of fusion • Off seam weld • Pipe aging • Residual stress (i.e., bending stress, denting) • Pipe material properties • Lack of procedures • Weld material
	Installation errors	• Poor practices
	Construction design	• Inadequate construction procedure
Equipment/component failure	Equipment/component failure	• Control equipment/electronics failure • Measurement device failure • Pipe body failure • Piping component/fitting failure • Pig barrel/receiver failure
		• Prime mover failure • Rigging failure • Valve failure • Weld failure • Equipment malfunction
Overpressure/High temperature	Incorrect Operations (e.g. Human and/or Equipment Factors)	• Equipment malfunction • Third party • Design • Incorrect operation • Unknown • PRV failures (pressure protection system failures) • PCV Failures (pressure control system failures) • Incorrect operation mitigation actions • Incorrect operation company procedures

mechanisms that could threaten the integrity of the pipelines. Operators then perform a susceptibility analysis for each individual segment for the threats identified. The minimum effort level of this process should include explicitly the threats outlined in the ASME B31.8S and CSA Z662 standards:

• design hazards, construction hazards, operator error, material failure, pipeline degradation, third-party activities, natural hazards, and. The hazard identification process also identifies active (i.e., SCC defect growing into time), inactive (i.e., dormant ERW imperfections) and potential (i.e., soil movement).

Practically, companies should maintain and use detailed information and records in designing, constructing, operating, and maintaining, and abandoning pipelines systems for making informed decisions that support the process of pipeline hazard and threat identification.

The hazard identification is a company-specific process that has typically the following common benefits and characteristics:

a. Integrate company specific, industry, standards and regulatory information and procedures into a formalized hazard identification process,

b. Analyze pipeline system on a sectioned/segmented basis to accurately identify and define hazards for unique and specific conditions,

c. Assess active and potential hazards on a regular and continued basis to ensure all damage mechanisms are included for risk evaluation through a specific defined process,

d. Implement a defined and consistent process to evaluate for interaction of anomalies, hazards and threats (e.g., SCC, dents, metal loss, manufacturing related features, third party/ mechanical damage, etc.) incorporating study outcomes (e.g., insufficient depth of cover in a river crossing, coating condition associated to SCC features) and uncertainty in information,

e. Share thoroughly hazard identification process and information throughout company departments and external organizations and industry,

f. Implement a hazard identification and report culture within the company to empower company staff, contractor, third party, and stakeholders to identify and report pipeline hazards, near misses, incidents and non-compliances to take necessary preventative measures and actions,

g. Validate the hazard identification process throughout pipeline inspections to determine the severity of the hazards,

h. Integrate hazard identification outcomes into decision-making process at every stage gate within the organization.

Threat Assessment practically encompasses all above activities in identifying and analyzing pipeline hazards and threat for the evaluation of the likelihood of the threat to fail the pipeline. This process typically addresses three main risk questions:

1. Hazard—What can be wrong? (i.e., coating disbondment in wet area)
2. Threat/Damage—How can be failing? (i.e., Pipeline leak due to pin-hole—Corrosion)
3. Likelihood—How frequent or what is the chance might it happen? (i.e., high probability of failure of pipeline due to Manufacturing threat)

Effective hazard identification depends on availability of information (see Section 3.5.3—Pipeline Integrity Threat Assessment), and having a systematic method and organizational controls for identifying hazards. Information that should be available to identify the existence of a hazard includes:

• Product transported (i.e., level of sour gas leading internal corrosion)
• Technical information along the pipeline (i.e., soil movement, pipeline design)
• Technical process (i.e., event tree technique)

The latter factor is to ensure that knowledge is applied effectively to the task of identifying hazards and of defining safe process conditions (i.e., missing procedure to start or shut-down a pump station booster that may lead in pipeline pressure surge). A clear defined process and techniques to identify threats and hazards helps pipeline operators to understand the level of pipeline operational risk. Description of techniques is outlined in the Section 4.5.3 of this chapter.

4.5.1 Hazard and Threat Identification and Analysis Process

Figure 4.5 illustrates an example of a detailed flowchart to identify hazards and threats. This approach includes a mix of creative/ imaginative, experience-based, and logical/rational techniques to identify and assess hazards. The approach includes elements such as:

• Gathering information from previous incidents and root-cause analysis of those incidents,
• Evaluating the listed hazards based on industrial practices, and
• Performing hazard identification techniques such as Event Tree Analysis, Brainstorming technique, Hazard and Operability Analysis (HAZOP) and Failure Mode and Effect Analysis (FMEA) for the failure modes and failure mechanisms associated with the identified hazards.

The main steps illustrated in the Figure 4.5 are description of pipeline segments or system to identify hazards, identification of threats through previous hazard identification studies, investigation analysis, or performing root-cause technique such as an event tree analysis. Hazard analysis step is to determine if identified hazard are associated with pipeline material, normal and abnormal operations, products, pressure, temperature, work environment, management system, and Human activities, etc., establish hazards interaction and estimate the level of hazard susceptibility and severity. Ranking of the identified hazards and documentation of the hazard identification, root-cause analysis, and hazard assessment are the final steps of the process for a continuous improvement of risk assessment.

Thorough hazard identification and threat assessment processes are the core elements in the Risk Assessment. The core concepts of the hazard identification are [4]:

• The aims of a robust hazard identification process are to ensure that the pipeline operator knows about existing hazards which could lead to major events at their pipeline system, and that new hazards might be recognized before they are introduced.
• Once hazards have been identified, the operator of a pipeline system will be able to take action to properly manage them.
• It is important to choose a hazard identification technique, or techniques, which provide an adequate depth of analysis.
• Hazard identification should provide sufficient knowledge, awareness, and understanding of the hazards that could lead to major incidents to be able to prevent and mitigate undesirable outcomes.
• During hazard identification, pipeline operators may wish to incorporate identification of all health and safety related hazards as the safety management system (SMS) must provide for all hazards and risks, not just risks of major events.
• Hazard identification provides a basis for identifying, evaluating, defining and justifying the selection (and rejection) of control measures for reducing risk.
• The full range of hazard and event types should be considered and the outputs of the hazard identification process fully documented.
• Identified hazards should not be ignored or discounted simply because control measures are, or will be, in place.
• The hazard identification process should consider all operating modes of the pipeline, and all activities that are expected to occur. It should also consider human and system issues as well as engineering issues.
• The hazard identification process should recognize that combinations of failures can occur, even though these may appear highly unlikely.
• The hazard identification process should be ongoing and dynamic.

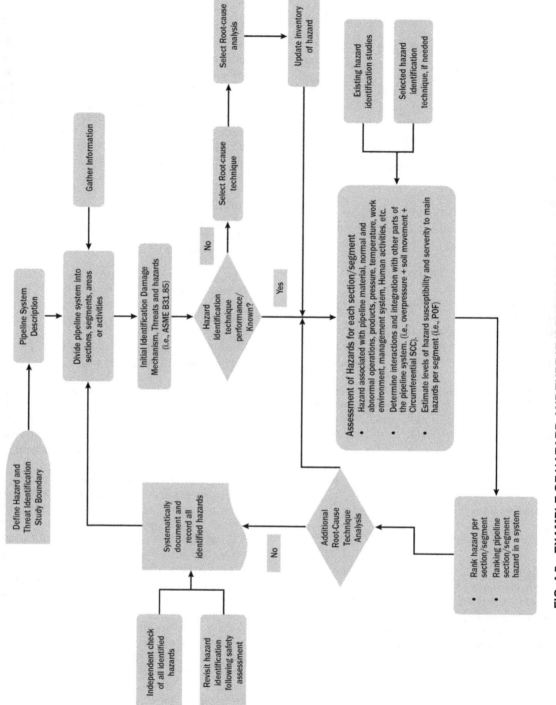

FIG. 4.5 EXAMPLE OF HAZARD AND THREAT IDENTIFICATION STEPS AND PROCESSES

It should not just be carried out during development of the safety case, but also in a range of defined circumstances, such as when there is a pipeline system modification, after any major event or dangerous occurrence, if a control measure deficiency is identified, and at defined intervals.

4.5.2 Outcomes of Hazard Identification and Analysis

The outcomes of the identification hazards process are to:

- Identify all hazards to the environment, health and safety of people, and pipeline integrity at or near the pipeline system;
- Identify the associated events and outcomes and rank them based on risks;
- Show clear links between hazards, causes and the potential events;
- Identify hazards can lead to major events or incidents;
- Provide the pipeline operator and the workforce with sufficient knowledge, awareness and understanding of the hazards to be able to prevent and deal with accidents and dangerous occurrences;
- Estimate the severity and level of susceptibility of the identified hazards for a hazard and pipeline section ranking;
- Provide a systematic record of all identified hazards which may affect health and safety of people, environment and pipeline integrity at or near the pipeline system, and in particular those which may lead to major incidents/events, together with any assumptions; and
- Provide a basis for identifying, evaluating, defining and justifying the selection (and rejection) of control measures for eliminating or reducing risk.

4.5.3 Hazard Identification Technique

There are wide ranges of hazard identification techniques available to use in this process to identify integrity hazards and threats. It is important to choose the most appropriate identification tool to provide the appropriate level of detail related to a specific integrity area of concern. The hazard identification techniques are structured processes to identifying fault conditions that lead to hazards, and reduce the chance of missing hazardous events. They all require considerable experience and expertise. The following example illustrates the most common hazards identification and analysis techniques/tools.

An Example of Hazard Identification Techniques:

- Brainstorming,
- Questionnaires,
- Pipeline industry benchmarking,
- Based-risk scenario analysis,
- Risk assessment workshops,
- Pipeline incident investigation through Fishbone Diagram,
- Auditing and inspection,
- Fault tree/Event tree analysis,
- Check list and What if Analysis,
- Failure Investigation and Root cause Analysis,
- Failure Mode and Effect Analysis (FMEA),
- Hazard and Operability Analysis (HAZOP),
- Process Hazard Analysis (PHA) [5],
- Others formal risk analysis tool in common use by pipeline industry.

HAZOP studies, scenario-based fault tree, FMEA, failure investigation analysis, and event tree analysis are approaches categorized as "*Scenario-Based Risk Method.*" These techniques are useful for examining specific situations, and often they are used with other techniques.

4.5.3.1 Event Tree Analysis (ETAs) Event tree analysis begins with an initiating event and work towards the outcome. This method provides information on how a pipeline failure can occur and the probability of occurrence.

An event tree is a graphical representation of potential incident pathways, and outcomes or scenarios, following a "hypothetical" initiating event. The event tree model depicts the logical inter-relationships between potential system successes and failures, or dependent events, as they respond to the initiating event.

The ETAs is usually completed in five stages:

1. Identify an initiating event of interest or concern.
2. Identify controls, safeguards, and barriers in place, and the safety functions designed to deal with the initiating event.
3. Construct the event tree beginning with the initiating event and proceeding through failures of the safety functions.
4. Describe the resulting potential pipeline incident event sequences.
5. Identify the critical failures that need to be addressed.

Advantages of Even Tree Analysis

- It is a structured, rigorous, and methodical approach.
- It can be effectively performed on several processes of pipeline lifecycle phases such as design, operation, and pipeline modifications.
- Permits probability assessment.
- The nature of the results is both quantitative and qualitative.

Disadvantages of Event Tree Analysis [6]

- An Event Tree Analysis can only have one initiating event, therefore multiple ETAs will be required to evaluate the consequence of multiple initiating events.
- Partial successes/failures are not distinguishable.
- Requires an analyst with some training and practical experience.

There are some rules when doing PoO calculations. To calculate the probability of occurrence of each final outcome (i.e., Figure 4.6 has 4 outcomes and Figure 4.7 has 9 outcomes), multiply along the branches, travelling from left to right from initiating event to final outcome. Thus, from the diagram depicted in Figure 4.6, the probability of occurrences of the initiating event (i.e., Leak event of a natural gas pipeline) happening AND systems 1 (i.e., *Immediate ignition*) AND system 2 (i.e., *Delay ignition*) AND system 3 (i.e., *Condition for explosion*) working properly is:

$$PoO_{outcome1} = I \times PoO_{S1} \times PoO_{S2} \times PoO_{S3}.$$
$$= 4.5 \times E\text{-}4 \times 0.05 = 2.25 \times E\text{-}5$$

The sum of each success/failure (Yes/No) probability pair, at each specific node adds up to 1. So, for instances, $PoO_{S1} + PoO_{F1} = 1$; $0.05 + 0.95 = 1$. This means that only given the value for success probability of a particular system, it is easy to calculate the failure probability for that same system, because the two will add up to give 1.

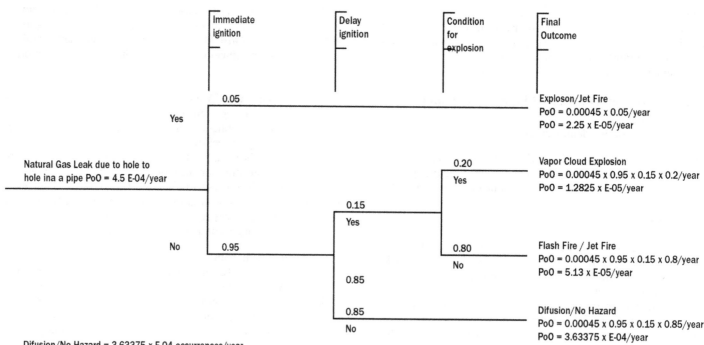

Difusion/No Hazard = 3.63375 x E-04 occurrences/year
Explosion/Flash Fire / Jet Fire = 2.25 x E-05 + 5.13 E-05 = 0.0000738 occurrences/year
Vapor Cloud Explosion = 1.2825 x E-05 occurrences/year

FIG. 4.6   EXAMPLE OF EVENT TREE ANALYSIS OF A NATURAL GAS PIPELINE LEAKAGE

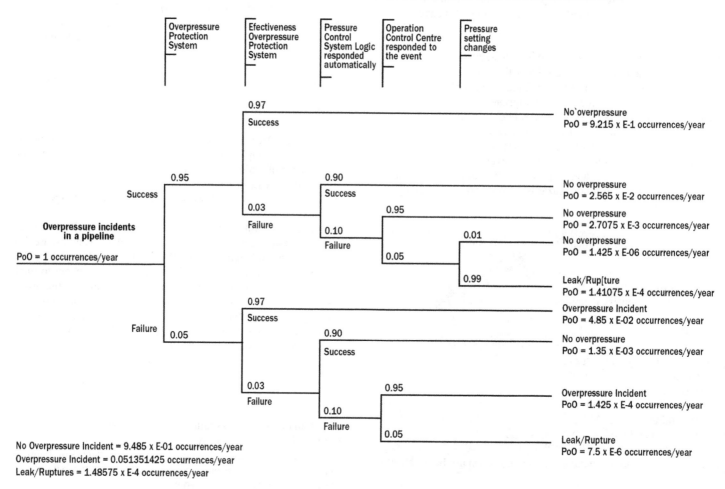

No Overpressure Incident = 9.485 x E-01 occurrences/year
Overpressure Incident = 0.051351425 occurrences/year
Leak/Ruptures = 1.48575 x E-4 occurrences/year

FIG. 4.7   EXAMPLE OF EVENT TREE ANALYSIS OF OVERPRESSURE INCIDENT

The sum of all the final outcome frequencies or probability of occurrence (for each outcome) will add up to equal the frequency of the initiating event.

4.5.3.2 Fault Tree Analysis (FTA)

Fault Tree Analysis (FTA) for pipeline is one of the logic and probabilistic technique used in the risk analysis and pipeline system reliability assessment. The FTA method start with a well-defined pipeline incident (i.e., pipeline rupture in a landslide zone), or the top event, following a deductive approach for identifying several ways and scenarios that can cause the pipeline incident. For instances, the top event of the Figure 4.9 is "*Pipeline Failure due to Equipment Impact.*" Deductive process is a reasoning "top-down" logic process that links clear hypothesis based on existing theory with true conclusions to reach a logically certain-deducting conclusion. The fault tree is also a graphical representation analysis answering the question "What can cause this failure?", and it essentially is used to identify failure modes and its root causes, and their causal relationship leading to a specific system failure mode. FTA is a formal deductive approach for resolving the basic causes of a given undesired event and is commonly used to illustrate events that lead to a failure consequently the failure can be prevented.

As illustrated in Figure 4.8, the methodology uses logic gates to show all credible paths from which the undesired event could occur. The fault tree is developed from the top down (i.e., from the undesired top event to the primary events, which initiated the failure) and the logic gates indicate the passage of the fault logic up the tree. The event should be traced back until it cannot be developed further, either due to lack of knowledge or because no other causes can be identified.

The logic gates predominantly consist of AND and OR gates (or modified versions) to indicate if the preceding event requires either one or a number of failures to occur. Once the fault tree has been fully developed frequencies/probabilities can be designated to each primary event, and by following the logic in the diagram the risk associated with the top event can be calculated.

The analysis is usually completed in six stages:

1. Identify the FTA's objectives, which define the purposes of the analysis (i.e., Identify root-causes of the pipeline failure due to mechanical damage from the maintenance/repair at the site-specific).
2. Familiarization with process of the threat, which defines the threat and its process. For instances, construction activities, maintenance, rehabilitation, pipeline repairs, or third party excavation can lead to mechanical damages. Mechanical damage threat is defined in the context as wrinkle, bends or buckle, denting, and broken pipe.
3. Identify the top event, which identifies the immediate, or basis cause that can be mitigated and preventive through pipeline operator processes. Additionally, identify "*what integrity threat occurs*" and "*when threat can occur.*" For instances, pipeline failure occurred due to equipment impact during company contractor and third party excavations.
4. Construct the fault tree analysis, which define FTA boundaries, ground rules, and breakdown of the top event to identify the root causes. The intent is to narrow the scope or basic cause insuring that all key components, elements, and systems required to capture the basis cause are included.
5. Analyze FTA, which determines the frequency and probabilities of the primary cause/event and estimates the probability of failure. For instances, the probability that a protection measure (i.e., concrete slab) protect the pipeline hit from third party is 95%. Check if basic events of historical failures have been included in the Fault Tree and to see if probabilities make sense.
6. Document FTA, which includes results, process, assumptions, parameters, methodology, and inputs.

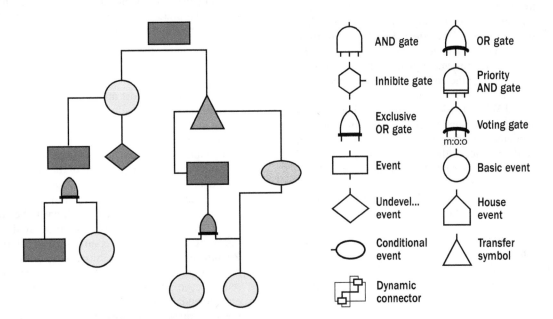

FIG. 4.8 EXAMPLE OF STANDARD SYMBOLS USED IN FAULT TREE ANALYSIS (FTA) (*Source:* Microsoft Corporation, Microsoft Visio 2007)

Advantages of Fault Tree Analysis [7]

- It is able to produce quantitative results. Probabilities or frequencies can be allocated to the initiating conditions, and using the logic present in the developed tree the probability or frequency of the event can be calculated.
- Shows a logical representation of the sequence of events. A pictorial representation of the failure path is produced which indicates the logical sequence of events leading to realization of the hazard.
- Can be used to assess a wide range of failures. Process failure (i.e., leak due to cracking) can all be easily incorporated into the logic of the process.

Disadvantages of FTA

- Require time consuming and expensive for complex systems. Each event has to be broken down to its initiating conditions, values for these conditions are then required to be identified and the logic followed to quantify the hazard.
- An experienced assessment team is required or errors in the logic can be made. The connection between the initiating conditions is required to be properly identified or errors can occur in the logic and from that to the quantification of the hazard.
- Some top events might be missed. Time and effort is required to identify all the top events that are required to be studied. A preliminary study might need to be performed to fulfill this criteria, as this technique cannot easily identify these events.

Example: Fault tree analysis for a pipeline failure due to mechanical damage caused by conducting authorized excavation for a pipeline repair or unauthorized excavation-by-excavation equipment such as excavator, backhoes, ploughs, etc. is illustrated in the Figure 4.9

4.5.3.3 What-If Analysis [8]   This technique is a systematic, team-based study to identify risks. The facilitator and team use standard 'what-if' type questions to investigate how a system, plant item, organization, or procedure will be affected by deviations from normal operations and behavior. This technique is widely applied to systems, plant items, procedures, and organizations generally. In particular, it is used to examine the consequences of changes and the associated risks thereby altered or created.

The industry-accepted What-if risk assessment methodology is used primarily to determine requirements for events which may result in loss of containment leading to harm to plan personnel, contractor personnel, the public, or the environment. The What-if analysis should be conducted by a team of technically qualified personnel with the intent of identifying hazardous scenarios and qualitative determining the risk scenarios, for instances. Staff required to conduct a What-If analysis include:

- Operations Team Leaders
- Control Room Operators
- Plant/Field Operation Personnel
- Maintenance personnel
- Security Staff
- Safety Engineers
- EHS, Supply Chain
- Engineering Staff
- Others based on specific scenario

Advantages of "What if?" Analysis

- It is easy to apply. The principle behind the technique is simple and therefore can be easily applied to a process.

Disadvantages of "What if?" Analysis

- Experienced assessors are required or hazards can be missed. The principle is simple though experience is required to ask all the appropriate questions or hazards might be overlooked.
- Require time consuming for complex processes. Complex processes will contain many items that are required to be assessed. Each one needs to have the appropriate questions applied to it and the results need to be recorded with associated hazards and consequences.

4.5.3.4 Hazard and Operability Analysis (HAZOP) [8]   HAZOP is a structured and systematic technique for examining a defined system, with the objective to:

- Identify potential hazards in the system. The hazards involved may include both those essentially relevant only to the immediate area of the system and those with a much wider sphere of influence, e.g., some environmental hazards; and
- Identify potential operability problems with the system and in particular identifying causes of operational disturbances and production deviations likely to lead to nonconforming products.

Advantages of HAZOP

- It is a systematic and comprehensive technique. A detailed plan for performing the technique is available which systematically applies guidewords and parameters to all the pipes and equipment in the process.
- Examine the consequences of the failure. Thought should be given by the assessment team to the consequences of the deviations identified. This aids in the production of recommendations for methods to minimize or mitigate the hazard.

Disadvantages of HAZOP

- This technique is most appropriate for plants. However, this technique might be useful for overpressure, and control room operation analysis.
- Require time consuming and expensive. Most systems contain a large number of pipes and equipment each of which need to be examined by the application of the various guidewords and parameters.
- Require detailed design drawing to perform the full study. To fully perform the study the process has to be designed to such a level that all the pipes and equipment are detailed with their operating conditions, and control instrumentation.
- Require experienced practitioners. Experienced team members are required to identify all possible causes and consequences of the deviations, as well as producing realistic recommendations.
- Focus on one-event causes of deviation only. Only the hazards associated with single deviations can be studied. Hazards that are caused by two or more separate deviations cannot be identified by the technique.

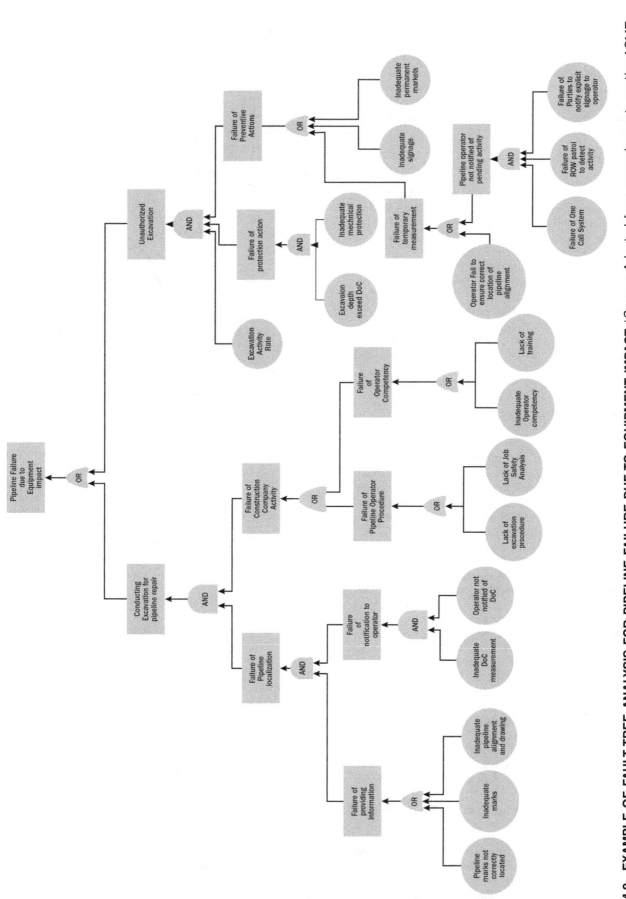

FIG. 4.9 EXAMPLE OF FAULT TREE ANALYSIS FOR PIPELINE FAILURE DUE TO EQUIPMENT IMPACT (*Source:* Adapted from several reports and/or ASME IPC papers prepared by C-FER Technologies www.cfertech.com authored by Playdon, D., Stephens, M., Chen, Qishi, C. Nessim, M. and ASME IPC2006-10433, Modeling Damage Prevention Effectiveness on Industry Practices and Regulatory Framework authored by Chen, Q., Davis, K. and Parker, C.)

4.5.3.5 Failure Modes and Effect Analysis (FMEA)

[8] Failure Modes and Effect Analysis (FMEA) is a systematic procedure, proactive tool, technique, and quality method for failure analysis of a system to identify the potential failure modes, their causes, and effects that could lead to undesirable consequences. The FMEA can be a qualitative technique, but it may be put on a quantitative basis when mathematical failure rate models are combined with a statistical failure rate mode.

Advantages of FMEA

- It is a systematic and comprehensive technique. A documented uniform method of assessing potential failure mechanisms, failure modes and their impact on system operation, resulting in a list of failure modes ranked according to the seriousness of their system impact and likelihood of occurrence.
- It is the basis for hazard determination.
- Identification of critical areas of the pipeline system (Root Cause Analysis).
- Captures the collective knowledge of a team.
- Improves the quality, reliability, and safety of the process.
- Provides historical records and establishes pipeline baseline.
- Enhances teamwork, understanding and cross-functional working relationship.

4.6 THREAT SUSCEPTIBILITY ASSESSMENT

This is the first level of the spectrum of risk analysis methods to specify if threat has actually been assessed, threat is potentially present, or no indication of threat in the pipeline. Threat screening or susceptibility analysis evaluates known relationship between data that indicate a threat does or does not exist, or is susceptible to possibly deterioration of pipeline integrity.

Firstly, Pipeline susceptibility assessment is practically applicable to a pipeline section that has not yet a susceptibility assessment performed, which the uncertainty of contribution of susceptible factors are considerable high. Secondly, when the threat probability has changed during condition monitoring, performance a threat susceptibility assessment can be considered. Finally, determination of the susceptibility assessment should be performed when a pipeline section has been previously assessed but that pipeline has been found not to be susceptible. Conversely, Probability of Failure estimation is applicable to threat when the following decision points are particularly associated with a level of risk:

1. Pipeline section is susceptible to the identified threat or possibly deterioration of pipeline integrity.
2. Threat does exist or threat is present.
3. Threat severity is evaluated by determining the threat acceptability and conducting characterization of failure modes (i.e., leak or rupture).
4. Interacting threats to pipeline integrity are known and related.

A typical first level of the threat susceptibility assessment might start with criteria and factors presented in the ASME B31.8S, or other industry study. This level requires minimal data and it is not considered a risk analysis method. Pipeline segment information is essential to perform this analysis. Where the pipeline data is missing, conservative assumptions should be used when performing the analysis or, alternatively, the segment might be prioritized as a high susceptible to the threat.

Example: Fatigue has played an important role in both liquid and gas pipelines. Every operator should evaluate its system to determine if it has pipeline segments susceptible to fatigue. It may be possible that time-dependent growth is occurring. For instances, linear indication due to lack of Fusion in the manufacturing process may grow and become hook crack due to pressure fluctuation. In order to manage systems safely, pipeline operators should be proactive, update facility records after pipeline inspections, and integrate data from operating pressure history. However, there are circumstances when pressure cycling and other forces can act upon resident subcritical imperfections in pipelines, and potentially lead to fatigue. For example:

- Pipelines that experience relatively high and frequent pressure cycles—specifically those locations without a historical pressure test in excess of normal operating pressure
- Facilities that experience a high magnitude of vibration
- Above-ground facilities such as spans that are subject to cyclical external loading such as vortex shedding and thermal expansion
- Liquid pipelines that experience flow direction changes
- Product service change
- Operating pressure unsteady

A threat susceptibility process involves several permutations and combinations of hazards to determine whether they have the capacity to create a threat to a pipeline. Therefore, for each threat evaluated (i.e., see Table 4.1 of this Chapter) a list of hazards that give capacity to damage to a pipeline can be created. Data integration also provides great value in the determination of susceptibility to a given threat. For instances, correlation of pipeline inertial data (i.e., strain), geotechnical surveys and line walk patrolling may indicate areas with hazards such as potential for displacement, settlement, or geometric damages resulting in threats such as ovalization, denting, or wrinkling.

Once the integrity threat has identified and the susceptibility has been reviewed, in the previous steps, selection of one or more integrity assessment methodologies can be conducted based on the priorities determined in the threat susceptibility analysis and which pipeline segment is susceptible. The most common integrity assessment methods used by pipeline operators are in-line inspection, pressure testing, and direct assessment. A general process for determining an ILI program or other integrity assessment is outlined in Figure 4.11.

4.6.1 Susceptibility Classification

Classification of the susceptibility levels provides guidelines for selecting the integrity assessment method to confirm the threat existence and review its category. Once the credibility or susceptibility is determined, threat categorization based on concepts/levels such as *"non-susceptible threat," "potential-susceptible threat," "susceptible and identified threat,"* and *"threat presented/assessed"* has been the practice. Another categorization based on threat stability behavior is *"Stable/Dormant/Inactive," "Moderately stable," "Unstable/Active,"* and *"Critical Threat."* Table 4.3 provides examples of levels of threat susceptibility.

Example: As illustrated in Figure 4.10, SCC Susceptibility Analysis methodology consists of comparing the data elements to the criteria. For instances, segments of pipeline having stress concentration (i.e., corrosion or residual such as bending), coating disbondment, high operating stress level, poly-tape coating type, low Charpy-V-Notch value and 40 years in operation, might have high probability to develop SCC anomalies. Figure 4.10 show an example of a flow chart for SCC Susceptibility Analysis.

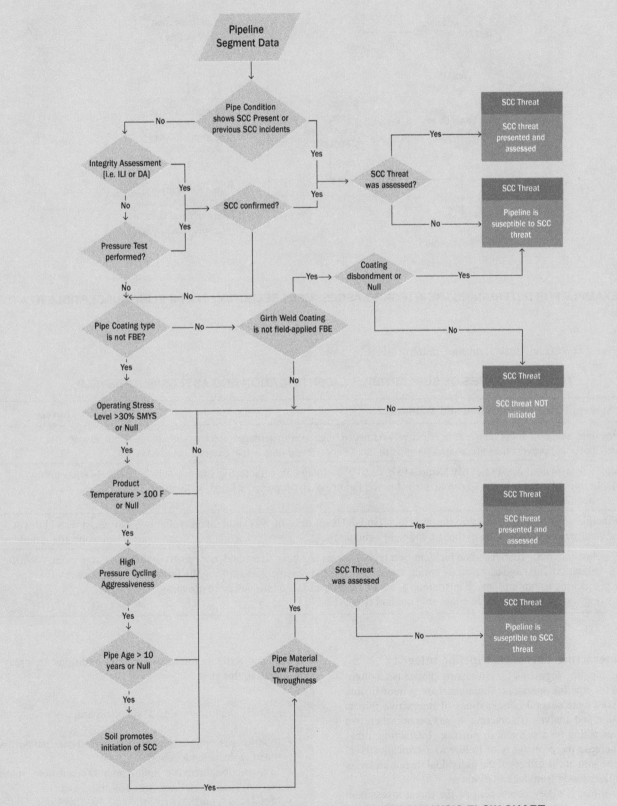

FIG. 4.10   ILLUSTRATION OF SCC SUSCEPTIBILITY ANALYSIS FLOW CHART

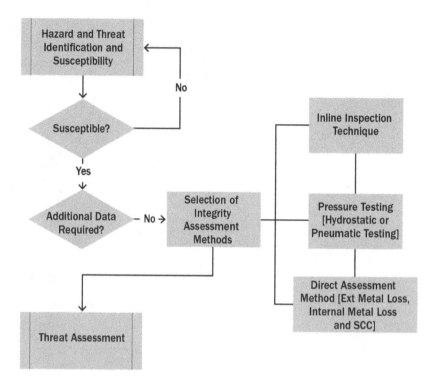

FIG. 4.11 EXAMPLE FOR DETERMINING THE INTEGRITY ASSESSMENT SELECTION IN A PIPELINE SUSCEPTIBLE TO A THREAT

Example: Threat classification based on susceptibility levels:

TABLE 4.3 CASES OF SUSCEPTIBILITY, IDENTIFICATION AND ASSESSMENT LEVELS

Cases for Susceptibility, Identification and Assessment	Levels
CASE 1: Pipelines with no integrity hazards or effectively managed (e.g. coating damage, effective cathodic protection) that could prevent the initiation and/or growth (active) of a threat (e.g. active external corrosion)	Not susceptible
CASE 2: Pipeline with confirmed integrity hazards (e.g. pre-1970 vintage manufacturing and pressure cycling), but integrity threat has not been identified in the field (e.g. ILI-reported without validation, hydrostatic testing failure)	Likely susceptible
CASE 3: Pipeline with confirmed integrity hazards (e.g. landslides on right-of-way) and threats reliably identified (e.g. wrinkle on the pipe in the ILI and field validated)	Susceptible and threat present: Identified
CASE 4: Pipeline that has experienced an incident or integrity threat has been identified, confirmed in the field (e.g. validation digs) and assessed by conducting engineering assessment identifying the failure mode (e.g. cracking to rupture), mechanism (e.g. pressure cycling-induced fatigue) and estimating criticality (e.g. safety factor) with an applicable method (e.g. API 579 Level 2)	Threat present: Identified and assessed

4.6.2 Interacting Threats to Pipeline Integrity

Monitoring pipeline segments for interacting threats is an ongoing challenge for pipeline operators. Threats that are systematic and repeatable may create several combinations of interactive threats for investigation and analysis. Interacting threats occurs when two or more threats acting on a segment or pipeline. Interacting integrity threats increase the probability of failure to a criticality level greater than the sum of the effects of the individual threats and may change the failure mode from leak to rupture.

Interacting threat is a key component of the threat assessment process. Typical scenarios in which an interacting condition may exist, and best practices used by pipeline operators to assess, manage, and mitigate interacting threat conditions are areas to be considered in this step.

Example: Typical Interacting Threat Scenarios

- Pipeline susceptible to both internal and external corrosion threat at the same location
- Construction threat (i.e., crack on long seam weld) interacting within unstable land (i.e., soil movement)
- Metal loss (gouge, corrosion, etc.) and mechanical damage (i.e., dent)

- Long seam weld (i.e., preferential corrosion) and cracks (i.e., Lack of Fusion)
- Girth weld anomalies (linear planar crack, lack of fusion, lack of penetration, etc.) and natural forces (soil movement)

The focus is practically on understating, and developing an appropriate methodology for evaluating the probability of failure (PoF) associated with a super-imposed set of threats occurring coincidentally called *"co-existing threats or interacting threats."*

Currently, USA regulation and supporting standards (i.e., ASME B31.8S) mention that threat interactions should be considered and addressed. However, there is limited industry knowledge on the evaluation of interaction of threats, and determination on how those interactions influence the overall PoF of a pipeline segment. Current methods may only simplistically evaluate and look at single threat failure instead of combined threats.

Example: Challenging PoF assessments and evaluation for interacting threats.

- Preferential corrosion on long seam weld with a hook crack (i.e., corrosion fatigue modeling)
- Failure pressure and remaining life of interacting threats as opposed to analyzing individual threats one at the time

Table 4.4 describes an example of likelihood of occurrence for interacting threats. The matrix outlines the relationship between different combinations of threats. Management process for interacting threats should be developed, which each particular active combination of threat is managed predominantly for pipeline segment. Each company may develop a qualitative matrix or likelihood of occurrence table depending on historical observation of how those threats have been interacted in different locations of the pipeline systems. Interacting treat matrix/table may differ from system to system and region to region. Overlaying inline inspection data from different technologies is a good start for identifying interactive threat to identify potential threat interactions. However, the interaction cannot determine the severity of defects.

Example: Threat Interaction using ILI
MFL and Caliper ILI data correlation may determine that some metal loss features are interacting with mechanical damage anomalies as well as external and internal metal loss.

Limitations for determining, evaluating, and assessing "Interacting Threats"

- Limited models and methodologies for assessing the interacting threat PoF and determining failure modes (i.e., leak or rupture)
- Data sources and contributors to interacting threats
- Rules of interaction for multiple integrity threats
- Limited growth modeling for interacting threats to assess interactive threat failing pressure and remaining life

Integrity threats could coexist in many cases, but confirmation of interacting threats is needed because increases the criticality and it should be properly characterized and managed.

Qualitative analysis of interacting threats such as "should be considered" or "should not be considered" may be helpful, but it does not provide a relative ranking of the most severe interactions per pipeline segment or sub-segment. The interacting threat analysis should address the following questions:

1. Identification—What process should a company use to identify interacting threats and how should they be incorporated into the pipeline integrity assessment?
2. Likelihood—Could it be incorporated into the pipeline integrity assessment?
3. Risk—Can interacting threats result in a significant/unacceptable risk levels to the pipeline when a single threat risk is at "acceptable" level?
4. Process—Can a process or methodology be employed to continuously monitor threat interactions and identify concerns at defined thresholds of risk?
5. Mitigation—How should threat interactions be mitigated? Which combinations of threats are most important to be controlled and mitigated?
6. Controlling—Which combinations of threats are most important to understand and control?

Example: Likelihood of Occurrence for some interacting threats for a selected system

TABLE 4.4   EXAMPLE OF QUALITATIVE MATRIX OF INTERACTIVE THREAT COMBINATION

Existing threats	Co-existing threats and/or failure mechanisms	Interacting threat likelihood
External corrosion Internal corrosion Microbiologically induced corrosion (MIC) Selective ERW seam corrosion Soil corrosion	Manufacturing wall thinning	Moderate
	Cracking	Moderate
	Geotechnical and hydro technical	Unlikely
	Mechanical damage	High Likely
	Material/manufacturing defects	High Likely
	Construction defects	Likely
	Equipment/component	Unlikely
	Incorrect operations	Moderate
	Weather and outside forces	Unlikely

(Continued)

TABLE 4.4   EXAMPLE OF QUALITATIVE MATRIX OF INTERACTIVE THREAT COMBINATION (Continued)

Existing threats	Co-existing threats and/or failure mechanisms	Interacting threat likelihood
Stress corrosion cracking (SCC) Sulfide stress cracking (SSC) Hydrogen-induced cracking—HF Mechanical damage delayed cracking Corrosion fatigue cracking Fatigue cracking	Corrosion	Moderate
	Geotechnical and hydro technical	Unlikely
	Mechanical damage	Moderate
	Material/manufacturing	Moderate
	Construction	Unlikely
	Equipment/component	Unlikely
	Incorrect operations	Moderate
	Weather and outside forces	Unlikely
Landslides/mass movement-displaced pipe Inland flood-spanned pipe Scoured-pipe Washout/erosion-spanned pipe Undermining subsidence-strained pipe Seismic-displaced pipe	Corrosion	Likely
	Cracking	Moderate
	Mechanical damage	Likely
	Material/manufacturing	Moderate
	Construction defects	High Likely
	Equipment/component	Unlikely
	Incorrect operations	Moderate
External interference Denting Wrinkle, buckle, and ripple	Corrosion	High Likely
	Cracking	High Likely
	Geotechnical and hydro technical	Moderate
	Material/manufacturing	Moderate
	Construction	Moderate
	Equipment/component	Moderate
	Incorrect operations	Moderate
	Weather and outside forces	Moderate
Weld cracking Defective pipe body Defective long seam weld	Corrosion	Moderate
	Cracking	Moderate
	Geotechnical and hydro technical	Moderate
	Mechanical damage	Unlikely
	Construction	Unlikely
	Equipment/component	Unlikely
	Incorrect operations	Moderate
	Weather and outside forces	Moderate
Construction design Installation errors Defective girth weld	Corrosion	Moderate
	Cracking	Moderate
	Geotechnical and hydro technical	High Likely
	Mechanical damage	Moderate
	Material/manufacturing	Unlikely
	Equipment/component	Unlikely
	Incorrect operations	Moderate
	Weather and outside forces	High Likely
Equipment/component failure	Corrosion	Likely
	Cracking	Likely
	Geotechnical and hydro technical	Unlikely
	Mechanical damage	Moderate
	Material/manufacturing	Likely
	Construction	Likely
	Incorrect operations	Moderate
	Weather and outside forces	Moderate

TABLE 4.4 EXAMPLE OF QUALITATIVE MATRIX OF INTERACTIVE THREAT COMBINATION (Continued)

Existing threats	Co-existing threats and/or failure mechanisms	Interacting threat likelihood
Incorrect Operations (e.g. Human and/or Equipment Factors)	Corrosion, wall thinning	Moderate
	Cracking	Moderate
	Geotechnical and hydro technical	Moderate
	Mechanical damage	Moderate
	Material/manufacturing	Moderate
	Construction defects	Likely
	Weather and outside forces	Likely

Example: PoF of Interacting Threats

Consider the integrity threats A, B, and C. The probability of failure from these threats is PA, PB, and PC respectively. If the threats are considered interacting then P(A+B+C) > PA + PB + PC

The distinction that makes threats interacting is that the resulting probability of failure from the Interactive Threat is GREATER THAN the sums of the probabilities of failure of the individual threats. For instances, consider threats A, B, and C. The probability of failure from these threats is PA, PB and PC respectively. Threats are considered interacting if P(A+B+C) > PA + PB + PC.

4.6.3 Threat Severity and Significance

Operating condition should be considered when evaluating the severity of a single threat. Features identified in the field through investigation digs or ILI report are assessed using defect assessment methodologies to estimate the severity or conduct pressure sensitivity analysis that indicates whether features are acceptable.

Defect assessment methodologies are outlined in detail in Chapter 12. Characterization, identification, and sizing of features provide support to determine what the failure modes and help to answer the following questions:

How could the pipeline fail? (i.e., leak or rupture).

What is the failure mechanism? (i.e., cracking, corrosion, denting, mechanical damages, or a combination of them).

Why could the pipeline fail? (i.e., high-pressure fluctuation, overpressure, equipment impact by third party, etc.).

What is the failure pattern? (i.e., whether anomaly grows over time. Threat is time-dependent or non-time dependent).

What is the growth mechanism? (i.e., growth by fatigue, environment, or both pressure and environment. For instances, Corrosion-fatigue cracks, SCC colony anomaly, or pure fatigue such as crack-like anomaly on long seam weld).

Anomaly tendency to have a failure mode such as leak (i.e., short and deep) or rupture (i.e., longer and shallower) can be determined using graphs to classify the severity of the anomaly. Figures 4.12 and 4.13 show examples of critical defect graphs

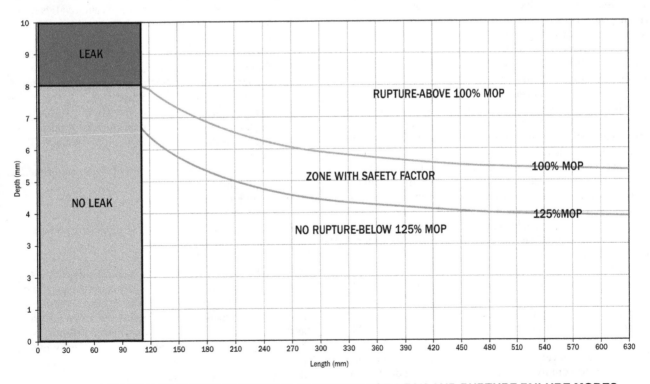

FIG. 4.12 EXAMPLE OF MAXIMUM CRITICAL FLOW SIZE FOR LEAK AND RUPTURE FAILURE MODES

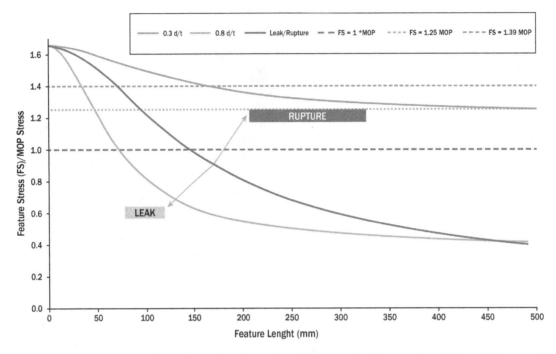

FIG. 4.13 EXAMPLE OF FLOW STRESS-DEPENDENT EQUATION GRAPH FOR LEAK AND RUPTURE BOUNDARIES

using for classification of failure modes (i.e., leak or rupture) and feature severity.

4.7 PIPELINE THREAT ASSESSMENT

Once the integrity threat existence has been confirmed threat in the pipeline segment, a pipeline threat assessment is the following process of the threat management. The first step in the threat assessment process is to conduct a detailed review of the available information since its pipeline inception such as design and material characteristics, manufacturing and construction records, and pressure testing parameters, methods and results.

The objective of the threat assessment is to assess the site-specific hazards exposure to a given threat, or new integrity threat, and evaluate if existing threat has vanished or became inactive to establish the most effective preventive and mitigation actions and address these relevant hazards in the most effective manner. In doing so, the threat assessment estimates additionally the safety level inherent in the options of mitigation action.

Pipeline operators typically used levels of threat assessment to evaluate conformances and compliances and it goes from a simple pipeline data review to a performance of detailed quantitative analysis. The following example helps to understand in detail levels of the threat assessment approach. Identification, assessment, ranking, treatment and mitigation (Chapter 11) of the integrity threats are steps the threat assessment process.

Primarily, the objective of the first level of the threat assessment is to review the attributes for all potential threats to a pipeline system in consideration of the status of design, material, fabrication, construction, and operational parameters that are associated with the pipeline system of interest. It is practically a review of the parameters and attributes prior to in-service pipeline.

Typically, different levels of threat assessment can be planned and implemented to comply with regulations and standards. The following example depicts the threat assessment methodology.

Example: Threat Assessment methodology to demonstrate compliance that it may involve three levels of evaluation, as follow:

Level A: Review of the design and construction (Prior to in-service) records to confirm whether the existing piping complies with standard and regulation requirements for the specific pipeline- location integrity concern (i.e., leak due a Microbiological Induced Corrosion and SCC). The review may comprise:

a. Review design and material specification and characteristics,
b. Verification of the maximum allowable stress and wall thickness,
c. Review pipeline construction procedures, methods and failures,
d. Review pressure-testing parameters, method, records and results at the construction time,
e. Verification of the stress analysis of all road crossings including imperfection as specified in standards,
f. Pipe manufacturing procedures and records (i.e., LSW process),
g. Review the depth of cover and clearance of the location of pipeline,
h. Evaluation of imperfections and its repairs,
i. Welding field review,
j. Coating condition review for pipe body and girth weld,
k. Valve spacing review, etc.

Level B: Identification of potential integrity threats and hazards (during in-service) to the pipeline at the specific pipeline-location integrity section. Assessment may comprise to check, review, and evaluation of the identified potential integrity threats at the specific pipeline-location segment to identify significant hazards according to pipeline operator's integrity management process for further detailed assessment. Aspects and threats to be assessed can include the following:

a. Leak and Ruptures incidents and Root Cause Analysis,
b. Integrity Assessment (ILI, Post-construction Hydrostatic testing, Direct Assessment),
c. Actual repairs and mitigation for all threats,
d. Operational Changes (e.g., pressure, temperature, flow, cycling),
e. Mitigation (Repairs),
f. Prevention (Damage Prevention, CP Surveillance, Actual pipeline coating),
g. External corrosion,
h. Internal corrosion,
i. Stress corrosion cracking,
j. Corrosion fatigue,
k. Mechanical damages,
l. Geotechnical threat,
m. Manufacturing and construction threat,
n. Overpressure threat,
o. Equipment malfunction, and
p. Fatigue cracking.

Level C: A qualitative or quantitative assessment and probabilistic methodologies are performed to rank the identified threats in previous levels and ensure preventions and mitigation actions, where necessary, could result in a safety level that meets or exceeds industry standards and regulations. Safety levels acceptability or criticality should be analyzed and supported. Pipeline operators are primarily concerned with undertaking operations and maintenance to reduce the likelihood (threat estimation) of an event to limit the potential for consequences. Effective methods of reducing the threat level and its failure rate shall be identified and examined for the assessed specific site to ensure acceptability of safety levels. The review and assessment may consider:

a. All hazardous contributing factors,
b. Indexing methodology for ranking threat,
c. Quantitative Risk Assessment (QRA),
d. Several scenarios,
e. Mitigation methods for the most dominant threat(s) (i.e., Concrete Slabs installation for Mechanical Damages threat, and DCVG, CIS surveys, repair external corrosion anomalies and ILI MFL tool run for Metal Loss threat),
f. Consequences reduction analysis,
g. Likelihood reduction analysis.

Probability of failure for a given threat (i.e., corrosion) can be calculated using different methodologies. Monte Carlo simulation is a technique used to calculate the probability of failure based on simulations, how ranges of estimates are created, and how likely the resulting outcomes are. The Monte Carlo simulation technique is usually used for corrosion, SCC, crack-like, and denting threats. A typical Monte Carlo simulation calculates the model hundreds, thousands or millions times, each time using differently randomly selected values. Calculation of Probability of Failure and Reliability of the pipeline system from a Fault tree is other technique and an example explaining the calculation method is described in the following example. Fault Tree Analysis calculation is typically used for Time-Independent threats (i.e., Mechanical Damages, Geotechnical and weather and outside forces threats). Probability of Exceedance (POE) is a probability approach to ranking inline

inspection metal-loss anomalies by predicted failure pressure (i.e., rupture failure mode) and depth dimensions (i.e., leak failure mode) from the tool vendor's pipe tally. The POE approach is explained in detail in the Section 4.7.2.

4.7.1 Probability of Failure (POF) Technique Using Fault Tree Analysis

The probability of occurrence of pipeline failure due authorized or unauthorized excavation due to excavation equipment impact can be estimated using the fault tree example shown in the Figure 4.9. The complexity and accuracy of the calculation procedure depends on the characteristics of the fault tree. Output event probabilities can be calculated directly from the basic principles of probability theory of AND gate and OR gate using probability of occurrence from the probabilities of each basic event. Probabilities values of basic events can be taken from pipeline operator failure frequency or industry data.

Based on these principles the outcomes for he AND gate can be calculated using the following equation:

$$Po = \prod_{n=1}^{n} Pn = P1 \times P2 \times P3 \times \ldots \ldots \times Pn$$

Where, Po is the outcome of the Probability of occurrence, and the P1 P2, P3, ..., Pn are the probability of the each input event to the associated "AND" gate. For an "OR" gate, the outcome probability of occurrence is given by the following equation:

$$Po = 1 - [(1 - P1) \times (1 - P2) \times (1 - P3) \ldots \ldots \ldots \times (1 - Pn)]$$

Figure 4.14 depicts a Fault Tree defining different paths requiring different mathematical calculation (and & or gates) for estimating their associated probability of failure.

4.7.2 Probability of Failure of In-Line Inspection Anomalies Using POE

The Probability of Failure (PoF) of anomalies reported by In-Line inspections (ILI) such as metal loss and cracking can be estimated using a methodology named Probability of Exceedance (POE). The "Probability of Exceedance" (POE) analysis is an engineering methodology to assess threats identified by in line inspection and rank metal loss and cracking anomalies by determining the probability of leaks and ruptures. The POE method helps preparing remediation (e.g., mitigation, hydrostatic testing) and re-inspection interval assessments for leak (i.e., depth) and rupture (i.e., predicted burst pressure)-dependent anomalies accounting for the ILI performance. Remediation and associated re-inspection scenarios can be assessed using Present Value (PV) analysis. POE does not account for the variability of pipe material properties and uses the growth models (e.g., corrosion, cracking) as input for re-inspection estimation [9–11].

The probabilistic methodology (POE) analysis determines the current and future integrity status of pipeline system using both deterministic and probabilistic approaches. POE analyses provide a realistic integrity and serviceability assessment for pipeline systems.

4.7.2.1 ILI Tolerance, Certainty or Confidence Level and Probability Distributions POE uses both the ILI-reported dimensions and the ILI performance accuracy (e.g., detection,

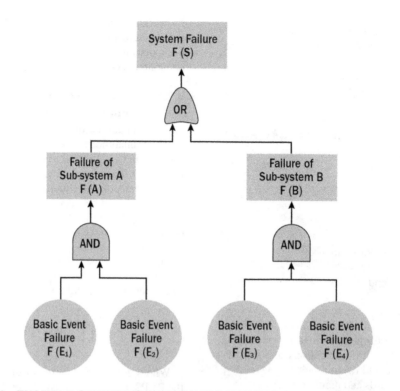

FIG. 4.14 EXAMPLE OF PROBABILITY OF FAILURE FROM A FAULT-TREE ANALYSIS

identification, and sizing). The ILI performance accuracy of the ILI is typically expressed by ILI vendors as the tool tolerance (e.g., +/−10% of the wall thickness) and the certainty or confidence level (e.g., 80% or 90% of the time) that can be graphed as a "natural" frequency distribution.

As illustrated in Figure 4.15, if ILI validation dig data is available, the natural frequency distributions are built determining the differences between the ILI-reported depths and predicted burst pressures and their corresponding values determined in the field by a given Non-Destructive Examination (NDE) technique (e.g., pit

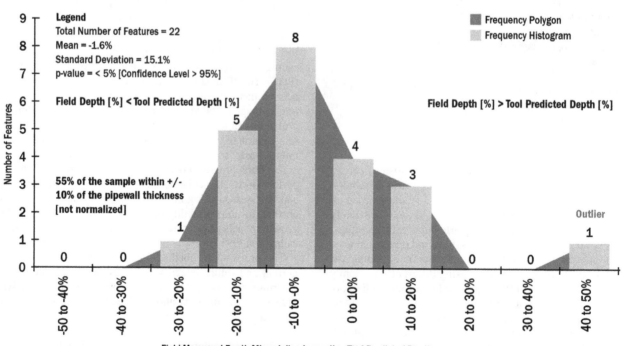

FIG. 4.15 FREQUENCY DISTRIBUTION: DIFFERENCES BETWEEN ILI REPORTED AND AS-FOUND DEPTH

gauge, ultrasonic, eddy current, phased array). Consideration should be made to evaluate the NDE performance (i.e., field tool tolerance and certainty) to be captured in the PoF calculation, if significant.

As illustrated in Figure 4.16, a natural frequency distribution of the field—ILI differences (i.e., depth and predicted burst pressure) can be typically converted into a Probability Density Function (PDF) expressed in terms of coefficients such as mean and standard deviation for normal and lognormal distributions. The statistical analysis determines the level of representativeness of the PDF distribution (e.g., confidence level $\geq$ 95%) and whether outliers need to be removed from the distribution for a specific analysis.

If ILI validation dig data is not available yet, the PDF distribution can be "simulated" using a Monte Carlo simulation of selected ILI and field data and expected ILI performance accuracy (e.g., vendors or company experience). Alternatively, the distribution coefficients (e.g., mean and standard deviation) can be estimated based on the selected tolerance and certainty (see example below).

Based on experience, the most common type of probability distributions for ILI performance sizing are Normal, Lognormal, or Weibull. Further to determining the confidence level from the ILI-field differences, the estimation of *p-value* or *significance* of the ILI-field regression can also be used for assessing their statistical representativeness and outliers to be removed for special analysis. The regression provides an equation (i.e., coefficients) for adjusting the ILI-reported values and a Standard Error as the standard deviation of the regression equation.

In some cases, ILI detection performance may have an important contribution increasing the Probability of Failure (PoF) of an ILI-reported anomaly. Therefore, the POE_{leak} and $_{rupture}$ due to sizing of a given ILI-reported anomaly should be factored (i.e., increased) using the field distributions of the Non-detected (or non-reported) anomalies and the worst non-reported anomaly expected to be found in the field. ILI identification/characterization performance can also be factored using the distribution of the ILI anomalies (e.g., crack—potential to grow) not properly characterized as found in the field (e.g., notch—low potential to grow).

4.7.2.2 PoF for ILI Leak-Dependent Anomalies (Depth)
Following the discussion in Chapter 12 Fitness-For-Service Assessment, the leak-dependent anomalies are more likely to fail as a leak due to their depth. However, maximum critical flaw size graphs should be considered in determining whether the ILI-reported anomalies and associated distribution fall within the leak-dependent region for a selected pipeline operating pressure (% MOP).

As illustrated in Figure 4.17, the probability of an ILI-reported anomaly depth actually exceeds a defined limit, in the field, which could cause a leak is named Probability of Exceedance for Leak (POE_{Leak}). The variability of the depth found in field for a given ILI-reported depth can be represented as a distribution, shown as a normal distribution in the figure for different depths. The dashed ovals represent the chance or probability of ILI-reported features (e.g., 60%, 70%, or 80% depth) can exceed the defined limit of 80% depth in the field.

In the Figure 4.17 [10], 80% depth has been defined as a limit at which the anomaly may experience a "sudden break" or leak due to its remaining 20% of the wall thickness for onshore pipelines. For thin pipelines (e.g., telescopic design, NPS schedule < 40), the limit may be defined as to be 60% to 70% depth; whereas, for very thick pipelines (e.g., offshore pipelines, special crossings), the limit may be defined as to be 90% depth. Maximum critical flaw size graphs accounting for wall thickness, material properties and operating pressure should be used for determining the POE_{leak} limit.

Example: PoF (leak) of ILI anomaly using POE

The POF of a pitting corrosion (type) with a 60% depth (size) as reported by an ILI with a sizing accuracy of +/−10% of the wall thickness (tolerance) at 80% of the time (certainty, $z = 1.28$) for normally distributed field-tool differences and a *p*-value < 5% (confidence level > 95%)—outliers removed.

The $POE_{leak\ (reported\ 60\%\ deep)}$ is calculated as a probability to exceed 80% of the wall thickness (leak condition) in the field.

Using Excel, the formula for POE (leak) = 1 − NORMDIST (Max, Value, Standard Deviation, True)

Where:

> *Max = depth limit at which leak may start occurring (e.g., 80% nominal wall thickness)*
> *Value = ILI-reported depth (e.g., value is recommended to be adjusted based on the selected field-tool regression equation)*
> *Standard deviation = measurement of field-tool difference spread (e.g., standard error from regression should be used if adjusting equation is used)*
> True = cumulative calculation
> POE (leak) = 1 − NORMDIST(0.8, 0.6, 0.1/1.28, True) = 5.2×10^{-3}

4.7.2.3 PoF for ILI Rupture-Dependent Anomalies (Predicted Burst Pressure)
Following the discussion in Chapter 12 Fitness-For-Service Assessment, the rupture-dependent anomalies are more likely to fail as a rupture due to their depth and length. However, maximum critical flaw size graphs should be considered in determining whether the ILI-reported anomalies and associated distribution fall within rupture-dependent region for a selected pipeline operating pressure (% MOP).

As illustrated in Figure 4.18 [10], the probability of an ILI-reported anomaly burst pressure actually exceeds a defined limit, in the field, which could cause a rupture is named Probability of Exceedance for Rupture ($POE_{Rupture}$). The variability of the burst pressure estimation determined in field for a given ILI-reported depth can be represented as a distribution [12]. The dashed ovals represent the chance or probability of ILI-reported features (e.g., 80%, 100%, or 125% MOP) can exceed the defined limit of 100% MOP estimated in the field.

Example: PoF (rupture) of ILI anomaly using POE

The POF of a general corrosion (type) with a safety factor of 125% MOP as reported by an ILI with a sizing accuracy of +/−12% MOP (tolerance) at 80% of the time (certainty) for normally distributed field-tool differences and a *p*-value < 5% (confidence level > 95%)—outliers removed. The $POE_{rupture\ (reported\ 125\%\ MOP\ predicted\ burst\ pressure)}$ is calculated as a probability to exceed 100% of the MOP (rupture condition) in the field.

Using Excel, the formula for POE (rupture) = 1 − NORMDIST (Max, Value, Standard Deviation, True)

where:

> *Max = pressure limit at which rupture starts occurring (e.g., 100% MOP)*
> *Value = ILI-reported predicted burst pressure (e.g., value is recommended to be adjusted based on the selected field-tool regression equation)*

Field Validation Data Distribution for a Tool Accuracy of +/- 10% of the Pipe Wall Thickness

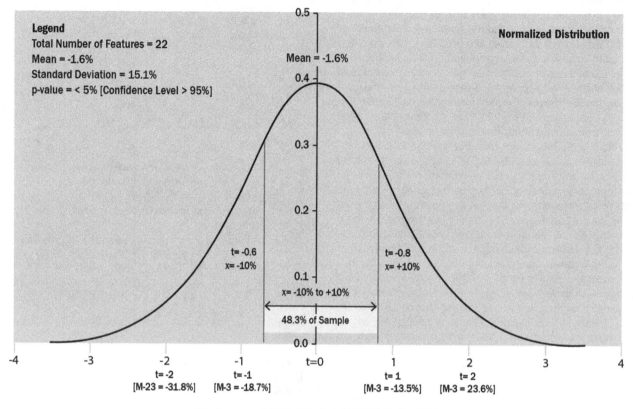

Tool Accuracy of 30% of the Data Validated in the Field

Tool Accuracy for 30% of the Data Validated in the Field

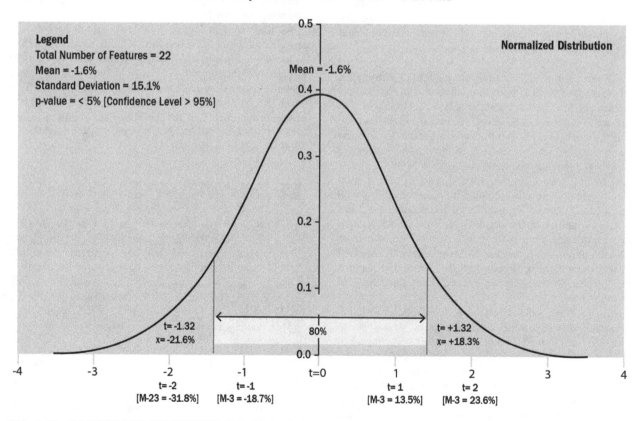

FIG. 4.16   ILI PROBABILITY DENSITY FUNCTION DESCRIBING DIFFERENT ILI TOLERANCE AND CERTAINTY

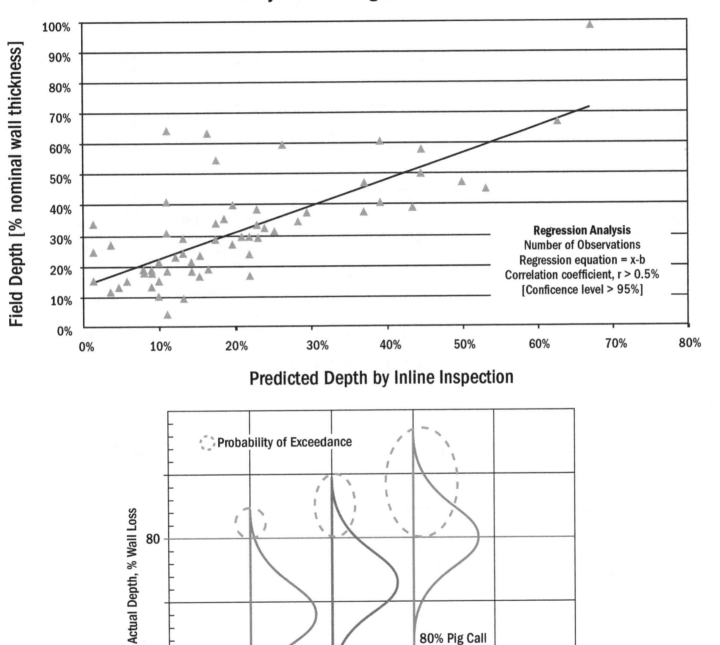

FIG. 4.17   FIELD VERSUS ILI PREDICTED DEPTH AND PROBABILITY OF EXCEEDANCE FOR LEAK

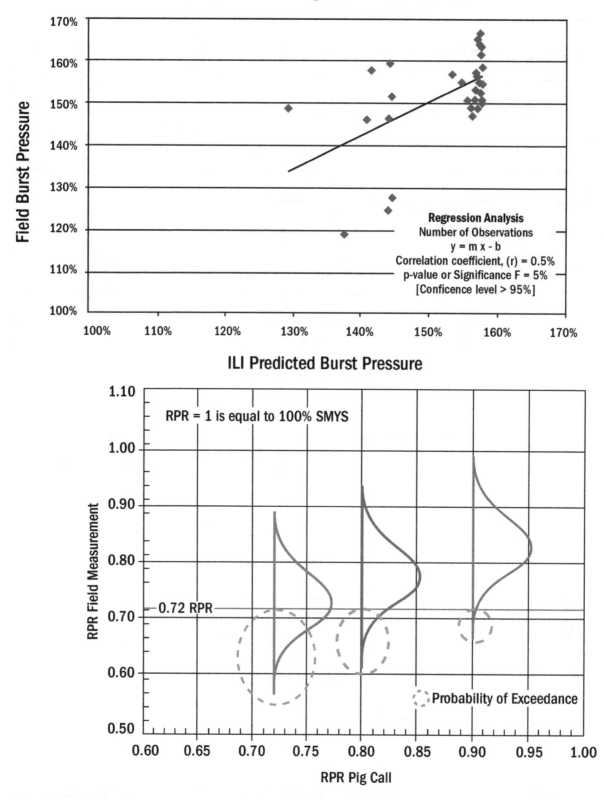

Field versus ILI Predicted Burst Pressure
Unity Plot and Regression Coefficients

Regression Analysis
Number of Observations
y = m x - b
Correlation coefficient, (r) = 0.5%
p-value or Significance F = 5%
[Conficence level > 95%]

ILI Predicted Burst Pressure

RPR = 1 is equal to 100% SMYS

0.72 RPR

Probability of Exceedance

FIG. 4.18 FIELD VERSUS ILI PREDICTED BURST PRESSURE AND PROBABILITY OF EXCEEDANCE FOR RUPTURE

Standard deviation = measurement of field-tool difference spread (e.g., standard error from regression should be used if adjusting equation is used)
True = cumulative calculation
POE (rupture) = NORMDIST(650, 812.5, (0.12*650/1.28), True) = 3.8 × 10^{-3}

4.7.2.4 PoF Criteria for ILI Excavation Programs PoF limits are defined for both leak (e.g., depth-based) and rupture (e.g., predicted burst pressure-based) conditions; hence, two (2) conditions need to be met by the selected ILI excavation program. The mitigation effects of the ILI excavation program would determine a safe and reliable period of the remaining anomalies until the next ILI re-inspection, discussed in the next section.

The following are some criteria for selecting a PoF limit to be used in determining the extent of ILI excavation programs identifying the type, location, and number of anomalies for field investigation.

1. *Improvement of Historical Failure Frequency*

    Failure frequency history of a pipeline system or network can be used as a reference for determining an acceptable PoF for the selection and prioritization of ILI excavation programs. The selection of the acceptable PoF should be made with the intent to improve on (e.g., smaller over time) over the historical trend. This PoF should be a value lower than unity such it can be used as a PoF. As failure frequencies reach very low values (e.g., 1 × 10^{-3} leaks/km-year), they may be considered numerically equivalent as a Probability of Failure (PoF) per kilometer per year [13].

2. *Maximum Probability of Failure (PoF) Derived from Individual and Societal Risk*

    Some regulators and industry standards have identified individual and societal risk levels for land-use planning near industrial hazards (e.g., pipeline). The individual risk only focuses on the likelihood that a particular person (e.g., normal, vulnerable) could be affected by a dangerous dose or exposure (e.g., HSE: fatality with one or less chances per million per year or 1 × 10^{-6} cpm per year). Furthermore, societal risk focuses on the likelihood for one incident to affect a large number of people (e.g., HSE: 50 fatalities with 200 chances per million per year or 2 × 10^{-4} cpm per year).

As illustrated in Figure 4.19, pre-defined acceptable societal risk can be transformed into acceptable PoF (failure/km-year) for a given consequence (i.e., number of fatalities). In the case of CSA-Z662 Annex O acceptable limits, the natural gas transmission pipeline reliability targets (1 – POF) for Ultimate Limit State (i.e., large leak and rupture) have accounted for both individual and societal risk [14].

When high ILI uncertainty and high density of anomalies are reported, a review of the assessment length for calculating the Cumulative PoF is recommended. The review may require an iteration process to anticipate cumulative PoF values would not numerically reach non-existing equivalent defects. The following is a recommended conceptual review process of the assessment length:

1. Determine the PoF limits based on pre-defined defect criteria (e.g., depth ≥70%, Predicted Burst Pressure ≤ 110% MOP).
2. Identify the pipeline segments with the highest density anomaly areas.
3. Determine whether defects or near-defects (anomaly + tolerance =~ defect) are present. If present, please remove the defects and near-defects from the sampled areas.
4. Evaluate whether the ILI performance (e.g., tolerance and certainty) is outside of normal expected ranges. Please note that the ILI uncertainty is very high, the

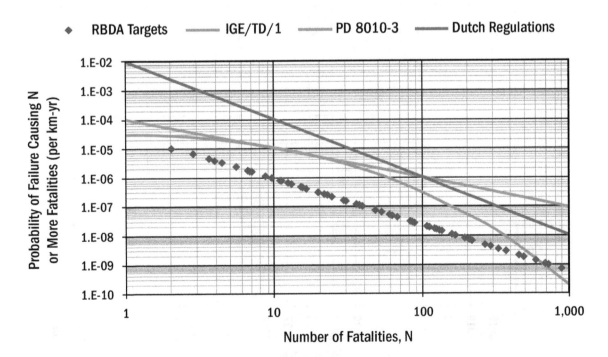

FIG. 4.19 SOCIETAL RISK LEVELS FROM MULTIPLE INDUSTRY STANDARDS AND REGULATIONS (*Source:* Nessim, M., Stephens, M., Adianto, R., 2012, Safety Levels Associated with the Reliability Targets in CSA-Z662, IPC2012-90450, ASME)

assessment length most likely would need to be customized for high-density areas, if any.

5. Calculate the cumulative PoF per km for the sampled high density areas without the defects and near-defects.

6. Compare Cumulative PoF values for the sample against the PoF limits. If the $PoF_{Cumulative}/PoF_{Limit} > 2 \times 10^{-1}$, please consider reducing the assessment length for calculating the Cumulative PoF until reaching a minimum Cum/Lim PoF ratio of 2×10^{-1}.

Example: Anomaly density and ILI uncertainty causing nonexisting critical Cumulative PoF

A high density of shallow anomalies (e.g., 450 reported anomalies with a depth < 40% and a low ILI performance >+/−15% at 80 of the time) within an assessment length (e.g., 1 km) can produce a large value for cumulative PoF per km per year equating to a non-existing large critical defect (e.g., >70% deep)

---

3. *Maximum Probability of Failure (PoF) Referenced to Deterministic Criticality*

An acceptable PoF for the selection and prioritization of ILI excavation programs can be correlated to the worst anomaly "acceptable" to remain in the pipeline at a given time.

Example: Acceptable Deterministic Criticality

• The worst anomaly with a depth always shallower than 60% at any given year
• The worst anomaly with a safety factor always greater than 1.25 MOP at any given year

---

The deterministic criticality of the worst anomaly can be converted to PoF using POE by defining the anomaly size and type, and their associated expected ILI performance (e.g., sizing, detection, identification).

Acceptable POE limits for leak and rupture can now be compared to against the cumulative POE evolving over the years for the only purpose of the selection and prioritization of ILI excavation programs. The verification of the residual PoF for leak and rupture should be conducted for ensuring the expected risk at the end of the program is considered acceptable.

The equation (1) provides the formulation for determining the POE cumulative per year "accumulating" all selected anomalies within a selected assessment length in a given year.

$$POE_{Cumulative} = 1 - \prod_{i=1}^{i=n}(1-Pi) \qquad (1)$$

where

Pi = Probability of Exceedance per feature (leak or rupture)
$\prod$ = Factorization of the selected POE values (1 − Pi)
n = Number of the selected metal loss features

As illustrated in Figure 4.20, cumulative POF per km per year is typically determined "accumulating" individual probabilities for leak and rupture estimated for a selected year (i.e., grown) and an assessment length. When comparing a selected acceptable POF level with the Cumulative PoF/km-year curves from the excavation scenarios, the excavation scope of work (i.e., selected anomalies for repair or removed from the Cumulative PoF) and the maximum time-to-repair alternatives can be obtained.

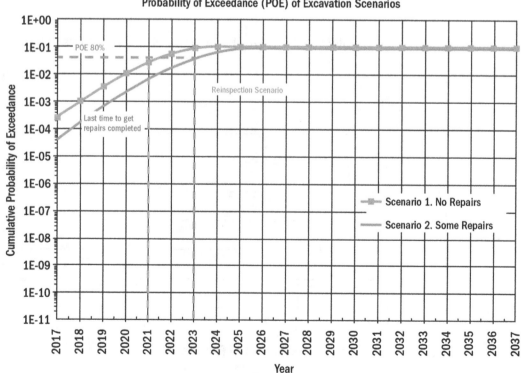

FIG. 4.20 CUMULATIVE POE LEAK BASED-CRITERIA FOR SELECTING ILI EXCAVATION PROGRAM SCENARIOS

4.7.2.5 Verification of Residual PoF and Re-Inspection for Risk Evaluation Residual risk verification is a typical practice for ensuring that the actions (or to be) taken provide an acceptable risk condition at the end of the process. Similarly, residual PoF verification provides the assurance of achieving the expected acceptable PoF through the selected mitigation (e.g., ILI excavation programs).

Furthermore, companies and regulators with diverse assets (e.g., pipelines, plants, stations, and vessels) are required assessing risk that can be compared among each other asset to make consistent decisions (e.g., risk significance, priorities, and schedule) across the organization and/or jurisdiction. The use of risk matrices based on individual or single scenarios enables management and engineers to evaluate and compare risks adequately and consistently.

Example: Individual and Cumulative PoF comparison among assets

A pressure vessel with a PoF_{leak} of 1×10^{-5} could not be compared with a Cumulative PoF_{leak} of 1×10^{-5}/km-year for a crude oil pipeline. The rationale is that the conceptual and mathematical meaning of a specific equipment PoF (e.g., single/individual scenario of pressure vessel failing) and cumulative PoF per km in a pipeline are different. Furthermore, a cumulative PoF per km per year do not reflect a credible failure scenario as the "accumulated" anomalies within 1 km would not physically fail altogether, but it can used for excavation ranking purposes.

As opposed to the cumulative PoF, a residual PoF for a pipeline, resulting from removing a selected excavation alternative, can adequately be compared to other residual PoF(s) from different assets. The residual PoF residual for a pipeline is determined by defining a credible failure scenario (e.g., leak-dependent, rupture-dependent) with the worst remaining ILI-reported anomalies and their associated conditions (e.g., abnormal operations, ILI performance, coincidental anomalies).

Example: Residual PoF comparison among assets

The PoF_{leak} of a pressure vessel of 1×10^{-5} (residual PoF after repairs) can be compared to the remaining worst leak-dependent anomaly (residual PoF after repairs) with a PoF of 4×10^{-6} within the same risk matrix for making decisions based on the significance of the risk.

4.8 REFERENCES

1. API 579-1/ASME FFS-1 2007 Fitness-For-Service.

2. Bunch, C., Cameron, G., Mora, R.G., *Guidelines to Conducting Threat Assessment of Pipeline Prior to Reactivation* IPC2010-31585, International Pipeline Conference (IPC), American Society of Mechanical Engineers (ASME), Calgary, AB, Canada.

3. Canadian Standard Association, CSA Z662-11 Annex H, Oil and Gas Pipeline Systems.

4. National Offshore Petroleum Safety and Environmental management Authority—Guidance Note N-04300-GN0107 Rev 4, December 2011.

5. USA, 40 CFR Part 68, and OSHA's Process Safety Management (PSM) standard, 29 CFR 1910.119.

6. Gould, J., Glossop, M., Ioannides, A., 2000, Review of Hazard Identification Techniques, Health and Safety Laboratory, UK Health and Safety Executive, United Kingdom http://www.hse.gov.uk/research/hsl_pdf/2005/hsl0558.pdf.

7. UK HSE, Review of Hazard Identification Techniques, United Kingdom Health and Safety Executive http://www.hse.gov.uk/research/hsl_pdf/2005/hsl0558.pdf.

8. TSSA, 2010, *Guidelines for the Implementation of the Level 2 Risk and Safety Management Plan*, Technical Standard Safety Authority, Ontario, Canada.

9. DOT PHMSA, 2003, *Hazardous Liquid Integrity Management: Frequently Asked Questions (FAQs)—# 7.19*, USA http://primis.phmsa.dot.gov/iim/faqs.htm.

10. API, 2013, *Managing System Integrity for Hazardous Liquid Pipelines, RP 1160, Section 7.2 Developing a Risk Assessment Approach*, American Petroleum Institute (API), USA http://www.americanpetroleuminstitute.com/publications-standards-and-statistics/standards/whatsnew/publication-updates/new-pipeline-publications/api_rp_1160#sthash.Vin5xGRm.dpuf.

11. Mora, R.G., Parker, C., Vieth, P.H., Delanty, B., 2002, *Probability of Exceedance (POE) Methodology for Developing Integrity Programs based on Pipeline Operator-Specific Technical and Economic Factors*, IPC02-27224, International Pipeline Conference (IPC), American Society of Mechanical Engineers (ASME), Calgary, AB, Canada.

12. Mora, R.G., Murray, A., Paviglianiti, J., Abdollahi, S., *Dealing with Uncertainty in Pipeline Integrity and Rehabilitation*, IPC2008-64612, American Society of Mechanical Engineers (ASME), Proceedings of the International Pipeline Conference, Calgary, Canada.

13. CSA Z662, 2015, *Oil and gas pipeline systems - Annex O Reliability-based design and assessment (RBDA) of onshore non-sour service natural gas transmission pipelines*, Clause O.1.5.3. Leakage limit sates, CSA Group, Toronto, Ontario, Canada.

14. Nessim, M., Stephens, M., Adianto, R., 2012, *Safety Levels Associated with the Reliability Targets in CSA-Z662*, IPC2012-90450, American Society of Mechanical Engineers (ASME), Proceedings of the International Pipeline Conference, Calgary, Canada.

CONSEQUENCE ASSESSMENT FOR PIPELINES

5.1   INTRODUCTION

Chapter 5 is intended for technicians, technologists, engineers, and managers working within the pipeline integrity area to gain a better understanding of the categories, processes, and content and format results of pipeline consequence assessments. Even though some generic formulation, conceptual guidance, and references are provided, qualified personnel in this area should be used in conducting and validating consequence assessments.

Chapter 5 starts by providing the terminology used in consequence assessments followed by the consequence categories and associated affected receptors and outcomes. As illustrated in Figure 5.1, the consequence assessment process is structured using a PLAN-DO-CHECK-PLAN management system approach integrated with engineering core elements associated with the consequence *analysis, classification, and validation.*

The *management system* sections describe the goals, objectives, targets of the consequence assessment in conjunction with planning, scheduling, verification, and management of change and performance review.

The *consequence analysis* section describes the pipeline discharge and dispersion phases for hazardous liquids, gas and two-phased fluids (i.e., High Vapor Pressure—HVP) including conceptual processes for approaching the modeling of consequence extent and severity. Pipeline case and model-specific must be considered for achieving reliable results. Dispersion modeling is described for the non-ignition, ignition, and explosion including Boiling Liquid Expanding Vapour Explosion (BLEVE) and toxicity at a conceptual level.

The *consequence classification* section describes multiple methodologies and their approach for classifying consequence areas. Methodologies approaching consequence from the safety and health viewpoint such as dwelling count and population density are described. Moreover, methodologies also factoring the environment accounting for the release extent are outlined supported by dispersion modeling for hazardous liquids, non-sour gas, and other fluids. Public consequence is defined in terms of community and regulatory consequences, as well as how media is factored as a measure. Financial consequence is also defined as the direct cost and in terms of how it differs from business consequence. Business consequence is not detailed in this book.

Validation of consequence classification section provides some guidance about the consideration to be made for confirming the type of consequence designated to a given pipeline segment, as well as some of the methods to conduct it. An additional section provides some recommended content and format for presenting the consequence results in terms of mapping and databases.

5.2   PIPELINE CONSEQUENCE CATEGORIES

Pipeline releases create conditions (e.g., product vapor cloud in air, emulsion in water and spread on land) that may cause singular and multiple types of consequences. For management purposes, consequences can be grouped in categories based on the type of affected receptors or generated outcomes, as follows:

a. *Safety and Health*: employees, landowners in proximity, and public in general
b. *Environmental*: air, land and water including habitat (e.g., flora and fauna)
c. *Community*: transportation (e.g., road, rail, navigable water), utilities (e.g., water, electricity, gas)
d. *Regulatory*: penalties, fines and enforcement
e. *Financial*: repair and clean-up cost, insurance premium or direct cost
f. *Business*: reputation (e.g., media), market trust (e.g., share), downtime (e.g., loss of expected income), contractual (e.g., customer penalties due to supply interruption), access to credit or indirect cost
g. *Human Talent*: engagement and hiring rotation
h. *Internal supply*: equipment parts, fuels and contracting
i. *Technological*: software, hardware and processes

This chapter only focuses on the safety, health, environmental, community, regulatory and financial consequences. However, the other categories may need to be assessed and managed by other processes.

5.3   CONSEQUENCE ASSESSMENT PROCESS

Some industry standards and regulations require pipeline operating companies to conduct consequence assessments for emergency response preparedness, identification of areas of high consequence, and determination of risk. Risk prioritization reveals the sequence of pipeline segments for conducting risk reduction including probability and consequence reduction within a given period. Even though consequence reduction may be harder than probability reduction, reducing drain and blow down volume releases as well as the dispersion extent and severity are measures to protect employees, the public and environment.

As illustrated in Figure 5.1, the conceptual process for conducting consequence assessments using a management system approach PLAN-DO-CHECK-ACT can be divided in the following stages:

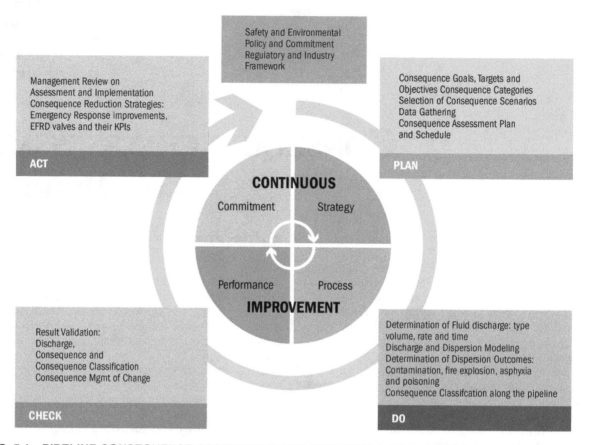

FIG. 5.1 PIPELINE CONSEQUENCE ASSESSMENT PROCESS WITH A MANAGEMENT SYSTEM APPROACH

5.3.1 Goals, Objectives, and Targets

Society, government, regulators, and special agencies, pipeline industry, special groups of interest and individuals likely have different expectations on pipeline consequences. Therefore, each stakeholder should identify goals, objectives, and targets before conducting a consequence assessment such the reduction measures are aligned to their expectations.

5.3.2 Selection of Consequence Scenarios

The selection of consequence scenarios (e.g., worst, most likely case) should factor the following aspects:

- Pipeline operating conditions and response at a given time
- Location of the release (e.g., distance from pumps/compressors, elevation, proximity to receptors and valves)
- Failure mode (e.g., corrosion, cracking) or discharge types (e.g., leak, large leak or rupture)
- Duration of the discharge from the pipe and dispersion into environment (e.g., time intervals for monitoring)
- Expected dispersion outcomes (e.g., overland flow, stream tracing, vapor cloud, fire, explosion)
- Data uncertainties

The varying capabilities of the following processes should be factored in identifying and quantifying the consequence scenarios:

1. Leak/rupture detection and isolation (e.g., hardware and software)
2. Control room and alarm management (e.g., human factors, abnormal operating conditions)

3. Emergency response in minimizing extent of consequence (e.g., accessibility, availability of personnel and equipment deployment)

5.3.3 Data Gathering

The following are some of the key data required for consequence assessment:

- Pipeline product characteristics (e.g., viscosity, density, molecular weight, vapor pressure)
- Weather conditions
- Emergency response time to contain release
- Pipeline characteristics (e.g., diameter, wall thickness, centerline, pressure)
- Selected release locations
- Road, rail and hydrology network (e.g., flow barriers, polylines, polygons)
- Pipeline mapping (e.g., Digital Elevation Model—DEM)
- Land use/cover (e.g., surface flow friction)
- Water crossing velocity and associated precipitation
- Regulatory framework (e.g., consequence definitions, response expectations)

5.3.4 Consequence Assessment Plan and Schedule

A plan and schedule should be prepared factoring the following:

- Scope definition agreed with operations and SMEs
- Data gathering: internal (e.g., company) and external (e.g., Public)
- Consequence analysis

- Consequence classification
- Consequence validation: modeling/office and field
- Report review and approvals
- Recommended path forward based on the results

5.3.5 Consequence Analysis

Determine the extent of the consequence, the receptors reached, and their grade of affectation using data gathering, field recognizance, mathematical formulation, computational algorithms, and desktop verifications.

5.3.6 Consequence Classification Along the Pipeline

Evaluate the results of the consequence analysis to determine the classification either qualitative or quantitative of the consequence along the pipeline based on predefined consequence criteria. Please refer to the next section in this book for details.

5.3.7 Consequence Results Validation

The results of consequence modeling used in the analysis as well as the classification should be tested, as follows:

- Were the selected method/software and assumptions applicable as expected?
- Were the consequence analysis results credible?
- Were the consequence classification results verified in the field?

Please refer to the next sections in this book for details.

5.3.8 Consequence Management of Change

The results of a pipeline consequence assessment may lead to changes associated with system modifications (e.g., additional valves, telecommunications for detection and isolation), procedure improvements (e.g., response time), redefining locations for emergency response (e.g., new worst release/consequence cases) and personnel training and qualification (e.g., response effectiveness).

5.3.9 Management Review on Assessment and Implementation

Management may determine establishing consequence reduction strategies involving multiple teams such as operations, projects, and emergency response. Those strategies would require monitoring via KPIs for planning, implementation, and effectiveness review.

5.4 CONSEQUENCE ANALYSIS: FLUID DISCHARGE AND DISPERSION

The consequence analysis focuses on understanding how the fluid and how much is discharged out of the pipeline as well as how the fluid is dispersed in the environment causing consequence. The following are some of the key aspects for conducting consequence analysis:

- Fluid type (e.g., gas, liquid) and properties (e.g., viscosity, flashpoint)
- Pipeline conditions (e.g., pressure, temperature) and discharge characteristics
- Environment characteristics such as topography, geography, soil properties (e.g., infiltration), and weather (e.g., temperature, wind)

- Receptors and their proximity/sensitivity to the fluid dispersion outcomes (e.g., fire, explosion, toxicity, contamination)

Fluid discharge focuses on the fluid release from the pipeline. The discharge volume and rate depend on the fluid conditions inside and outside of the pipeline due to the breach that may cause the fluid phase to change (e.g., gas, liquid, two-phase).

Whereas, the fluid dispersion focuses on the extent of the mixture or fluid mixed into the environment (e.g., air, water, and land). Dispersion may occur at open or confined/obstructed spaces in which some dispersed fluids may create smoke or product vapor restricting emergency or escape routes. The mixture may affect humans and the environment due to its toxicity. In addition, the mixture may encounter ignition sources creating either an immediate or a delayed ignition, which cause different types of outcomes such as fire and explosion.

5.4.1 Determination of Fluid Discharge: Type, Volume, Rate, and Time

The first step in determining fluid discharge is to understand how the fluid is released in terms of pipeline opening/orifice (e.g., pinhole) and discharge type (e.g., leak). This step is followed by estimating the volume of the fluid discharged in their steady state condition (e.g., one or two-phase), the calculation of the discharge rate and time factoring the emission type (e.g., instantaneous, continuous).

5.4.1.1 Qualification of the Type of Fluid Discharge: Leak, Large Leak, or Rupture The fluid discharge from the pipeline can be qualified as a leak, large leak, or rupture. The fluid discharge out of the pipeline into the environment can occur in any of three (3) states: (1) gas (2) liquid or (3) two-phase such as pipeline transporting HVP (e.g., Butane, Propane, or other HVP liquid).

Pipeline leaks are discharges of the pipeline fluid causing partial loss of containment that may not be detected by operations or increased time before it is detected. Hence, the pipeline may continue "partially" transporting flow throughput. Leaks are typically detected by the public, farmers, right-of-way surveillance and sometimes, later on, by leak detection systems.

Leaks result from small openings (e.g., corrosion, seam, or girth weld cracking) going through the pipe wall. Hole sizes for small leaks are typically less than 10 mm in diameter; however, other jurisdictions (e.g., UK) define pinhole equal or smaller than 25 mm and small holes between 25 mm and 75 mm for modeling purposes.

Hole sizes for *large leaks* are typically longer than 10 mm and less than 50 mm [1]; however, other jurisdictions (e.g., UK) define large holes greater than 75 mm up to 110 mm for modeling purposes.

Conversely, *pipeline ruptures* are discharges of the pipeline fluid causing loss of containment immediately detected as it impairs or stops the operation of a pipeline. The hole sizes are typically longer than 50 mm [2]; however, other jurisdictions (e.g., UK) define rupture releases from holes greater than 110 mm for modeling purposes.

Liquid spills can be classified based on the discharge rate and duration [3], as follows:

Instantaneous Spills: entire volume discharged in a very short duration

Continuous Spills: volume with a specified finite discharge rate for a long duration

Finite Duration Spills: finite volume discharged over a given finite duration

Gas discharge can be classified based on the discharge rate and duration [4] accounting for their buoyancy:

1. *Instantaneous (Puff)*—Neutrally Buoyant Gas
2. *Continuous*—Positively Buoyant Gas (e.g., Momentum Jet)
3. *Time-Varying Continuous*—Dense Buoyant Gas

5.4.1.2 Quantification of the Discharge: Volume, Rate and Time The quantification of the *discharge volume* resulting from an unintended release is comprised of the *operational volume (V_o)* and the *drain (i.e., liquids) or blow (gases) down volume (V_d)*.

5.4.1.2.1 Discharge Volume. Discharge volume is comprised of the operational volume and the drain down for liquid phase or blow down volume for gas phase. The *operational volume* is the release occurred from a leak or rupture event up to the time when the valves are closed by operations. Gin, G [5] divided the operational volume into three (3) stages; each of the three (3) segmented volumes most likely would have a different discharge rate for a given time:

a. Initial ramp up volume as the pipe opened to release;
b. Maximum flow rate volume until the controller detected and responded by starting valve closure;
c. Volume until the valve has reached full close position

The *drain down volume* is the release occurred after the upstream and downstream valves have been closed discharging flow by accounting for the pipeline elevation profile.

The *blow down volume* is the release occurred after the upstream and downstream valves have been closed until the pipeline segment is fully depressurized.

In pipelines with a mixture of motor operated valves, manual valves, and check valves, the transition from operational volumes to drain or blow down volumes can be more complex, as the isolation segment changes with time.

The total discharge volume *(V_r)* can be calculated with the following formula:

$$\text{Discharge Volume } (V_r) = V_o + V_d$$

Where;
V_o is the Operational Volume = Operational Flow rate × Time (t_i)
V_d is the Drain Down or Blow down Volume

As illustrated in Figures 5.2 and 5.3, the elevation would affect the *drain down volume (V_d)* for pipelines transporting LVP (e.g., crude oil, gasoline, diesel, kerosene) as some of the volume between valves will stay in the pipeline and not be released.

5.4.1.2.2 Discharge Rate. The discharge flow rate can be considered [6] either

a. *Stationary* when the discharge is controlled by the upstream and downstream pressure making the discharge rate to be mainly constant, or
b. *Non-Stationary* or the discharge rate is changing over time because the conditions upstream are changing.

In most of the pipeline release cases, the flow discharge rate *is* not constant or non-stationary as the rate is reduced over time. However, for pipelines with small diameter and short length, the error of assuming constant discharge rate is minimal. For flammable, explosive, and toxic fluids, the constant discharge rate makes the vapor cloud to be over-predicted estimating higher volume and shorter duration than actual.

The International Association of Oil and Gas Producers (OGP) [7] recommends some considerations for determining discharge rates from underground pipelines as their discharge rate may be reduced due to the location on the circumference of the pipeline (clock position) and the size of the pipeline break/opening, as follows:

• For modeling ruptures, both high and low discharge rates as well as low release rates with a delayed ignition should be estimated for determining the envelope of the consequences.
• For modeling large leaks, discharge scenarios from the top, side, and bottom of the circumference of the pipeline should

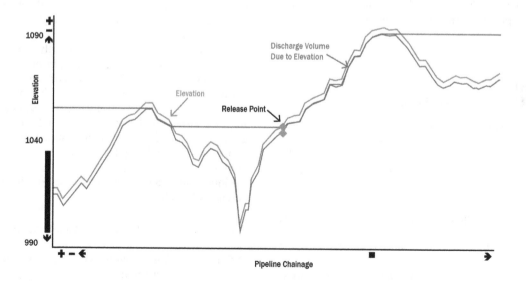

FIG. 5.2 DISCHARGE OR RELEASE VOLUME AFFECTED BY ELEVATION (*Source:* New Century Integrity Plus)

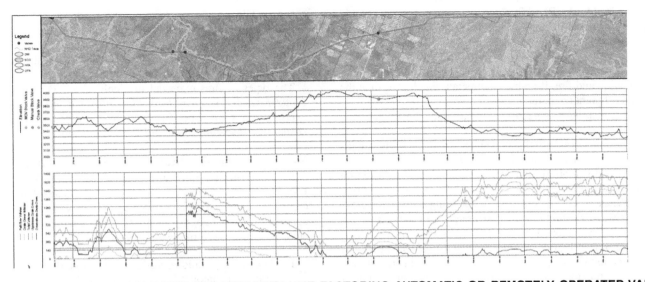

FIG. 5.3   EXAMPLE OF DRAIN DOWN AND TOTAL VOLUME FACTORING AUTOMATIC OR REMOTELY OPERATED VALVES
(*Source:* New Century Integrity Plus)

be understood. These scenarios would help for considering the effect of soil obstruction during the discharge of the fluid.

• For modeling small leaks, the normal discharge rate should be calculated in addition to the scenario of low pressure (e.g., 10% MOP) with a discharge rate reduced by the effect of soil, but increasing the hole size to obtain the normal discharge rate. For instance, scenarios such as:

  • a pressure of 1 bar (100 kPa or 14 psi) for operating pressures greater than 10 bar;
  • a pressure of 0.1 bars (10 kPa or 1.4 psi) for operating pressures less than 10 bar.

Discharge rate of a steady state liquid pipeline at ambient conditions can be determined by using the Bernoulli and Torricelli formulation (Perry and Green, 1984): where it is assumed a constant pipeline diameter between valves as well as fluid density and velocity during discharge. The frictional loss is represented by a discharge coefficient, C_d (Crowl and Louvar, 1990) [8, 9]. Hole area is calculated based on the type of expected discharge. The following are recommended hole diameters for determining the area of the orifice [2]:

Leak	10 mm
Large Leak	50 mm
Rupture	Full diameter

C_d is function of orifice shape and the velocity of the fluid through the hole. From Beek and Mutzall [10], the following average values can be determined:

Sharp and steep edges	0.62
Straight edges	0.86
Rounded edges	0.96
Full bore rupture	1.00

Note: For pipelines transporting steady-state liquids that at ambient conditions transform into gas, the discharge rate should be calculated as a gas.

Discharge rate of a gas pipeline can be determined by using the formula for liquids, but multiplying it by the expansion factor due to gas compressibility [11]. This approach assumes ideal gas, no heat transfer and no external shaft work.

The sonic or chocked flow for gas occurs when discharge rate reaches the maximum value becoming only dependent on the upstream pressure (P_1). For the sonic (choked) and subsonic flow, the discharge rate at that condition can be calculated [11].

Discharge rate of a Liquefied Petroleum Gas (LPG) pipeline can determined by using Tam & Higgins [12] methodology by figuring out the correlation mass decay (kg) in a Liquefied Petroleum Gas pipeline (LPG such as Butane, Propane).

LPG discharge rate at a given time can be deducted from their empirical single-node slip flow model, which determines the remaining mass inventory at a given time as a function of the mass decay constant. The decay constant depends on the diameter of the pipe and orifice, and temperature.

Newer numeral simulations for predicting discharge rates of condensable hydrocarbon mixture following a pipeline rupture can be found in this reference [13].

5.4.1.2.3 Discharge Duration. The time required for discharging the volume in the pipeline after the valves are closed depends on the discharge rate and volume, which can be mathematically expressed as follows:

Van der Bosch, C.J.H., Weterings [6] provides the numerical procedure for calculating the blow down time of pipelines through small holes for sonic and subsonic condition.

The summation of the individual times $_{(n)}$ resulting from the individual volumes and their associated discharge rates provide the total discharge duration of the total estimated volume released to the environment.

5.4.2   Dispersion Outcomes and Model Types for LVP, Gas and HVP

Dispersion outcomes are translated into consequences affecting people and environment and causing damage to property. Selection of a representative dispersion model is essential to estimate the extent and severity of the affected areas.

5.4.2.1 Liquids (LVP) Liquids (Low Vapor Pressure) would extend forming pools and may infiltrate into the soil, partially evaporate into the air, mix with water and further produce:

- *If high flammability (e.g., gasoline) with ignition*: pool contamination and fire, if ambient temperature is higher than the fluid flashpoint temperature.
- *If low flammability (e.g., crude, diesel)*: pool contamination.

The three dimensional (3D) geographical extension of the liquids on land can be estimated using *Overland Flow modeling* factoring evaporation and soil absorption. Whereas, the liquid extension on water can be estimated using *Tracing on Water modeling* factoring floating product, sinking product (e.g., crude aging and specific gravity exceeds water's), fluid + water mixture or emulsion, which can be partially left behind on the water crossing banks or stream bed.

The extension of liquids on land and water is reduced by environmental conditions (e.g., retaining topography and permeability, evaporation) and emergency response from the pipeline company and nearby community first responders.

5.4.2.2 Gas (Sour and Non-Sour) Gases (Sour and Non-Sour) would extend as a spray, aerosol, or flash forming vapor clouds that could produce:

- *If no ignition occurred*, potential asphyxia and poisoning causing injuries or death
- *If ignition occurred*, fire creating thermal radiation and potential explosion causing injuries, death and property damage. If gas is liquefied via pressure in a container and exposed to boiling conditions, it may cause a Boiling Liquid Expanding Vapor Explosion (BLEVE)

The extension of vapor clouds traveling with a mixture of fluid and air can be estimated using *Momentum Jet* and *vapor dispersion modeling*. However, the model selection depends on fluid material and thermodynamic properties (e.g., vapor pressure, specific gravity and associated temperature, buoyancy) and release rates.

5.4.2.3 Two-Phase Release (HVP) HVP fluids would extend as a fluid + air mixture flowing on top of the ground and creating vapor clouds in the air that could produce:

- *If no ignition occurred*, potential for asphyxia and poisoning causing injuries or death
- *If ignition occurred*, fire creating thermal radiation and potential explosion with an overpressure blast causing injuries, death or property damage. If HVP is liquefied via pressure in

a container and exposed to boiling conditions, it may cause a Boiling Liquid Expanding Vapor Explosion (BLEVE)

The extent of a HVP mixture vapor cloud can be estimated using *Heavier-than-Air* vapor dispersion modeling.

The types of fires expected in pipelines are *Pool Fire, Vapor Cloud Fire, Jet* or *Spray Fire, Flash Fire* and *Fireball*. Fireball, overpressure blast, and BLEVE are discussed in the section *Dispersion Modeling with Ignition and Explosion*.

5.4.3 Dispersion Modeling With No Ignition

5.4.3.1 Overland Flow and Tracing on Water Modeling for Liquids (LVP) As illustrated in Figure 5.4, liquids (flammable or combustible LVP) may experience up to four (4) dispersion phases depending on the topographical, soil, water, temperature, wind, and environmental conditions, as follows:

1. *Liquid pool*: fluid spreads on the ground following topographical contours
2. *Infiltration*: fluid may be partially absorbed by soil and mix with ground water
3. *Evaporation*: fluid may evaporate as it changes its buoyancy from internal stable to ambient turbulence
4. *Emulsion*: fluid may encounter a water body (e.g., creek, river, lake, and ocean) mixing and following its currents

The overland flow and tracing on water modeling are conducted to determine the extent of liquid spills flowing on land and becoming an emulsion while traveling on water bodies (e.g., lakes, ponds, streams, rivers, and springs). The extent or dispersion of the liquid spills depends on factors such as [14]:

- *Fluid Characteristics*: viscosity, evaporation, miscibility, solubility
- *Pipeline Characteristics*: diameter, elevation/slopes (e.g., Digital Elevation Model), release hole size
- *Pipeline Operational Conditions*: flow rate, pressure, detection and isolation time, separation between automatic or remotely-operated valves, and emergency response
- *Environment and Weather Characteristics*: soil permeability, vegetation, water body flow and speed, precipitation, groundwater and drain tiles

Digital Elevation Models (DEM) [15] provide a 3D representation of the terrain surface describing the topography, elevation, water body shapes and paths that are used for developing spill modeling. DEMs are typically represented by equally-sized cells containing location coordinates and elevation attributes. DEMs can

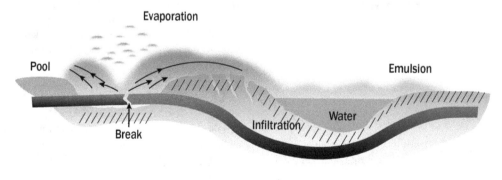

FIG. 5.4 PHASES OF LIQUID PIPELINE DISPERSION (NO IGNITION)

be developed by methods such as Light Detection And Ranging (LIDAR), optical satellite imagery and aerial survey-based stereo photogrammetry. DEM resolutions are typically available ranging from 30 m to 3 m. DEMs with a resolution of five (5) meters provide excellent results balancing processing time and precision.

As illustrated in Figure 5.5, the overland flow models can be developed by integrating multiple platforms of software and data (e.g., spreadsheets, databases, Geographical Information Systems, consultant-specific) or programming extensions/scripts in GIS for determining the path and extent of a liquid spill. The conceptual process of an overland flow and tracing on water can be described, as follows [5]:

a. Setup the pipeline centerline and valves on a GIS Digital Elevation Model (DEM) using raster data emphasizing the variations in topography (e.g., road beds, roadside ditches, culverts), and confinement (e.g., buildings, storage tanks, containment)

b. Define release points along the 3D DEM pipeline

c. Select the release scenario for the LVP fluid (e.g., leak, large leak, or rupture with/without fire, explosion) identifying whether liquid pool vapors could trigger fire and/or explosion

d. Identify fluid characteristics and operational conditions for calculating the release volume

e. This should factor the release detection and isolation times for valve closure and places/time where the product would be contained by the emergency response plan. Alternatively, a leak not detected for some time can also be used for estimating the volume

f. Define the environment conditions and weather characteristics identifying potential receptors

g. Select the dispersion model (i.e., overland flow, tracing on water, liquid pool vapor) identifying their transitions/combinations

FIG. 5.5 SAMPLES OF OVERLAND FLOW AND TRACING ON WATER MODEL RESULTS (*Source:* New Century Integrity Plus)

h. Determine the spill path and direction on the 3D DEM identifying liquid spread, pooling (i.e., Overland Flow Model) limited by the emergency response time and/or spill volume

i. Determine liquid transient rate for infiltration into the soil and evaporation (i.e., Overland Flow Model). If the soil permeability is high, the groundwater can be affected and emulsion or dissolved contaminants (e.g., benzene moving with the water) may be transported

j. Once point of contact with water bodies (e.g., lakes, rivers) is reached by the Overland Flow model, the volume at that time is transferred to the Tracing on Water model. If the fluid is miscible and soluble, the emulsion volume will increase over time on the water

k. Identify the flow dynamics of the contacted water body differentiating between channel flow (e.g., creeks, rivers) and body masses (e.g., ponds, lakes)

l. Determine the spill path on the water body accounting for its flow rate, wind, bank retention and evaporation and identifying whether continental or marine/oceanic weather is affecting the spill behavior/path

m. Identify the affected receptors, their sensitivity and grade of recovery to the release effects

n. Conduct quality assurance and control reviews verifying the process has been completed for all spill points and ensuring that the interfaces overland + tracing models are continuous in time and volume. Results should be initially validated via simulations and experience ensuring the models reflect the reality in the field

o. Prepare maps and GIS files with the spill points, time-varying spill extent (e.g., 4, 8, 12 hours) for land and water indicating the intersected consequence areas for communicating the results to the appropriate stakeholders (e.g., emergency responders, integrity, regulators)

5.4.3.2 Vapor Cloud Modeling for Gas (Non-Sour) and Two-Phase Release (HVP)

As illustrated in Figure 5.6, for gases (sour and non-sour), the dispersion phases of onshore gas pipelines start with a rapid release going into the atmosphere creating a vapor cloud mixture with air. The vapor cloud expands and flows with wind decreasing concentration until it loses all its release energy. Gas fluids may go up to four (4) dispersion phases along the topographical contours:

1. *(Jet) Initial Turbulent Expansion*: continuous or instantaneous release with rapid expansion and fluid dilution losing concentration. If pipeline is underground, consideration should be made to factor the soil resistance related to the pipe clock position of the leak (i.e., top, lateral or bottom side); similarly, if pipeline rupture, the flow from the other side of the rupture impinging the flow rate coming onto the ground.

2. *(Cloud) Spreading and Mixing*: cloud creation expanding, moving faster than wind and mixing with turbulence driven by fluid buoyancy (e.g., heavier, neutral or lighter than air). The cloud may touch down on the ground at this phase.

3. *(Cloud) Slumping Dense Plume*: cloud touches down on the ground slowly reducing mixing rate. Concentration is near uniform in the vertical plane. Cloud flows vertically down by ambient buoyancy (i.e., gravity) of the mixture and horizontally by the wind.

4. *(Cloud) Passive Dispersion*: cloud on the ground with no release energy driven by wind.

Depending on the type of fluid and operational and environmental conditions, dispersion stage may go directly from the first (i.e., turbulent jet) to last phase (i.e., passive) or in-between phases [7].

The two-phase or dense gases (HV) disperse by flowing downwards to the surface and spreading radially driven by gravity. The shallow cloud on the surface expands horizontally developing a head at the cloud perimeter with a strong Vorticity. Once the radial spread is finished, the vapor cloud goes onto the passive phase.

As illustrated in Figure 5.7, for a two-phase fluid (e.g., HVP), the dispersion phases of dense gases are described in Hunt, Bosch, and Weterings [6]:

a. *(Jet) Flashing*: creating a vapor mass fraction and temperature in the jet stream

b. *(Cloud) Gravity-Spreading*: vapor cloud starts mixing and rapidly slumping due to gravity

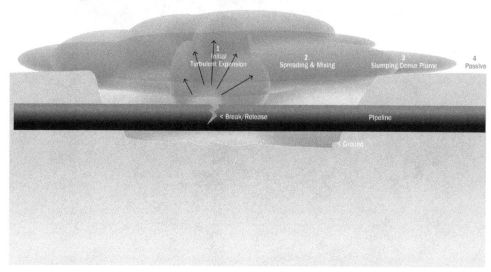

Dispersion Phases of an Underground Pipeline Release (no ignition)

FIG. 5.6 DISPERSION PHASES OF AN UNDERGROUND GAS RELEASE (NO IGNITION)

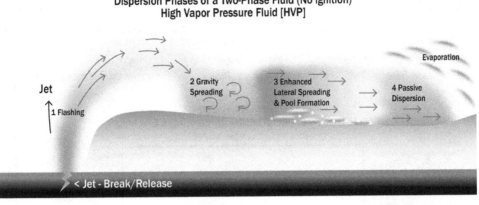

Dispersion Phases of a Two-Phase Fluid (No ignition)
High Vapor Pressure Fluid [HVP]

FIG. 5.7   DISPERSION PHASES OF A TWO-PHASE FLUID (HVP) RELEASE (NO IGNITION)

c. *(Cloud and Pool) Enhanced Lateral Spreading and Pool Forming*: vapor cloud is further mixed by both atmospheric turbulence and gravity. Droplets are formed from entraining air creating a (evaporating) liquid pool

d. *(Cloud and Evaporation) Passive Dispersion and Pool Evaporation*: mixture moved by atmospheric turbulence as the difference between vapor cloud density and ambient air becomes small. Evaporation from the liquid pool is combined with the cloud

The vapor cloud dispersion models can be developed by integrating multiple platforms of software (e.g., Figure 5.8) and data (e.g., excel, databases, Geographical Information Systems, consultant-specific) or programming extensions/scripts in GIS for determining the path and extent of a liquid spill. Once the pipeline characteristics and operation, and geographical data are prepared, the vapor cloud modeling with no ignition can be conducted with the following conceptual process:

a. Setup pipeline centerline and valves on a GIS Digital Elevation Model (DEM) using raster data accounting for variations in topography (e.g., road beds, roadside ditches, culverts), confinement (e.g., buildings, storage tanks, containment) and potential receptors that could be affected

b. Select the release scenario for gas (e.g., vapor cloud) or HVP (e.g., pressurized jet) identifying whether fire and/or explosion could occur

c. Define release points along the 3D DEM pipeline with their associated fluid characteristics and operational conditions, release volume limited by the valve closure times, and the flow rate versus time correlation

d. Define the environment conditions including the roughness length of the area of interest

e. Define the weather characteristics including the stability of the atmosphere and friction velocity, if any

f. Select the dispersion model (i.e., Momentum Jet, Neutrally Buoyant or Heavier-than-Air or Dense) identifying the transitions/combinations (e.g., momentum jet to heavier-than-air) as well as the nature of the release (i.e., instantaneous, continuous, time-dependent). Check applicability and limitations of the selected models

g. Identify the affected receptors, their sensitivity and grade of recovery to the release effects

h. Plot maps with the vapor cloud extent in form of shapes such as circular and conical

5.4.4   Dispersion Modeling with Ignition: Fire and Thermal Radiation Scenarios

5.4.4.1 Thermal Radiation   In the context of this chapter, heat would be understood as it is emitted from a fire of flammable gas, High Vapor Pressure fluids—HVP—(e.g., Butane, Propane) or liquid (e.g., gasoline) being transported by a pipeline. Heat is transferred as *thermal radiation, conduction,* and *convection*. Thermal or infrared radiation is a type of electromagnetic radiation generated by thermal motion of atomic particles that does not need a specific medium for heat transfer such as object (e.g., heat conduction) or liquid or gas (e.g., convection).

Thermal radiation is transmitted for a distance from the ignition source affecting receptors (e.g., injury, fatality, and damage) that can be reduced via absorption by water vapor and carbon dioxide in the air. The thermal radiation can be estimated using semi-empirical, field, and integral models.

Semi-empirical models are typically used by the pipeline industry in estimating the thermal radiation at a given distance (i.e., receptor) with a few simplified factors. Thermal radiation modeling of pool and jet fires is approached by assuming fire emitters from either *a point source* (i.e., single or multiple) or *a surface emitter*; whereas, the thermal radiation modeling of a fireball is typically approached by assuming a surface emitter using the fraction of the generated combustion energy and the release pressure.

Using a surface emitter model, the calculation of the thermal radiation (TR) at a given distance (i.e., receptor location) requires the following three (3) conceptual components [16]:

1. *Surface Emissive Power (SEP)*: heat flux rate from a radiating surface of the flame expressed in $J/(s.m^2)$ or W/m^2. The SEP is determined from the *combustion-burning rate*, the *heat of combustion of the material*, and *the surface area of the flame*. The *Yellow Book* [6] differentiates the combustion burning rate per type of fire, as follows:
   - *Pool Fire*: the burning rate equates to the evaporation rate from the pool.
   - *Jet Fire*: the burning rate under stationary conditions equates to the release rate.
   - *Fireball*: the burning rate equates to the total amount of flammable material divided by the duration of the fireball.

2. *View Factor (F_{view})*: a geometric relationship between emitting and receiving surfaces depending on the dimensions, shape of the flame, the distance to and the orientation of the

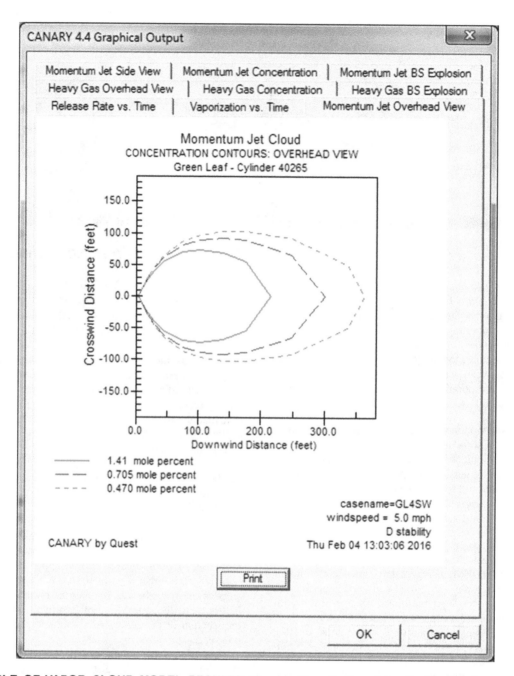

FIG. 5.8   EXAMPLE OF VAPOR CLOUD MODEL RESULTS (*Source:* New Century Integrity Plus, Software: Canary by Quest Consultants)

receiving receptor. View factor is equal to 1, if flame entirely fills the field of view of the receptor surface.

3. *Atmospheric Transmissivity* (τ_a): factor accounting for the absorption by the air composition (i.e., water vapor and carbon dioxide) between the fire and the receptor. τ_a is typically calculated as a function of the product water vapor pressure (p_w) and distance from the flame surface to receptor.

Each type of fire (i.e., *Pool Fire, Vapor Cloud Fire, Jet* or *Spray Fire, Flash Fire,* and *Fireball*) creates different thermal radiation areas; however, sometimes a fire may evolve from one type of fire

to another (e.g., from fireball to jet fire). The type of fire can be characterized by the following factors:

a. Discharge direction (e.g., vertical, inclined, and horizontal)
b. Obstruction of the discharge feeding the vapor cloud
c. Buoyancy of the vapor cloud holding or reducing flammable conditions
d. Timing for ignition (i.e., immediate or delayed) and
e. Place (i.e., local, remote) of the ignition related to the source

5.4.4.2 Pool Fire   A pool fire is a turbulent diffuse flame resulting from the combustion of a fluid evaporating from a layer of liquid with no or low initial momentum (Figure 5.9). The fire and the pool are interdependent as the fire transfers heat to the flammable fluid creating its own evaporation and changing the pool size (if not continuously fed). This will end up affecting the fire source.

The USA FEMA Handbook [3] proposes a method for calculating the maximum thermal radiation distance (radius) from the center of a pool fire by using a *point source emitter model* and creating a thermal radiation circle. This circle contains a thermal radiation exposure of 5 KW/m² or 1600 BTU/hr-ft² during 45 seconds able to affect unprotected (e.g., no fire retardant clothing) and unsheltered (no dwelled) people by causing them the effects of 2nd degree burns. Then, the receptors are referenced related to the thermal radiation circle to identify whether they are affected.

Furthermore, the distance to injury (e.g., 5 KW/m²) or death (e.g., 10 KW/m²) levels are parameterized as a function of the selected emissive power (KW/m²) and the pool radius (meters).

Using *surface emitter models*, the thermal radiation from a pool fire $TR_{(Pool\ fire)}$ on land at a given distance can be calculated determining the fire dimensions and properties, its heat capacity, emitter versus receptor spatial relationship, and the thermal reduction effects from the surrounding environment [6]. The thermal radiation from a pool fire using the surface can be determined with the following conceptual process:

Fire Dimensions and Properties

1. *Equivalent Liquid Pool Diameter (D)* as the actual pool area is irregular, unless confined. The diameter of the pool fire depends upon the discharge mode (i.e., instantaneous, continuous, or finite discharge), discharge quantity and rate, and the burning rate. The worst-case equivalent liquid pool diameter can be estimated by assuming instantaneous discharge with the minimum spill thickness, which creates the maximum pool area and the associated liquid pool diameter *(D)* of a circle.

2. *Flame Characteristics accounting for wind velocity*:
   - Average Flame Length (*L* = height from ground)
   - Flame Tilt (i.e., vertical angle Θ)
   - Flame Shape (e.g., cylindrical, conical and cylindrical with elongated flame base diameter)
   - Elongated Flame Base Diameter (*D'*) that may decrease the 'reach' distance to receptor

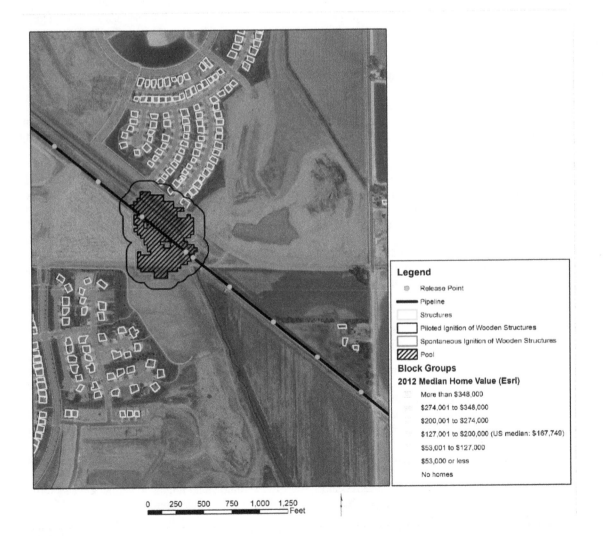

Legend

- Release Point
- Pipeline
- Structures
- Piloted Ignition of Wooden Structures
- Spontaneous Ignition of Wooden Structures
- Pool

Block Groups
2012 Median Home Value (Esri)

- More than $348,000
- $274,001 to $348,000
- $200,001 to $274,000
- $127,001 to $200,000 (US median: $167,749)
- $53,001 to $127,000
- $53,000 or less
- No homes

0   250   500   750   1,000   1,250
Feet

FIG. 5.9   EXAMPLE OF A POOL FIRE CONSEQUENCE EXTENT (*Source:* New Century Integrity Plus)

3. *Combustion burning rate* (m"): typically equates to the evaporation rate from the pool
4. *Fraction of heat radiated from the surface of the flame* (F_s)

Heat Capacity
5. *Surface Emissive Power* (SEP) calculated from steps 1 to 4: as a function of D, $m"$, L, F_s

Emitter and Receptor Spatial Relationship
6. *View Factor* (F_{view})

Surrounding Environment (Thermal Radiation Reduction)
7. *Atmospheric Transmissivity* (τ_a)

End Result: Thermal Radiation at a given distance from the Fire to the Receptor
8. *Thermal radiation from a pool fire surface $TR_{(Pool\ fire)}$ at a given receptor*: as a function of SEP, F_{view} and τ_a

5.4.4.3 Vapor Cloud Fire   A *Vapor Cloud Fire* occurs during the dispersion of the cloud that reached an ignition source propagating through the space occupied by the cloud. Fire can only occur if the vapor mixture is within the flammable limits and the flame travels back to the cloud origin as a laminar flow. If the flow of the flame is turbulent, an explosion would occur producing an overpressure blast.

The thermal radiation from a vapor cloud fire at a given receptor can be determined with the following conceptual process [3, 6, 17]:

1. Determine the discharge rate and the fluid state at the dispersion condition (e.g., HVP is liquefied during transportation in the pipeline, but dense gas during dispersion) such as
   a. Buoyant (i.e., floating/emerging in the air)
   b. Neutrally buoyant (i.e., floating in the air at given height), and
   c. Heavier-than-Air or Dense (i.e., slumping on the ground)
2. Identify whether the dispersion would occur in a open field (e.g., rural) or congested (i.e., town, cities) area related to the potential obstacles for dispersion
3. Determine the atmospheric stability based on the wind speed, day or night cloudiness, which can be estimated by the Pasquill stability classes from A (extremely unstable) to G (extremely stable)
4. Identify the Lower Flammability Limit (LFL) of the fluid as the volume percentage (%) of its concentration in the air or milligram/liter at a given pressure and temperature
5. Calculate the distance from the fluid source to the vapor cloud reaching the Lower Flammability Limit over which the fluid will ignite causing injuries and potential deaths. The width of the vapor cloud is typically 16% to 50% of the vapor cloud length. Thus, the fire area generating thermal radiation can be determined by the length and width of the vapor cloud at the LFL position
6. Determine the Surface Emissive Power, View Factor and Atmospheric Transmissivity
7. Calculate the thermal radiation at a given distance/receptor considering the short duration of the fire (e.g., tenths of a second) and the temperature of the flame

5.4.4.4 Flash Fire   Flash fire [18] is a sudden fire of short duration, high temperature and high speed, less than sonic velocity,

causing thermal radiation, smoke and low overpressure. Flash fires can cause asphyxiation due to smoke in confined spaces and/or tissue damage to the lungs. Thermal radiation is calculated following the Vapor Cloud Fire methodology considering that Flash Fires produce short thermal hazards with negligible overpressure blast.

5.4.4.5 Jet or Spray Fire   A jet or spray fire is a turbulent diffusion flame resulting from the combustion of a fuel continuously released with some significant momentum (Figure 5.10) in a particular direction or directions [6]. Jet fires can arise from releases of gaseous (e.g., natural gas, HVP), flashing liquid (two phase) and vaporization of liquid. The burning rate for pipeline equates to the release rate.

For underground pipelines experiencing rupture, the Jet fire is impinged by the opposite jet flow and the crater created by the rupture [19]. This impinging effect reduces the momentum of the jet and widens up the area affected by contamination and fire.

Three (3) types of models are used for Jet fire:

1. Single point source (i.e., API RP 521, Brzustowski and Sommer, 1973)
2. Multiple point source on a center line trajectory (i.e., Cook, 1987) to simulate the flame size and shape. This type of model is applied in computer applications such as WHAZAN, FLARESIM, and THORIN
3. Geometric solid flame using an angular conical frustum shape (i.e., Chamberlain, 1987) to simulate the thermal radiation with a uniform surface emissive power. This type of model is applied in computer applications such as FLARE, TORCH, MAJESTIC SHELF2 PIPEFIRE MAJ3D, and SHELL THORNTON/PHAST

The thermal radiation from a Jet fire at a given receptor using the conical frustum shape can be determined with the following conceptual process [6]:

1. Calculate the exit velocity of the expanding jet considering molecular weight and thermodynamic properties of the gaseous fluid, pressure at hole and temperature in the jet, and mass flow rate evaluating whether choked (i.e., expanded diameter) or unchocked (e.g., orifice diameter) flow
2. Determine the combustion effective source diameter representing the throat diameter of an imaginary nozzle releasing air of density ρ_{air} at a mass flow rate m'
3. Determine the surface area of the conical frustum solid (flame) including end conical discs: length, width of the frustum base and tip, angle between hole axis and the horizontal in the direction of the wind and lift-off of the flame
4. Determine the Surface Emissive Power, View Factor (i.e., Conical shape) and Atmospheric Transmissivity
5. Calculate the thermal radiation from the Jet fire at a given distance/receptor

5.4.5   Dispersion Modeling with Ignition and Explosion

5.4.5.1 Vapor Cloud Explosion, Overpressure Wave, or Blast   A Vapor cloud explosion occurs when a liquid vapor, gas or HVP fluid cloud ignites sometime after its formation (i.e., delay ignition) with either a release or obstacles interacting with expansion flow ahead of the flame and consequently, creating an overpressure wave or blast. As the explosion can happen at any location, the blast distance is typically added to the thermal radiation distance (Figure 5.11) in order to define the potential consequence areas [18].

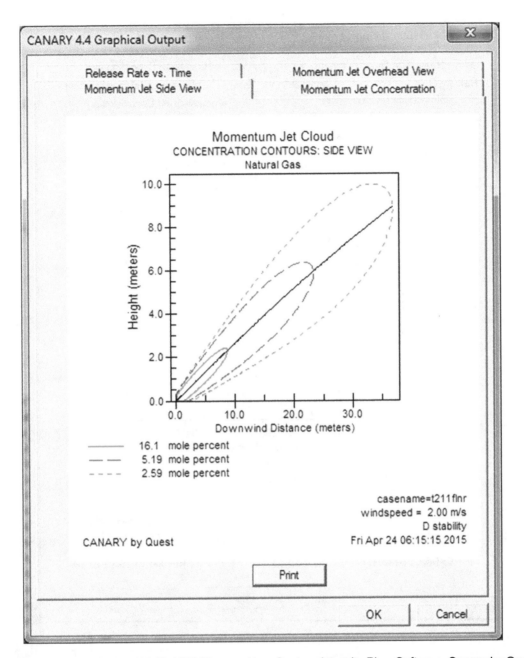

FIG. 5.10 MOMENTUM JET CLOUD AND FLARE (*Source:* New Century Integrity Plus, Software: Canary by Quest Consultants)

Vapor cloud explosions require special conditions to trigger the overpressure wave or blast. Otherwise, a vapor cloud fire, Jet fire, or fireball would occur. The following are some of those special conditions [6]:

- Fluid under pressure and temperature conditions enabling the creation of a vapor cloud mixture within the Lower and Higher Flammability/Explosion (LFL/HFL) Limits
- Turbulence of the vapor cloud with a delayed ignition resulting from either flame accelerating fast through the vapor cloud (e.g., congested facilities and structures) or high-momentum release generating a detonation. Flame propagation speed through the vapor cloud increases the overpressure blast effects

According to the flame propagation speed, the vapor cloud explosion can be categorized as either a deflagration or detonation. Deflagration occurs when the flame travels at a speed less than the speed of sound. Conversely, detonation occurs when the flames travel at a supersonic speed.

There are multiple models for calculating the Vapor Cloud Explosion overpressure blast. Those models can be classified as either TriNitroToluene (TNT) equivalency or Fuel-Air Blast (FAB) models.

The TNT equivalency models determine the peak overpressure of a vapor cloud explosion in terms of the equivalent mass of TriNitroToluene charge required to produce a similar explosion energy causing the same property damage at a given distance. The TNT hemispherical surface burst graphs established by Kingery and Bulmash in 1984 and adapted by Lees in 1996 are used to correlate the Hopkinson-scaled distance to the peak side-on overpressure, side-on impulse, positive phase duration, and time of arrival of the overpressure wave for explosion touching the ground or only air explosion. The Hopkinson-scaled distance depends on the real distance to the receptor and the TNT charge mass equivalent to the mass and combustion heat of the fuel causing the vapor cloud

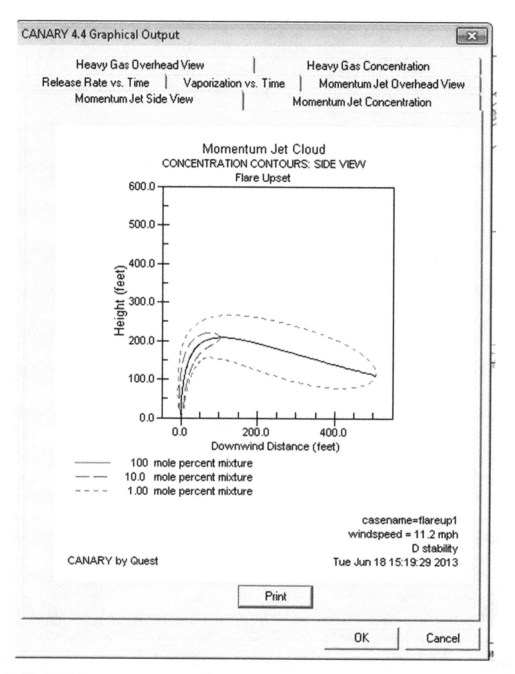

FIG. 5.10 (CONTINUED) MOMENTUM JET CLOUD AND FLARE (*Source:* New Century Integrity Plus, Software: Canary by Quest Consultants)

explosion; for which, conversion factors from chemical (e.g., combustion heat) to mechanical (e.g., blast) are needed.

The TNT equivalency methods are simple and easy to use focusing on high amplitude and short duration shock-waves causing property damage, but not accounting for the fuel combustion rate. Conversely, vapor cloud explosions produce blast-waves of lower amplitude and longer duration. TNT equivalency method overestimates the pressure blast near the explosion source and underestimates the pressure blast far from the source.

The Fuel-Air Blast (FAB) methods determine the blast overpressure at a given location accounting for the vapor cloud explosion mechanism and boundary conditions including confinement and obstacles. The following are some of FAB methods developed for predicting vapor cloud explosion overpressure: the Multi-Energy

Method (MEM), Baker-Strehlow-Tang (BST), Confinement Assessment (CAM) and Numerical Methods.

The Multi-Energy Method (MEM) has provided an improvement over the TNT equivalency methods as the approach considers the possibility of multiple blast or explosion sub-sources within the same vapor cloud that may have different confinement or obstruction conditions. Incident and experimental explosions have demonstrated that the total quantity of fuel in the cloud (i.e., combustion energy) has no correlation to the actual blast effects concluding that only portions of the vapor cloud actually exploded.

In the MEM, each explosion sub-source should be evaluated determining the explosion/blast strength associated to the level of confinement (i.e., 1 to 10) considering that this method does not account for directional blast effects. United Kingdom (UK) Health

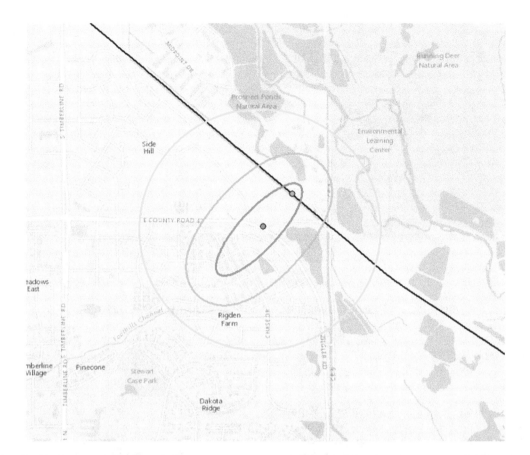

FIG. 5.11 VAPOR CLOUD ½ LFL (RED), THERMAL RADIATION (ORANGE), AND OVERPRESSURE (YELLOW) (*Source:* New Century Integrity Plus, Software: Canary by Quest Consultants)

and Safety Executive (HSE) has defined confined volume as to anything within 5 meters of obstacles. The interpretation of multiple explosion sub-sources causing overpressure, which may be overlapping or occurring at different times, is challenging.

The Baker-Strehlow-Tang (BST) method is based on the MEM, but using flame speed instead of explosion strength. Thus, the overpressure graphs are correlated to flame speed. Both Multi-Energy methods based on either explosion/blast strength or flame speed have challenges in their cloud assumptions (e.g., homogeneous, hemispherical) and definition of blast sub-source areas avoiding non-representative fragmentation or lumping [20].

The Congestion Assessment Method (CAM) model determines the overpressure at a given location from the ignition location using pressure decay curves and accounting for the type of fuel, level of confinement (e.g., walls) and congestion (e.g., obstacles) [21].

Numerical methods use Computational Flow Dynamics (CFD) models simulating the physics of the phenomena through thousands of cells. Those cells follow the Navier-Stokes equation for determining the velocity, pressure, temperature, and density of a moving fluid applying the principles of conservation of mass, momentum, and energy. In addition, these equations are interconnected with models for simulating turbulence and combustion energy.

Reviews of multiple VCE methodologies identifying advantages and limitations recommend using multiple methods ensuring applicability to the specific conditions is confirmed and the results can be compared and interpreted [22].

5.4.5.2 Fireball Fireball is a burning fuel-air cloud in the form of somewhat spherical radiant heat [18], if the release ignites immediately; otherwise, a jet fire would develop. While the inner core

of the cloud consists almost completely of fuel, the outer layer consists of a flammable fuel-air mixture. Due to the buoyancy forces of hot gases and discharge momentum, the burning cloud rises, expands, and takes a spherical-like shape dispersing quickly (Figure 5.12).

The thermal radiation of a Fireball depends on the size of the fireball (i.e., mass, diameter), the distance to the receptors (horizontal distance and height or lift off), their orientation to the flame (i.e., View Factor), the Surface Emissive Power (SEP) and the atmospheric transmissivity (i.e., water vapor pressure and ambient temperature).

Thermal radiation from Fireballs is modeled determining the correlations between fireball mass (i.e., mass consumed is equal to the flow from pipeline) and duration, typically less than 30 seconds, accounting for the stoichiometric combustion substance specific factor. The fireball radius can be determined by its mass assuming a spherical shape. The lift-off height of the fireball is typically calculated by a factor (i.e., ≥2) of the fireball radius. Then, the distance from the center of the fireball to the receptor is a diagonal distance resulting from the vertical lift-off height and horizontal projection on the ground to the receptor [23].

The thermal radiation from a fireball at a given receptor using a spherical shape can be determined with the following conceptual process:

1. Calculate the mass of the flammable material as a function of the flow rate
2. Establish the heat of combustion and vaporization contribution to net heat of radiation
3. Determine the radius and lift-off height of the fireball and associated time/duration

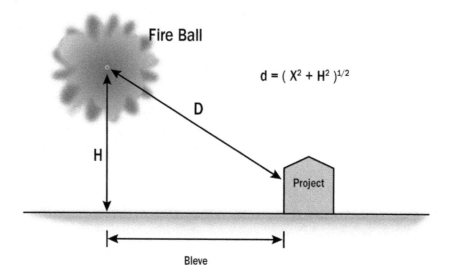

Fire Ball

$$d = (X^2 + H^2)^{1/2}$$

D

H

Project

Bleve

FIG. 5.12   FIREBALL DIMENSIONS AND DISTANCE TO RECEPTOR

4. Determine the Surface Emissive Power, View Factor (i.e., spherical shape) and Atmospheric Transmissivity
5. Calculate the thermal radiation from the Fireball at a given distance/receptor

5.4.5.3 BLEVE Boiling Liquid Expanding Vapor Explosion   Pipeline rarely contains pressure vessels, which originate BLEVE; however, some high level information is provided for complementing the fire modeling section. BLEVE is an explosion as the consequence of the catastrophic rupture of a pressure vessel containing a liquified gas. The sudden depressurization due to the rupture will lead to an explosive vaporization inside the bulk of the liquid being heated under pressure reaching a boiling point. Blast wave and even shock wave can be generated to have destructive impact on the surroundings and human bodies as well as the projectiles. If the liquid were flammable, a jet fire, pool fire, and fire ball would cause a fire hazard [24].

5.4.6   Toxic Effects during Dispersion: Dose and Lethality

The harm from toxic substances being dispersed depends on the concentration of the fluid + air mixture (e.g., chemical composition), exposure duration, and the receptor vulnerability. The exposure to the chemical can cause acute, chronic or no effects. Acute effects result from one time accidental exposure at high concentrations for a short period; whereas, chronic effects result from repeated exposures at lower concentrations for a longer period [7, 9, 25].

Receptor vulnerability is determined by consequence models of population accounting for the effects of the discharge and dispersion of the release via inhalation, ingestion and body contact (e.g., skin or the mucous linings of the eyes, mouth, throat, and urinary tract) causing asphyxiation and intoxication that may lead to affecting the nervous system and causing injuries and death. Receptor vulnerability may be higher in sensitive populations such as elderly, children, and people with respiratory and cardiovascular difficulties [26].

The toxicity effects of a given dose can be represented by the distribution of the percentage of affected people (y-axis = frequency) versus their response to a given dose (z-axis = vulnerability); however, multiple experiments are required to capture a population response to a multiple dosage of a given toxic chemical.

For predictability purposes, dose-response behavior can be parametrized using either the Probability Plot equation (Probit Function) or the Determination of Harmful Dose such as the Specified Level of Toxicity (SLOT) or the Significant Likelihood Of Death (SLOD).

The Probit represents the trend equation of data points obtained experimentally for a given toxic chemical within a range of doses causing a response in a selected population. The equation is comprised of toxic chemical coefficients and the dose. The Probit coefficients are available on the public domain for the majority of chemicals and can experimentally be obtained from the following conceptual process (Figure 5.13):

1. Determine the response distribution of affected people for a given dose
2. Obtain the response mean and standard deviation for a range of doses by repeating step (1)
3. Construct the Mean+/–Standard Deviation Response versus Dose curve
4. Transform Dose-Response distribution to Cumulative and Percentage Response using a Logarithmic scale
5. Use the Percentage to Probit converter to determine (Probit, Log-Dose) data points
6. Plot the (Probit, Log-Dose) data points determining their trend equation in terms of $Y = k_1 + k_2 \ln (c^n t)$

Conversely, the consequence (i.e., population response) of a toxic chemical dose can be determined identifying the coefficients of the appropriate Probit equation, generically expressed as $Y = k_1 + k_2 \ln (c^n t)$; where the coefficients k_1, k_2, and n can be obtained from multiple sources; and the dose is expressed as the toxic chemical concentration (c) and the exposure time (t). Once the Probit value (Y) is calculated, the percentage of affected people is calculated using the Probit versus Probability (i.e., Percentage) correlation developed by Finney in 1971. This conversion can also be expressed for use in spreadsheets to determine the percentage of people affected by a given Probit value Y.

The UK HSE uses the concept of Toxic Load (A) required to cause a level of harm. Toxic load is expressed as a percentage of lethality typically representing deaths in highly susceptible individuals to the toxic chemicals, as well as individuals requiring medical attention and individuals experiencing severe distress.

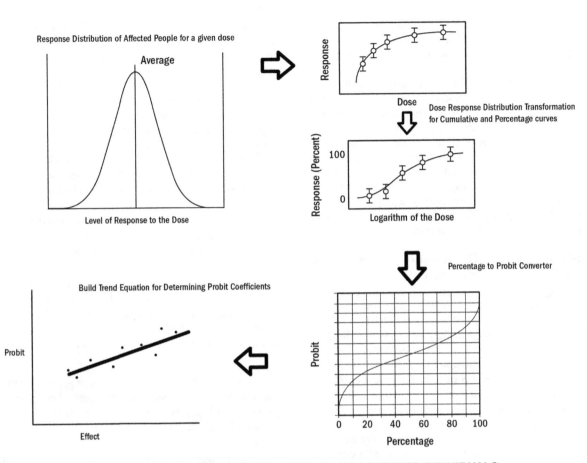

FIG. 5.13 TOXIC DOSE AND PERCENTAGE OF AFFECTED INDIVIDUALS

The toxic load can be defined for specific levels of harms, as follows:

- *Specified Level Of Toxicity* (SLOT) for a toxic load causing a 1 to 5% of lethality (i.e., LD_1 or LD_{1-5})
- *Significant Likelihood Of Death* (SLOD) for a toxicity load causing 50% of lethality (i.e., LD_{50})

5.5 CONSEQUENCE CLASSIFICATION ALONG THE PIPELINE

The extent of the release resulting from the discharge and dispersion analysis requires evaluation to determine the consequence classification based on the affectation criteria defining the type of receptors affected and their level of severity or impact. In the context of Section 5.3 of this book, the following five (5) types of consequences have been used for defining methodologies to classify consequence along the pipeline:

1. *Safety and Health*: employees, landowners in proximity and public in general
2. *Environmental*: air, land and water including habitat (e.g., flora and fauna)
3. *Community*: transportation (e.g., road, rail, navigable water), utilities (e.g., water, electricity, gas)
4. *Regulatory*: penalties, fines and enforcement
5. *Financial*: repair and clean-up, insurance premium (i.e., direct cost)

The following methodologies and their combination used by some pipeline industry standards and regulations focus on one (1) or more type(s) of consequences in classifying consequence along the pipeline:

a. Dwelling Count and Concentration: *Safety & Health*
b. Population Density and Concentration: *Safety & Health*
c. Potential Impact and Release Extent: *Safety and Health, and Environment*
d. Public Impact: *Community and Regulatory*
e. Financial Impact: *Direct Cost*

5.5.1 *Safety and Health:* Classification by Dwellings Count and Concentration

This methodology uses an indirect Safety and Health approach in assessing consequences by counting dwellings (e.g., houses, buildings intended for human occupancy) within a window defined by a length along the pipeline (e.g., 1 mile or 1.6 km long) and a fixed width on either side of the pipeline centerline (i.e., 0.2 km or 1/8 mile from centerline) [27]. In addition, the method uses a fixed concentration of people of 20 persons or greater with an occupancy frequency defined as either "normal use" [1] or a number of days a week per a given period (e.g., PHMSA CFR 192.5).

The window was arbitrarily defined based on the view reach of a fixed wing airplane and its operator spotter's ability to count dwellings within the predefined window; however, the use of aerial and satellite photography has facilitated a more accurate counting. The dynamic window is moved along the centerline until it captures the highest frequency of dwelling count to identify the worst case class location bucket or designation that pipeline point belongs to (e.g., 10 or less dwellings for Class 1; 11 to 45 for Class 2; 46 or greater for Class 3; and buildings with 4+ stories for Class 4).

Some standards recommend considering places of public assembly with high people concentration evacuating at slower rates such as churches, schools, nursing homes, casinos, multiple apartment complexes that may require higher class location than the designation determined by counting. Higher class location designation means lower maximum allowed operational stress or higher safety factor.

The indirect nature of the dwelling calculation for estimating the population density does not necessarily capture the extent and severity of the pipeline fluid release on air, land, and water beyond the fixed window width. The method is not applicable for liquid pipelines crossing elevated areas (e.g., hill, mountains) and water crossings as well as large diameter and/or high pressure liquid and gas pipelines. In addition, the method does not capture the impact on the public associated to their transportation via roads, railways and navigable waters, and their utilities supply (e.g., potable water, electricity, gas and sewer) and their need to evacuate due to other potential hazards (e.g., fire, explosion, air poisoning).

5.5.2 *Safety and Health:* Classification by Population Density and Concentration [28]

This methodology uses a direct Safety and Health approach in assessing consequences by determining the population density within a predefined area. The UK IGEM TD/1 and CSA-Z662-11 Annex O techniques for calculating population density are herein described.

In the United Kingdom, the IGEM TD/1 *Steel Pipelines and Associated Installations for High Pressure Gas Transmission* defines three (3) types of Areas: R or *Rural*, S or *Suburban*, and T or *Town*.

IGEM TD/1 provides the formulation for calculating population density by first defining the area as the multiplication of the length of 1.6 km strip centered on the pipeline centerline by the assessment width of 8 times the minimum *Building Proximity Distance* (BPD) for a Type R (i.e., rural designation is used as a default). Then, the population within the calculated area is determined from Census including the counting of persons within dwelling.

Example: for a pipeline with a diameter of 323.9 mm (NPS 12) and a Maximum Operating Pressure (MOP) of 10.03 MPa (100 bar, 1456 psig), the population density (ρ) is calculated using IGEM TD/1, follows:

- The IGEM/TD/1 *Steel Pipelines and Associated Installations for High Pressure Gas Transmission*, the minimum BPD for the given diameter and MOP is 50 m
- The Area (A) for assessment is calculated based on IGEM/TD/1 section 6.7.2
- A = (1600 m)(8)(50 m) = 640,000 m² (64 Ha 0.25 mile²)
- Then, the population (P) is based on the last Census indicating no multi-storey buildings, dense traffic, or underground services for the given area of 64 Ha
- P = 330 persons
- Then, the population density (Pd) is calculated as the average calculated from the total number of people (P_A) within Area (A) divided by the Area (A)
- Pd = P_A/A = 330 persons/64 Ha = 5.15 persons/Ha
- As the population density exceeds 2.5 person/hectare and does not exhibit characteristics of an area Type T or Town, this area can be classified as a Type S or Suburban

In regards to concentration of people, the IGEM TD/1 defines "Sensitive Locations" as locations with difficult evacuation or increased sensitivity to thermal radiation such as hospitals, convalescent homes, nursing homes, seniors' homes, sheltered housing, schools, colleges, buildings with five or more storeys, large

community and leisure facilities and large open air gatherings. These sensitive locations may trigger a change in the Type of Area (e.g., R, S) as well as increase in the protection of the pipeline and Right-Of-Way (e.g., external interference).

In Canada, the CSA-Z662-11 *Oil and Gas Pipeline Systems* Standard—Annex O defines population density within the context of consequence assessment area for the purpose of calculating the expected number of people potentially affected by a natural gas pipeline release that ignited.

CSA-Z662 Annex O.1.5.2.3 clause provides the formulation for calculating population density by first defining the assessment area as the lesser of the two (2) following calculations:

1. A rectangle with a length of 1.6 km and the assessment width along the pipeline centerline
2. A square with sides equal to the assessment width

The assessment width seeks to represent the impact circle diameter caused by a ignited rupture as a function of the pipeline pressure (P) and diameter (D) with the formula of $0.158\sqrt{(PD^2)}$. Please refer to CSA-Z662-11 clause O.1.5.2.3 clause for identifying the advantages and limitations of each calculation.

Example: for a pipeline with a diameter of 323.9 mm (NPS 12) and a Maximum Operating Pressure (MOP) of 10,038 kPa (100 bar, 1456 psig), the population density (ρ) is calculated using CSA-Z662 Annex O, as follows:

- The Assessment Width (Aw) in meters is calculated using the formula of $0.158 \sqrt{(PD^2)}$
  Aw = $0.158 \sqrt{(10.03\text{MPa})(323.9 \text{ mm})^2}$ = 162 m
- The Area (A) for assessment is calculated using two (2) techniques:
  Rectangular, A1 = (1600 m)(162 m) = 259,200 m² (25.92 Ha, 0.10 mile²)
  Square of Aw side, A2 = (162 m)(162 m) = 26,244 m² (2.62 Ha, 0.01 mile²)
- Then, the population (P) is based on the last Census indicating no multi-storey buildings, dense traffic, or underground services for the given area of 25.92 Ha (A1) or 2.62 Ha (A2).
  P_{A1} = 134 persons in an area of 25.92 Ha
  P_{A2} = 24 persons in an area of 2.62 Ha (average affected by concentration in one Building)
- Then, the population density is calculated as the average calculated from the total number of people (P_A) within Area (A) divided by the Area (A)
  $Pd_1 = P_{A1}/A1$ = 134 persons/25.92 Ha = 5.15 persons/Ha
  $Pd_2 = P_{A2}/A2$ = 24 persons/2.62 Ha = 9.16 persons/Ha

The CSA-Z662 Annex O suggests selecting the lesser of the two (2) values defining the population density to be 5.15 persons/Ha. Table O.2 correlates average population density to Class location that would allocate the calculated population density greater than Class Location 2 average (3.3 persons per Ha) and smaller than Class Location 3 average (18 persons per Ha).

The population density calculated using the technique #2 or Pd_2 shows the effects of concentration of people when the area estimation is focused in segments with buildings allocating multiple persons.

5.5.3 Safety and Health, and Environment: Classification by Potential Impact and Release Extent

This methodology uses a combined Safety, Health and Environmental approach in assessing consequences by determining the extent of the release on the air, land and water affecting all organic and inorganic matter and living organisms as well as their interacting natural systems.

Fluids during a pipeline release may behave by changing its phase from liquid phase, liquid-gas phase and gas phase, which makes the previous classification for natural gas and hazardous liquids not entirely either complete or applicable. As discussed in Section 5.4, potential for overpressure and toxicity shall be reviewed based on the fluid characteristics (e.g., dose and lethality), pipeline operation (e.g., temperature, pressure), Discharge-Dispersion transition (e.g., fluid + air mixture, flashing, concentration, dispersion turbulence), and environment (e.g., confinement, population proximity).

In the context of this book, the difference between a *Pipeline Consequence Assessment* (PCA) and an *Environmental Impact Assessment* (EIA) is that the PCA focuses on an accidental release during the operation of the pipeline. Whereas, the EIA can focus on the entire pipeline life cycle (from design to abandonment) identifying changes to the environment, human socio-economic conditions, heritage and use of lands for traditional purposes by Aboriginal people [29].

Furthermore, PCA uses pipeline fluid discharge and dispersion modeling in conjunction with emergency response for determining the areas affected by hazards such as fire, explosion, contamination, poisoning, and asphyxiation. Both EIA and PCA seek mitigative and preventative measures for their own focus or scope.

Pipeline mapping on a Geographical Information System (GIS) platform containing population and environmental data facilitates the process of determining potential impact and release extent. The USA PHMSA has developed a *National Pipeline Mapping System* (NPMS) providing the public, pipeline operators and government officials with information about pipeline location, transported fluid, status (e.g., in-service, idle, out-of-service), and operator information (e.g., company name, contact address, phone, email). NPMS also collects digital geospatial data capturing the location of pipeline centerline with a minimum accuracy of +/−150 m (500 feet) from all regulated pipelines in the USA.

The Canadian Federal regulator, the National Energy Board (NEB), has also built a Geographical Information System with all regulated pipelines for internal use; moreover, The NEB [30] produced an interactive map for graphically presenting the NEB incident database by fluid and event type.

The UK HSE [31] has determined distances from the Major Hazard Pipeline (MHP) that allows specific development (e.g., houses, schools, buildings) identifying where they are located within the zones (e.g., inner, middle or outer) for planning and emergency response purposes.

Reductions in the impact of a release can be achieved by natural and man-made processes and structures. Some natural options for reducing dispersion are biodegradation, fluid evaporation (e.g., photolysis) and mixing (e.g., hydrolysis) that may decrease the harmful fluid volume and concentration of the release. Man-made options for impact reduction can be made during the design, construction, and operation of the pipeline. For instance, volume discharge can be reduced by shorter valve separation and faster valve closure times (i.e., automatic shutoff valves); while the installation of containment ponds and physical barriers can reduce dispersion of spills and effects of fire and explosion.

5.5.3.1 Natural Gas Pipelines

As described in Section 5.4, a release of a natural gas pipeline would disperse as a vapor cloud that depends on fluid and pipeline characteristics as well as operational and weather conditions. The vapor cloud may cause asphyxia along its trajectory; moreover, its exposure to an ignition source would trigger fire and explosion causing safety and health consequences.

The classification methodology for natural gas pipelines estimates the *Potential Impact Circle* affecting primarily human; however, the affectation of the environment or damage of property could also be estimated within this area as significant. This classification methodology uses either a simplified formula (i.e., C-FER/GRI) or a complex computational solution based on case-specific pipeline characteristics and area conditions such as topography and weather.

C-FER developed for Gas Research Institute a simplified model for sizing High Consequence Areas (HCA) that could potentially be affected by a natural gas pipeline failure. The model is based on a dominant hazard of thermal radiation from a sustained jet or trench fire. The model proposed the formula, later accepted by USA DOT PHMSA, for estimating the radius of potential impact with an assumed threshold heat intensity for natural gas of 15.8 kW/m^2 (5,000 BTU/hour-ft^2).

The Figure 5.14 illustrates the potential impact circle for gas pipelines and the model assumptions within the context of San Bruno failure including secondary fire-damage on property resulting from delayed valve closure.

As illustrated in Figure 5.15 [19], the simplified C-FER formula was validated by comparing the proposed HCA radius versus burn extent of actual pipeline failures in the USA and Canada.

Furthermore, the USA DOT PHMSA included in the CFR 192 the concept of *identified site* in an attempt for capturing areas or structures with concentration of people (i.e., 20 persons) and a given occupancy frequency as well as persons confined with limited mobility and/or difficult to evacuate. The occupancy frequency is typically 50 or more days within 12 months, 5 days a week for 10 weeks within 12 months.

In regards to affectation of the environment, the industry standard IGEM TD/1 suggests conducting an Environmental Impact Assessment (EIA). IGEM TD/1 under some regulations (i.e., The Public Gas Transporter Pipe-line Works (Environmental Impact Assessment) Regulations 1999) is required for specific range of diameters (e.g., >800 mm), (e.g., 40 km) and/or pressure (e.g., 7 bar) and accounting for whether the pipeline could affect the public and environment.

5.5.3.2 Hazardous Liquid Pipelines

As described in Section 5.4, a release of a hazardous pipeline would disperse as a liquid and/or vapor depending on fluid and pipeline characteristics as well as operational and weather conditions. The fluid disperses on land, water or air contaminating the environment, evaporates (e.g., air) causing asphyxia and pollutes drinking water affecting human consumption; moreover, its exposure to an ignition source may cause fire and explosion also causing safety and health consequence.

The classification methodology for hazardous liquid pipelines estimates the *Release Extent* applied on either *Direct* or *Indirect* consequence areas (Figure 5.16). The direct areas are along the Right-Of-Way (ROW) affected by the release; whereas, indirect areas are not in vicinity of the ROW, but affected by fluid

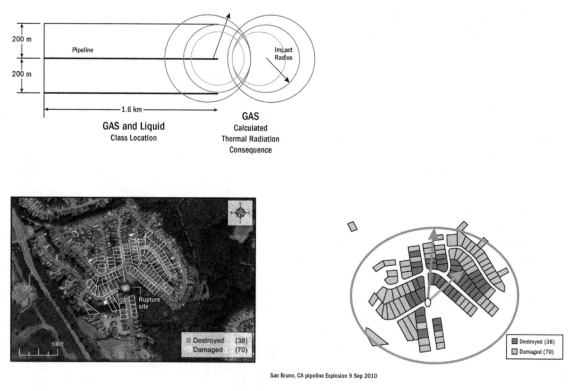

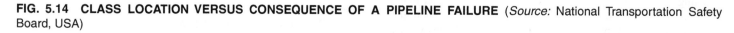

Post Rupture Image

FIG. 5.14 CLASS LOCATION VERSUS CONSEQUENCE OF A PIPELINE FAILURE (*Source:* National Transportation Safety Board, USA)

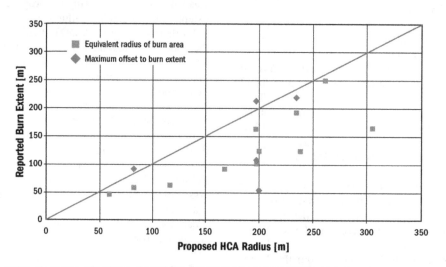

FIG. 5.15 PROPOSED HCA RADIUS VERSUS ACTUAL PIPELINE FAILURES IN THE USA AND CANADA (*Source:* Stephens, M.J., Leewis, K., Moore, Daron K., 2002, A Model for Sizing High Consequence Areas Associated with Natural Gas Pipelines, ASME International Pipeline Conference, Paper IPC 2002-27073, Calgary, AB, Canada)

transported to far away distances via water bodies or geographical channels. This classification methodology uses a computational solution based on case-specific pipeline characteristics and area conditions such as topography and weather.

Populated or environmentally-sensitive areas are determined as a *Direct Consequence Area* if they are intersected by a pipeline meaning that they may be directly affected by a potential pipeline release. Direct consequence areas can be denominated as high,

medium, or low depending on the criteria used for each one by the regulatory jurisdiction or in absence of definition, by the company.

For instance, from the population standpoint, the USA PHMSA defined High Consequence Area (HCA) as the urbanized area that contains 50,000 or more people and has a population density of at least 1,000 people per mile^2. Furthermore, HCA included the Other Populated Area (OPA) concept defined as a concentrated population, such as an incorporated or unincorporated city, town, village, or other

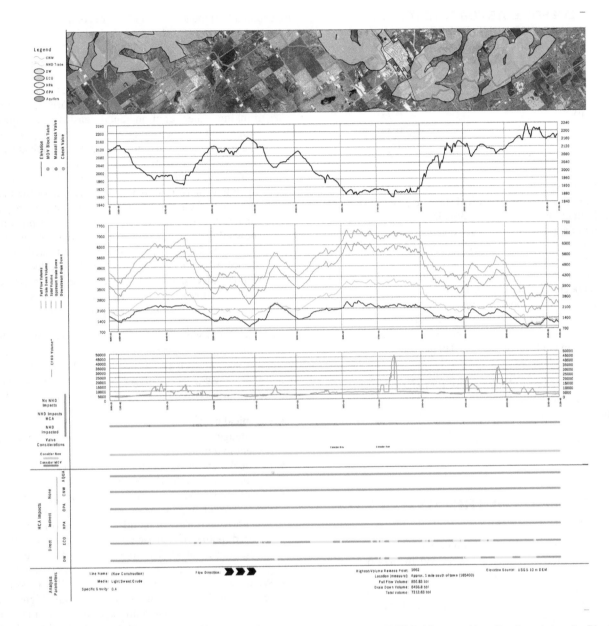

FIG. 5.16   SAMPLE OF DIRECT AND INDIRECT CONSEQUENCE AREAS (*Source:* New Century Integrity Plus)

designated residential or commercial area. For instance, from the environmental standpoint, the High Consequence Area (HCA) definition included the unusually sensitive areas (USA, e.g., drinking water), National Fish Hatcheries, and commercially navigable waterways.

Indirect Consequence Areas are determined by computational modeling tracing the fluid spill on water bodies or channels (e.g., farm drain tiles, irrigation systems, rain water) until the location where emergency response team is able to reliably stop it. The categorization of high, medium or low depends on the area type (e.g., HCA) reached by the transported fluid.

5.5.3.3 Other Fluid Pipelines   USA DOT RSPA OPS retained Baker [32] to conduct a study for determining the Potential Impact Circle formula for gases other than natural gas (i.e., methane). Even though formulae were provided for typical gases (i.e., Ethylene, Hydrogen, rich natural gas and synthesis gas) transported in the

USA, the lack of validation either full scale testing or incident for those gases did not allow for confirming their validity, but requiring more study. Table 5.1 depicts some coefficients for different types of gases that can be used the Potential Impact Radius formulae [32].

5.5.4   *Public Impact Consequence*: Community and Regulatory

The public as a community can be impacted directly and indirectly from a pipeline release. Communities may need to be evacuated or their access roads to be closed and re-routed due to potential or actual hazards (e.g., fire, explosion), environmental effects (e.g., air to breath, water to drink), and human outcomes in the short (e.g., panic, violence, vandalism) and long term (e.g., post-trauma).

In the context of this book, the extent of the public impact consequence can be measured by the level of disruption (e.g., evacuation, road blockage, human outcomes) and its duration (e.g., days,

TABLE 5.1   COEFFICIENTS FOR POTENTIAL IMPACT RADIUS FORMULAE FOR SOME GASES (*Source:* Courtesy of USA Department of Transportation)

Fluid	PIR Coefficient
Natural gas (lean)	0.69
Natural gas (rich)	0.73
Hydrogen	0.47
Ethylene	1.04

weeks, months, years) in every given community. Determination of the level of media coverage at the critical stage, as fades away over time, may help in confirming the estimated level of public impact. Quantification of the public impact consequence is challenging and sometimes can be only assessed qualitatively such as low, medium, high, and extreme.

Regulatory consequences from a pipeline release come from letters of warning and notices of violation, corrective action orders, pressure restrictions, administrative monetary fines and penalties, and augmented regulatory scrutiny (e.g., information requests, inspections, applications to new/modified installations) on the pipeline operator due to the increase in the level of risk posed to the public.

5.5.5   *Financial Impact Consequence*: **Direct Cost**

In the context of this book, financial impact is represented by the direct cost paid by the pipeline company responsible for the pipeline release. The direct cost includes the costs for emergency response, repairing the pipeline, cleaning-up the spill, restoration of the environment and associated monitoring, and lawsuits fees and damages.

5.6   VALIDATION OF THE CONSEQUENCE CLASSIFICATION

Validation of the results of the consequence assessment process including analysis and classification, described in previous sections, recognizes that the content and quality of the data varies from source to source. Accuracy variation in the data may result in incorrect and/or incomplete extent and classification of High, Medium or Low consequence areas [33].

Dispersion models may reach wrong locations due to incomplete assumptions on geographical barriers (e.g., culverts, roads) as well as emergency response times or spill on water flows (e.g., faster under worse storm/rain conditions). Maps with consequence areas (e.g., water intakes, campgrounds, endangered species, farm drain tiles) may not be up-to-date or not properly delineating the true boundaries due to update cycle or licensing issues.

Methods for validation of the consequence classification assigned to an area include field visits, private vendors with higher resolution aerial and satellite imaging, and remote sensing and geospatial technology such as Light Detection and Ranging (LIDAR), radar or electromagnetic radiation reflected or emitted by an object (e.g., Landsat).

5.7   CONSEQUENCE ASSESSMENT RESULTS: CONTENT AND FORMAT

The consequence assessment results can be presented in multiple formats ensuring that their input is dated identifying the source for validation and future updates, the methodology and assumptions are detailed for repeatability and future comparison, and the results presented according to the engineering, integrity and regulatory requirements.

The integrity requirements include the following:

1. Pipeline listing should identify the start and end of each consequence area either direct or indirect area (e.g., population, drinking water, environmental sensitive, concentration of people and water crossings).
2. The listing should also include reference points such as stations, valves, crossings, and landmarks.
3. Maps with a minimum scale of 1:25,000 centered at the pipeline centerline showing the envelope of the plume/spill areas up to their maximum extent (e.g., emergency response capability). An overview map of the entire pipeline should show all the individual maps for reference purposes.
4. Listing should list all release points simulated in the consequence assessment indicating the pipeline location, released volume, affected receptors, and the time for the release to reach them. As a minimum, a listing of the worst-case scenarios per consequence area or pipeline segment should be prepared.

The USA Environmental Protection Agency (EPA) published the *"Risk Management Program Guidance for Offsite Consequence Analysis"* providing a list of information to be submitted as a result of an offsite consequence analysis for the worst-case scenarios. The list issued in March 2009 has been included for reader's benefit.

- Chemical name;
- Percentage weight of the regulated liquid toxic substance (if present in a mixture);
- Physical state of the chemical released (gas, liquid, refrigerated gas, gas liquefied by pressure);
- Model used (OCA or industry-specific guidance reference tables or modeling; name of other model used);
- Scenario (gas release or liquid spill and vaporization); Quantity released (pounds); Release rate (pounds per minute); Duration of release (minutes) (10 minutes for gases; if you used OCA guidance for liquids, indicate either 10 or 60 minutes);
- Wind speed (meters per second) and stability class (1.5 m per second and F stability unless you can show higher minimum wind speed or less stable atmosphere at all times during the last three years); Topography (rural or urban); Distance to endpoint (miles, rounded to two significant digits); Population within distance to endpoint (residential population rounded to two significant digits);
- Public receptors within the distance to endpoint (schools, residences, hospitals, prisons, recreation areas, commercial, office or industrial areas);
- Environmental receptors within the distance to endpoint (national or state parks, forests, or monuments; officially designated wildlife sanctuaries, preserves, or refuges; Federal wilderness areas); and
- Passive mitigation measures considered (dikes, enclosures, berms, drains, sumps, other).

5.8   REFERENCES

1. Anon., 2015, *CSA-Z662 Oil and Gas Pipeline Systems*, Canadian Standard Association, Ontario, Canada.

2. Anon., 2000, *API 581 Risk-Based Inspection, Base Resource Document*, American Petroleum Institute, USA.

3. Federal Emergency Management Agency (FEMA), 1990, *Handbook of Chemical Hazard Analysis Procedures*, Maryland, USA.

4. Chandrasekaran, S., 2012, *Dispersion Models—Atmospheric Pollution—Module 2: Lecture 5*, Department of Ocean Engineering, Indian Institute of Technology Madras, Chennai, India.

5. Gin, G., Davis, J., 2012, *Spill Impact Assessment—Crude Oil Pipeline ERCB #3353-1*, ASME International Pipeline Conference, Calgary, AB, Canada.

6. Van der Bosch, C.J.H., Weterings, R.A.P.M., 2005. *"Yellow Book" Methods for the Calculation of Physical effects due to releases of hazardous materials (liquids and gases)*, TNO—The Netherlands Organization of Applied Scientific Research, The Hague, Netherlands.

7. International Association of Oil & Gas Producers (OGP), 2010, *Consequence Modeling*, OGP Risk Assessment Data Directory Report 434-7, England and Wales, United Kingdom.

8. Tweeddale, M., 2003, *Managing Risk and Reliability of Process Plants*, Gulf Professional Publishing—Imprint of Elsevier Science, Burlington, MA, USA.

9. CCPS, 1999, *Guidelines for Consequence Analysis of Chemical Releases*, Center for Chemical Process Safety (CCPS), American Institute of Chemical Engineers, USA.

10. Perry, R., 1997, *Perry's Chemical Engineering Handbook*, 7th edition, McGraw-Hill, USA.

11. Covello, V.T., Merkhoher, 2005, *Risk Assessment Methods: Approaches for Assessing Health and Environmental Risks*, Plenum Press, USA.

12. Tam, V.H.Y., Higgins, R.B., 1990, Potential impact simple transient release rate models for releases of pressurised liquid petroleum gas from pipelines, *Journal of Hazardous Materials*, Elsevier, Volume 25, Issue 1, pp. 193–203, NY, USA.

13. Economou, I., Mahgerefteh, H., Oke, A., Rykov, Y., 2003, Potential A transient outflow model for pipeline puncture, *Chemical Engineering Science*, Elsevier, Volume 58, Issue 20, pp. 4591–4604, NY, USA.

14. Muhlbauer, 2015, *Pipeline Risk Management: The Definitive Approach and Its Role in Risk Management*, Clarion Technical Publishers, Houston, TX, USA.

15. Cote, P., *GIS Manual—GIS Tutorials and Resources*, Harvard University Graduate School of Design, http://www.gsd.harvard.edu/gis/manual/dem/.

16. Steenbergen, R.D.J.M., van Gelder, P.H.A.J.M., Miraglia, S., Vrouwenvelder A.C.W.M., 2013, *Safety, Reliability and Risk Analysis: Beyond the Horizon*, CRC Press, Florida, USA.

17. Environmental Protection Agency (EPA), 2009, *Risk Management Guidance for Offsite Consequence Analysis*, Washington, USA.

18. CCPS—AIChE, 2010, *Guidelines for Vapor Cloud Explosion, Pressure Vessel Burst, BLEVE and Flash Fire Hazards*, 2nd Edition, Center for Chemical Process Safety, New York, USA.

19. Stephens, M., 2000, *A Model for Sizing High Consequence Areas for Natural Gas Pipelines*, C-FER Technologies for Gas Research Institute #8174, AB, Canada.

20. Ryder, N.L., Schemel, C.F., Mannan, S., 2009, *Part 1 Design Requirements: A Comparison of Vapour Cloud Explosion Models and the Importance of Properly Assessing Potential Incident Impact*, Proceedings of the Mary Kay O'Connor 2009 International Symposium, College Station, Texas, USA.

21. Puttock, J.S., 1995, *Fuel Gas Explosion Guidelines—The Congestion Assessment Method*, Shell and Thornton Research Centres, ICHEME Symposium Series No. 139, Institution of Chemical Engineers (IChemE), Rugby, United Kingdom.

22. Lea, C.J., Ledin, H.S., 2002, *A Review of the State-of-the-Art in Gas Explosion Modelling*, UK Health and Safety Executive, Report HSL/2002/02, United Kingdom http://www.hse.gov.uk/offshore/strategy/hazard.htm#Modelling.

23. Kinsman, P., Lewis, J., 2002, *Report on a second study of pipeline accidents using the Health and Safety Executive's risk assessment programs MISHAP and PIPERS*, Research Report #036, Health and Safety Executive, United Kingdom.

24. Mengmeng, X., 2007, *Thermodynamic and Gas Dynamic Aspects of a BLEVE*, Faculty of Applied Sciences Delft University of Technology, The Netherlands.

25. UK Health and Safety Executive (HSE), *Methods of approximation and determination of human vulnerability for offshore major accident hazard assessment* http://www.hse.gov.uk/foi/internalops/hid_circs/technical_osd/spc_tech_osd_30/spctecosd30.pdf.

26. Withers, R.M.J., Lees, E.P., 1985, *The Assessment of Major Hazards, Journal of Hazardous Materials*, Department of Chemical Engineering, Loughborough University of Technology, Loughborough, Leicestershire, Great Britain.

27. Anon., 2012, *ASME B31.8 Gas Transmission and Distribution Piping Systems*, American Society of Mechanical Engineer, ASME Press, NY, USA.

28. Goodfellow, G., Haswell, J., 2006, *A Comparison of Inherent Risk Levels in ASME B31.8 and UK Gas Pipeline Design Codes*, IPC2006-10507, Proceedings of the 2006 International Pipeline Conference, American Society of Mechanical Engineer, Calgary, AB, Canada.

29. Government of Canada, 2012, *Canadian Environmental Assessment Act*, http://lois.justice.gc.ca/eng/acts/C-15.21/index.html.

30. National Energy Board, 2015, Interactive Incident Map, https://www.neb-one.gc.ca/sftnvrnmnt/sft/dshbrd/mp/index-eng.html.

31. UK HSE, 2007, *HSE Land Use Planning Zones for Major Hazard Sites and Pipelines*, UK Health and Safety Executive (HSE), United Kingdom.

32. Baker Inc., C-FER Technologies, 2005, *Potential Impact Radius Formulae for Flammable Gases Other Than Natural Gas*, Technical Task Order Number 13 (TTO 13), USA Department of Transportation (DOT) Research and Special Programs Administration (RSPA) Office of Pipeline Safety (OPS), USA.

33. Freeman, B., 2005, *HCA Validation Optimizes Pipeline Integrity*, Geospatial Information & Technology Association (GITA), USA, http://www.freemangis.com/written-works/.

RISK ASSESSMENT FOR PIPELINES

Chapter 6 provides the concepts, elements, process, and methodologies for assessing the integrity risks in pipelines. The chapter starts defining risk assessment and control as elements of pipeline risk management allocating its processes within a PLAN-DO-CHECK-ACT or management system framework. Core risk concepts such as perspectives toward risk, uncertainty, and conservatism, individual and societal risk, and risk aversion are also explained to holistically understand the pipeline risk results.

The purpose and objectives of the risk assessment is described following by some examples of applications and outcomes. The types of methodologies for assessing risk such as Qualitative, quantitative and semi-quantitative are explained describing their characteristics, advantages, limitations and some graphical examples.

The overall process of risk assessment including their components and some guidance for planning, data gathering, and integration are described. Furthermore, sensitivity analysis for identifying risk reduction options, risk assessment frequency, and documentation are explained.

Each of the risk assessment components (i.e., analysis, estimation and validation, and evaluation), is explained following their sequential order along the chapter till to the end.

6.1   INTRODUCTION TO PIPELINE RISK MANAGEMENT

Risk management is an integrated management process identifying, assessing, preventing, monitoring, controlling, and minimizing or eliminating an unacceptable risk from hazards and threats and associated probability of an event and its associated consequences. Pipeline risk management is necessary for timely and effectively managing pipeline system assets (e.g., pipeline, components, equipment and any appurtenances) to achieve a safe, reliable and sustainable pipeline operation protecting the environment.

Example: Decision-Making Processes Applying Pipeline Risk Management Principles

- Design
- Construction
- Operation
- Maintenance and integrity
- Class location changes
- Incidents
- Deactivation, reactivation and abandonment

Regulatory bodies, government agencies, pipeline safety advocates, and pipeline operators have promoted the continued development and implementation of pipeline risk management. It assists both regulatory and pipeline industry goals in improving public safety and environment protection by optimizing the aspects of pipeline design, operation and maintenance. Because of this, the risk management programs have been tailored specifically to satisfy those needs.

Managing pipeline integrity risks is an integrated and iterative process. The management process should also ensure that measures are timely integrated into company's daily operation. These processes are aimed to continuously

- Identify hazards (conditions) and integrity threats
- Analyze likelihood of failure and potential consequences
- Estimate and evaluate risk
- Determine the measures for reducing risk
- Verify the effectiveness of applied measures
- Measure identifying continuous improvements

Risk (Figure 6.1) increases when either the probability of a failure increases or the consequences of an event (magnitude of the potential loss) increases. Pipelines are subject to different hazards, threats, and damage mechanisms through all phases of the pipeline life. Pipeline integrity threat assessment is discussed in more detail in Chapter 4. The integrity threats associated with pipelines may induce risks. Pipeline failures will also have impacts in the company on both direct costs and revenue, and indirect costs such as company's public perception, reputation and the regulatory scrutiny.

The threat probability may be reduced by implementing effective assessment, monitoring, prevention, and mitigation plans.

Example: Probability Reduction Alternatives

- Reduce operating pressure
- Repair anomalies
- Pipeline re-routing/lowering
- Change operational procedures
- Install protections or system modifications

The consequence is typically increasing as opposed to decreasing. Change of the class location designation (e.g., population increase) as to going into a higher-class designation is not controlled by the pipeline operator. Pipeline consequence assessment is discussed in more detail in Chapter 5.

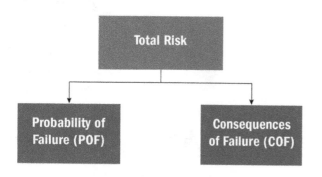

FIG. 6.1 RISK: PROBABILITY AND CONSEQUENCE

Example: Consequence Increase Cases

- New campground area
- New dwellings (e.g., houses, offices, factories)
- Development of industrial areas

Conducting only mitigation actions cannot eliminate risk. Implementation of an effective risk management process can assist in reducing pipeline failures by timely identifying high risk pipeline segments (risk assessment) and defining measures (risk control). Those measures are discussed in Chapter 11 followed by the Fitness-For-Service assessment in Chapter 12.

Risk management reinforces the need for pipeline operators to propose risk reduction alternatives, estimate risk values, evaluate (for each alternative) risk profiles, make safety decisions aimed to control unacceptable risk, and implement corrective actions.

The resulting risk reduction programs applied across an entire company allows integrity staff and senior management manage local and enterprise risks through a process that is employed with fidelity. The management review process should verify that the measures have achieved an acceptable residual risk meeting the expectations of all stakeholders.

6.1.1 Risk Management Process

Risk management is a process of strategizing appropriate risk reduction measures and implementing them in the on-going management of the pipeline activities following with a continuous improvement process.

As illustrated in Figure 6.2, the risk management process can be framed with the Quality Management System (QMS) operational principle of PLAN-DO-CHECK-ACT. The continuous improvement process on risk involves defining strategies and processes, setting up objectives, goals, and specific targets with well-defined risk performances, and establishing high level of commitment from senior management and all level of the pipeline organization.

Risk assessment process is an overall process of hazard and threat identification, risk analysis and risk evaluation to increase reliability of a pipeline system. Risk analysis is a structured process for estimation of risk determined from the likelihood and consequence of pipeline failure associated to identified hazards and receptors, respectively. Risk evaluation is the process to review acceptability of risk based on comparison with risk standards or criteria, and the trial of various risk reduction measures. Risk control is a process of reducing risk by preventing, mitigating and monitoring hazards associated to consequence impacts, and communicating risk management [1].

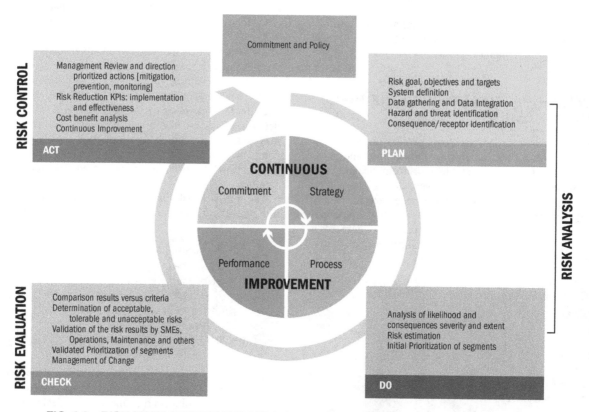

FIG. 6.2 RISK MANAGEMENT PROCESS WITH A MANAGEMENT SYSTEM APPROACH

As illustrated in Figure 6.3, risk management essentially embraces two main processes, risk assessment and risk control.

1. **Risk Assessment** is conceptually comprised the following components,
   a. **Risk Analysis and Estimation**
      Planning
      • Risk goal, objective, and target definitions
      • System definition
      • Data gathering
      • Data integration
      • Hazards and threats, and consequence/receptor identification
      Implementation (Do)
      • Analysis of likelihood and consequence severity and extent
      • Risk estimation
      • Initial prioritization of the pipeline segments
   b. **Risk evaluation**
      Verification (Check)
      • Comparison of risk results versus acceptance criteria
      • Determination of acceptable, tolerable and unacceptable risks in the systems
      • Validation of the risk results from multiple cold eye review approaches such as Risk Subject Matter Experts (SMEs), pipeline operations and maintenance and industry performance
      • Validated prioritization of the pipeline segments
      • Management of change
2. **Risk control** includes the following,
   Management Review (Act)
   • Management review and direction
   • Prioritized actions: mitigation, prevention and monitoring
   • Cost benefit analysis
   • Risk reduction KPIs: implementation and effectiveness
   • Continuous improvement

6.1.1.1 Perspectives Toward Risk   Perspectives of risk such as likelihood, consequences, outcomes, significance, casual scenario, and population may differ from case to case. These differences depends on level of tolerability, acceptability, perception, analysis, evaluation, pipeline systems location, stratification, jurisdictional regulations, and modeling including data interpretation, accuracy and completeness of risk assessment. As illustrated in Figure 6.4, these perspective differences are sources of debate among society,

special interest groups, individuals, government, regulators and agencies and pipeline companies.

6.1.1.2 Uncertainty and Conservatism   Uncertainty and conservatism during the estimation and evaluation of risks are challenges in the risk assessment that require a reasonable balance. Risk assessment incorporating a high or low level of conservatism will contribute to overestimate or underestimate the risk. Some collected data (i.e., ILI-reported anomaly sizing, detection and characterization, field measurement of flaws, operating over-pressures, etc.) would contain uncertainty that needs to be estimated or quantified (e.g., probability density functions) building a reasonable conservatism in estimating risk.

Uncertainty arises from risk factors such as limitations of ILI tools, pipe strength, wall thickness measurements, depth of cover, pressure reading, cathodic protection readings, pipe coating conditions, growth rate modeling. Uncertainty in the pipeline degradation and changes of pipeline conditions over time make risk professionals more conservative when estimating risk. Quantifying uncertainty using frequency distributions of events within the risk algorithm may increase the risk assessment credibility.

It is paramount to determine the level of conservatism and uncertainty suitable to pipeline condition using substantiated assumptions to enhance the risk assessment. The approach is not to report a larger magnitude of unrealistic risk and apply intentional and non-substantiated bias in the process of risk assessment.

6.1.1.3 Individual Risk   Individual risk is the measure of risk as perceived by a specific individual that might be located near of the pipeline during a pipeline incident assuming that the individual is present 100% of the time. CSA Z662 bases the individual risk calculation on the annual probability of the fatality of an individual located within the pipeline hazard zone during an incident.

The US Federal regulation 49 CFR 192 [2] and the ASME B31.8S [3] standard describe the potential impact zone, which is considered as a surrogate for consequence as a function of the outside diameter, actual maximum operating pressure and a coefficient for natural gas pipelines.

The Major Industrial Accidents Council of Canada (MIACC) issued risk-based land use planning guidelines, which provide procedures on the use of risk assessment in respect to the development of land use plans for municipalities and industry. MIACC proposes acceptable levels of individual risk in terms of acceptable land uses. The recommended acceptable levels of public location risk

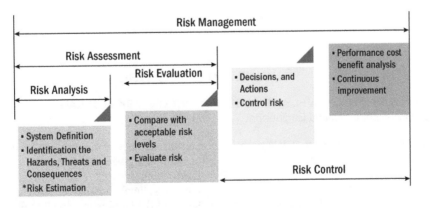

FIG. 6.3   RISK MANAGEMENT: RISK ASSESSMENT AND CONTROL

FIG. 6.4 FIVE (5) PERSPECTIVES OF RISK

Annual Individual Risk
Chance of fatality per year

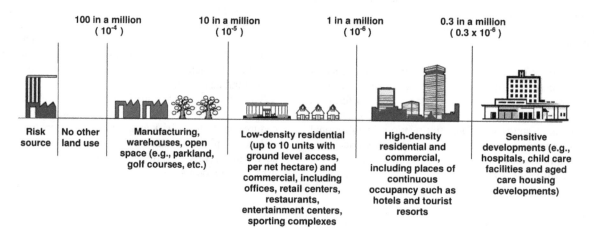

Allowable Land Uses

FIG. 6.5 RECOMMENDED ACCEPTABLE LEVELS OF PUBLIC LOCATION RISK FOR LAND USE (*Source:* Chemical Institute of Canada [CIC] and Canadian Society of Chemical Engineering [CSChE])

for land use based on the MIACC proposed by the CIC/CSChE Process Safety Management division are illustrated in Figure 6.5.

6.1.1.4 Societal Risk Societal risk measures the overall risk where consequences considered is a function of the expected of fatalities occurring due to a pipeline failure. Societal risk is defined as the relationship between frequency of number of fatalities (F)

and the number of fatalities (N) suffering from a specified level of harm in a given population from the consequence of hazardous fluids being released from a pipeline. Examples of sites that could present societal risk include natural gas pipeline, liquefied petroleum gas (LPG) pipeline, and liquified natural gas (LNG) pipelines.

Societal risk evaluation is focused on the estimation of the casualties of more than one individual being harmed simultaneously by a pipeline incident. Societal risk usually takes into account the actual population density sites to characterize the number of individuals at a given time that could suffer a specified injury or fatality due to a pipeline incident. The societal risk is evaluated as the annual probability of failure in an evaluation length.

Example: The evaluation length could be determined as any area in Class 2, Class 3 or Class 4 location where the potential impact radius (PIR) is greater than 200 meters and/or any area within a potential impact containing 20 or more buildings intended for human occupancy. The evaluation length could also be any 1600-m length successive window of a class location-designation area along the pipeline segment.

Societal risk can be depicted graphically, in the form of F-N curves or numerically, or in the form of a risk integral. F–N diagram is used to represent the historical record of incidents [4]. *N*, is the predicted number of fatalities or persons harmed associated with the event while *F*, is the predicted frequency of occurrence.

Mathematically, the equation for an *F-N* criterion curve may be presented as:

$$R = F \times N^a$$

Where,

   F = the frequency of N or more fatalities
   N = the number of fatalities
   a = aversion factor (often between 1 and 2)
   R = constant

A pipeline operator may adopt societal risk criteria issued by governmental bodies at the country, regional (e.g., provincial or state) and local level, or criteria developed by industry. Similarly, pipeline operator companies may establish company-specific societal risk criteria based on historical data, if available.

The UK developed risk criteria for advising on land use planning to local planning authorities to address both individual risk and societal risk. *Guidelines for Developing Quantitative Safety Risk Criteria* book provides additional guidance for worldwide risk criteria developed by governmental bodies including societal and individual risk criteria for process management decisions. Figure 6.6 provides UK HSE maximum tolerable societal risk and *As Low As Reasonably Practicable* (ALARP) criteria [4, 5].

ALARP is accepted in some jurisdictions not all as a principle for managing risk. In some cases, risk is reduced regardless of the high level of resources, efforts, and costs invested in. Then, pipeline companies may need to identify the best benefit-cost scenario (i.e., pipeline replacement, repair and/or pipeline operation at reduced pressure) to attain their goals and objectives (i.e., the point where the costs exceed benefits) to achieve a level of acceptable risk.

The above figure illustrates three (3) main risk regions:

1. *Unacceptable,*
2. *Tolerable if ALARP*, and;
3. *Broadly Acceptable* region.

The unacceptable risk region contains the level of risk that is not tolerable requiring mitigation measures at any cost to continue operation. The ALARP region is the level of risk that can be tolerated, but can be further assessed to find a risk level as low as reasonably practicable considering the benefits and costs of the additional mitigative measures. The Figure 6.6 shows an ALARP region between risk levels of 1×10^{-3} (workers) and 1×10^{-6} (ALL). The ideal risk level is within the broadly acceptable region where the risk levels are negligible or so small that they can be managed by routine maintenance activities (i.e., ILI, Cathodic Protection Surveys, Right-of-Way Surveillance) and no additional risk mitigation measures are required.

6.1.1.5 Societal Aversion   Societal aversion reflects the social concerns and reactions to risk. People are more averse to single events with larger consequences than multiple events with lower fatalities. Society would think differently between the risk of one person being killed every year and the risk of 100 people being killed every 100 years in a kilometer of pipeline. Risk aversion is characterized by the society attitude and sensitivity to catastrophic or large consequence outcomes. It is not the same across societies and the perception can be different from country to country. (Assumed risk versus risk imposed upon the society, i.e., flying in an airplane, driving a car versus living next to a refinery or near a pipeline.)

Societal aversion captures the society perception of a low probability incident causing a large number of fatalities represents a higher risk than a higher probability incident causing a proportionately lower number of fatalities. Figure 6.7 shows the slope of the best-fit line from the correlation between the probability of an incident and the associated number of fatalities that can be implied as the degree of aversion [6] of failure consequences from a sample of onshore natural gas pipelines using the Reliability-Based Design and Assessment (RBDA) methodology adopted by Canadian industry standard CSA-Z662. Furthermore, the RBDA targets are compared with other regulatory and industry jurisdictions.

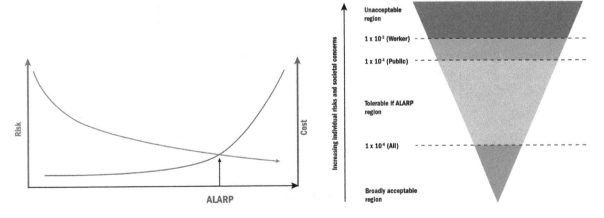

FIG. 6.6   ALARP—AS LOW AS REASONABLY PRACTICABLE

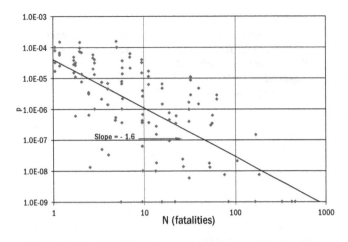

FIG. 6.7 SOCIETAL RISK LEVEL COMPARISON

The following examples illustrate the difference between societal and aversion risks. The societal risk associated with a number of fatalities or injured people are quantified in the examples with and without aversion factor assumed to be 1.5.

Example: Using No Aversion Factor

Case A: Risk associated with a higher probability of failure causing a proportionately lower number of fatalities
Probability of a pipeline failure due to external corrosion and mechanical damage = 1.25×10^{-4} failure/km-year and consequences of failure in a class 1 location within an impacted pipeline length of 800 m.
Three (3) dwelling units with a population density of 2 persons per dwelling within a class 1 location during normal use equates to six (6) expected number of fatalities/failure.

Case A: Risk of Failure = $(1.25 \times 10^{-4}$ failure/km-year) $\times$ (6 fatalities/failure)

= 7.5×10^{-4} fatalities/km-year for the class 1 location

Case B: Risk associated with a low probability of failure causing a large number of fatalities
Probability of a pipeline failure due to mechanical damage = 1.25×10^{-6} failure/km-year and consequences of failure in a class 4 with an impact pipeline length of 800 m.
For 80 single dwelling units, 2 buildings with 4 stores, a motel intended for human occupancy and two playground occupied by 20 persons during normal use with a total population density of 600 persons equates to 600 expected number of fatalities/failure.

Case B: Risk of Failure = $(1.25 \times 10^{-6}$ failure/km-year) $\times$ (600 fatalities/failure)

= 7.5×10^{-4} fatalities/km-year for the class 4 location

When no aversion factor is applied in Case A and B, the risk estimate has the same calculated number.

Using Figure 6.8 to plot the expected number of fatalities (N) and their frequency (F) from the above example, the risk for Case A and Case B are considered unacceptable because it is above to the maximum tolerable risk societal estimated using BSI PD-8010-3 standard: 2009 [7]. However, the values which might be fallen above the ALARP criteria line then mitigations scenarios can be evaluated.

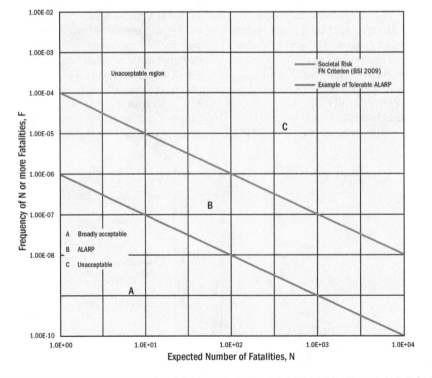

FIG. 6.8 F-N DIAGRAM—EXAMPLE OF SOCIETAL RISK ACCEPTABILITY USING BSI PD 8010-3

Example: Using Aversion Factor greater than one (1) a = 1.5

The Table 6.1 shows that the risk increases exponentially with the number of fatalities 10 times more with an aversion factor of 1.5 (i.e., an aversion factor greater than 1), which means that a lower probability of failure of pipeline causing higher number of fatalities represent a higher risk values than higher probability of failure causing proportionately lower number of fatality. This reflects the risk aversion. Recognizing risk aversion associated with multiples fatalities, societal risk acceptance criteria are usually expressed in term of F-N relationship.

Social media and TV have an impact on society when a pipeline incident occurs. However, society has the tendency to ignore a single fatality or pipeline rupture distributed over time or space, while the sum of these cause a societal response. For instance, people paid more attention to the Pacific Gas and Electric Company (PG&E) ruptured in a residential area in the city of San Bruno, California on September 9, 2010. The accident killed eight (8) people, injured many more, and caused substantial property damage [8]. Conversely, people tend ignore a pipeline rupture with no fatalities or environmental damages.

Furthermore, Figure 6.9 shows the reliability targets used by the Reliability-Based design and Assessment method [9] adopted by CSA-Z662 that can be used to make design and operational decisions that meet specified target reliability levels accepting a given risk level. The benefits of these methods include:

- Consistent and demonstrable safety levels,
- Integration of design and operational decisions to reach lowest cost solutions,
- Well suited to unconventional pipelines such those in arctic regions, and those using high strength steel.

TABLE 6.1   EXAMPLE OF SOCIETAL RISK ESTIMATION WITH AND WITHOUT AVERSION

Case	Frequency of fatality (F) (Failure/km-year)	Number of fatality (Fatality/failure)	Risk of failure® (Fatality/ km-year)
A	1.25×10^{-4}	$6^{1.5}$	1.83×10^{-3}
B	1.25×10^{-6}	$600^{1.5}$	1.83×10^{-2}

Example: Reliability-Based Design Assessment (RBDA) Applicability:

- Design of new pipelines especially those involving novel loading, new technologies and high consequences,
- Developing cost effective plans to deal with changes in original design conditions such as class location upgrades or pressure increase,
- Planning of defect repairs and in-line inspection,
- Optimization of damage prevention activities, and
- Reliability-Based assessment for river crossing.

6.2   RISK ASSESSMENT PURPOSE, APPLICATIONS, FUNCTIONS, AND OUTCOMES

The purpose of the pipeline risk assessment is to quantify and estimate the risk levels of pipelines. Some regulations also require companies using the risk assessment to prioritize the integrity assessments developing baseline and continual reassessments and determining additional preventive and mitigative measures.

Example: Risk Assessment Applications

- Optimization of pipeline maintenance, inspections and replacement
- Evaluation of change of service (e.g., type of product being transported), operating condition changes (e.g., pressure, temperature)
- Upgrade of class location or consequence (i.e., population, environment)
- Increment of existing pipeline reliability
- Assessment of safety of the reliability-based pipeline design with the aim of demonstrating the structural adequacy and increasing the availability and reliability of a pipeline

Developing a functional risk assessment approach that fits company business model is a challenge in the form of expertise, resource limitations, and data constraints. Risk analysis and results

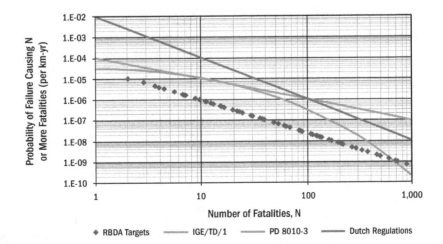

FIG. 6.9 ULTIMATE LIMIT STATE RELIABILITY TARGET FOR DESIGNED CLASS LOCATIONS 2 TO 4 [9]

cannot be interpreted as an exact expression of the structural condition of the pipeline, its fitness for service, or any other pipeline condition. Assumptions, accuracy of information, risks methodology used and the level of completeness of the assessment of hazards, threats and consequences should play a key role in the company making-decision process.

Essential functions are needed to establish a basic risk assessment process with logical and structured approach to integrate efficiently all available information into a robust risk assessment. The following examples of objectives and expected outcomes provide the foundations to build a comprehensive risk assessment program or enhance an existing risk model verifying meaningful risk estimates are determined:

Example of Essential Risk Assessment Functions:

- Measures risk in verifiable units, (i.e., consequences: $/failure, fatality/failure, Spilled volume/failure),
- Calculates failures probabilities grounded in engineering principles, (i.e., probability of failure calculation based on Monte Carlo simulation),
- Fully characterizes consequences of failure,
- Profiles risk along a pipeline,
- Integrates pipeline knowledge,
- Promotes more accurate decision making,
- Controls the bias, and
- Verifies proper aggregation.

Example of Expected Outcomes:

- Efficient and transparent risk modeling,
- Accurate, verifiable and complete risk results,
- Improved understanding of actual risk,
- Risk-based inputs to guide integrity decision making—true risk management,
- Optimized resources allocation leading to higher level of public safety,
- Appropriate levels of standardization facilitating smoother regulatory audits, and
- Expectation of regulators, the public, and management fulfilled.

6.3   RISK ASSESSMENT TYPES

The selection of the optimal method depends on several factors, such as the following:

- Expectation/Purpose of the risk assessment
  - Examples: pipeline segment ranking for ILI, analyzing cost-effective risk reduction, selection of mitigation actions to reduce risk
- Objective of the IMP
- Available resources and number of pipeline segments to study
- Level of effort in gathering and processing data
- Complexity of the analysis and time to conduct the assessment (e.g., months, years)
- Nature, quality and availability of information and data

Risk assessments are typically conducting using the following methodologies:

1. Qualitative
2. Quantitative
3. Semi-quantitative

Depending on the selected methodology, the following techniques may be useful in supporting the risk analysis:

- Event trees
- Fault trees
- Scoring or indexing
- Probabilistic

6.3.1   Qualitative Risk Assessment

Qualitative risk assessment methodology is a relative assessment method ranking systems and components relative to each other based on subject matter expertise and score or index modeling. Section *"Risk Algorithm"* illustrates an example of indexing modeling calculation. Qualitative Risk Assessment is also a method for screening analysis of hazards, threats, and receptors for onshore and offshore pipelines such as decision-tree model. The following are some of the application of qualitative risk assessments:

- Where the probability of failure and the consequences cannot be easily quantified,
- Identification of dominant hazards and threats,
- Integrity planning including segment ranking and optimization of maintenance activities,
- Scheduling of hydrostatic, inspection intervals and repairs,
- Pipeline pressure increase and class location designation change,
- Performance of drill-down and data analysis for contributing risk factors,
- Prioritization of integrity mitigation, prevention and monitoring measures, and
- Input for developing pipeline integrity Key Performance Indicators (KPI).

This methodology has a more limited use, and its limitation is based on type of pipeline segment, and the availability of data to conduct the analysis. Qualitative risk assessment requires the estimation of relative probabilities and consequence severities in broad groups such as very high, high, medium, and low, as illustrated in Figure 6.10 and the example below. Although you can use any number of groups, you will probably not be able to assign with sufficient confidence more than five failure probability and consequence severity groups. This approach is often compelled to use subjective likelihood based on intuition and expertise given the partial or non-representative data. These constitute the major source of uncertainty in the risk profile and evaluation.

Example: Priority Categories Shown in the Qualitative Risk Matrix

VERY HIGH or *"Unacceptable Risk"*: It should be mitigated immediate with engineering and/or administrative controls to a risk ranking of Low or Very Low. For instance, line replacement for a pipeline with a leak or pressure reduction for a pipeline with anomalies with a depth greater than 80% wall thickness.

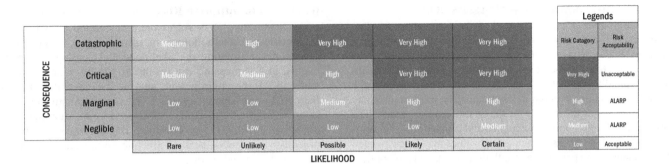

FIG. 6.10 EXAMPLE OF QUALITATIVE RISK MATRIX AND SIGNIFICANCE

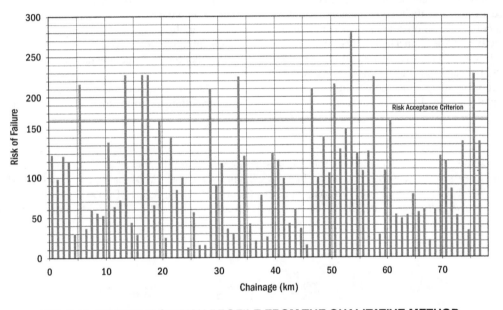

FIG. 6.11 EXAMPLE OF RISK PROFILE FROM THE QUALITATIVE METHOD

HIGH: It is should be mitigated with engineering and/or administrative controls to a risk ranking of Low or Very Low with a specific time frame, such as between 6 months and a year. *ALARP approach can be applied to reach a tolerable risk.* For instance, pressure reduction and perform a scouring analysis in a pipeline exposed under a creek, or conduct a HDD under a river crossing for an exposed pipeline under a river containing cracking anomalies.

MEDIUM: It should be mitigated with engineering and/or administrative controls to a risk ranking of Low or Very Low with a specific time frame, such as between one (1) and three (3) years depending on mitigation action difficulty. For instance, repair features failing within a time period of two years. *ALARP can be applied.*

LOW or "*Acceptable Risk*": Should be monitored and verified on a continuous basis that procedures or controls are in place. No mitigation required for risk reduction. However, prevention actions might consider improving process. For instance, revise integrity procedures every two years or running an ILI tool for corrosion monitoring with an interval of five years.

Qualitative Risk Assessment advantages over other methods include the following:

- Use subjective weights, scores, and measures in risk estimation.
- Relies on qualitative data from other studies, expert opinions, and personal experience.
- Both probability and consequences is not a measurable quantity.
- Criterion relies on subjective scores and weights.
- Easy to perform and helps eliminate non-issues.
- It is less accurate and lacks reproducibility and consistency.
- Terminology used is barrier to communication: probable, unlikely, extremely unlikely.
- Vulnerable to criticism and undermines effort to standardization.

Figure 6.11 is a risk level profile from an indexing modeling called relative risk assessment, a matrix plot representing the likelihood and consequences indexing levels, and the SCC index levels showing the SCC drives (variables) that contribute to the SCC index results [10].

6.3.2 Semi-Quantitative Risk Assessment

Semi-quantitative is a term that describes any approach that has aspects derived from both the qualitative and quantitative approaches. It is geared to obtain the major benefits of the previous two approaches (e.g., speed of the qualitative and rigor of the quantitative). Typically, most of the data used in a quantitative approach is needed for this approach but in less detail. The models also may not be as rigorous as those used for the quantitative approach. The results are usually given in consequence and probability categories rather than as risk numbers but numerical values may be associated with each category to permit the calculation of risk and the application of appropriate risk acceptance criteria.

As illustrated in Figures 6.13 and 6.14, description of probability of failure may be estimated in a quantitative approach and consequences might be expressed in a relative ranking. In a semi-quantitative approach, different scales are used to characterize the likelihood of adverse events and their consequences. Analyzed probabilities and their consequences do not require accurate mathematical data. The objective is to develop a hierarchy of risks against a quantification, which reflects the order that should be reviewed and no real relationship between them. The combination of the two models can be a solution in some cases, combining the specific advantages of each and decreasing their disadvantages.

6.3.3 Quantitative Risk Assessment (QRA)

A Quantitative Risk Assessment is defined as a formal and systematic risk methodology of identifying potentially hazardous events, estimating the likelihood and consequence of those events, and expressing the results as risk to people (employee and public), the environment, and the business. It is a probabilistic risk assessment approach. This probabilistic approach is more scientific, statistical, technical, formal, quantitative, and objective. Ideally, the probabilistic risk assessment is based on objective likelihoods such as pipeline failure rates inferred from statistical data and theories.

Population risk or risk of fatalities or injuries is the quantification of the risk to population that might be impacted by a certain hazard surrounding the pipeline. Quantitative risk assessment uses two (2) concepts: individual and societal risk to quantify the risk as measured by the combination of probability and consequence of failure with the number of people affected by a hazardous phenomenon (e.g., fire, explosion, toxicity).

Quantitative Risk Assessment advantages and disadvantages over other methods include the following:

- Require parameter data with probability density function distribution, deterministic and stochastic approaches

---

Example: A typical risk profile for a liquid pipeline transmission is illustrated in Figure 6.12 showing the variation of risk units' percentage for four main threats (i.e., ERW-Manufacturing threat, Mechanical Damages [MD] threat, External Corrosion [EC], and Stress Corrosion Cracking [SCC] threat), consequence impacts, and the sum of all risk in terms of Total Risk Units. Such a plot identifies high-risk segment due to two threats, SCC and ERW threats, as well several segments of pipeline with joints exceeding the pipeline risk criteria.

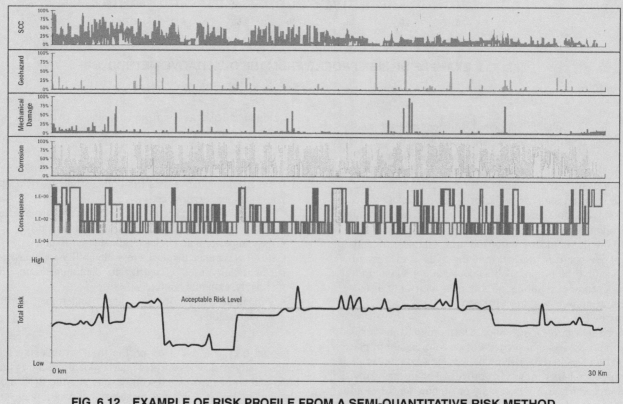

FIG. 6.12 EXAMPLE OF RISK PROFILE FROM A SEMI-QUANTITATIVE RISK METHOD

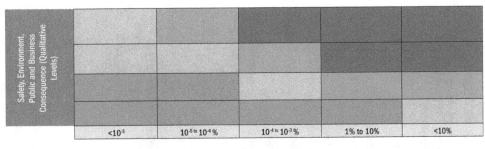

Quantitative Probalility Range

FIG. 6.13   EXAMPLE OF SEMI-QUANTITATIVE RISK MATRIX

Public

Death or severe health effects. Large community impact and evacuation
Injury or moderate health effects. Medium community impact
Small injury or health effects. Medium community impact
No injury, minor injury, or health effects. Temporary closure of side road

Safety

Death or severe health effects
Serious Injury or moderate health effects. Medium community impact
Small Injury or health effects. Medical treatment required
No injury, minor injury, or health effects. Require First Aid

Environment

Significant adverse effects and major emergency. Affecting water sources
Intermediate Emergency Response
Minor adverse effects
No adverse effects or close proximity require First Aid

Business Downtime

More than six months
Between one and six months
Between one week and one month
Less than one week

Capital Loss and Property Damage

Significant amount
Between significant and high amount
Between high and medium amount
Minimum amount

Loss of Market Share and Reputation

Significant
High
Medium
Low

FIG. 6.14   EXAMPLE OF CATEGORIZATION OF CONSEQUENCES

- Both probability and consequences must be a measurable quantity
- It is the process of using objective or quantitative measures to estimate a risk
- Provides risk scenario(s) and relies on quantitative data from other studies, and expert opinions for the risk model
- Provide relatively accurate estimates and it is defendable to the stakeholders
- Quantify risk associated with defined scenarios (i.e., line replacement vs repairs)
- Easy to quantify a cost/benefit analysis for risk reduction
- Requires thorough scientific review. Quantitative risk assessment
- May be too complex for stakeholders
- Requires data that not always available
- Relies on comprehension of mathematical methods

Table 6.2 compares the purpose, type of method, typical process used, and acceptance criteria for determining the Scenario-based, Index/Relative and Probabilistic/Quantitative Risk Assessment Methods.

6.4   RISK ASSESSMENT PROCESS

6.4.1   RA (Plan)

Planning phase is the preliminary step to conducting the risk assessment by establishing the scope and context for risk assessment aligned with the company policies, goals and objectives as well regulations and codes. This phase may involve an information sharing and consultation process with risk advisors, decisions-makers, risk consulting companies, personnel of different company divisions, and various stakeholders prior to conducting the actual risk assessment.

Companies clearly ought to define risk assessment goals, objectives, and targets to determine its success in identifying the risk levels and the actionable measures. Companies also should identify failure the integrity threats and pipeline release type (i.e., leak, large leak, and rupture) causing a consequence applicable to the selected pipeline system factoring the safety margins (i.e., 1.25% of MAOP), risk limits (i.e., 0.3×10^{-6} fatality/year for sensitivity institutions) and expected cost effectiveness (e.g., benefit/cost ratio).

Definition of the pipeline system, segments, data gathering and integration, risk criteria and acceptability, and hazard-threat identification processes are required in this phase. The Hazard and Threat

TABLE 6.2 DIFFERENTIATING SCENARIO-BASED, INDEX/RELATIVE AND PROBABILISTIC/QUANTITATIVE RISK ASSESSMENT METHODS

Concept	Scenario-Based Model Risk Assessment	Index/Relative Risk Assessment	Probabilistic/Quantitative or Probabilistic Risk Assessment
Main Objective	As per API 1160-2013 section 7.2, Scenario-based model uses sequences of events that lead to the risk of release, which can be calculated either by fault tree or event tree methodologies.	As per API 1160-2013 section 7.2, Relative Risk Assessments use algorithms with variable, attributes and weighting factors to provide scores for ranking purposes. The method uses all applicable integrity threats and consequences to schedule integrity assessments, mitigation and re-inspections	As per API 1160-2013 section 7.2, Probabilistic Risk Assessment determines the probability of occurrence and consequence (e.g. cost) of integrity-related events (e.g. damage or release). Probability of Exceedance is typically used for mitigating threats following an In-Line Inspection.
Type of method Process	Qualitative or semi-quantitative • Identify hazard (e.g., flammable, combustible, explosive, pressure, human factors, toxic) • Hazard evaluation: (e.g., loss of power, corrosion, overpressure, overflow) • Hazard controls (e.g., safety devices, design, or procedures) • Define risk scenarios (credible and specific) • Determine consequences • Determine probability fault or (event tree) • Determine risk using the matrix • Assess the risk for the defined scenarios • Define preventative and mitigation measures	Qualitative or semi-quantitative • Identify threats (e.g., corrosion, dents, cracking) and associated hazards/conditions (e.g., coating damage, ROW activities, pressure cycling) and consequences • Assign qualitative weights for threats and consequence types based on subject matter experts • Assign scores to attributes per Hazard (e.g., coating damage) leading to a threat (e.g., corrosion) • Calculate likelihood per threat and consequence • Determine risk drivers • Define ranking based on risk, likelihood and consequence • Validate with SMEs • 2 Standard deviation of likelihood per threat and risk score frequency • Risk significance matrix	Quantitative • Identify Hazards/Threats • Determine Probability and Frequency of Failure • Quantify Consequences in monetary terms • Determine Individual Risk • Determine Societal Risk • Compare against Acceptance Criteria • Identify whether the calculated risk is broadly acceptable or tolerable • Determine Cost Benefit Analysis for each mitigation alternative of the ALARP region risks • Make decisions
Acceptance criteria	• Likelihood and consequence levels • Risk categories (significance)		• Leak, Rupture, Serviceability and Fatigue Targets or Industry Failure Frequency • Acceptable Risk (numeric)

identification are detailed in the next section. W. Kent Muhlbauer, in his book Pipeline Risk Assessment [11], provides also risk assessment techniques in the *"Risk Assessment Building Block"* section, which describes various ideas how to understand and measure risk.

6.4.2 Pipeline System Definition
The following are some recommended elements for the definition of the pipeline system:

1. Identify the pipeline systems including subsystems and parallel loops
2. Characterize the pipeline sections using physical boundaries (e.g., class location changes, valves, pump stations)
3. Define the pipeline segmentation methodology (e.g., fixed pipe length, pipeline attribute changes) ensuring credible failure scenarios are not diluted due to over-segmentation
4. Anticipate the available data when segmenting the pipeline while keeping in mind the objective of the risk assessment (e.g., In-Line Inspection schedule prioritization). Avoid either excess modeling or excess data collection not related to the objectives.

6.4.3 Data Gathering
The next step is performing an inventory of the existing pipeline system data related to the risk assessment objectives. The amount and type of data to support risk assessment will mainly vary depending on the integrity threats (e.g., metal loss, cracking, incorrect operations) being assessed while the consequence data may tend to be the same for multiple types of risk assessment.

Data gathering can be accomplished by use of *Geographical Information Systems* (GIS) tools capturing databases, spreadsheets, and external sources. The *Pipeline Open Database Standard* (PODS) [12] provides industry recognized-database architecture for storing pipeline and integrity data that can be accessed by software using a GIS platform. The *ArcGIS Pipeline Data Model* (APDM) is a geo-database model derived from other database architectures such as PODS; APDM is intended to be a template, not a standard.

Gathering is also conducted by extracting data from maps, alignment sheets, pipeline spreadsheets sourced by companies and pipeline industry. In some cases, interviews with stakeholders, company personnel from operations, contractors, maintenance department, and other resources are useful in gathering data.

6.4.4 Data Integration
Data integration requires the definition of a common pipeline location reference that allows data from various sources to be accurately associated and integrated. For instance, in-line inspection (ILI) data is referenced to the distance traveled along the inside of the pipeline (i.e., chainage from ILI wheel count/odometer) versus the right of way surveys such as rectifiers, close interval survey (CIS) and ground patrolling that may be referenced to stationing. The following are some of the considerations for data integration:

a. Data models for upload and process for sustainment
b. Location referencing
c. Multiple data alignment
d. Quality control for data migration
e. User data analysis and validation
f. User data update and sharing
g. Data retention and maintenance

Multiple data alignment and location referencing are accuracy critical for data upload and migration of pipeline systems. Layering of data using the same referencing method enables the integrity professional to conduct integrated analysis leading to better criticality and risk assessment. A key in the analysis is spotting changes and entering new data into the assessment process. Integration can be accomplished in many different ways, i.e., manual, manual within GIS, or automatic within GIS. Pipeline operators are usually employing GIS methods. The integration helps reviewing whether areas of interest may be producing incorrect analysis by data cross-reference.

An integrated data management provides the following benefits:

- Quality control
- Standard data format
- Standardization of process
- Same answer regardless of who performs the analysis
- System-wide implementation of upgrades and changes
- Ability to measure performance
- Facilitates communication among teams
- Efficient and consistent decision making

Example: Data Type and Categories for Risk Assessment

- Pipeline hierarchy, sections and segments
  - Maps, alignment sheets, digital photos, stationing data
- Pipe design attributes
  - Pipeline and facility specifications and operating information (e.g., pipe grade, diameter, pressure limits, elevation profile, leak detection system)
- Construction and Installation: hydrostatic failure, stresses and duration of tests
  - Quality assurance reports
- Pipeline operation: fluid composition, pressure and temperature spectrums, history
- Maintenance and Integrity
  - Release and repair history; pipeline inspections
  - Class location changes, right-of-way encroachments
  - Integrity threat susceptibility and identification reviews
  - Corrosion control history: internal and external
  - Inline inspection and hydrostatic results
  - Cathodic protection: rectifier readings, soil resistivity readings, Direct Current Voltage Gradient (DCVG) records
  - Repair and maintenance records back from pipeline inception
  - Mitigation, monitoring and prevention programs
- Future operation (e.g., new/changes in the integrity threats)
- Expected consequence changes (e.g., new neighborhoods/developments)
- Industry information related to pipeline failures

6.4.5 Hazard, Threat, and Consequence Identification
The identification of hazards (e.g., coating damage) helps in either anticipating or further assessing integrity threats (e.g., corrosion) to the pipelines. Similarly, the receptors (e.g., populated areas, water bodies) need to be identified prior to establishing the potential extent and severity of the consequences of a failure [13].

Sections 6.5.1 and 6.5.2 start discussing the assessment of integrity threats and consequences. Furthermore, Chapters 4 and 5 explain in detail the Hazard, Threat and Consequence and Assessment process.

6.4.6   RA (Do)

6.4.7   Risk Analysis

Risk analysis is the first process that determines the probability of failure of the integrity threats and their consequences (e.g., leak, large leak, or rupture) per segment to determine the risk per pipeline segment and along the profile of the pipeline. This qualitative risk analysis can be based on the relative or indexed risk algorithms.

Risk factors per integrity threat and consequence are defined and data requirements mapped out to start the data gathering process. The risk factor data is expected to have an intrinsic uncertainty (e.g., low, medium, high) that need to be understood for providing context to the risk analysis results. Even within the same integrity threat category (e.g., corrosion), differences in uncertainty and timing of the input data (e.g., accuracy and year of In-Line Inspection –ILI-, Non-Destructive Examination –NDE-, cathodic protection and coating surveys) may not accurately describe a physical phenomenon (e.g., corrosion growth) impacting the representativeness of the risk analysis.

If the *risk factor data uncertainty* is quantified and timing synchronized, risk may change in some pipeline segments over others shifting more appropriately the risk ranking. This may cause higher estimated risk value in some pipeline segments as a result of the applied uncertainty, but causing the user to seek for better data to reflect their perception, but not necessarily for the need to conduct risk reduction. It is a sound practice to separate higher risk segments with high data uncertainty or unknowns from the segments with higher risk values due to actual conditions. These two (2) cases may have different risk management strategies (e.g., risk reduction, risk data gathering, and monitoring).

Risk drivers are the risk factors with the highest contribution to the estimated risk for a given pipeline segment. For instance, SCC risk factors such as coating type (e.g., tape disbondment), cathodic protection (e.g., ineffective), stress level (e.g., >60% SMYS) may be the factors with the highest contribution to the estimated risk for a given segment becoming the risk drivers. Sometimes, true risk drivers consistently identified by the pipeline operator over time do not show up on the initial risk model results requiring its iterative optimization.

6.4.8   Risk Estimation

Risk estimation provides the outcomes of probability and consequences analysis and estimated risk results, and requires validation of the risk model formulation (i.e., algorithm) and input data creating make-sense and meaningful results. The validation of risk model reviews the following aspects:

a. Formulation applicability and the completeness of the variables
b. Data input used in the model identifying the level of uncertainty
c. Make-sense of the estimated risks

The frequency for evaluating risk of pipeline segments can be defined on either fixed interval (e.g., every 5 years) or driven by change in the conditions (e.g., new consequence areas, higher/lower pressure, and service/product change or pipeline acquisitions).

The results of the risk assessment should be documented and presented indicating the risk methodology, sources of data, assumptions and challenges, pipeline segment ranking by risk, probability, and consequence indicating their risk drivers, and options identified

for risk reduction. Decisions about selecting options to reduce risk will be part of the risk management process.

6.4.9   RA (Check)

6.4.10   Validation of Risk Results

Validation of the risk assessment should be conducted by pipeline-specific personnel and Subject Matter Experts (SME) from multiple disciplines such as risk, integrity, operations, leak detection, right-of-way inspection, and surveillance. This validation may involve running multiple iterations adjusting the risk model and their results; particularly, cases where the risk values and ranking are not representative of the actual reality portraying the risk to be unrealistically too high or too low, and an inverse ranking. The validation process should ensure that the model and its results captures to the extent possible, the phenomena causal and posing barriers of risks (e.g., design, construction, operations, and environment). The validation should be able to reaffirm, modify, and reject risk results whether their mathematical rationale is logical.

6.4.11   Risk Evaluation and Sensitivity Analysis

Risk evaluation process identifies the pipeline segments meeting and exceeding the risk acceptance criteria and proposing alternatives for risk reduction. The evaluation process uses the ranking of the pipeline segments by risk, probability, and consequence to identify the risk drivers of those highly ranked segments for sensitivity analysis purposes.

Sensitivity analyses can be conducted by increasing (causal) or decreasing (barrier) risk drivers. Furthermore, the effects of the uncertainty from the input data on risk can be understood and simulated by dynamically changing (up and down) the values of the risk factors within an expected variability range and determining the range of change in risk. Sensitivity analysis can also help fine-tuning the data input associated to educated opinions or weighting factors used in the model.

6.4.12   RA (Act)

The selection of a risk reduction effective plan is the next step that involves balancing the costs and efforts against their benefits. Regulatory, legal, and other requirements such as safety of people, and the protection of the natural environment should be considered in selecting the most appropriate preventive and mitigation actions to reduce risk.

The evaluated risk may lead to a decision to perform further analysis or not to treat the risk in any way other than maintaining controls of existing risks. This decision would need to be tested against company's values, goals, and objectives.

6.5   RISK ANALYSIS

Risk analysis is a structured process used to identify both the likelihood of integrity threats and extent of their consequences. Risk analysis answers four fundamental questions [14]:

a. What can go wrong?
b. How likely is it?
c. What are the consequences?
d. What is the level of risk?

Risk analysis provides an input for risk evaluation and control providing the basis for making decisions involving different types and levels of risks.

As illustrated in the example in Figure 6.15, a risk analysis could involve:

- Identification of hazards, integrity threats such as scenarios associated with pipeline materials (i.e., low Charpy V-Notch values for pre-1970 vintage pipelines), system, process and facilities (i.e., overpressure in a refined-product pipeline).
- Determination of undesirable consequences to human, environmental, and economic impacts such as thermal radiation flux for fires (jet fire, flash fire, or fireball) in a propane terminal or pipeline ROW nearby a playground factoring both the existence and location of receptors.
- Estimation of possible risk scenarios
- Identification of potential safeguard options (e.g., mitigation, prevention, monitoring) needed to minimize the possible consequences

Category	Data
Attribute data	Pipe wall thickness Diameter Seam type and joint factor Manufacturer Manufacturing date Material properties Equipment properties
Construction	Year of installation Bending method Joining method, process and inspection results Depth of cover Crossings/casings Pressure test Field coating methods Soil, backfill Inspection reports Cathodic protection (CP) installed Coating type
Operational	Gas quality Flow rate Normal maximum and minimum operating pressures Leak/failure history Coating condition CP system performance Pipe wall temperature Pipe inspection reports OD/ID corrosion monitoring Pressure fluctuations Regulator/relief performance Encroachments Repairs Vandalism External forces
Inspection	Pressure tests In-line inspections Geometry tool inspections Bell hole inspections CP inspections (CIS) Coating condition inspections (DCVG) Audits and reviews

FIG. 6.15 EXAMPLE OF DATA FOR INTEGRATION (*Source:* from ASME B31.8-2014)

Existing controls and their effectiveness and efficiency should also be taken into account. The way in which consequences and likelihood are expressed and the way in which they are combined to determine a level of risk should reflect the type of risk, the information available and the purpose for which the risk assessment output is to be used. These should all be consistent with the risk criteria. It is also important to consider the interdependence of different risks and their sources [1].

The confidence in determination of the level of risk and its sensitivity to preconditions and assumptions should be considered in the analysis, and communicated effectively to decision makers and, as appropriate, other stakeholders. Factors such as divergence of opinion among experts, uncertainty, availability, quality, quantity, and ongoing relevance of information, or limitations on modeling should be stated.

Risk analysis can be undertaken with varying degrees of detail depending on the level of consequence, the purpose of the analysis, and the available information, data and resources. The analysis can be qualitative, semi-quantitative, or quantitative. Risk estimation is to estimate the total risk by determining the Probability of Failure (POF) by the Consequence of Failure (COF).

6.5.1 Integrity Hazard and Threat Assessment

Hazard and Threat Assessment is the first element in the process of risk analysis. Its prime purpose is to identify hazards leading to threats that may become failure modes (causes) requiring the estimation of the Probability of Failure (POF).

Anomalies or defects causing modes of failure are introduced into a pipeline at any point during manufacturing or construction process or developed by contributing factors such as corrosion or cracking during in-service. They may worsen up over time under specific operational or environmental conditions via failure mechanisms such as fatigue-induced or environmentally-assisted cracking causing in-service pipeline failures. Table 6.3 provides examples of pipeline failure mechanisms. Please refer to Chapter 4 Integrity Hazard and Threat Assessment of this book for details.

6.5.2 Consequence Assessment

Consequences can be determined by modeling the outcomes of an event or set of events, or by extrapolation from experimental studies or from available data. Consequences can be expressed in terms of tangible and intangible impacts. In some cases, more than one numerical value or descriptor is required to specify consequences for different times, places, groups, or situations.

Leak and rupture scenarios are caused by different failure modes and mechanisms. Pipeline release scenarios could be leak, large leak and rupture causing different types of consequences. A leak scenario may lead to have a higher Probability of Failure; whereas, a rupture scenario may lead to higher consequences depending on the detection capabilities of the system or public discovery. The rupture scenario could drive the risk outcomes as a worst-case scenario; however, leaks may go undetected for a longer period of time causing a large consequence.

The consequences of failure can be expressed in terms of population density and/or concentration, the probability of ignition and the size of the hazard area. The hazard area can be defined as the area within which people would potentially be exposed to a lethal heat dosage [13]. A range of mathematical models is available to assist in estimating the consequences. Consequences models commonly used in the oil and gas pipeline industry are overland flow modeling, source term modeling, jet fire, BLEVE, flash fire, vapor cloud explosion, and toxic gas dispersion. Please refer to Chapter 5 Consequence Assessment of this book for details.

TABLE 6.3   EXAMPLE OF FAILURE MECHANISMS

Corrosion	Cracking	Geohazards	External damage	Material manufacturing	Others
External corrosion	Stress corrosion	Scouring and bank erosion	First-party damage	Defective long seam weld	Fire
Internal corrosion	Hydrogen induced	Mass wasting (e.g., landslides)	Contractor Second-party damage	Defective girth weld	SCADA malfunction
Sulfide stress corrosion	Delaying cracking (mechanical damages)	Subsidence	Third-party damage	Wrinkle, Buckle Ripple	Human errors
Microbiological induced corrosion (MIC)	Immediate cracking (mechanical damages)	Seismic	Vandalism	Defective pipe body (i.e., lamination, slivers)	Lightning

Example of Input for Consequence Modeling

- nature of fluid (e.g., flammable, toxic, reactive, etc.),
- pipeline design, buried-or-aboveground topography,
- environmental conditions,
- size of hole or ruptures,
- overland flow modeling,
- dispersion of fluid,
- probability of ignition and explosion following ignition,
- toxic effects,
- ground/water pollution.

6.6   RISK ESTIMATION

Risk estimation measures the level of effect on safety, environment, and business for a specific integrity threat estimating the risk values for the purposes of identifying whether they are *broadly acceptable, tolerable, or unacceptable.*

6.6.1   Risk Factors and Drivers

Risk factors (e.g., hazards, threats, or consequences) represent some conditions related to risk, and these are estimated for a particular segment of pipeline. Selecting appropriate and realistic risk factors will drive to clear risk outcomes avoiding subjective scores. Identification of adequate risk factors on both parts of the risk equation (i.e., probability and consequence) will determine representative risk estimation. Some risk factors may contribute higher to risk and considered as "risk drivers," which can also used as barriers to help in reducing the risk.

Example of Risk Drivers

- Depth of Cover (DoC) may be a risk driver directly contributing to higher probability for third-party mechanical damage; however, deeper DoC may become the barrier needed for reducing the chance to access the pipeline by 3rd parties.
- Fracture toughness material property may be a risk driver directly contributing to higher probability of seam weld cracking; however, higher Charpy-V-Notch (CVN) values may become the barrier needed for reducing the cracking growth rate and risk.

6.6.2   Estimated Risk

Estimated risk is the exposure determined from the frequencies of occurrence and consequences of the identified integrity threat. The estimated risk is calculated by multiplying the Probability of Failure (POF) by the Consequence of Failure (COF). The POF is based on the *probability or likelihood* that an event or condition will result in failure, tempered by the confidence in the pipeline data. Developing a balanced ROF component requires more data, susceptibility and identification analysis; whereas, COF is somewhat more straightforward relying on geographical data, modeling, and validation.

6.6.3   Comparison Criteria for Risk Evaluation

6.6.4   Risk Algorithm

Risk Algorithm is a set of logical rules that uses data and information involving threats, hazards, variables, and attributes of variables in a qualitative risk approach to estimate a meaningful risk score. Developing a risk algorithm that reflects pipeline company's needs is a challenge. The pipeline industry has been introducing changes to risk algorithms in order to achieve risk results reflecting closely reality.

Risk algorithms can range from very simple screening tools and relative algorithms to an enormously complex set of logical rules, equations, and combination of them. Those algorithms can be created using a simplistic spreadsheet or vendor's solutions, which provide configurable software solutions for running complex quantitative risk and qualitative risk models. Depending on the complexity of the risk models, the results of estimated risk should be easy to understand and defend, and successfully pass regulatory scrutiny providing transparency and justification to the risk results.

Model building, application, evaluation, and understanding are therefore the foundation of the risk assessment. New algorithms are now more intuitive, easier to configure with built-in data that can dynamically integrate spatial, geographical, and tabular data at runtime to maximize the goal of data integration. New algorithms have the ability and flexibility to show risk results in either relative terms for qualitative models or absolute terms for fully quantitative risk modeling.

The completeness and appropriateness of a risk algorithm can be achieved by

- Using all relevant and available data
- Eliminating hides of effects

- Removing blinded issues
- Making appropriate assumptions
- Understanding unmitigated hazards and threats that a pipeline can be exposed
- Considering all potential mitigation effectiveness, and
- Recognizing the pipeline resistance

This approach definitely leads to accurate risk outcomes for better risk management decisions preventing releases and enhancing the ability to comply with regulations, codes, and standards. Subject Matter Experts (SME) facilitate the development of risk algorithms by interacting with the company personnel. The meetings are designed to assist the operator in setting up and customizing the risk algorithm to arrive at a tailored set of evaluation that reflects operator's risk management philosophies.

Generic algorithms as part of a software standard configuration do not represent any industry minimum standards nor should it be used as recommended risk conditions. However, they can be used as a reference for customization and identifying basic data requirements such as the following:

- Industry incident data
- Industry studies, papers and researches (e.g., PRCI, ASME International Pipeline Conference papers and industry association studies such as CEPA, INGAA, AOPL)
- International standards, regulations, and codes

The foundation of a successful risk assessment analysis is dependent on company's ability to accurately define an algorithm that captures its success and failure experiences specific to design, construction, operation, maintenance, and integrity practices.

Throughout an algorithm development process for a qualitative risk approach, the following terminology is recommended to describe key risk facets:

6.7   RISK EVALUATION AND SENSITIVITY ANALYSIS

6.7.1   Risk Criteria and Acceptability

Risk criteria refer to elements and definitions used for evaluating the risk significance of a given estimated risk related to the company, industry, or regulatory tolerance. Typically, the risk matrix provides the categorization of risk (e.g., acceptable, tolerable, not acceptable) portraying the risk criteria that reflects the company values, goals, objectives and timing to respond according to company's risk management policy. Each company develops their own risk criteria accounting for legal, regulatory and stakeholder requirements.

The way how probability and consequences of failure are expressed to determine the outcomes of risk assessment should always be consistent with the risk criteria outputs. Comparing the level of risk in the risk evaluation step with risk criteria established should be comparable.

Example: Consistency Expressing Risk

If probability of failure is expressed as "numerical," the criteria for that probability should not be expressed as "low, medium, or high."

The risk acceptability criteria established and compared with risk outcomes will influence the company decision and ultimately the risk attitude and actions. Risk criteria are very dynamic over time and should be revised in the continuous improvement process once mitigation actions are implemented and risk level is reduced.

Example of a Quantitative Risk Algorithm (Indexing Modelling Risk Algorithm)

Calculation of Risk of Failure (RoF) can be accomplished using the indexing modeling risk algorithm shown below. PoF and CoF indexes can be calculated given weights to the threats, and receptors, respectively. Threat and impact indexes can be estimated given weights to variables and score contributions to the attributes as illustrated. For instance:

$$RoF = PoF \times CoF$$

$$PoF_{(index)} = (\% \times EC_{index}) + (\% \times IC_{index}) + (\% \times TPD_{index}) + (\% \times Cracking_{index}) + (\% \times MFG_{index}) + (\% \times EQP_{index}) + (\% \times CON_{index}) + (\% \times IO_{index}) + (\% \times SCC_{index})$$

$$PoF_{(index)} = (0.15 \times EC_{index}) + (0.1 \times IC_{index}) + (0.15 \times TPD_{index}) + (0.15 \times Cracking_{index}) + (0.03 \times MFG_{index}) + (0.08 \times EQP_{index}) + (0.08 \times IO_{index}) + (0.09 \times CON_{index}) + (0.12 \times SCC_{index})$$

$$CoF_{(index)} = (\% \times People_{index}) + (\% \times Env_{index}) + (\% \times Employee_{index}) + (\% \times Financial_{index})$$

$$CoF_{(index)} = (0.35 \times People_{index}) + (0.30 \times Env_{index}) + (0.15 \times Employee_{index}) + (0.20 \times Financial_{index})$$

$$EC_{index} = \Sigma \, (External\ Corrosion\ Variable\ Weight \times Score\ Contribution)$$

$$EC_{index} = (0.04 \times Soil\ Type\ score) + (0.02 \times Environment\ score) + (0.02 \times pH\ Value\ score) + (0.03 \times MIC\ location\ score) + (0.03 \times Peak\ Temp\ score) + (0.03 \times Coating\ Type\ score) + (0.03 \times Coating\ age\ score) + \ldots\ldots\ldots + (0.02 \times Repair\ score)$$

Impact of People (IOP)$_{index}$ = Σ (IOP Variable Weight x Score Contribution)

LOF (Threat Types)	Weightings
External Corrosion	15%
Internal Corrosion	10%
Third Party Damage	15%
Cracking (LoF)	15%
Manufacturing	3%
Equipment	8%
Incorrect Operations	8%
Construction	9%
Stress Corrosion Cracking	12%
Total Percent	**100%**

External Corrosion Index : 15%

Variable	Variable Weight
Soil Environment	2%
Attributes	**Scores**
Stable	0
Soil Contamination	8
Exposure to Salt	8
Exposure to chemicals	8
Moisture cycling	5
Poorly Drainage	7
High Soil Stress	10
No Data	10

COF (Receptor Types)	Weightings
People Impact	35%
Environment Impact	30%
Employee Impact	15%
Financial Impact	20%
Total Percent	**100%**

Threat	Var #	Variable Name	Variable Weight
EC	1	Soil Type	4.00%
EC	2	Environment	2.00%
EC	3	pH Value	2.00%
EC	4	MIC Location	3.00%
EC	5	Peak Product Temperature	3.00%
EC	6	Coating Type	8.00%
EC	7	Coating Age	3.00%
EC	8	Type of CP	1.00%
EC	9	Operating Pressure vs. PS	4.00%
EC	10	Wall Thickness	5.00%
EC	11	Pipe Age (Ins)	5.00%
EC	12	Seam Type	1.00%
EC	13	Years of No CP	2.00%
EC	14	Years of Questionable CP	1.00%
EC	15	Years of Inadequate CP (Calc.)	1.00%
EC	16	Coating Condition	6.00%
EC	17	Pipe Condition	5.00%
EC	18	ILI or VI Metal Loss	2.00%
EC	19	ILI or VI RPR	5.00%
EC	20	Incident Age	5.00%
EC	21	Casing	1.00%
EC	22	Casing Short	2.00%
EC	23	DA Severity Index	6.00%
EC	24	Completed Assessment Age	8.00%
EC	25	Metal Loss Anomaly Density	5.00%
EC	26	# Years to 80% Wall Loss	5.00%
EC	27	TP vs PS	1.00%
EC	28	EC Failure Frequency	2.00%
EC	29	Repair/Mitigation Method	2.00%
		Total	**100.00%**

Impact	Var #	Variable Name	Variable Weight
IOP	1	Identified Site	17%
IOP	2	Proximity, (m) to Identified Site	7%
IOP	3	Class Location	25%
IOP	4	Potential Property Damage	4%
IOP	5	Potential for Secondary Failure	4%
IOP	6	Crossing Type (Topographic)	4%
IOP	7	Crossing Proximity, (m)	4%
IOP	8	Barriers to Dispersion	4%
IOP	9	Potential Impact Radius, (m).	5%
IOP	10	Program Status - Public Education	2%
IOP	11	Incident Response Time, min	3%
IOP	12	Incident Type By Threat	3%
IOP	13	Pressure Control Systems	8%
IOP	14	Failure Mode	10%
		Total	**100%**

Threat weights, variable weights and attribute scores are given based on experiences or Subject Matter Experts (SME) and brainstorm meetings. Interpretation, analysis, and assessments of the threat and impact index contributions to RoF, PoF, and CoF depends on the maximum and minimum values given for each variable weights and attribute scores.

Risk acceptability refers to the acceptable risk limits or ranges providing the reference to identify whether a pipeline segment achieved the risk goals and objectives of the company. Risk acceptable limits can be individually established for each integrity threat, consequence and associated risk. The probability acceptability can be either a "fixed" or "ranged numerical value" depending on consequence or resulting from a formulation (e.g., acceptable, critical region).

Normal distribution is a representation of the risk, probability and consequence level values that are used for calculation the comparison criteria in an indexing modeling risk assessment.

Figure 6.16 depicts comparison criteria calculation typically used on qualitative risk assessments.

The coefficient **n** may be selected as to 1 (67%), 1.28 (80%), or 2 (95%) based on the level of aversion desired by the company.

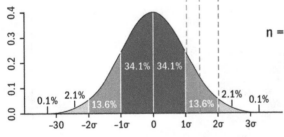

n = 1 (green), 1.28 (orange), 2 (red)

Comparison Criteria (CC) = Mean (σ) + n Standard Deviation

FIG. 6.16   EXAMPLE OF COMPARISON CRITERIA

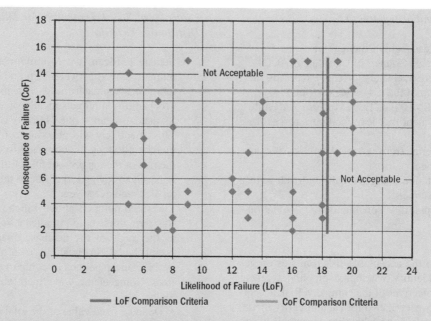

FIG. 6.17 EXAMPLE OF LIKELIHOOD AND CONSEQUENCE ACCEPTABILITY CRITERIA FROM A QUALITATIVE RISK METHOD

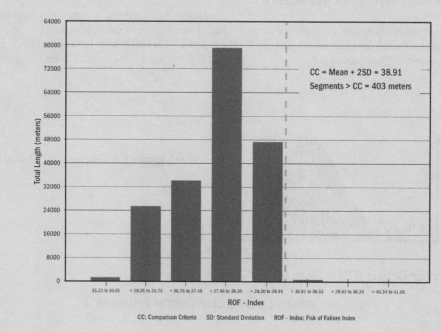

FIG. 6.18 EXAMPLE OF RISK ACCEPTABILITY CRITERIA USING A QUALITATIVE RISK METHOD

When the risk analysis process has been completed, it is necessary to compare the estimated risks against risk criteria established by the company. The risk criteria may include elements such as associated costs and benefits, legal and regulatory requirements, socioeconomic and environmental factors, concerns of stakeholders. Risk evaluation is used to make decisions about the significance of risks to the company and whether each specific risk should be accepted or treated.

6.7.2   Risk Drivers and Drill Down Analysis

There are several ways that the results from a risk assessment can be viewed for evaluation purposes. The basic output is in a grid format, which provides basic functionality for sorting and filtering.

As illustrated in Figure 6.19, graphical outputs are also available for displaying the risk analysis results. The results can be also displayed as linear graphs, GIS, and pivot tables. Drill down reports are also used, which can display the variables that are contributing the most to the overall risk score as well as the potential reduction in risk score contribution for each of these variables.

6.7.3   Sensitivity Analysis: Risk Reduction and Benefit/Cost Ratio

Benefit-cost analysis provides additional elements for ranking decision alternatives. This analysis can evaluate various alternatives as a function of time. Benefit-cost analysis on existing pipeline systems with time-dependent damage mechanisms can assist,

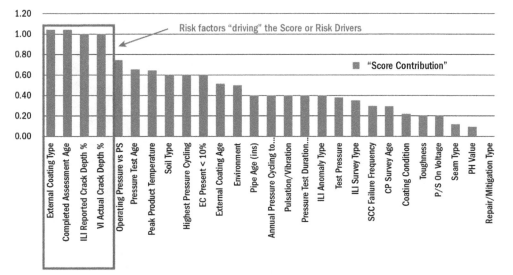

FIG. 6.19   EXAMPLE OF SCC FACTORS AND DRIVERS FROM A QUALITATIVE RISK METHOD

as one of the tools, in identifying the most suitable strategy to mitigate risk and extend their remaining life.

Benefit-cost alternatives depend on each pipeline system case depending on its integrity condition, location, consequences, probability of failure, mitigation options, threats and risk outcomes, and other aspects such as company's risk aversion and policy.

The general methodology for a benefit-cost analysis follows the following steps [15]:

- ΔRisk: Calculate the change in risk through applying each mitigation alternative to the probability of failure and consequence of failure models;
- ΔCost: Estimate the cumulative costs involved with the implementation of each alternative;
- Benefit-cost: Calculate the benefit-cost ratio.

To evaluate the benefit-cost ratio of each alternative, a common base alternative can be taken to compare to the enduring alternatives and implementation cost values against is required. As the purpose of a benefit-cost evaluation is to determine the effectiveness of applying various mitigation alternatives, the base case should represent the pipeline in its unmitigated state or common base alternative. The associated risk of failure is a function of all known damage features and mechanisms.

6.8   REFERENCES

1. *ISO*, Anon., 2009, *Risk management—Principles and Guidelines*, ISO 31000, International Organization of Standardization, Geneva, Switzerland.

2. USA Department of Transportation, 2015, *49 Code Federal of Regulations, Part 192 Transportation of Natural Gas and Other Gas by Pipelines*, U.S. Government Printing Office, Washington, USA.

3. Anon., 2012, ASME B31.4—2012 *Pipeline Transportation Systems for Liquids and Slurries*, American Society of Mechanical Engineers, New York, USA.

4. CCPS, 2009, *Guidelines for Developing Quantitative Safety Risk Criteria*—Appendix A, Center for Chemical Process Safety, New York, USA.

5. Talbot, Julian, ALARP (As Low As Reasonably Practicable), http://www.jakeman.com.au/media/alarp-as-low-as-reasonably-practicable.

6. Nessim, M., Zhou, W., Zhou, J., Rothwell, B., McLamb, M., N., 2004, *Target Reliability Levels for Design and Assessment of Onshore Natural Gas Pipelines*, IPC04-0321, American Society of Mechanical Engineers (ASME), Proceedings of the International Pipeline Conference, Calgary, Canada.

7. BSI, *Pipeline systems. Steel pipelines on land. Guide to the application of pipeline risk assessment to proposed developments in the vicinity of major accident hazard pipelines containing flammables*, PD 8010-3, British Standard Institution, London, United Kingdom (UK).

8. NTSB, *Pacific Gas and Electric Company Natural Gas Transmission Pipeline Rupture and Fire San Bruno, California on September 9, 2010, Accident Report*, National Transportation Safety Board, NTSB/PAR-11/01, PB2011-916501, Washington, DC, 2011.

9. Nessim, M., Stephens, M., Adianto, R., 2012, *Safety Levels Associated with the Reliability Targets in CSA-Z662*, IPC2012-90450, American Society of Mechanical Engineers (ASME), Proceedings of the International Pipeline Conference, Calgary, Canada.

10. Cote, E. I., Ferguson J., Tehsin, N., 2010, *Statistical Predictive Modelling: A Methodology to Prioritize Site Selection for Neutral pH Stress Corrosion Cracking*, American Society of Mechanical Engineers, Proceedings of the 8th International Pipeline Conference, Calgary, Canada.

11. Muhlbauer, 2015, *Pipeline Risk Management: The Definitive Approach and its Role in Risk Management*, Clarion Technical Publishers, Houston, TX, USA.

12. PODS, *What is the PODS Pipeline Open Data Model?*, http://www.pods.org/pods-model/what-is-the-pods-pipeline-data-model/.

13. Gin, G., Davis, J., 2012, *Spill Impact Assessment—Crude Oil Pipeline ERCB #3353-1*, International Pipeline Conference, American Society of Mechanical Engineers, Calgary, AB, Canada.

14. CSA Z662, 2015, *Oil and gas pipeline systems*—Annex B Guidelines for risk assessment of pipeline systems, CSA Group, Toronto, Ontario, Canada.

15. Blackwell, C., Cote, E., Gagne, C., *Decision Making Methodology for Cost Effectively Managing Pipeline Risk*, American Society of Mechanical Engineers, Proceeding of IPC 2012, 9th International Pipeline Conference, Calgary, Alberta, Canada.

PIPELINE INTEGRITY ASSESSMENT METHOD SELECTION AND PLANNING

7.1 INTRODUCTION

Chapter 7 describes the process for selecting the type(s) of integrity assessment methods needed to identify, characterize and assess or simply verify the integrity condition of a pipeline segment. In addition, the planning process supplemented with lessons learned in each section.

The chapter begins with introducing the pipeline integrity assessment methods (i.e., in-line inspection, pressure testing, direct assessment, and alternative methods) focusing on their purpose and technology/technique used, which are discussed in detail in the following chapters within this book. Subsequently, a 10-step process is proposed for selecting the pipeline integrity assessment method that starts with choosing a given integrity threat(s) and reviewing how they can be adequately and effectively detected by all methods. The process explores the feasibility of conducting ILI as a preferable method particularly when the pipeline crosses areas of high consequence.

The selection of individual and combination of integrity assessment types are discussed providing references to recognized worldwide industry standards. Then, rationale are provided in deciding when to perform additional integrity assessment method based on (a) the first selected integrity assessment method limitations; (b) or performance issues (e.g., detection, sizing, and uncertainty, re-inspection interval prediction); or (c) the potential high consequence. Some examples/ cases of performing additional integrity assessment methods are also provided.

Benefits of integrating results from multiple integrity assessment methods and other integrity data are presented with recommended centerline reference methods for data integration. In closing the 10-step process, some guidance on professional and assessment principles are provided developing requirements for conducting an independent quality assurance and control review.

The chapter closes with considerations in planning integrity assessment activities that includes associated lessons learned. From asset management perspective, the planning process is explained in three (3) stages: (1) setting the plan goals and objectives (*What do we want to achieve with the plan?*); (2) Determining the current condition for planning (*Where are we now to plan?*); and (3) Planning enablers and controls (*How can we achieve the plan?*).

7.2 SELECTION OF PIPELINE INTEGRITY ASSESSMENT METHOD TYPE

7.2.1 Pipeline Integrity Assessment Method Types

The pipeline integrity assessment methods can be mainly classified in four (4) types:

1. In-Line Inspection (ILI)
2. Pressure Testing
3. Direct Assessment (DA)
4. Alternative methods

ILI is a non-destructive inspection technique that can be used for integrity assessment of pipelines. The type of ILI survey performed is dependent upon the type of integrity threat that is being assessed. In-Line Inspection pigs are distinguished from other pigs in that they have sensors for collecting and recording information about the condition of the pipeline. In-Line Inspection tools (smart pigs) have sensors for collecting and recording information on the condition of the pipelines, which are distinguished from other types of tools for cleaning or repairs (i.e., utility pigs). In-Line Inspection is explained in detail in the Chapter 8.

As illustrated in Table 7.1, In-line inspection tools can be classified by integrity threat type and technology principle including some examples of the threats listed in the ASME B31.8S:

Pressure testing is a destructive testing technique to detect/eliminate (by failing) the largest defect in the pipeline at the time of the testing that can fail due to internal pressure (i.e., pressure-dependent defect). The test is performed at a pressure higher than the proposed pipeline operating pressure providing a pressure-dependent safety margin to the normal operation. Regarding pressure testing limitations, remaining anomalies of both short and deep dimensions (i.e., leak-dependent defects) or with subsequent growth can cause failure (i.e., leak or rupture) following a successful pressure testing. Pressure testing is explained in detail in the Chapter 9.

As illustrated in Table 7.2, each type of pressure testing has a specific purpose that should be considered in selecting or combining with the integrity assessment method(s).

Direct Assessment (DA) is a non-destructive assessment technique for classifying pipeline regions with common characteristics (i.e., *Pre-Assessment*) that may be experiencing the selected integrity threat (e.g., external corrosion, internal corrosion, or stress

TABLE 7.1   INTEGRITY THREAT TYPE AND IN-LINE INSPECTION TECHNOLOGY PRINCIPLE

Integrity threat type	In-Line Inspection technology principle
Metal loss (e.g., external and internal corrosion, manufacturing-related)	• Magnetism • Ultrasound • Current density [*] • Remote field eddy current
Deformation (e.g., damage such dents, ovalities; construction or weather-related such as wrinkles, buckles, ripples)	• Mechanical fingers • Eddy current • Magnetic • Ultrasound
Cracking (e.g., manufacturing-related in the seam, welding-related in the girth weld or environmentally-assisted such as SCC)	• Magnetism • Ultrasound, phased array • Electro-magnetic acoustics • Self-excited eddy current • Remote field eddy current
Movement (e.g., weather-related/geohazards such as hydro-geotechnical, subsidence, upheaval, earth movements, seismic, bending strain)	• Inertial measurement (i.e., strain, curvature and centerline trajectory)

Note [*]: Current density is used in cathodic protection in-line inspection tools.

TABLE 7.2   PRESSURE TESTING TYPES AND PURPOSE

Pressure testing types	Purpose
Strength	Establish the operating pressure limit of a pipeline segment based on the required safety factor
Leak-tightness	Determine that a pipeline segment does not show evidence of leakage at the time of the testing
Spike or yielding	Improve or restore the pipeline integrity condition (e.g., cracking removal, yielding or remediation) at higher stress levels for a short duration to minimize unnecessary growth and test breaks

TABLE 7.3   DIRECT ASSESSMENT TYPES AND PURPOSE

Direct assessment types	Purpose
External Corrosion (ECDA)	• Buried pipelines with external corrosion
Internal Corrosion (ICDA)	• Dry gas • Wet gas • Liquid petroleum
Stress Corrosion Cracking (SCCDA)	• High-pH SCC • Near-neutral pH SCC

corrosion cracking). The integrity information about those regions is supplemented with field surveys (i.e., *Indirect Inspection*) and validated through excavation (i.e., *Direct Examination*). Provided a good representation of the pipeline condition has been achieved, the results are used for evaluating the effectiveness of the assessment and estimating the reassessment interval (i.e., *Post Assessment*). Direct Assessment is explained in detail in the Chapter 10.

As illustrated in Table 7.3, each type of direct assessment and associated purpose is listed that should be considered in selecting/combining the integrity assessment method(s).

Alternative methods such as inferred condition technologies and data analytics may be used when operational (e.g., pipeline system configuration), technological (e.g., small diameters) and environmental (e.g., water availability and disposal) conditions do not permit the other three (3) main types of pipeline integrity assessment methods.

7.2.2   Process for Selecting Pipeline Integrity Assessment Type(s)

Figure 7.1 provides a 10-step process for selecting one or more types of integrity assessment methods considering their *limitations,*

potential pipeline consequences, feasibility for implementation of the method that may require *combining multiple technologies and integrity assessment methods.*

The 10-step process is described in details below:

{1} *Select a given Integrity Threat*

The 10-step process starts with the *list of applicable integrity threats* resulting from the threat assessment described in Chapter 4. At this time, the selection of an integrity threat is only required to start the process and iterative cycles until all applicable threats are worked through the flow process.

The selection of an integrity assessment method may be different depending on whether the integrity threat has been identified (e.g., field examination) or simply suspected to be in the pipeline (e.g., susceptibility analysis: coating damage, pressure cycling on pre-1970 vintage pipe). This differentiation in the certainty of the presence of the integrity threat may generate different specifications of the selected integrity assessment method(s). Integrity threat susceptibility results would likely provide less detailed information than the integrity threat identification processes about the threat morphology (e.g., dimensions, shapes), failure mechanisms (e.g., growth) and failure classification (e.g., leak or rupture-dependent) making difficult to accurately specify the ILI tool and hydrostatic testing to be used.

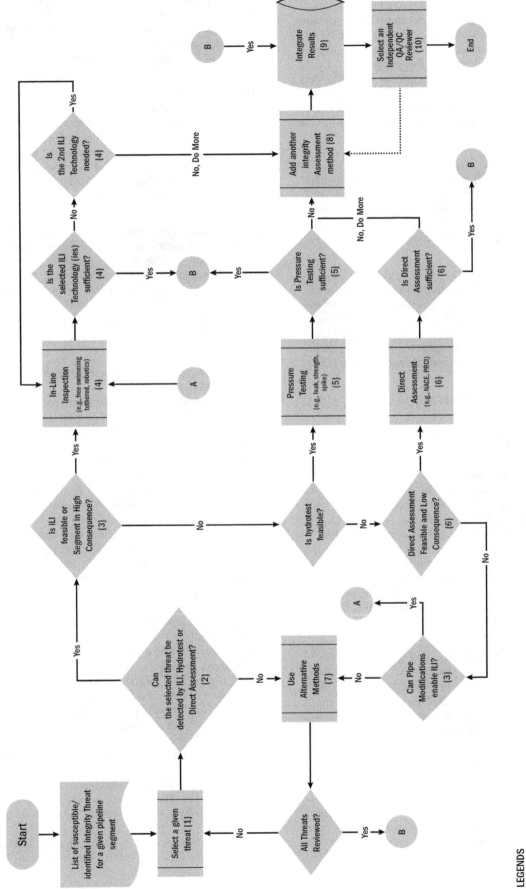

FIG. 7.1 PROCESS FOR SELECTING THE PIPELINE INTEGRITY ASSESSMENT METHOD TYPES

LEGENDS
Activity # explained within this chapter

Lack of morphology information about a susceptible integrity threat (e.g., short and deep versus long and shallow) may affect the selection of the technology for detection and sizing resolution (e.g., ILI super high, high, and standard resolution). Similarly, this information may affect the selection of the stress level (e.g., higher or standard) to be applied to hydrotest for integrity purposes. Every method for assessing or verifying the integrity of a pipeline has its own strengths and weakness. Some methods are focused on a specific type of integrity threat (e.g., corrosion) requiring operational requirements (e.g., flow speed, medium, pressure reduction for excavation).

Example: The ultrasonic wall measurement (UTWM) in-line inspection method may be strong in detecting long and narrow corrosion in natural gas pipelines; however, its limitation of requiring liquid medium (e.g., diesel, water) for running the UTWM ILI tool may need to be considered.

Example: Pressure testing may be suitable in testing multi-diameter Jet A-1 product pipelines over in-line inspection. However, hydrotest has the limitation of introducing water in high quality liquid (e.g., Jet A-1) and gas (e.g., natural gas) pipelines that would require a stringent requirement for water removal to achieve a very dry state at low points after the hydrostatic testing.

{2} *Can the selected integrity threat be detected by ILI, Hydrotest or Direct Assessment?*

Every integrity assessment has limitations related to the type (e.g., corrosion) and/or size of anomaly able to detect (e.g., pinhole versus general corrosion). Some integrity threats may require multiple methods (e.g., Hydrotest and ILI, ILI and Direct Assessment, Hydrotest and Direct Assessment) and/or multiple technologies (e.g., ultrasonic and magnetic flux leakage) complementing each other to reach expected detection capabilities.

In-Line inspection technologies are not yet available for directly measuring integrity threats such as incorrect operations causing overpressure, pipeline equipment and some weather-related (e.g., river bed scour); however, if the damage from the integrity threat can be quantified (e.g., crack growth from overpressure, pipe strain from flooding at a water crossing), these measurements may partially assist as an input for the assessment of those threats.

Example: Dual crack in-line inspections (e.g., EMAT, UTCD or CMFL) on pre-1970 vintage pipelines have considerably advanced, but injurious cracks non-reported by ILIs are still being found in the ditch. Spike or hydrostatic testing higher than 1.25 MOP is helping in removing undetected manufacturing without significantly growing subcritical cracks. Dual crack ILIs and spike testing are jointly working in the detection and removal of cracking increasing the reliability of old pipelines.

Hydrostatic testing is not capable of providing adequate and effective verification of geometry anomalies (e.g., mechanical damage), weather-related threats (e.g., scour, settlement) and incorrect operations unless they contain or produce stress concentrators that could leak under the pressure testing.

Example: Dents even with stress concentrators have passed hydrostatic testing levels even exceeding 100% SMYS without showing a sign of leakage at the time of the testing. However, hydrostatic testing up to burst levels has shown the first releases coming off their stress concentrators.

Direct Assessment (DA) may not be able to provide confidence in assessing the integrity condition of the entire pipeline segment containing threats other than corrosion-related features. The selection of representative excavation sites (i.e., Direct Examination) is an essential step in confirming the analysis and assumptions required for understanding the condition of the pipeline [1]. In some cases, the pipeline sites examined in the field may become the only sites that the assessment can fully rely on.

Example: The NTSB Safety Study in 2015 identified limitations of the DA methodology and its application in preventing failures when used as the only integrity assessment method. Direct Assessment (DA) is limited to the corrosion-related threats being looked for and the areas investigated in the field. Caution is recommended in only selecting DA for corrosion identified as a high relative risk or within areas of high consequence.

{3} *Is ILI feasible or Can Pipe Modifications enable ILI? Is the segment in High Consequence?*

The first recommended option for conducting an integrity assessment method is the use of In-Line Inspection technologies individually or in combination with other methods. However, the feasibility of the in-line inspection may be limited by the current pipeline configuration (e.g., diameter restrictions) and operation (e.g., flow rates). Even if the feasibility to conduct an ILI is deemed to be low, senior management effort should make every effort possible to conduct in-line inspections in areas of high consequence.

The configuration of the pipeline may not have traps for launching and receiving in-line inspection tools or the trap's lengths are not sufficient for longer tools. The pipeline may have configuration that would not allow a standard in-line inspection tool to go through without get damage or get stuck. The following are some of the common restrictions to look for to be resolved:

- Cleanliness of the pipe (e.g., sediments)
- Mitre bends
- Short bend or elbow radius (e.g., <3.0 diameter radius)
- Diameter changes (e.g., NPS 12 pipeline with NPS 10 valves)
- Valve types (e.g., piston, deep or long seat)
- Heavy wall (e.g., drastic changes getting the tool speeding up)
- Excessive girth weld protuberance (e.g., damaging tool caps)
- Unbarred Tee of similar diameter to the pipeline (e.g., tool getting into the tee)
- Dents (e.g., tool getting stopped and may be damaged)

Operational complications may also reduce the feasibility of conducting an ILI. The following are some of the common complications to look for to be resolved:

- Very low or high flow speed (e.g., for ILI data collection)
- Intermittent flow (e.g., for ILI data collection)

- Very low or high pressure (e.g., for ILI propulsion or instrumentation limits)
- Slack flow (e.g., for ILI displacement, couplant medium)
- Very low or high temperature (e.g., for ILI instrumentation limits)
- Illegal tapping presence (e.g., tool damage)

The ILI plan must consider an emergency plan in case the tool is stuck in the pipeline. The tool shall have a transmitter device to find it in case it is stuck in the pipeline.

Sometimes special modifications to the ILI tool may be possible; however, disadvantages associated to reduction in the ILI data collection and resolution as well as residual risk of the ILI tool damage may still exists.

{4} *Selecting In-Line Inspection*
For In-Line Inspection, API RP 1160-2013 Table 1—In-line Inspection Tools and Capabilities provides guidance about the Magnetic Flux Leakage (MFL), Ultrasonic (UT), caliper and inertial measurement unit (IMU) tool capabilities [2]. They are compared against their ability to detect and size anomalies such as corrosion (e.g., internal, external), cracking (e.g., selected seam corrosion, SCC), geometry (e.g., dents, wrinkles, buckles), manufacturing (e.g., expanded pipe, laminations, hard spots) and movement (e.g., strain). For more details, Chapter 8 of this book explains the ILI technologies, how they work, impacts on their performance, and their selection by pipeline integrity threat.

ILI cracking technologies have shown significant progress. However some challenges in detecting or accurately sizing relevant anomalies due to

- Variety of morphology (e.g., crack opening, shape, dimensions and discontinuous flaws)
- Adjacent to pipe geometry shapes (e.g., weld corners/caps)
- Manufacturing imperfections causing noise or "shading" responses on the ILI raw data

The use of a 2nd or a 3rd ILI technology in the same pipeline being run back-to-back has helped "complementing" the detection capabilities of individual technologies. In addition, multiple ILI raw datasets of different ILI technologies on the same pipe joints provides the opportunity for capturing multiple "views" of the same anomaly better characterizing its morphology and may be revealing coincidental anomalies (e.g., dents with cracking, cracking with corrosion under disbonded coating). Furthermore, having a good understanding of sizing, detection, and identification capabilities of the technology using multiple ILIs with "ILI raw data overlay" may be able to eliminate false critical anomalies (e.g., deep cracking with high amplitude, but corresponding to a non-injurious seam weld thickness change from thicker to normal thickness).

Example: Circumferential Magnetic Flux Leakage (CMFL) technology has been identified to detect axially-oriented crack-like manufacturing anomalies. CMFL has shown performing better on manufacturing cracks with opening (crack width) than tight cracks (fatigue-induced) requiring other ILI technologies such Ultrasonic (e.g., UTCD, EMAT, Phased Array) for increasing the detectability success.

{5} *Selecting Hydrostatic Testing*
For hydrostatic testing, CSA-Z662 [3], ASME B31.4 [4] and B31.8 [5], and AS 2885 [6] industry standards provide guidance about the pressure testing process, types (i.e., strength, leak, spike), test medium and minimum test duration. Chapter 9 of this book explains the role of pressure testing in pipeline integrity management, objectives and scope of the different types of pressure testing, pressure test failures and other considerations such as test lengths, elevation, backfill and restraint effect and crossings.

Example: Finding girth weld defects during in-service inspections is challenging. Under the uncertainty of unknown remaining defects or their growth sources, hydrotesting has provided an option for checking whether the safety factor over the operating pressure can be withstood at the time of testing (no leakage or rupture) and the operation can be restarted. Most likely, monitoring of growth of girth weld cracks may be needed with either retesting intervals or in-line inspection.

The following are some of the common environmental challenges in conducting hydrostatic testing to be evaluated:

- Water quality and chemical components (e.g., may induce internal corrosion)
- Water availability
- Water disposal after the test (e.g., contaminated with product)

{6} *Selecting Direct Assessment*
For Direct Assessment, NACE SP0502 [7], SP0206 [8], SP0110 [9], SP0208 [10], and SP0204 [11] provide guidance to identify locations where corrosion integrity threats can be formed (i.e., new) or growing (i.e., existing corrosion) in the future driven by conditions that are reviewed at the locations selected for field investigation. NACE SP0210 [12] also provides guidance to validate previous corrosion assessments verifying if the upcoming reassessment interval is still appropriate or may need to be reduced.

Example: Here are some typical conditions reviewed during the direct assessment. For *internal corrosion*: water/entrained electrolytes, flow velocities and patterns, residue accumulation, temperature profile. For *external corrosion*: cathodic protection effectiveness, coating condition; and *near-neutral pH Stress Corrosion Cracking (SCC)*: coating type and age, operating stress, r-ratio.

Some of the common issues to be considered for direct assessment are:

- Land permit to perform the excavations
- Environmental restrictions
- Information alignment quality and certainty
- Capability and tolerances of the tools to be use

{7} *Selecting Alternative Integrity Assessment Methods*
Some regulators and industry research associations have identified a few alternative methods to the three (3) previously discussed for assessing the integrity of pipelines.

Example: Guided Wave or Long Range Ultrasonic Technology (GW/LR UT) has been used for short non-piggable pipeline segments (e.g., road crossings, pipe with internal/external insulation) to identify larger metal loss anomalies where high sizing accuracy is not required; however, GW/LR UT is highly dependent on data interpretation with limited performance close to appurtenances.

Any alternative integrity assessment method is intended to cover the overall pipeline segment for integrity verification purposes; otherwise, the use of any alternative method on a sample of the pipeline segment length would become a step of a Direct Assessment process (i.e., direct examination) [13].

{8} Add another Integrity Assessment Method

The following are some of the reasons that may lead to adding another integrity assessment method:

1. Detection—when the primary method has limitations in detecting the threat(s) objective of the integrity assessment
2. Sizing—when the primary method may not be able to accurately size or dimension of anomalies that may become critical
3. Uncertainty—when the uncertainty of the results from the primary method is high enough that would not allow preventing a threat(s) from becoming critical
4. Higher consequence—when the consequence can reach levels high enough that would make the risk intolerable, even if the probability of failure of the threat may be low
5. Re-inspection interval prediction—when the primary method and existing integrity information may not be able to predict a safe and reliable re-inspection interval (i.e., before a critical threat fails). The key inputs for prediction are current condition and growth
6. Failures occurrence—when failures have occurred even though some ILI technologies have already been used

7. Modification of operational requirements—increase the operational pressure, change of flow direction and or abandonment
8. Large pipeline modifications—variants or realignments

Table 7.4 provides examples of three (3) cases that may illustrate the focus in adding a secondary integrity assessment method to complement the primary method based on identified/susceptible threats.

{9} Integrate Results

The following are some of the key benefits obtained from integrating the results from integrity assessments:

- Better understanding of the criticality of the integrity threat providing additional information such as morphology (e.g., shape, extent, dimensions), coincidental anomalies and adjacent features. The integrity criticality may increase due to the integration.
- Repeatability of the integrity threat providing confidence of their type (e.g., crack), morphology, and location.
- Identification of potential conditions or hazards contributing to the initiation (e.g., coating damage) or growth of the integrity threat.
- Prioritization of pipeline sites based on the consequence assessed at the identified integrity threat requiring a sooner remediation response.
- Optimization for resource allocation as result integration increases the effectiveness of the integrity work.

The integration requires using a common reference for comparing datasets, which can be achieved pipeline chainage or stationing ensuring that girth welds have been matched obtaining the same joint length among the datasets. This process may require stretching or shrinking the pipeline chainages in conducting the matching; it also requires a more detailed analysis when some segments of the pipeline are changed. Global Positioning System (GPS) coordinates of the pipeline centerline. Geographical Information Systems

TABLE 7.4   EXAMPLE OF ADDING INTEGRITY ASSESSMENT METHODS

Example	In-Line Inspection (ILI) focus	Pressure Testing (PT) focus	Direct Assessment (DA) focus
Case A: seamless pipeline with multiple threats	***Primary Method*** External corrosion; dents with stress concentrators	If pipeline not susceptible to girth weld defects, PT may not be needed	***Add Secondary Method*** Coating condition and cathodic protection effectiveness
Case B: pre-1970s vintage seam welded and tape coated pipeline with multiple threats	***Primary Method*** External corrosion; selective seam weld corrosion	***Add as a Secondary Method*** Hydrostatic testing at higher stress level and short duration (e.g., spike testing) followed by leak tightness testing	Tape coating disbondment may not allow for reliable cathodic protection information disabling the DA method
Case C: pre-1980s vintage seam welded and multiple diameter pipeline with multiple threats	**Note**: *If areas of high consequence, pipe modifications are strongly recommended to do ILI as soon as possible*	***Primary Method*** Hydrostatic testing at higher stress level and short duration (e.g., spike testing) followed by leak tightness testing	***Add Secondary Method*** Internal Corrosion Direct Assessment

(GIS) are used as a platform referencing the pipeline datasets (e.g., latitude, longitude, elevation).

The following are some examples of integrity analyses derived from the integration of the results of the integrity assessment methods:

Example: In a hazardous liquid pre-1970 vintage pipeline, integrating two (2) different types of In-Line Inspection technologies (ILIs) such as cracking and metal loss may reveal a mechanically-driven crack in a corrosion area. The crack growth may be faster due to the environmental effect of corrosion. The crack may be partially growing by pressure cycling-induced fatigue and also be exposed to environmentally-assisted corrosion growth. The overall area of both coincidental anomalies may also be larger than the individual anomalies showing the contributing hazard factors such as coating damage and areas of ineffective cathodic protection.

Example: In a High Vapor Pressure (HVP) butane post-1980 vintage pipeline, integrating a caliper ILI with the pipeline listing from another ILI may reveal a dent or deformation on or in proximity to seam/girth weld. The weld could have been impacted during the occurrence of the deformation worsening up the condition of the weld anomalies by cold hardening, inducing micro cracking and reducing their remaining life. If mechanical damage occurred, coating damage would likely create conditions for corrosion on the dent and weld surface.

Example: In a natural gas pipeline post-2000 vintage pipeline, integrating Internal Corrosion Direct Assessment (ICDA) findings with metal loss ILI in a natural gas pipeline may differentiate shallow internal corrosion from internal manufacturing anomalies. ICDA for natural gas pipelines identifies potential stagnant liquid locations where internal corrosion could occur based on the pipeline slopes, gas specific gravity, and flow speed. Those ICDA susceptible locations matching ILI-reported internal metal loss may have a higher likelihood of being internal corrosion.

{10} *Select an Independent Quality Assurance and Control (QA/QC) Reviewer—Best Practice*

Pipeline industry and regulators have learned that activities with higher change sensitivity (e.g., ILI-reported crack sizing) on risk drivers (e.g., leak or rupture) require a higher level of confidence in the input and assessment to make safer integrity management decisions in preventing releases. The selection of an independent quality assurance and control reviewer(s) has demonstrated adding value in achieving integrity assessment results.

Governments, regulators, and industries have developed guidelines for conducting independent verification of programs (e.g., safety, environmental) that outline principles and processes that may assist the pipeline integrity professionals and managers in identifying the applicable attributes of a verification process, the roles and responsibilities and an expected verification output.

For instance, the Australian Government expects that verifiers adopt the following *professional principles* when performing *Generator Efficiency Standards* (GES) independent verifications [14, 15] that could be applied for pipeline integrity reviews:

- Independence
- Ethical Conduct
- Fair Presentation
- Due Professional Care

Furthermore, the Australian Government has also identified the following *assessment principles* drawn from ISO/DIS14064 Parts 1–3 that should be expected from the plans and reports subject to verification:

- Completeness
- Consistency
- Accuracy
- Transparency
- Relevance
- Conservativeness

The independent QA/QC reviewers are typically focused on the review of following aspects of the integrity assessment:

1. Representativeness of the **input** (e.g., data completeness and quality),
2. Adequacy and effectiveness of the **processes** (e.g., ILI raw QA/QC review), and
3. Missing or incorrect **output** and interpretation (e.g., ILI detected anomalies, but not processed).

The following are some activities recommended to be included in the scope of an independent quality assurance and control review:

Example: Review of Integrity Assessment Method Selection

- Review of the pipeline questionnaire for validity and accuracy
- Due diligence review of the ILI feasibility study (e.g., pipeline modifications, consequence)
- Effectiveness review of the selected ILI technologies and tools for a given integrity threat(s)
- Review of pressure testing design and plan for adequacy minimizing threat growth
- Feasibility review of using combined integrity assessment methods

Example: Review of Integrity Assessment Results

- QA/QC Review of ILI raw data for unreported and misclassified anomalies
- Review of actual pressure testing performance and potential secondary effects
- Representativeness review of the selected dig sites for DA direct examination

7.3 INTEGRITY ASSESSMENT PLANNING: PROCESS AND LESSONS LEARNED

Once the methods have been selected for conducting the integrity assessment (e.g., multiple ILIs, hydrostatic testing, DA and combination), a process for planning their execution capturing lessons learned from either internal or external sources is recommended. The process can be comprised of setting the goal and objectives, determining the current condition for planning and ensuring enablers and controls are in place to start implementation. Lessons learned are described as examples in each subsection.

7.3.1 Setting Plan Goal and Objectives: *What Do We Want to Achieve with the Plan?*

The goal and objectives of an integrity assessment execution plan are typically defined by the following factors:

a. Personnel safety in executing the integrity assessment method
b. Avoid impacts to the operation including flow, pressure and temperature conditions
c. Integrity assessment method setup, operations and field verification ensuring prevention of time and cost losses
d. Quality of the integrity assessment data collection, processing and interpretation
e. Delivery time of integrity assessment results

Example: Integrity Assessment Method Plan Goal and Objectives
Goal: *Safe, efficient and cost effective in-line inspections required for successfully dimensioning the extent and severity of the identified integrity threats*
Objectives

- Safe ILI field and office execution (e.g., zero incidents)
- Successful ILI runs collecting complete and quality data (e.g., zero re-runs due to either pipeline conditions or ILI tool malfunctioning)
- Timely delivery of the field, preliminary and final reports (e.g., zero delays)
- Cost effective implementation of the ILIs (e.g., right cost for the right job, zero budget exceedance)

7.3.2 Determining Current Condition for Planning: *Where Are We Now?*

At the highest level, the understanding of the current condition for developing the plan can be started by identifying the corporate expectations associated with the integrity assessment execution. Senior management expectations for safety, quality, operational requirements, and timing would drive the plan requirements. In reciprocity, senior management would provide the commitment necessary (e.g., resources, authorizations, timely responses) for achieving the plan objectives.

Looking at the internal context, previous internal events such as incidents, overspending, lack of qualifications or delays may influence the focus of the plan as the corporation would be looking at actions to prevent reoccurrence in the plan being developed.

Looking at the external context, compliance with new or existing regulations, technology improvements, and company standards will define the framework, and sometimes, the schedule to be met. For instance, planning requirements for the first (i.e., baseline) versus the recurrent (i.e., continual) integrity assessments may differ in the review process (e.g., previous critical anomalies, increase of consequence areas, risk changes).

Example: Current Condition for Planning
In this hypothetical case, senior management is concerned of the effectiveness of the crack identification, mitigation and monitoring of pre-1970 vintage pipelines.

- Pipeline incidents have caused safety, environmental and property damage consequences as well as supply restrictions to refineries and communities deriving in societal and regulatory higher risk aversion (e.g., internal or external context)
- ILI technology limitations in adequately detecting cracking as well as reliable methodologies for criticality assessment is requiring faster implementation of applied research and redundancy technological inspection and hydrostatic testing for MOP verification (e.g., external context)
- Shortage of qualified personnel is limiting a faster development (e.g., internal and external context)
- Increased funding supported by resources need to be in place for faster implementation (e.g., internal context)

A baseline assessment plan is a program that provides the first complete information of the condition of the pipeline. Continual assessment plan provides an understanding of the current condition and the changes related to the previous assessment enabling for projecting future activities based on growth.

Baseline Assessment Plan typically uses integrity assessment method(s) such as ILI tools, pressure test, direct assessments, other alternative technologies or a combination of them. Some jurisdictions require companies to develop a Baseline Assessment Plan (BAP) focusing on the following:

- Identify all potential threats and threats, and supporting information,
- Select methods to assess the integrity of the pipeline,
- Schedule for completion the integrity assessment of all segments, including all contributing risk factors,
- Plan direct assessment depending on the threats to be addressed, if applicable, and
- Develop a procedure to ensure that the BAP minimizes environmental and safety risk

---

Example: Baseline and Continual Assessment Plan Process

7.3.3 Planning Enablers and Controls: *How Can We Achieve and Control the Plan?*

An *effective organizational structure with clear accountabilities, responsibilities and roles* for the integrity assessment method (IAM) planning and execution would provide the first planning enabler in demonstration of senior management commitment [16].

A plan enabled by *activities prioritized by risk* would provide timely identification, assessment, mitigation, prevention, and monitoring of risks. These activities would work towards to the senior management goals and objectives associated with safe operation and risk reduction.

A plan enabled by *effective coordination* including *communication, participation and consultation* with integrity, engineering, field implementation, product supply and operations, right-of-way and lands, vendors and contractors personnel.

Example: Coordination via Communication, Participation, and Consultation

The *coordination with integrity engineering* could focus on the participating in the development of integrity assessment method planning specifications and requirements for processes such as pipeline readiness (e.g., temporary or permanent traps, testing heads), technology customization (e.g., spike test instrumentation, ILI tool selection), operational conditions (e.g., flow speed, cleanliness, medium) and run implementation and acceptance.

The *coordination with field implementation* could focus on communicating and participating in the verification of the feasibility/practicality of the proposed integrity engineering requirements. This feedback would also provide solutions to the anticipated issues and increase ownership for a teamwork implementation.

The *coordination with product supply and operations* could focus on consulting to timely synchronize the flow throughput and rates, potential shutdowns, special operations during launching and receiving ILI tools or hydrostatic testing or pressure reductions during the DA excavations. Furthermore, control rooms need to be timely informed of any upcoming activities on the pipeline and right-of-way.

The *coordination with right-of-way and lands* could focus on communicating the plans and schedule to initiate dialogues, notifications and potential negotiations with landowners for confirming right-of-way access to sites as well as agreement upon potential land remediation resulting from activities such as ILI run tracking, hydrostatic testing failures and water disposal, and DA excavations.

The *coordination with vendors and contractors* could focus on determining their personnel, equipment, and material availability for the proposed schedule as well as communicating the specifications and requirements. Cost savings may result from longer term planning, coordination-based increased volume and market competition, which all require enough time in advance to get worthwhile cost effectiveness.

A *Management System (MS)* to execute, monitor and continuously improve the plan for achieving the integrity goals defined by the Integrity Management Program (IMP). MS processes such as training, qualification, and competency are enablers for executing the plan with the quality and time expected.

Example: Training, Qualification, and Competency

The *training* program could be developed for pipeline integrity engineers in areas of planning and/or implementation requiring both technical and project management curriculum. Integrity technical curriculum could focus on understanding of integrity hazards and threats (i.e., initiation and growth), defect assessment (i.e., criticality and acceptability), inspection, and repair methodologies. Integrity project management curriculum could focus on planning, scheduling, and cost control as well as purchasing and contracting, vendor performance and project close-out.

Example: Training, Qualification, and Competency

The *training* program could be developed for pipeline integrity engineers in areas of planning and/or implementation requiring both technical and project management curriculum. Integrity technical curriculum could focus on understanding of integrity hazards and threats (i.e., initiation and growth), defect assessment (i.e., criticality and acceptability), inspection, and repair methodologies. Integrity project management curriculum could focus on planning, scheduling, and cost control as well as purchasing and contracting, vendor performance and project close-out.

The *qualification and competency* program could be defined by oral, written, and/or on-the-job verification by a senior integrity engineer or manager. Activities such as defect assessment, in-line inspection and repair method selection can be conducted by the trainee with the guidance and supervision (or approval) of the senior personnel.

A plan enabled by *document and records management* providing up-to-date procedures and records for planning and implementation of the integrity activities. The following are some examples of key integrity activities requiring detailed procedures created, updated, and approved prior to their implementation and previous records:

Example: Procedures and Records

- *Safety procedures for* each step (e.g., hazard identification and safety analysis, check-list, permits, previous incidents and lessons learned)
- *In-Line Inspection* (e.g., prior to shipping review, field verification, tool launch, station by-pass, tracking, tool receive and run approval, previous ILI results and field findings)
- *Hydrostatic Testing* (e.g., stress levels and duration, water collection and disposal, failure-repair-response, successful test criteria, pipe manufacturing and hydrotest failure history)
- *Excavation and Repair plan* (e.g., site preparation, excavation length and selected anomalies for NDE, proposed repair method and integrity support to the field, site remediation, previous repairs)
- *Non-Destructive Examinations*—NDE (e.g., data gathering for metal loss, geometry and cracks and comparison to ILI or DA, NDE interaction with integrity, acceptance criteria and reporting)
- *Fitness-for-Service Assessment* (e.g., expert qualifications, objectives and deliverables, field verification of assumptions, mitigation, prevention and monitoring)
- *Pressure Reduction* (e.g., criteria for initiation and lifting pressure reduction)

A plan with *process controls* of resource leveling, schedule and costs should be developed. The process controls should be focused on timely measuring the progress of the integrity activities with the purpose of early identify variances from the plan and/or anticipate future deviations that could be prevented. Process controls supported by Key Performance Indicators (KPIs) provide management with the assurance, visibility, and opportunity for timely actions.

The following are some examples of KPIs for process controls of integrity activities:

Example: Process Controls KPIs
Input KPI (e.g., integrity plan resources)

- *Number of work/purchase orders (WO/PO) planned versus issued*

Progress KPIs (e.g., integrity plan activities)

- *Percentage of WO/PO planned versus completed ()*
  - *(%) On Time*
  - *(%) Delayed*
    - *Average and maximum of delayed days)*
  - *(%) On Budget*
    - *($) Exceeding Cost and Percentage (%) over the planned Budget*

Output KPIs (e.g., integrity plan)

- Number of Quality Non-Conformances
- Number of successful in-line inspections, digs, cathodic protection surveys completed

Outcome KPIs (e.g., failure prevention objective)
- Number of critical defects repaired
- Number of integrity responses (e.g., ILI, Dig) conducted before deadline (e.g., regulatory, engineering assessment-driven)

A plan enabled by *Management of Change (MoC)* continuously identifying the changes in processes and procedures as well as conditions (e.g., pressure reduction). Some of the benefits of creating an MoC are informing the parties involved to either assess/verify or get prepared prior to the change takes place.

Example: Management of Change for a Pressure Reduction

The MoC should describe the reason, rationale or justification (e.g., anomaly reported by ILI) for the pressure reduction (e.g., 20% of the highest pressure experienced during the last 60 day period) affecting some of the station-to-station sections.

The change would likely affect the planned flow throughput and delivery to customers requiring communication to the supply team as well as operations for setting up the new maximum pressure set limits. They should be identified in the MoC for review, endorsement, and/or approval. Field implementation team may also be informed for starting the planning of the repairs.

Senior management may also be participating in the MoC to ensure a safe start-up and/or lifting the pressure back to MOP, when applicable.

7.4 REFERENCES

1. NTSB, 2015, *Integrity Management of Gas Transmission Pipelines in High Consequence Area—Safety Study*, NTSB/SS-15/01, PB2015-102735, National Transportation Safety Board, USA http://www.ntsb.gov/safety/safety-studies/Documents/SS1501.pdf.

2. API, 2013, *Management System Integrity for Hazardous Liquid*, RP 1160-2013, Second Edition, American Petroleum Institute, USA.

3. Anon., 2015, *CSA-Z662 Oil and gas pipeline systems*, Canadian Standard Association, Ontario, Canada.

4. Anon., 2012, ASME B31.4—2012 *Pipeline Transportation Systems for Liquids and Slurries*, American Society of Mechanical Engineers, New York, USA.

5. Anon., 2012, *ASME B31.8 Gas Transmission and Distribution Piping Systems*, American Society of Mechanical Engineer, ASME Press, NY, USA.

6. AS, 2012, *Pipelines—Gas and liquid petroleum—Design and construction, AS 2885.1*, Australian Standards, Australia.

7. ANSI/NACE, 2010, *Pipeline External Corrosion Direct Assessment Methodology (ECDA)*, NACE SP-0502, National Association of Corrosion Engineers—International, USA.

8. NACE, 2006, *Internal Corrosion Direct Assessment Methodology for Pipelines Carrying Normally Dry Natural Gas (DG-ICDA)*, SP0206, National Association of Corrosion Engineers—International, USA.

9. NACE, 2010, *Internal Wet Gas Internal Corrosion Direct Assessment Methodology for Pipelines (WG-ICDA)*, SP0110, National Association of Corrosion Engineers—International, USA.

10. NACE, 2010, *Internal Corrosion Direct Assessment Methodology for Liquid Petroleum Pipelines (ICDA)*, SP0208, National Association of Corrosion Engineers—International, USA.

11. NACE, 2015, *Stress Corrosion Cracking (SCC) Direct Assessment Methodology (SCCDA)*, SP0204, National Association of Corrosion Engineers—International, USA.

12. ANSI/NACE, 2010, *Pipeline External Corrosion Confirmatory Direct Assessment (ECCDA)*, SP0210, National Association of Corrosion Engineers—International, USA.

13. USA PHMSA, 2012, *Hazardous Liquid Integrity Management: Frequently Asked Questions*, U.S. Department of Transportation—Pipeline and Hazardous Materials Safety Administration (PHMSA), Pipeline Technical Resources FAQ # 6.23, USA https://primis.phmsa.dot.gov/iim/faqs.htm.

14. Australian Government, 2006, *Independent Verification Guidelines*, U.S. Department of the Environment and Heritage, Australian Greenhouse Office, Australia.

15. EQE International Ltd, 1996, *Guidelines for Preparing a MODU Verification Scheme*, International Association of Drilling Contractors, North Sea Chapter, Report No 107-12-R-1, Australia.

16. BSI, *Asset Management, Part 2: Guidelines for the application of PAS 55-1*, PAS 55-2: 2008, British Standard Institute, The Institute of Asset Management, United Kingdom.

IN-LINE INSPECTION

8.1 INTRODUCTION

A generic term for a device inserted into a pipeline and then freely traverses the line is a "pig." Pigs come in many shapes, materials, colors, and functions. From a functional perspective, there are cleaning pigs, inspection pigs, and repair pigs. The focus of this chapter is on smart pigs that perform in-line inspections (ILI) surveys.

In-line inspection is a non-destructive inspection technique that can be used for integrity assessment of pipelines [1, 2]. Smart pigs are distinguished from other pigs in that they have sensors for collecting and recording information about the condition of the pipeline. There are many different types of in-line inspection types. The type of ILI survey performed is dependent upon the type of integrity threat that is to be addressed and assessed [3, 4]. Since vendor's initial product launch of the ILI tools, several hardware generations (ILI vehicle, electronics and sensors), and ILI software and data analysis developments and enhancements have resulted from operator's needs and the engineering evaluation. For instance, an inline inspection service capable of continuously identifying and differentiating the pipe grade was lately developed to support the process of validating the MOP (maximum operating pressure).

Pipeline operators typically contract high resolution in-line inspection technologies, which in conjunction with specialized analytical software result in location, detection, identification and sizing of anomalies, appurtenances, features, and defects related to metal loss, cracking, mechanical damage, and manufacturing threats [5].

For instance, Magnetic Flux Leakage (MFL) tools are typically used to detect and size both internal and external metal loss features. Straight beam ultrasonic (UTWM) tools also have the ability to detect and size internal and external metal loss with the added ability to detect and size deformations and laminations that occur in the wall of the pipe. Shear wave ultrasonic devices (UTCD) are best used for detection of crack-like features. Electro Magnetic Acoustic Transducer (EMAT) tools were developed to detect and size axially aligned crack anomalies in dry gas pipelines. Geometry tools are used primarily for detection of deformation features (e.g., dents, wrinkles, buckles, etc.). There are also inertial mapping tools that are able to map the route of the pipeline, detect land movement and identify locations of strain in the pipeline. In various applications, eddy current technology (EC) is used (e.g., mounted to some geometry ILI tools) for a touch less determination of ovalities or dents. In other cases, eddy current sensors are used to differentiate between internal and non-internal anomalies or to scan the internal pipeline surface for shallow corrosion or cracking. Remote Field Technology (RFT-MFL) has been recently developed for detecting corrosion and girth weld cracking in pipelines with internal coating. There are nuisances to all of these inspection techniques that we will cover in subsequent sections of this chapter.

The following is a summary of how in-line inspection technologies that can be used to assess threats or hazards.

- Metal Loss—MFL, UTWM, EMAT, EC, RFT
- Deformation—MFL, UTWM, EMAT, EC
- Cracking—MFL, UTCD, Phased Array, EMAT, EC, RFT
- Movement/Mapping—Inertial
- Cathodic Protection—Current Mapper

As illustrated in Figure 8.1, this chapter follows a Management system approach, as follows: PLAN, DO, CHECK, and ACT. There is brief discussion of the options available for ILI. Following the identification of ILI technologies, there will be a discussion on pipeline threats that the ILI tool can detect and size. In some cases, ILI tools will have the ability to detect a feature but not size the feature.

Terminology this chapter aims to use common pipeline industry terminology when referring to ILI. API 1163 has been selected as the consistent standard for terminology [6]. The most relevant terms are listed below. Additional terms can be found in API 1163.

Anomaly: An unexamined deviation from the norm in pipe material, coatings, or welds, which may or may not be a defect

Appurtenance: A component that is attached to the pipeline (e.g., valve, tee, casing, instrument connection)

Characteristic: Any physical descriptor of a pipeline (e.g., grade, wall thickness, manufacturing process) or an anomaly (e.g., type, size, shape)

Defect: A physically (e.g., field, laboratory) examined anomaly with dimensions that exceed acceptable limits

Detect: To sense or obtain an indication from a feature

Feature: Any physical object detected by an ILI system

Identification: Generally understood to be the delineation of the type of feature (e.g., classification); however, this may be extended to include secondary characterization of an anomaly where mitigation decisions hinge around such characterization

Imperfection: An anomaly with characteristics that do not exceed acceptable limits

Indication: A signal from an ILI system

Probability of Detection (POD): The probability of a feature that has been detected by an ILI tool

Probability of Identification (POI): the probability of a detected feature that has been correctly classified (e.g., as metal loss, dent)

Service Provider: Any organization or individual providing ILI services

Sizing Accuracy: The accuracy with which an anomaly or characteristic measurement is reported

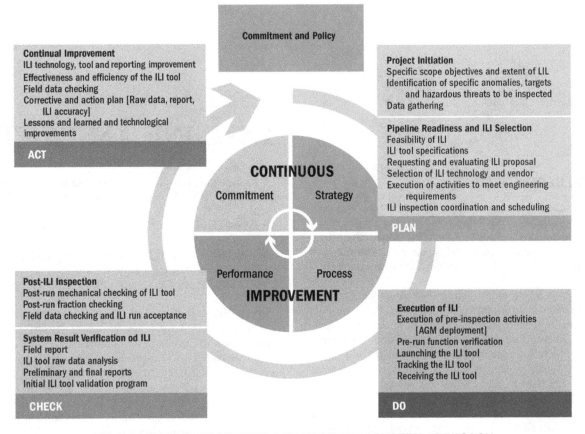

FIG. 8.1 ILI PROCESS WITHIN A MANAGEMENT SYSTEM APPROACH

8.2 IN-LINE INSPECTION (ILI) PROCESS

The pipeline industry typically relies on non-destructive inspection to ensure the serviceability of the pipeline transportation infrastructure. Pipeline operators primarily develop the inline inspection process for conducting ILI inspection activities, having knowledge of pipeline condition, and performing structural pipeline analysis. ILI service companies dedicated to providing solutions that overcome the physical limitations of the pipeline structure mainly perform the ILI inspection.

A general process of an ILI program is outlined in the Figure 8.1. The objective of the ILI process in this context is to ensure that all holistic, systematic, and coordinated activities of an inline inspections program follow the principles of management system PLAN-DO-CHECK and ACT. This process applies to all inline inspection technologies such as MFL, UTCD, EMAT, Caliper, UTWM, Inertial, EC, and a combination of them.

Some aspects of the ILI process include:

- Selecting of the most suitable ILI tool on tool capability
  - Anomaly locating, detecting, identifying and sizing
- Determining the performance specifications for ILI tools for the threat being assessed
  - Probability of detection (POD), probability of identification or characterization (POI), sizing (POS) accuracy and their confidence level
- Understanding requirements for a successful pipeline system operational verification
  - Minimum engineering requirements for: pre-inspection, inspection, and post-inspection.

- Performing an in-line inspection activity systematically with a detailed cost control
- Identifying the piggability of a pipeline
- Verifying the reported inspection results to confirm that the minimum performance requirements of the ILI tool is met
- Agreeing on data analysis methods for evaluating reportable features (e.g., Modified B31G method for metal loss anomaly, API 579 for seam weld cracking, rules for anomaly interaction)
- Developing a growth rate analysis on dimensional changes for matched anomalies (e.g., ILI consecutive run comparison) and associated re-inspection interval

8.3 SELECTION OF IN-LINE INSPECTION TECHNOLOGY BASED ON PIPELINE THREAT

Integrity assessment by in-line inspection (ILI) is very common in the pipeline industry. ILI tools are selected based on the types of features that they are able to either detect and size or just detect. The particular operational conditions of the pipeline are considered in the selection process. This section will provide an overview of the ILI technologies that are available in industry today and the types of features the tools can assess. There are a number of documents available in the pipeline industry that provides guidance on the performance of ILI tools by pipeline threat. These documents include:

- American Petroleum Institute's (API) Recommended Practice 1160—Managing System Integrity for Hazardous Liquid Pipelines (API RP 1160)

- API Recommended Practice 579—Fitness for Service (API RP 579)
- ASME B31.8S—Supplement to B31.8 on Managing System Integrity of Gas Pipelines (B31.8S)
- API Recommended Practice 1163—In-Line Inspection Systems Qualification (API RP 1163)
- API Recommended Practice for Assessment and Management of Cracking in Pipelines (API RP 1176)
- Specifications and Requirements for Intelligent Pig Inspection of Pipelines—Pipeline Operators Forum (POF 2009) [7]
- Interstate Natural Gas Association of America's Report to the National Transportation Safety Board on Historical and Future Development of Advanced In-Line Inspection Platforms for Use in Gas Transmission Pipelines (INGAA)
- NACE International's Standard Practice 0102 In-Line Inspection of Pipelines (NACE SP0102)
- NACE International Publication 35100 In-Line Inspection of Pipelines (NACE 35100)

When discussing the ability of ILI tools to detect and size or just detect pipeline features, we will primarily reference API RP 1160. Table 1 of the API 1160 document shows a listing of pipeline threats with the corresponding assessment ability of ILI tools. API's table has an "S" when a tool is able to detect and size a feature and a "D" when a tool is just able to detect and not size.

Table 8.1 helps pipeline operators to select in-line inspection tool capturing some of the latest experiences in technology and data analysis developments; however, they continue to evolve requiring the reader updating the table over time. Limitations in sizing, detection, and identification anomalies are primary challenges for vendor and operators for selecting the most suitable tool. Tool characteristics and performance in the selection process are pertinent to the inspection objectives. NACE SP0102 standard—Section 4 details also the types of ILI tools and inspection purposes. Some ILI vendors offer a situational selection process to support pipeline operator in finding the appropriate inspection strategy.

8.3.1 ILI Technologies and Detection Capabilities

8.3.1.1 Magnetic Flux Leakage
MFL surveys were the earliest form of internal inspection of pipelines. The first MFL tool was an Axial Magnetic Flux Leakage (AMFL) tool. The original technology has evolved significantly over the course of its history. Some of the area that the tool has evolved:

- **Variable by-pass MFL Tools** that allowed for more consistent tool speed when performing inspection of high flow pipelines (e.g., >5 m/second). By allowing the self-adjust tool speed performing at an optimal speed, the pipe can be magnetized without speed distortions in the magnetic waves obtaining better data quality and resolution for processing.
- **Dual-diameter MFL Tools** allowed for inspection of pipelines with multiple diameters. Typically, dual diameter tools can inspect a pipeline with one nominal diameter difference in the same inspection survey; however, special designs have been made for bigger diameter pipelines (e.g., NPS 36 and 48).
- **Multi-diameter MFL Tools** today ILI tools can be designed much more complex. Multi-diameter designs are available with up to 8 inches difference in ID. These tools are characterized by special "umbrella" pull units towing the ILI tool thru the pipeline.

- **Combined MFL and Mapping Tools** allowed one survey to perform both the AMFL ILI survey and map the coordinates of the pipeline.
- **Circumferential MFL Tools** were developed to find features that have a narrow circumferential and a long axial component that were not optimal for the AMFL to detect and size.
- **Bi-Axial and Tri-Axial Sensors MFL Tools** were developed to measure the magnetic field vector in two or three spatial directions at the same time with an array of sensors (e.g., axial, radial, and circumferential). This allows a more precise sizing of the true dimensions of metal loss anomalies.
- **Residual or Low Field MFL Tools** were developed to identify cold working around dents, to detect hard spots in pipe and localized areas of strain. Recently, it has been used to aid determination of mechanical properties of the pipe steel.

It is worth mention that properties like variable by-pass, dual or multi-diameter option as well as the mapping feature are not exclusively used for MFL ILI tools. Other technologies (e.g., EMAT or caliper tools) can be configured in a similar manner.

8.3.1.1.1 Axial Magnetic Flux Leakage. AMFL ILI tools are the most common form of ILI. These tools were first introduced to the pipeline industry in the 1960s and developed because operators desired to do more to maintain the integrity of their pipelines in the wake of high profile pipeline failures. Internal inspection allowed for more detailed assessment then what hydrotesting and cathodic protection monitoring provided by them.

The 1970s saw a shift from low-resolution AMFL technology to high-resolution AMFL technology. Additional vendors of the technology entered the marketplace in the 1980s that then lead to further enhancement of the technology in the 1990s through research and development.

The major advancement in the 2000s was the combination of tool technologies in the same survey. This allowed multiple threats (e.g., metal loss and deformations) to be assessed at the same time. This was significant because a caliper and AMFL tools could be run simultaneously and the data could be overlaid more easily. This allowed operators to respond promptly to potentially high consequence features such as deformations with metal loss that result from third party damage.

Early versions of the AMFL tool were considered Standard or Low Resolution. These early versions could say a metal loss feature was light (10% to 30% in depth), moderate (30% to 50% in depth), or severe (greater than 50% in depth). The tool could not however report if the feature was on the internal or external surface of the pipe. When high resolution AMFL tools were developed, they had significantly more sensors and could make more accurate reporting of number and absolute size of metal loss features and ultimately were able to report if a metal loss feature was internal or external.

Figure 8.2 below shows an AMFL tool. The tool in the figure has just inspected a gas pipeline and is an equipped with by-pass valve at the front. The liquids and residuals visible on the tool originate from condensate and compressor oil present inside the pipeline.

According to API 1160 AMFL tools are best used to detect and size external metal loss and internal metal loss. AMFL tools are used for metal loss features volumetric in nature that extend in circumferential direction. AMFL tools can also detect without sizing: dents, wrinkles and buckles, laminations, bends and curvatures. AMFL tools are typically run with high-resolution caliper tools. High-resolution caliper tools are able to detect and size features that AMFL can only detect.

TABLE 8.1 ILI MFL, CALIPER, AND ULTRASONIC TECHNOLOGY CAPABILITY FOR DETECTION AND SIZING (2016)

ILI purpose	MFL technology		Mechanical finger	Ultrasonic technology		
	AMFL (Axial)	CMFL (Circumferential)	Caliper	Wall measurement	Angle beam and phased array[11]	EMAT
Propellant	Liquid or Gaseous Medium[1]		Liquid or Gaseous Medium	Liquid Medium Special Gaseous Medium[1]		Liquid or Gaseous Medium[1]
Corrosion External corrosion Internal corrosion Microbial internal corrosion	Detection[2] Sizing[3]		No detection[2]	Detection[2] Sizing	Detection[2] Sizing[3]	Detection[2] Sizing[3]
Narrow axial external corrosion (NAEC)	No detection[4]	Detection[2] Sizing[3]	No detection	Detection Sizing[2]	Detection[2] Sizing[3]	Detection[2] Sizing[3]
Crack like and crack-field anomalies (AXIAL)[8] Stress corrosion cracking, fatigue cracks, longitudinal seam weld anomalies, lack-of-fusion (LOF) cracks, weld anomaly, toe-cracks, hook cracks, HIC	No detection	Detection[2,5] Sizing[3,8]	No detection	No detection	Detection[2] Sizing[3]	Detection[2] Sizing[3]
Circumferential cracking (e.g., girth weld)	Detection[5] Sizing[5]	No detection	No detection	No detection	Detection[2] and sizing[3] if modified[6]	Detection if modified[6]
Dents wrinkle bends	Detection[4] Sizing[7]		Detection, Sizing	Detection[4], Sizing[7]	Detection[4] Sizing[7]	No detection
Buckles	Detection[4] Sizing[7]		Detection[4] Sizing	Detection[4], Sizing[7]	Detection, No sizing	No detection
Notches and gouges	Detection[8,9]		No detection[9]	Detection[9]	Detection[2] Sizing[8,9]	Detection[2] Sizing[8,9]
Laminations	Limited detection Sizing[7,8]		No detection	Detection Sizing	Detection[9] Sizing[3]	Detection[9] Sizing[3]
Previous repairs	Detection of steel sleeves and patches, others only with ferrous markers		No detection	Detection if welded or steel marked to pipe, may show as a thicker wall	Detection if welded to pipe, may show as a thicker wall	Detection if welded to pipe, may show as a thicker wall
Mill-related (non-geometric) anomalies	Limited detection[9]		No detection	Limited detection[9]	Detection[8,9]	Detection[8,9]
Bends	Detection[12]		Detection Sizing[3]	Detection[12]	Detection[12]	Detection[12]

(Continued)

8.3.1.1.2 Circumferential Magnetic Flux Leakage. Circumferential Magnetic Flux Leakage (CMFL) tools were first introduced developed in 1996 to assess pipelines for Narrow Axial External Corrosion (NAEC). CMFL was officially introduced to the pipeline industry in 1998 as an emerging technology. Vendors of the technology have trade names such as Axial Flaw Detection (AFD) or Transverse Field Inspection (TFI). CMFL can detect and size axially oriented metal loss that AMFL technology cannot detect. If there is

sufficient opening of a crack to cause a disturbance in the magnetic field, then cracks can also be detected and sized by CMFL.

Following a failure of an electric resistance weld (ERW) in 1997, the CMFL tool was first used to detect longitudinal seam weld features when a ultrasonic crack detection tool was not available [8]. The decision was made to use the technology in a new way to reduce the effort and cost involved when performing multiple hydrotests. The technology is known for producing a high number

TABLE 8.1 ILI MFL, CALIPER, AND ULTRASONIC TECHNOLOGY CAPABILITY FOR DETECTION AND SIZING (2016) (Continued)

ILI purpose	MFL technology		Mechanical finger	Ultrasonic technology		
	AMFL (Axial)	CMFL (Circumferential)	Caliper	Wall measurement	Angle beam and phased array[11]	EMAT
Ovalities	Limited detection[9]		Detection Sizing[3]	No detection	No detection	No detection
Coating (disbondment detection)	No detection		No detection	No detection	No detection	Detection[10]

[1] In *some gaseous mediums*, suitable sensor modification (e.g., angle, attenuation) may be possible provided that comprehensive testing and validation are conducted to confirm ILI capabilities. The inspection of a pipeline operated with a gaseous medium requires the preparation of a liquid batch

[2] Varies depending on anomaly morphology (e.g., depth, length, width). Specific morphologies may require specialized designed sensors and algorithm to be detected

[3] Sizing accuracy defined by *ILI vendor-specific* sensor technology and data analysis advancement

[4] Detection only if *sensor resolution* (i.e., smaller size; higher data density) is adequate to the target anomaly

[5] Require *minimum crack opening* for ILI detection (e.g., Reduced Probability of Detection [POD] for tight cracks). Not reliable for SCC detection and characterization

[6] Transducers to be rotated by 90 degree from axial (i.e., pipeline axis) to circumferential (e.g., diameter) direction

[7] Sizing may be possible based on ILI vendor-specific experience on *indirect measurement* supported by field validation; however, anomaly identification and characterization is a challenge

[8] Anomaly identification and characterization continues to be challenging as to the crack ILI technology and interpretation ability to differentiate between *manufacturing (e.g., stable, non-growing, non-injurious) and cracking (e.g., potential growth and injurious)*

[9] Detection, but *not reliable anomaly identification and characterization*

[10] Detection of Coating disbondment in the following types: Polyethylene, FBE, Tape wrap, Coal Tar Enamel, Asphalt

[11] Angle beam (e.g., shear-wave) and Phased Array ultrasonic technologies are integrated for educational purposes; however, they have differences associated with the *level of detection and sizing capabilities* based on the specific sensor technology and data analysis advancement used

[12] Detection when multiple odometer riding around the circumference on the bends to see odometer length differences

of magnetic disturbances that may or may not be actual anomalies. In early uses of the technology, a significant number of validation digs were required to determine the accuracy of the tool. Data interpretation is valuable in determining action required in response to the survey. The tool has increased in performance as more validation digs have been performed and the results were reviewed for better signal interpretation.

Figure 8.3 below shows a high resolution CMFL tool. The technology requires typically two measurement units. The second unit is covering the gaps created by the magnetic brushes of the first unit. The brushes on the side of the sensors induce the magnetic field in the circumferential direction into the pipe.

API 1160 states that CMFL tools are able to detect and size external metal loss and internal metal loss. CMFL specializes in detecting metal loss features that have axial length with minimal circumferential width. API 1160 shows that CMFL can detect without sizing: external selective seam corrosion internal selective seam corrosion, axially oriented crack-like anomalies, dents, wrinkles and buckles, laminations, bends, and curvatures.

CMFL can also be used as a double check for ultrasonic crack detection tool. Cracks with a sufficient air gap are detected by the CMFL tool even if they were not detected by the ultrasonic crack detection tool. The combination of both technologies provides a better differentiation of cracking from other similar anomalies (e.g., gouges, stringers, or laminations). High resolution caliper tools are also typically run with CMFL tools to assess predominate pipeline

threats. CMFL has become a valuable double check to both traditional AMFL tools and ultrasonic crack detection tools because the CMFL tool sees features differently. This gives operators confidence that they have found the most injurious features on the pipeline.

8.3.1.1.3 Helical Magnetic Flux Leakage. In traditional pipe making where there is one longitudinal weld seam, AMFL and CMFL are sufficient for detecting and sizing metal loss features. When spiral welded pipe began to be used in the pipeline industry, this presented a new challenge to the industry for assessment. Helical MFL was first offered in the 1990s as a "Triaxial" technology [9]. The "Triaxial" technology was able to report in the axial, radial, and transverse direction. Helical MFL has been available since 1995 according to INGAA. The main advantage of Helical MFL is that it provides more accurate reported sizing of features than traditional AMFL technology.

8.3.1.1.4 Residual or Low Field Magnetic Flux Leakage. Residual or Low field MFL data can be used to detect material property changes, residual stress, and areas of cold work [10]. This is important for pipelines where an accurate material test record (MTR) is not available. Residual or Low Field MFL surveys can be performed with corresponding verification excavations to determine a pipeline's mechanical properties if they are unknown. This is important for maintaining accurate reports and integrity verification.

FIG. 8.2   HIGH RESOLUTION AXIAL MAGNETIC FLUX LEAKAGE (AMFL) ILI TOOLS
(*Source:* ROSEN Group and Baker Hughes, respectively)

FIG. 8.3   HIGH RESOLUTION CIRCUMFERENTIAL MAGNETIC FLUX LEAKAGE (CMFL) TOOL (*Source:* ROSEN Group)

FIG. 8.4   DUAL (HIGH AND LOW FIELD) MAGNETIZATION AXIAL MFL ILI TOOL (*Source:* ROSEN Group)

This technology is used in conjunction with high resolution MFL. The two MFL technologies are on the same pig and the Low Field MFL measures the remaining magnetic signal in the pipe steel following passage of the high magnetic field MFL tool. The first application of this technology was performed in 2008. Figure 8.4 below shows a dual magnetization MFL tool. The high magnetic field is induced by the front unit. The trailing unit is generating a controlled low magnetic field. The tool is lifted from the pull-rig. In the background, various test pipe is visible. For the purpose of the pull test the front cup of the tool has been slotted. This cup would be exchanged prior to the actual pipeline inspection run.

8.3.1.2 Ultrasonic Wall Measurement   Ultrasonic wall thickness measurement (UTWM) technologies were first introduced in 1985. The UTWM was developed after MFL technology and provides a high resolution inspection of pipelines. As with all piezo electrical based ultrasonic survey technologies, a liquid couplant is required to perform the survey. The couplant is used to transport an induced sound wave from the ultrasonic sensor to the pipe wall and back. UTWM survey tools are available in most common pipeline diameters.

UTWM is used less frequently than MFL tools because a couplant is required and pipelines are required to be cleaner than an MFL survey. Gas pipeline operators can choose to use a liquid slug to perform the survey. Some operators choose not to use this option because it could introduce an internal corrosion threat if the pipeline is not completely dried following the inspection. Gas operators may choose to perform an MFL survey if they do not want to introduce water into their pipeline.

UTWM offers some advantages over MFL technologies. MFL surveys require large magnets to induce the required signal into the pipeline. MFL tools typically require a crane or other lifting equipment to get the tool into the launcher. UTWM are typically much lighter and can sometimes be handled by hand when lifting the tool into smaller diameter pipelines. UTWM is capable of locating and detailing areas of internal or external metal loss. UTWM is specially able to detect and size pitting along the axial direction of the pipe. Some vendors of UTWM technology can detect deformation features with the same sensors that detect corrosion features but may not specify the capability.

UTWM tools are also a primary assessment for internal and external corrosion. UTWM tools have a few advantages over MFL tools. UTWM are better able to detect and size pits and narrow axial external corrosion. UTWM are lighter and smaller and have a wide range of pipe designs that they can traverse. UTWM come at a higher price to perform surveys. If a pipeline has pitting or narrow axial external corrosion then the extra cost may be warranted. UTWM are also not susceptible to being fooled by corrosion products that have magnetic properties like MFL surveys can be. Prudent operators are encouraged to rotate between MFL and UTWM technologies to ensure a complete assessment of corrosion.

Figure 8.5 below shows a UTWM tool. This design is characterized by sturdy metallic carrier for the ultrasonic sensors. This supports a precise guidance of the sensors along the inner pipe wall.

8.3.1.3 Ultrasonic Crack Detection Tool and Phased Array   The first ILI crack detection tool was developed in 1970 based on eddy current technology. The resolution of this experimental tool was not sufficient. It took another 20 years before the first ultrasonic crack detection tool (UTCD) was developed in 1994 in Germany. As mentioned, this tool and its successors were developed to perform integrity assessment of crack-like features. Crack-like features can be introduced into the pipe during the manufacturing process (e.g., hook cracks) or in the operation of the pipeline (e.g., SCC or fatigue). UTCD tools were necessary because MFL and geometry tools of the day were not able to detect and size crack-like anomalies.

In combination with UTWM technology, the UTCD tools can also detect and size narrow axial external corrosion and laminations. In the case of laminations, the UTCD tool may be able to determine if the lamination has crack growth that is connected to a surface of the pipe. If the lamination is connected to a surface of the pipe, then it should be assessed as a crack. Laminations that do not grow to a pipe surface are typically considered non-injurious.

UTCD tools have a high density of sensors that produce a sound wave in the form of a Shear Wave. Shear waves are mounted at an

FIG. 8.5 ULTRASONIC WALL THICKNESS MEASUREMENT (UTWM) ILI TOOL (*Source:* ROSEN Group)

angle to the internal pipe surface such that the wave hits the surface at an angle and reflects at an angle for the sensor to receive again. The angle that the sensor is set to is typically 45 degrees.

The front unit is generating ultrasonic shear waves travelling counter clockwise in circumferential direction. The rear unit is operating in clockwise direction. Therefore, axial cracking in the pipeline is detected from two different sides. Both units provide full resolution straight beam wall thickness measurement to characterize metal loss and laminations.

Phased Array is a technology based on ultrasonic principle, but using a "virtual" sensor comprised of multiple stripe-like elements (aka. Array) electronically grouped to produce an angled or beam-shaped pulse wave. The focus and angle of the ultrasonic wave are created by varying or delaying the timing sequence (e.g., nanoseconds) of excitation or firing of the elements [11]. The Ultrasonic Phased Array tool for inspection pipelines was first introduced in 2005 providing higher resolution of ultrasonic crack detection data. Figure 8.6 below shows UTCD and Phased Array tools.

Figure 8.6 below shows UTCD and Phased Array tools.

8.3.1.4 Electro-Magnetic Acoustic Transducer Electro Magnetic Acoustic Transducer (EMAT) ILI technology was developed to detect and size SCC and axially aligned crack-like anomalies in gas pipelines. EMAT provides gas operators an advantage over UTCD tools because it does not require a liquid couplant to obtain an integrity assessment of the pipeline. EMAT is an evolving technology and validation of the results through excavations is required to ensure the assessment is accurate.

The concept of how the transducers are used in the EMAT tool has been around since the late 1960s and early 1970s. The use of the transducer continued to evolve through the 1980s and into the 1990s when the EMAT technology was deployed for external inspection of pipelines, pipeline facilities, pipeline tanks, and support features. The gas pipeline industry drove the development of the EMAT technology for use in internal inspection of pipelines. The commercial in-line EMAT surveys were performed in 2000s and the technology continues to develop into the 2010s. At that time, a framework had been developed by the industry to address SCC and cracking in gas pipelines using EMAT technology [12]. With further refinement of the technology, the tool will be able to detect and size hook cracks and other longitudinal seam features beyond what it is currently capable of detecting.

Figure 8.7 shows an EMAT tool waiting to inspect a gas pipeline. The tool is ready to be pulled into the launcher barrel. The tool has a by-pass valve installed at the front. The by-pass flow is guided thru the center of the tool.

API 1160 states that EMAT tools can detect and size: axially oriented SCC, axially oriented cracks, circumferential cracking (with special setup), and hard spots. It also can detect without sizing: external metal loss, internal metal loss, external selective seam corrosion, internal selective seam corrosion, axially oriented crack-like anomalies, dents, wrinkles and buckles, laminations, bends and curvatures. EMAT is a quickly evolving technology.

8.3.1.5 Geometry Geometry ILI surveys are performed to detect and size deformation features in pipelines. Geometry tools come in two forms: (1) Mechanical Caliper, and (2) High Resolution Geometry. Mechanical calipers can detect and size dents and bends but can only detect wrinkles/buckles and expanded pipe. High Resolution Geometry tools have shown the ability to detect and size dents, wrinkles/buckles, expanded pipe, bends and in some cases internal corrosion (API 1160).

Geometry ILI tools were first introduced in the 1970s. They have evolved continuously since that time. The first generation of the tool were single channel with an odometer attached that gave the change in inner diameter along the pipeline but were a construction validation exercise and not for integrity management [13]. The next edition of the geometry tool was used for integrity management and operated on the same principles as the first generation. This generation was limited in reporting ability for both location in the o'clock position and location along the pipeline. The third generation of geometry pigs switches to a paddle design that were able to provide better reporting but were not able to detect sharp dents that passed between the paddles. Successive editions of geometry

FIG. 8.6 ULTRASONIC CRACK DETECTION (UTCD AND PHASED ARRAY) TOOLS (*Source:* Baker Hughes and ROSEN Group; GE PII, respectively)

FIG. 8.7   ELECTRO MAGNETIC ACOUSTIC TRANSDUCER (EMAT) TOOL (*Source:* ROSEN Group)

tools found a balance between caliper spacing and size to provide profiles of deformations so accurate that they can be imported into Finite Element Analysis (FEA) software packages.

Geometry tools have evolved such that a robust integrity assessment of deformations can be made with one survey. The ILI tool performing the survey can include an MFL tool with the high-resolution caliper tool. Combining these tools together provides an assessment of metal loss and deformations and mechanical damage.

Figure 8.8 shows a high resolution caliper tool. The caliper arms are equipped with additional eddy current sensors to compensate any measurement error caused by lift-off or bouncing of the caliper arms. Two sensor planes provide 100% sensor coverage.

8.3.1.6 Inertial Mapping   Inertial mapping ILI tools were first introduced to the pipeline industry in the 1980s. These first inertial mapping tools were only able to provide the X, Y, and Z coordinates for a global positioning system (GPS). Since 1990s, inertial mapping ILI tools are used to detect locations of vertical and horizontal curvature strain and bends in the pipeline enabling to see potential areas of movement. These features can then be used to determine if there is an interaction threat with other pipeline threats. Older pipelines that were not subject to rigorous quality assurance during construction can be affected by strain. Overlaying girth welds with vintage criteria that are susceptible to failure with the strain information from the inertial mapping tool can have great benefit for managing the vintage girth weld threat.

8.3.1.7 Cathodic Protection Current Mapper   Cathodic Protection Current Mapper (CPCM) is an in-line survey technique that assesses the performance of a pipelines cathodic protection system, illustrated in Figure 8.9. CPCM is not an integrity assessment method by itself. The tool is an aid to an overall integrity management program. CPCM can be seen as equivalent to a close-interval survey (CIS).

CPCM was developed in the 2000s as a way to perform cathodic protection current measurement. CIS surveys require surveyors to

walk the length of the pipeline and take measurements approximately every three feet. CPCM that an electrical measurement of the cathodic protection system from inside the pipeline. The measurement is a true potential on the pipe and not subject to the influences that an above ground survey are prone to causing error in measurement. There is also a significant health and safety risk reduction by performing the assessment as an in-line survey and not as an above ground survey.

The CPCM tool is approximately six feet long and measures the voltage drop over that length from the inner surface of the pipe. The known resistance of the pipe is then used with Ohm's law (voltage equals current × resistance) to determine the current at the location. Changes in current at any point along the pipe indicate a change in the system.

The total output voltage of the CP system is the summation of all the voltage drops of the system. In a well-coated and well-protected pipeline segment, there should be a smooth voltage curve generated by current both leaving and returning to the rectifier.

CPCM has a number of benefits. The inspection provides a one hundred percent inspection of CP system and ensures minimal gaps in integrity inspection data. The survey is able to measure current achieved on the pipe by the CP system under coating systems that are generally considered to block CP (i.e., shield the CP from the pipe). CPCM is also beneficial in locations where access is difficult such as offshore, swamps, mountainous terrain and congested urban areas.

The signals shown by the CPCM tool have clear interpretations:

- Large gains or losses over a single point = Rectifiers, Bonds, Shorts, Anodes
- Large gains over several feet or meters = areas of poor coating or bare pipe—high current density
- Small gains over longer areas are ideal and evidence of good coating and well distributed CP
- Shallow positive (up left to right) slope across the zero line = midpoint between sources and good CP coverage
- Flat line on the "0" line = no discernible CP

FIG. 8.8   HIGH RESOLUTION CALIPER TOOLS (*Source:* ROSEN Group and Baker Hughes, respectively)

FIG. 8.9 CATHODIC PROTECTION CURRENT MAPPER ILI TOOL (*Source:* Baker Hughes)

8.3.1.8 MFL Eddy Current—Remote Field Technology As illustrated in Figure 8.10, Remote Field Eddy Current Technology (RFEC or RFT for short) [14] is similar to Magnetic Flux Leakage (MFL) technology in that both techniques rely on an axial magnetic field induced into the pipe wall, which is distorted by corrosion its and other defects. The distortion of the magnetic field is what is detected by in-line inspection (ILI) Tools.

In the case of MFL Tools, the magnetic field is induced by strong magnets that are positioned very close to the inside surface of the pipe, with detection coils or solid state detectors placed between the poles of the magnets. In RFT Tools, the axial magnetic field is induced by a large "exciter" coil. If the exciter coil is energized by DC current, then the ILI Tool will work in the same manner as an MFL Tool, but if energized by AC current, then the exciter coil will create circumferential eddy currents in the pipe wall as well as the axial magnet field.

This gives the RFT technique the advantage of being sensitive to transverse and axially oriented defects such as cracks. For instance, Stress Corrosion Cracks (SCC), which are axially oriented) are not easily detected by standard MFL Tools, because the magnetic field runs in the same direction as the cracks, and therefore the cracks do not distort it. In the RFT ILI Tools, the circumferential eddy current crosses the SCC at right angles, and is therefore distorted by the cracks and detected by the sensors on the Tool.

Other strengths of RFT Technology are:

- measure through liners and scale
- equal sensitivity to inside and outside flaws
- detection of planar defects in girth welds

- Detection of planar defects in ERW and sub-arc longitudinal welds
- Detection of local stress and strain areas
- Sensors do not have to contact the pipe wall

Current limitations of RFT technique:

- speed is limited by the inspection frequency
- practical thickness limit is 0.675"
- non-contact sensors limit threshold of detection (TOD) to 20% for 1/2" diameter pit

8.3.1.9 Other Inspection Methods—Material Properties Determination Investigations of major incidents have driven national regulations of ageing pipelines. Uncertainties in design requirements from the area of pipe line construction due to gaps in pipeline records enforce the industry efforts in the development of new approaches in the ILI technology.

The objective of the recently developed material property tools is to provide a better picture for

- MAOP, MOP Calculation—Diameter, wall thickness, minimum yield strength
- Risk Assessment—Variables to describe the ductility like fracture toughness

Conventional ILI technologies are capable of delivering relevant information for maximum allowable operating pressure (MAOP), maximum operating pressure (MOP) calculation, such as wall thickness and diameter. Measurements of pipe yield strength or

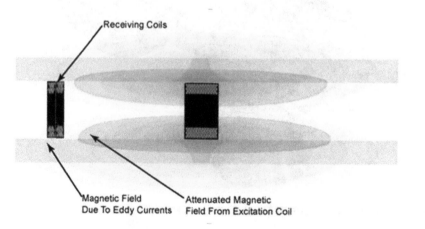

FIG. 8.10 MFL EDDY CURRENT—REMOTE FIELD TECHNOLOGY (*Source:* Russell NDE Systems)

toughness by ILI technologies was not possible in the past. Indirect correlation of measurements based on a combination of several technologies was the only way to get a qualitative impression of these material properties.

An Eddy Current approach was developed recently by one pigging vendor. This service uses an eddy current based pipe grade sensor (PGS) for the measurement of absolute values of tensile strength and yield strength. The use of ILI material property tools will increase the safety within the pipeline industry and will be an essential part in future risk assessment.

8.4   IN-LINE INSPECTION PLANNING

Planning is most important for the success of any project. Early determination of the expected outcome and identification of potential issues along the way are essential in planning. The goal of ILI performance is a meaningful integrity assessment in time to meet the needs of the operator. This section will guide the reader through the selection of an ILI technology and identify items to consider before the ILI vendor and field personnel arrive on-site to perform the survey.

8.4.1   Selection of In-Line Inspection Technology

Chapter 4 of this book showed the reader how to identify threats to pipeline systems. Chapter 6 then gave the reader methodologies for quantifying the risk to the pipeline. In-line inspection is one integrity assessment method to confirm the existence of the threat and to monitor the threat. Mitigation of threats identified comes through field excavation and repair of reported features by the ILI technology. Prevention, mitigation, and monitoring are discussed in Chapter 11.

The type of ILI technology that is used for integrity assessment on the pipeline must match the threats identified on the pipeline (NACE SP0102) and consider the following for a successful integrity assessment of the threat:

- Accuracy, detection and statistical confidence of the ILI technology
- Detection sensitivity
- Classification capabilities
- Sizing accuracy
- Location accuracy
- Requirements for defect assessment

Each of the six elements above is discussed in the follow subsections. These items are important for both determining the inspection technology to use and the contractual requirements with the vendor that performs the survey.

8.4.1.1 ILI Performance   When selecting an ILI technology or a combination thereof, the primary threat(s) should be defined and matched with the ability of the ILI tool to detect, identify/characterize, and size that feature as well as the ILI requirements to perform under the possible pipeline operation (e.g., cleanliness, flow, product response). To ensure integrity assessment the pipeline integrity engineer should match the vendors stated technology capabilities for probability of detection, classification, and sizing abilities with applied tolerances as well as the technology-pipeline requirements (e.g., magnetization or ultrasonic coverage versus wall thickness,

speed and product attenuation requirements). Table 1 of NACE SP0102 also provides detection and sizing abilities by tool technology and threat.

Some vendors back up the information given in their respective ILI performance specification in accordance with API 1163. This is a transparent process of demonstrating the tools ability by laboratory measurement, full-scale tests and field results from "historic" runs. This type of experience is reflected in two statistical parameters: the tolerance (e.g., +/−10% of the wall thickness) and certainty (e.g., 80% or 90% of the time) or confidence level. While the tolerance describes the accuracy of the measurement, the certainty captures the level of confidence of the selected tolerance. A detailed description can be found in API 1163. These parameters play an important role in the selection of the corresponding ILI technology and need close collaboration with the ILI vendor.

8.4.1.2 Detection Capability: More than Probability of Detection (PoD)   Tool detection specification may reference a minimum anomaly morphology (e.g., width, length, and depth) to be detected by a given ILI technology. This minimum anomaly morphology needs to be defined from what it is expected in the pipeline. The detection performance is measured by the ILI vendors as the Probability of Detection (POD) for contracting purposes; however, companies use additional performance indicators such as non-reported anomaly distributions (e.g., lognormal, Weibull) and largest anomaly missed (e.g., log-odds model, Probability of Inclusion) [15].

8.4.1.3 Identification or Classification Capabilities: Probability of Identification (PoI)   Classification means the ability to identify and differentiate one feature type from another. This ability to identify or classify features is important when performing an in-line inspection. Anomaly classification/identification helps differentiating between injurious (e.g., crack—potential to grow) and non-injurious (e.g., manufacturing—stable or low potential to grow) anomalies. Different types of anomalies may have different response criteria.

Depending on the type of anomaly, their mitigation timeline are different (e.g., laminations versus SCC). Therefore, ILI technology selection should also focus on the Probability of Identification (PoI) for the different targeted anomalies.

Example: Cracking Identification/Differentiation
Crack ILI technologies have been challenged with better differentiating the anomaly types for companies to timely prioritizing the mitigation of their effects (e.g., criticality and growth). Once detection capabilities are improved, sensor development/improvement and data analysis have been focused on reducing false "identification" calls (i.e., non-injurious anomalies). Sensor development has been oriented to miniaturization and higher data resolution for specific wall thickness. Combination of EMAT, MFL and Eddy Current sensors has been the latest trend, but ongoing validation is recommended.

8.4.1.4 Sizing Accuracy: Probability of Sizing (PoS)   The ILI tool selected should provide sufficient sensitivity to the size of feature that is being assessed. Sizing accuracies are provided in the

ILI tool specification sheet in terms of tolerance and certainty (i.e., confidence level). Field—ILI comparison should also be expressed in terms of both tolerance and certainty.

Example: Sizing Tolerance and Certainty Calculation

Based on a statistically representative sample investigated in the field, a normal distribution had been identified from field—ILI depth differences after the outliers were removed (separate analysis).

The Field-ILI depth normal distribution had a *p*-value < 5% achieving a confidence level of >95% becoming acceptable to be used for calculating ILI performance.

The following values were determined to estimate the tolerance and certainty for 80% certainty specified/promised by the ILI vendor.

Field-tool difference distribution description	Variable	Value (% Depth)
Mean of depth differences	u	+5%
Standard deviation of depth differences	*Stdev*	15%
Z value for 80% certainty of a natural normal distribution	z	1.28
x for upper limit for %80 certainty	*x at z(1.28)* = u + (z * stdev)	+24%
x for lower limit for %80 certainty	*x at z(1.28)* = u − (z * stdev)	−14%
Positive tolerance at 80% certainty from the mean	*x(upper)* − u	+19%
Negative tolerance at 80% certainty from the mean	*x(lower)* − u	−19%

Based on the field—ILI depth differences, the ILI performance for depth is +/−19% of the wall thickness (i.e., tolerance) for 80% of the time (i.e., certainty) from the mean (+5% wall thickness).

ILI sizing needs to be as close to the in-field anomaly as possible so depth and predicted burst pressure with their associated accuracy can be used for prioritization the mitigation and the re-inspection interval. Operators should have an anomaly response table prior to performing an ILI survey that is consistent with industry best practices, corporate requirements, and regulatory criteria. Accurate sizing, both tolerance and certainty, of reported features ensures that the injurious features are responded to in the appropriate timeframe preventing incidents.

8.4.1.5 Location Accuracy: Odometer and Inertial Features that are detected and sized accurately by an ILI tool must also have location accuracy so the reported feature can be mitigated as necessary through excavation. Location accuracy includes the odometer distance, the reported orientation, and the referenced joint number. Furthermore, the use of inertial measurement units installed with the selected technology for inspection can provide higher location accuracy in terms of longitude, latitude, and elevation of the pipeline and features. An inertial measurement unit (IMU) can be added to almost any ILI tool. There are tolerances to these locations as well in the data sheet.

As illustrated in Figure 8.11, for increasing the pipeline location accuracy, additional Differential GPS (DGPS) or GPS points tied to the inertial data via above ground markers need to be added. Typically, a DGPS leveling is conducted every kilometer or mile; however, the accuracy can be tailored based on the 1:2000 ratio.

Example: Increasing Accuracy via Inertial

A MFL inspection of a pipeline has been added with a 1:2000 accuracy inertial measurement unit. The company is expecting to get a one (1) m (3 ft) accuracy for locating the pipeline and detect potential areas of movement with the first one run; however, they know that two (2) inertial runs are needed for determining the differential displacement (e.g., axial, transverse and vertical).

The ILI and company have made the following actions aiming to the inspection objectives:

a. Schedule GPS tie point every 2000 m at kilometer post 2, 4, 6, and 8 in addition to the launcher and receiver traps to achieve the 1-m axial (along the pipeline chainage) accuracy.
b. If the inertial bias/drift correction can be applied using the converge of two (2) GPS tie points (upstream and downstream), there is a likelihood that the transverse and elevation of intermediate points may achieve a better accuracy than 1-meter (e.g., 0.5 to 1 m).
c. If the pipeline construction practices ensured that girth welds were not bent (e.g., straight pipe within 3 diameters of the girth weld), the vertical and horizontal curvature should be minimum or close to zero.
d. If item (c) is confirmed, the inertial data of the first (only run yet) would be reviewed at the girth weld locations for identifying "deviations." If vertical or horizontal curvature (i.e., "bending") is identified, those locations should be verified in the field to assess potential movement.

8.4.1.6 Requirements for Defect Assessment ILI tools report features that require defect assessment. When selecting ILI technologies, the resolution of the results plays a role in defect assessment. It is important to match the resolution capabilities of the ILI tool with the expected features on the pipeline. Tool selection for reassessment surveys can rely on the baseline assessment for determination if a tool with a higher resolution is required. For example, a dent that is reported by the ILI tool can be mapped and then placed in finite element software to analyze the local stresses and strains on the individual features.

8.4.2 Multiple In-Line Inspection Technologies

A fully robust integrity management program can be achieved through ILI assessments using multiple survey techniques that are selected based on the threats to the pipeline. The use of multiple technologies ensures that there are no blind spots in the integrity program. Performing a risk assessment on the pipeline will help prioritize which ILI multiple techniques are best for the pipeline.

There is no single ILI technology able to detect and size the most common pipeline integrity threats. All ILI technologies have strengths and weaknesses in detecting and sizing threats. When selecting the ILI technology, it is important to consider running multiple technologies. Multiple technologies can be selected from one survey to the next conducting repeatability analysis for understanding true morphology (e.g., length, width, depth) of the integrity threats. Multiple technologies can be used in the same inspection train.

FIG. 8.11 DIFFERENTIAL GPS OR TIE POINT TO THE INERTIAL DATA AT A SELECTED LOCATION (*Source:* ROSEN Group)

8.4.2.1 Single Threat Purpose Using Multiple ILI Technologies

Multiple technologies can also be performed in the same survey for the *same purpose or integrity threat.* For single threats such Stress Corrosion Cracking (SCC) and seam weld cracking, the comparison of raw signal from multiple ILI technologies (e.g., CMFL, UTCD, Phased Array) on the same anomaly locations has provided benefits in better characterizing the morphology (i.e., POI and POS) of the cracking and identifying additional cracking (i.e., POD) missed in the other dataset. *Example:* Single Threat ILI Purpose—Corrosion Multiple ILI Technologies

Even though Axial MFL technology is typically used for corrosion, Ultrasonic Wall Measurement technology may better detect and size longitudinal or channel corrosion and pinhole.

Example: Single Threat ILI Purpose—Seam Weld Cracking Multiple ILI Technologies

A pipeline may require crack in-line inspection such as both circumferential MFL and an Ultrasonic angle beam or phased array technologies to better detect anomalies in Double Submerged Arc Welding and Electric Flash Welded longitudinal seam welds. CMFL and ultrasonic technologies complement each other in capturing open-width and tight cracking as well as providing better input in determining the criticality of the anomalies.

8.4.2.2 Multi-Threat Purpose Using Multiple ILI Technologies Multiple ILI technologies in the same survey can be used for *multi-threat purpose* (metal loss and geometry anomalies) for data integration and wider anomaly classification spectrum (e.g., if laminations are identified by one technology with high confidence, this technology will support the identification of SCC by another technology). Furthermore, in some cases combination of the AMFL with UTWM has been made on one tool and type of integrity threat [16].

The most common type of combination survey is an MFL with caliper. This is the original combination tool and is the most common to the pipeline industry. Internal corrosion, external corrosion, and deformations are assessed in one survey when multiple survey techniques are performed together [17]. Vendors are able to overlay the data from multiple survey technologies based on the needs of the operator.

Performing surveys together save costs in multiple ways. Operators are not charged multiple mobilization fees when surveys are combined. Some survey techniques require changes in the pipe operation to ensure proper tool speed. By performing the surveys together, the pipeline through put is not altered multiple times. Figure 8.12 gives an example of the possible combination of geometry, IMU and AMFL and the corresponding benefits. Combination of corrosion and geometry surveys provide detailed information as a basis for strain and metal loss integrity management.

There are some safety benefits to performing multiple surveys together. Although performed frequently the operation of pig launchers and receivers in pigging operations are not necessarily routine. Therefore, by combining two or more such operations, significantly increase worker safety.

8.4.3 Pipeline Readiness

Pipeline inspection technologies are advancing rapidly. Pipeline elements that were considered unpiggable 5 years ago could now be considered piggable [18]. In the realm of pipelines, unpiggable is met with the old adage of "where there is a will, there is a way." Three categories are generally known to affect the ability of pipelines to be pigged.

1. Design Factors,
2. Operational Factors,
3. Economic Factors

The three factors that make a pipeline unpiggable are important for determining if a pipeline is ready for ILI performance. A

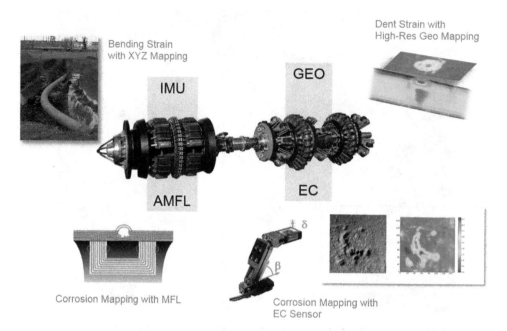

FIG. 8.12 EXAMPLE OF THE POSSIBLE COMBINATION OF GEOMETRY, IMU, AND AMFL TOOLS (*Source:* ROSEN Group)

pipeline must be designed in such a way as to allow for launching an ILI tool; allowing it to traverse the line collecting data without damage and be successfully received. The pipeline must also operate to allow for sufficient tool performance that the assessment has value to the overall integrity management program. The majority of design and operational factors can be overcome with investment in the ILI technology or investment in the pipeline asset. If the costs to make a line piggable or alter operation are too high, then another integrity assessment should be pursued.

On the operational side the characteristics of the medium are a key factor to understand the piggability. Cases like high temperatures, multi-phase medium, abrasiveness etc. can restrict a pipeline from being inspected [19].

8.4.3.1 Pipeline Questionnaire ILI vendors have questionnaires that they request pipeline operators fill out before entering into a contract to perform an ILI survey. Appendix A of NACE SP0102 shows an example of a questionnaire. Pipeline questionnaires focus on the design factors and operational characteristics of the pipeline. Operators are encouraged to be as thorough and honest when completing the pipeline questionnaire as possible. Identifying potential issues with the inspection as early as possible in the process will aid the likelihood of first run success. The ILI vendor will review the questionnaire together with the operator to start the customization of the ILI inspection equipment to match the particular objective of the inspection. This process is finalized some time ahead of the survey to ensure that issues that might seem minor to the operator are not significant to the vendor and therefore would compromise the success of the inspection program.

Pipeline questionnaires vary from vendor to vendor but they will all cover similar aspects. The ILI vendor will want to know what type of survey is being requested and where the service is being requested. Vendors will also request product characteristics (e.g., water, CO_2, H_2S, and wax content) and pipeline operating parameters (e.g., pressure, flow rate, temperature). It may also be important how these parameters vary along the pipeline. The questionnaire will also cover integrity and design aspects of the pipeline. It is helpful for the vendor to know when the last integrity assessment

was performed and if there were any issues with the survey. Design aspects include wall thickness, grade, diameter, and other aspects. There should also be a checklist for reviewing pipeline design, construction or past repair information to anticipate and resolve the challenges facing the assessment.

8.4.3.2 Modifications Driven by Pipeline Design Not all pipelines were originally designed to be piggable. Before an ILI can be performed for the first time, a detailed review of the pipeline design should be performed to ensure that there are not any obstacles to overcome. As-built drawings are a good source of information for design factors. It can also be helpful to interview field personnel that have local knowledge of the pipeline's design. Below are a few design elements that can make pigging difficult:

- Diameter—Not all ILI tools are available in all sizes and smaller sizes can be challenging to get the sensors compact enough to fit in the line.
- Bore restrictions—Some valves do not open the full diameter of the pipeline. When there is a restriction, it can be difficult for ILI tools to pass with speed.
- Tight bends—Radius of bend less than 1.5 times the pipe outer diameter can be difficult.
- Back-to-back bends.
- Mitered bends (greater than 10 degrees of bend).
- Unbarred tees—At tees in the pipeline if the diameters are unbarred and similar in size, the tool need to be equipped with special guiding elements. Also pipeline operation needs to stop the flow in and out of the tee.
- "Y" connections—Similar to unbarred tees, if the connection is not barred it the ILI tools have to be equipped with special guiding elements to control the path the ILI tool takes. Also the sealing length of the tool needs to be modified to ensure no unwanted by-pass occurs. "Y" connections can be symmetrical and asymmetrical.
- Diameter changes—Most ILI tools can traverse changes of one nominal pipe size. Technologies are advancing to allow for larger differences but it is difficult to maintain resolution.

- No Launcher or receiver—There are a number of workarounds for lack of launcher or receiver. Robotics and tethered tools can be used or temporary launchers and receivers can be brought in to perform he survey.
- Thick walled pipe—Wall thicknesses greater than 0.600 inch are considered heavy wall pipe. Pipe with a wall thickness greater than 1.0-inch can be difficult for MFL technologies to fully magnetize the pipe for inspection.
- Thin walled pipe—Wall thicknesses less than 0.188-inch can be difficult for ultrasonic technologies because the echoes and reflections of the signals overlap with one another.
- Deadlegs, crossovers, and laterals—Tethered or robotic tools can be used, depending on the configuration.
- Internal coatings—Thin flow coatings (<0.5 mm) are typically not affecting the inspection results. Thicker coatings may reduce the performance specification of the applied inspection technology because there is a separation between the sensor and the pipe steel. ILI tools also have the potential to damage internal coatings due to the sharp edges of brushes and sensors. Therefore, solutions are offered, where the ILI tools mechanics is supported by wheels.

8.4.3.3 Operational Similar to design characteristic, some pipelines do not have the operational makeup to allow for successful pigging. Below are a few operational characteristics that can make pigging difficult:

- Operational temperature above 150 degrees Fahrenheit—Operating temperature can cause problems with the drive cups and the sensors on the ILI tool. Some ILI technologies have a temperature package that can be used to allow for pigging.
- Operational pressure below 200 psig—ILI tools traverse pipelines using pressure differentials for propulsion. Low pressure is mainly an issue in gas pipelines where unwanted speed excursions of the ILI tool are caused by low pressure. In case of a combination of low pressure and low flow the risk of a stationary tool is increasing.
- Operational pressure above 2,500 psig—High pressures can damage the sensors and the electronics of the ILI tool. As with the high temperature operation, some vendors can retrofit their technologies to allow for high pressure pigging.
- Multi-phase flow—The gas phase in multiphase lines can prevent the usage of ultrasonic techniques but does not necessarily make the line unpiggable, since MFL, EC and EMAT methods still can be applied.
- Flow too fast/fast—ILI techniques have optimal speeds for inspections. Typically, this can be worked around by either (1) adjusting the flow rate of the pipeline or (2) the ILI vendor can modify the drive cups of the tool or add a by-pass valve to control tool speed.
- Cleanliness of the product—Product that has a high wax content or product that has rough particles in the stream can affect tool wear and sensor performance. Maintenance pigging history and results can be an indicator of what the ILI tool may encounter during inspection.

8.4.3.4 Economic Very few pipelines are truly unpiggable. If a pipeline operator is willing to invest the time and money to develop an ILI tool to overcome their unique obstacle, then the line can be pigged. Operators have limited financial resources to modify pipelines to make them piggable. The use of temporary launchers and receivers can be used to save money in the short term. The repeated

use of temporary launchers and receivers can have adverse effects too. If temporary launchers and receivers are used often enough, there could be a point in time where the total cost of using temporary traps exceeds the cost of installing permanent traps. When there are not permanent launchers and receivers, pipelines cannot be routinely maintenance pigged. Temporary traps are effective solutions when there is limited space available at locations where ILI tools will be launched or received.

8.4.3.5 Cleaning A successful inline inspection in collecting reliable data depends significantly of the cleaning process. The cleaning program should begin well in advance of the planned inspection date collecting historical cleaning and operational information. The effort for pipeline cleaning basically depend on the type of product transported within the pipeline, the frequency and type of cleaning pigs previously completed, and the ILI inspection technology to be used. Different types of debris are usually observed in pipelines depending on their service. The following Table provides a brief indication for the pipeline most commonly used in the oil and gas industry.

Pipelines can contain foreign material such as dirt, debris, and deposited solids such as salt. Solid deposits (i.e., salt, gasket parts, welding rods, hard scale, wax, sand, and others) can form an adherent, solid barrier that affects pig passage and adversely impacts ILI data quality, and can be very difficult to remove. Depending on conditions, pre-ILI cleaning can be an essential element in obtaining good quality integrity data. Such foreign materials can interfere with the sensors on instrumented ILI tools and affect the accuracy of geometry tools that may be run prior to the ILI. Cleaning can be accomplished by various methods including chemical and dry (scrapers, brushes, magnets). Although an ILI tool could be run in a dirty pipeline, the resulting data would be questionable thereby implying a "piggability" issue.

For pipelines that have not been maintenance pigged for an extended period of time and are suspected to have cleanliness issues, a progressive pigging program can be commenced. Progressive pigging programs begin with mild cleaning pigs and progress to more aggressive cleaning pigs until the pipeline is sufficiently clean for inspection. This progressive program manages the risk of getting any one pig stuck by cleaning in a staged manner. Solid particulate that is removed in cleaning should be tested in the field or sent to the lab testing to determine if there are any additional integrity concerns identified in the solids.

Some vendors offer intelligent data logger which can be mounted to any cleaning tool and do not need any sophisticated setup. These data loggers record time-base pressure, temperature, acceleration, and inclination of the cleaning tool. This allows the operator to identify girth welds, joint numbers, areas of considerable debris, or obstructions caused by dents without the need for intelligent pigging [20]. On the left of Figure 8.13, a pipeline data logger is depicted, which can be flanged to the body of the cleaning tool.

Handling of solids that are removed during the cleaning should be considered before the action is performed. There is a health, safety, and environmental risk when certain solids are removed from the line. Operators should consult with their safety departments before handling solids removed from the pipeline.

8.4.3.6 Gauge Sizing Before an in-line inspection is performed, a cleaning tool equipped with a gauge plate is utilized. Gauge tools do not typically have electronic sensors attached to them when they are used in pipelines. Gauge plates are a solid metal disc of a set diameter that is less that the inner diameter of the pipeline. The

Example: Typical pipeline debris found in the cleaning process before ILI inspection run

Pipeline service	Type of debris	Typical amount of debris	Effect on ILI
Refined products	Corrosion product, valve gasket parts and traces of valve grease	Little to significant amount	Low to no reduction in data quality
Crude oil	Hard and soft paraffin (wax), asphaltines, sand, valve gaskets parts, hard scale, corrosion product	Potentially large depending on product composition, crude velocity and temperature	Sensor contamination, Progressive data loss possible
Multi phase	Hard scale, sand, wax, corrosion product	Potential large depending on product composition	Sensor guidance affected. Reduced sensor sensitivity where occurring
Injection water	Hard scale, sand, corrosion product	Potential large depending on product composition	Sensor guidance affected. Reduced sensor sensitivity where occurring
Dry gas	Corrosion product, black powder, compressor oil, sand	Usually little if pipeline is regularly cleaned and no affected by black power	Low to no reduction in data quality
Sour gas service	Corrosion product, black power, small valve part, compressor oil	Large depending on corrosion severity and black power	Run behavior of ILI tools affected, progressive tool wear and coincidental reduction in data quality

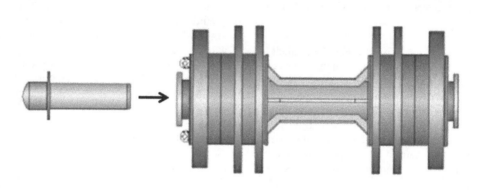

FIG. 8.13   SCHEMATIC VIEW OF A HIGH PERFORMANCE CLEANING TOOL WITH A DATA LOGGER (*Source:* ROSEN Group)

gauge plate is run from the beginning to the end of the pipeline to determine if there are any deformations preventing the passage of the ILI tool. For both the pipeline operator and the ILI vendor, it is required to find a deformation that could damage the ILI tool with a gauge plate prior to the actual ILI tool. Gauge plate tools can be equipped with the above-mentioned pipeline data logger. This will allow locating roughly the position of a deformation of pipe. Gauge plates will not identify sections of pipe that are expanded or overbore. Gauge plates are useful as post construction surveys to ensure that no construction defects exist in the pipeline.

8.4.4   In-Line Inspection Scheduling

NACE SP0102 Section 6 provides factors to be considered when scheduling an ILI survey. These factors included:

- Access to the site
- Throughput outage
- Manpower
- Inspection run time
- Land and access
- Environmental
- Procedural

Pipeline facilities are not always entirely owned by the pipeline company. Launchers and receivers can be collocated with other oil and gas assets and have site restrictions. For example, if a pipeline facility is on a refinery property, it may be necessary for any personnel performing the ILI survey to have refinery training prior to accessing the site. It is also important to schedule the access to the site at a time when there is no other work interfering with the ILI survey performance.

Throughput is considered during both the ILI performance and mitigation activities. A pipeline may require either additional or reduced flow rates to get a quality assessment. The inspection staff should communicate with the operational staff of the pipeline company to ensure that the flow rates for the ILI tool are achieved and do not adversely affect the business. Pipeline features that are reported by the tool may require mitigation. In some cases, the mitigation could involve a pressure reduction or shutting down of the pipeline to perform a replacement. Operational staff should be aware of any potential outages or contingencies before the run is performed.

Manpower includes both the number of people that will be required to be on site and also the equipment involved in the performance of the survey. With greater complexity in the tool, the more clarity that is required for roles and responsibilities of on-site personnel. Clear procedures should be in place and communicated prior to arrival on-site. This procedure will include operation of valves on the launcher/receiver and communication protocols.

Inspection run time should be considered when performing the final setup of the ILI tool. If the tool is expected to run for a long time then additional battery capacity may be required. When a large volume of features is expected, it may be necessary to increase the data capacity of the ILI tool. These items are best considered during the early planning phases of the ILI survey.

There are also regulatory deadlines for performing ILI assessments. Operators should schedule the performance of an ILI well ahead of the regulatory deadline for assessment. Allowing for more time in the schedule for assessment will mitigate some of the risk of schedule slippage. Schedules for ILI performance can vary based on operational needs of the business, availability of the ILI tool, and other factors.

8.4.5 Case Study #1
Context

- 50 mile long refined product pipeline and was installed in 1980
- 10-inch diameter X42 pipe with a 0.275-inch wall thickness
- Operating pressure is 1,200 psig and experiences aggressive pressure cycling
- The pipeline has AMFL and high resolution caliper surveys performed together every four years
- The line experience no post construction hydrotest failures
- The line has had a few in-service failures in the longitudinal seam weld that have shown signs of fatigue crack growth
- There have been no third party damage, external corrosion, or internal corrosion failures on the pipeline

Assignment

1. Select and justify the next ILI assessment methodology.

Results

In this case, it may be beneficial to perform an ILI survey with an UTCD tool. Part of pipeline integrity is ensuring that as many credible threats as possible are addressed through integrity assessment. If there have not been any failures due to deformations or corrosion, then these threats may be managed well on the pipeline. It would be prudent to understand if any other injurious features are present in the longitudinal seam weld. The pipeline has had fatigue failures in its history and currently operates in an aggressive manner. Performing a UTCD tool would be a logical course of action to prevent future failures of the longitudinal seam weld.

8.4.6 Case Study #2
Context

- 10-mile long crude oil pipeline
- Risk assessment has shown that there is the potential for internal corrosion due to low flow rates and intermittent flow
  - The lowest point of the pipeline is in the middle of a 100 foot wide river crossing
- The pipeline is 20 years old and has extruded polyethylene coating with shrink sleeves at the girth welds
- The pipeline is not currently piggable and a capital project is underway to make it piggable

- Temporary launcher and receiver option—$1,000,000 for piping modifications with an expected ILI cost of $500,000 for each survey
- Permanent launcher and receiver option—$2,000,000 for piping modifications and $250,000 for each survey
- It is expected that the line will need to be inspected every four years once the line is part of an ILI program
- There is a significant business and public reputational risk in the event of a loss of containment

Assignment

1. Select and justify whether you would select the permanent or temporary launcher and receiver option for the capital project.
2. Determine any strengths or weakness for each option.

Result

Either option would achieve the goal of performing an integrity assessment using ILI. Operators are constrained on the amount of capital project funds that are available for modifying existing piping for integrity assessment. Even though the permanent option may be cheaper in the long run (i.e., after four surveys), it may not be possible to spend all of the money required up front.

One additional advantage to the permanent trap option is that it would allow for routine maintenance pigging. Maintenance pigging that is optimized based on the results of the ILI survey and pipeline operation can aid in preventing or mitigating internal corrosion.

8.5 IN-LINE INSPECTION IMPLEMENTATION

Implementation of the plan to perform an integrity assessment with ILI is just as important as the technical planning and selection of the ILI tool prior to assessment. This section on implementation will aid the reader in determining things to look for during field operations to ensure a successful ILI survey.

8.5.1 Above Ground Marker Deployment

Above ground markers (AGM) are placed along the right-of-way at discrete points to track the progression of the tool during performance of the survey. AGMs should be placed at consistent locations for every survey to ensure comparison of data between surveys. The location of AGM can be marked with a permanent reference marker or be marked with GPS. GPS points can give consistent results for location of AGM placement; are not affected by encroachment on the right-of-way; and cannot be covered if the location is not maintained. Permanent reference stones give confidence that a consistent AGM location is used.

AGMs can be locations used to identify passage of the ILI tool and can serve as a reference point to locate reported features. The ILI tool and the AGM are synchronized before and after the run by GPS time. During the ILI run, the pipeline log-distance and GPS time are stored on the tool for every detected feature. The AGM GPS time can be used after the run to correlate the above ground position to the pipeline log-distance and in turn to the reported features. The depth of the pipe and location in relation to casings can affect the ability of the AGM to detect tool passage. If the pipe is too deep at an AGM location or is the location of a casing then the AGM will not be able to detect tool passage. AGM locations can also be used to confirm that the ILI vendor have met the location accuracy required in the contract. NACE SP0102 provides additional guidance on AGM concepts and provides requirements for pipe design to identify location of features reported by the tool.

8.5.2 Tool Verification Prior to Launching

Section 7.3 of API 1163 provides considerations related to checking the ILI tool before launching in the pipeline. These considerations are divided between functional and mechanical checks of the ILI tool. In this section, we will discuss why these checks are important.

In addition to the function and mechanical checks, an overall visual assessment and documentation of the tool's condition prior to survey should be performed.

8.5.2.1 Functional Checks of ILI Tool Prior to Launching
Functional tests ensure that the tool will perform to its stated capability. Functional tests of the ILI tool are typically the responsibility of the ILI vendor. Prior to arriving on site, be sure to check the contract with the ILI vendor as to what checks they are required to perform, how those records will be stored, and rejection criteria. This information could be important in the event of an unsuccessful survey through not collecting sufficient data.

Functional tests include:

- Checking power supply
- Checking sensors
- Checking data storage capacity
- Checking components are properly initialized
- Checking speed control is operable

Power supply and data storage can be an issue on longer ILI surveys. As battery and data storage capacity has improved, this has become less of an issue for pipeline assessment. It is still important to ensure that sufficient battery life and data storage is available. When a large number of features were reported in an earlier survey, data storage capacity should be checked to ensure all of the features detected previously and new features are stored.

Sensors and component settings need to be checked before performing the survey to ensure that the tool is able to collect data accurately throughout the survey. Failed sensors and inaccurate odometer readings can lead to failed surveys. First run success rate is important both for integrity and business performance. The operator needs to respond to threats in a timely manner and the business does not want to be affected multiple times due to reruns.

Speed control is an important element of ILI. MFL surveys need to be performed at a controlled speed to allow the tool to magnetize the pipeline. Without sufficient magnetization, the tool will not be able to be as accurate as possible.

8.5.2.2 Mechanical Checks of ILI Tool Prior to Launching
Mechanical checks of the tool include checking that the bolts and connection points are tight on the ILI tool. It is also important to check that all of the electronics packaging that need to be sealed from the external environment are intact. Mechanical failure of the tool can lead to a number of issues in the pipeline.

Losing the electronic capability of the tool due to product intrusion would lead to a failed survey because of lack of data collection. Mechanical failure of the tool from the device separating from itself could cause significant issues. Bolts and other metallic components that break or come loose from the tool could be passed through the system and cause problems with pumps, compressors and other equipment.

8.5.3 Launching the In-Line Inspection Tool

Procedures for launching and receiving an ILI tool should be discussed and clearly outlined before the ILI survey is performed. Any time pig traps are opened or closed there is a personnel safety risk. There are also personnel risks from handling of the ILI tool. Having clear procedures for loading and unloading, the tool and operations of the valves on the launcher and receiver can mitigate these risks. Review of Job site safety should be performed prior to any action on site.

8.5.4 Monitoring Progress of the Survey

It is important to monitor the progress of the ILI tool during the in-line assessment. Monitoring the progress ensures that the tool is moving along the pipeline at an appropriate speed and that it has not experienced any speed excursions. Tracking of the progress also ensures that the tool has not gotten stopped moving during the survey.

Field personnel are required to aid the monitoring of the ILI survey while it is being performed. These individuals need to be trained to use tracking equipment, perform line locating, calculate the time between progress points, and know the communication protocols to complete the project. A compass (mechanical or electronic) can be used to monitor when MFL tools have passed. Other tools are equipped with a low frequency (e.g., 20 Hz) transmitter, which can be received above ground.

The operating conditions of the pipeline also need to be monitored during the ILI survey. Consistent operational parameters are required for ILI tool performance. The operational conditions need to be within the contract specifications that the tool attains the best performance.

8.5.5 Receiving the In-Line Inspection Tool

Procedures for receiving the ILI tool are just as important as the procedures for launching the tool. There are more personnel safety considerations following the inspection. When the tool is exposed to the pipeline environment it can be exposed to conditions that are hazardous to people (e.g., Hydrogen Sulfide, Iron Sulfide, radioactive contamination, the product itself, etc.).

8.6 IN-LINE INSPECTION PRELIMINARY

The "check" step in the ILI process involves confirming that the tool performed acceptably when it performed the ILI survey. This step involves documenting the condition of the ILI tool when it exits the receiver and then assessing the functional systems on the tool and the data that was collected.

8.6.1 Tool Verification after Receiving

Section 5.2 from NACE SP0102 and Section 7.5 of API 1163 provide the user with information on what to look for upon completion of an ILI survey. The functional and mechanical checks performed before launching the tool should be repeated and compared to see if there were any discrepancies following assessment. The condition of the tool should be photographed. Tool damage, debris from the pipeline, and activity of the tool should all be documented.

8.6.2 Run Acceptance

Operators and ILI tool vendors should agree to contractual terms on what determines that an ILI survey was acceptable. This should be performed prior to commencing a survey so both parties know what to expect from an accepted and completed survey. NACE SP0102 provides guidance on what to consider in run acceptance criteria. Criteria to consider according to NACE are as follows:

- Physical damage to sensors after run
- Lost sensor channels on data
- Sensor noise
- Distance inaccuracy
- Velocity underruns or overruns

Damaged sensors and damage to the ILI tool in general can be a sign of the quality of data that was collected during the survey. ILI tools have a minimum number of sensors that are needed to collect data that is of acceptable quality. If a few sensors are lost then this may not constitute a failed survey if the other sensors were able to perform to specifications. Similarly, if there is sensor noise (i.e., disturbances in the sending and receiving of a signal) then the data may not be acceptable. Sensor damage and quality of data are essential contractual parameters prior to running the survey. Data collected should be continuous throughout the survey. Areas where there is an outage need to be evaluated differently than areas of full coverage.

Distance covered reported by the tool is generally reported as odometer distance in units of feet or meters. The user should compare the distance reported by the tool to the actual distance of the pipeline and the distance reported by previous ILI surveys. It has to be kept in mind though that the above ground distance is always different from the pipeline log-distance, due to the difference in elevation of pipeline and surface.

The velocity of the tool should be compares to the necessary parameters for quality ILI performance. Speed excursions by themselves may not be rejectable. The total length and location of speed excursions should be identified and compared to the other data. If the locations are in high consequence areas or areas that have never been assessed, then these may warrant rejection. Different ILI technologies are more sensitive than others to tool speed excursions. Consult with the vendor it there are any questions on tool speed.

8.6.2.1 MFL Considerations MFL tools require magnetic saturation of the pipeline wall to perform to stated capabilities. Carbon steel is saturated between 1.1 and 1.7 T. The user should check that the magnetization levels reported by the tool are to the specification provided by the vendor. As with other discrepancies in a survey, the location and number of occurrences out of specification will determine if the survey needs to be performed again.

8.6.2.2 Ultrasonic Considerations Ultrasonic tools should be checked for echo loss, which could be caused by loss of couplant or misalignment of the sensor. The percentage of echo loss that is acceptable should be agreed with the vendor prior to performing the survey. Excessive echo loss is a rejectable results.

8.6.2.3 Tool Rotation Considerations ILI tools rotate around their center axis when they traverse the pipeline. Some tools require rotation while others can obtain poor results if there is too much rotation. The user should determine how much the tool rotated versus the stated requirements of the tool. Inertial tools are required to rotate for inertial bias correction.

8.7 ILI REPORT REVIEW AND RESPONSE

Once the ILI survey has been performed and accepted, the results need to be acted upon. This action includes a review of the ILI Data (Section 8.8.1), verification of tool performance (Section 8.8.2), determine location of validation digs (Section 8.8.3). This process also includes the action to respond to reported features that meet regulatory, company, or industry best practice requirements.

8.7.1 In-Line Inspection Data Review

Once the inline inspection tool has captured the necessary data and approved the field report, Vendor's qualified analyst set to work to interpret it. Aided by specialized vendor's software, the analysis consolidates and correlates raw facts to generate meaningful insight for pipeline operators. ILI data analysts are qualified to the ANSI/ASNT ILI-PQ standard and have extensive experience in interpreting each ILI tool data type. Signals and analysis varies depending on inline inspection tool type.

A Quality Assurance of ILI Data (QAILID) process is a systematic and consistent approach that aims at improving and reliability in the application, interpretation, and evaluation of data acquired from inline inspection devices. This process involves field data check, checks the ILI data, repeatability process, and advanced ILI data analysis. The field data check task ensures that the ILI service provider has collected all necessary data for integrity assessment.

Example: Field Tool Report
The Field report describes the preliminary analysis of data collected by the ILI tool during the pipeline inspection. Additionally, the field report may provide the following information:

- Pre-launch inspections following the pre-run inspection procedures.
- Timing of launching and receiving.
- Debris collected during the tool travel.
- Tool condition as received in the receive trap indicating sensor and wheel damages.
- A preliminary evaluation of the captured data by the tool after the data was downloaded indicating the data suitability for integrity assessment and percentage of the data and number of sensors collecting data.
- An evaluation of the collected data by other systems such as odometers, and inertial functionality.
- Records of the tool trip timing at all AGM locations.
- Average velocity of the tool indicating the minimum and maximum too specification.
- A preliminary investigation of the sensor damages indicating the direct/immediate causes.
- Photos of the pre-run and post-run tool.
- General status of the tool.

Repeatability process checks that the new inline tool data set sees the majority of anomalies seen by the previous inspection tool, and records new anomalies identified between the two consecutive runs. This process is a continual process and it is conducted where a previous integrity assessment was performed. A detailed comparison of the two data sets is the basis for the repeatability analysis or Run-to-Run comparison. Changes in the estimated depth and length of features (i.e., metal loss and cracking anomalies) between the inspections could be attributed to anomaly growth or initiation of new anomalies.

Comparison of two sets of inline inspection data and obtain growth rate analysis involves activities such as direct visual inspection of all features identified (reported and non-reported features) in the previous data set, match the selected features to be investigated with the new data set and determine the differences to conduct the growth rate analysis. Additionally, this process helps to understand in detailed the potential misclassification of features and detectability of the inline inspection tool. Limitation of the resultant assessment is strongly influenced by the accuracy and completeness of the data recorded during the two comparative inspection operations and the comparing analysis practically is subject to obtaining, reviewing, and confirming the suitability of the two data sets for its purposes.

Understanding the validity of calculated growth rates in the context of inspection tool tolerances is driven by the estimation of the of the specific tool repeatability.

The advanced ILI data analysis comprises checking and reviewing raw data for anomalies that were reported by the inspection tool but are considered non-reportable anomalies. Non-reportable anomalies are those that are specified as not being able to be found with a given tool specification For instances, a crack reported by a UTCD tool with a depth of 1 mm and interlinking length of 20 mm. This anomaly is non-reportable feature because it does not meet tool specification of 1 mm in depth and 30 mm in length. However, there are features remaining in the pipeline that meet the tool specification and the ILI tool did not report it. Those features are typically called "False Negative" features.

The alignment of the two-inline data sets is conducted by matching girth welds locations along the entire pipeline, taking into consideration any changes in pipe joint lengths due to potential modifications of the pipeline between the two inspection runs. The Quality Assurance (QA) on the ILI raw data is to check any feature detected but non-reportable feature can be identified. This process involves checking the C-Scan and B-Scan captures from the UTCD software or client viewer software and database for the entire pipeline. The version of client viewer software usually does not provide the entire data set collected by sensors kits of the ILI tool. If the entire raw data set should be reviewed, the assessment shall be conduct laterally with the ILI provider using the analysis software.

Example: Selection criteria of the sample of SCC cracking features reported by UTCD tool

- Features with the highest Estimated Safety Factor (ESF) from the crack ranking of the both ILI data sets.
- Features with the deepest reported depth for both ILI dataset.
- Review and compare the feature profile to determine the differences in length and depth of features.

- Features associated with signal identified during the data analysis as visually changing between the inline inspection either in extend or severity.
- Features located in particular areas of interest as designated by pipeline operators (i.e., river crossings, road crossings, suspected growth areas, highest MOP in the system, populated areas, etc.).
- The assessment may account for any inaccuracy, incompleteness, and misclassification in the data set or the referenced ILI data.
- Other criteria selected by pipeline operators.

Example of the QA/QC process for the UTCD inline raw data for SCC and Crack-like features on the long seam weld and pipe body.

- The assessment is a qualitative and qualitative assessment for the inline inspection raw data. Qualitative assessment comprises reviewing, comparing and checking what the ILI tool reported to determining the misses of the tool (i.e., Probability of Detection), and the quantitative consist of a quantification of dimension deviation from the reported tool data (i.e., magnitude deviation for determining crack depth, probability of Sizing).
- The signal for each selected anomaly is directly compared across all recording sensors for a qualitative assessment to determine whether the feature shows any evidence of growth.
- Identification of features with high amplitudes (i.e., features with amplitude of the signal of 30 to 40 dB depending of the algorithm configuration) that do not meet the minimum length threshold with repetition in the next sensors (i.e., crack tip with length smaller than length threshold).
- For each ILI feature subjected to the quantitative assessment, differences on maximum depth, number of cracks, defective areas, remaining ligament area, length of longest indications with colonies, cumulative length and width, recorded amplitude, and status of the SCC individual crack (i.e., active or inactive cracks) are examined.
- Identification of faulty sensors, bad or noisy channels, overspeed areas, missing data, distance problems, AGM locations, and concerned areas to determining the continuity and discontinuity of information that may be associated with long seam weld position.

Reduction of the capacity of the pipeline structure due to incremental of anomalies over time can be called "Deterioration." This deterioration of lines is quantifiable through the advanced analysis to the inline inspection tool data (i.e., raw data). New anomalies can initiate or existing anomalies can grow between two ILI runs or even other anomalies can become inactive if operating conditions has changed. The growth of anomalies is the effect of pipeline degradation. Deterioration assessment can be performed having more than one consecutive inline inspection data set by evaluating changes in the signal recorded in both previous and next inspections. The following example shows a deterioration analysis due to cracking threat detected by various consecutive inline inspections that evaluate changes in the UT crack signal recorded by the Crack Detection tools.

Example: Deterioration analysis using UTCD raw data

The findings of the UT signals of the previous run are compared to the findings of the next run. Practically, four steps are considered and described as follow.

1. Conduct an analysis of the pipe joints having non-reportable (features with dimensions under ILI tool threshold) SCC or LOF anomalies either located in pipe body or associated to long seam weld in the most recent in-line inspection tool run to identify additional crack indications. It is noted that reviewed indications do not meet the criteria of the reported dimension in the ultrasonic tool specifications (i.e., crack features with a depth less than 1 mm and length less than 30 mm).

2. Perform an evaluation of the SCC and LOF growth rate between two consecutive runs to revise and update the SCC and LOF growth rate estimated previously. A detailed review of the B-Scan and C-Scan captures is meaningful to collect essential information in determining the crack signal and dimensions.

3. Determine the initiation period of the additional crack indications identified in the analysis. This is done by comparing UT signals reported in the second ILI tool data that were not reported in the first ILI tool data set.

4. Determine the dimensions (i.e., depth and length) of the crack indications identified in the analysis based on signal recorded by the tools. The estimation of the feature depth depends on the amplitude (dB) and information collected from B-Scan and C-Scan captures for each ILI tool algorithm.

5. Conduct statistical analysis of the dimensions. Estimate the Failure pressure, failure mode, and time to fail for the recorded and un-reportable features that do not meet the threshold of the ILI tool specification encountered in the analysis.

8.7.2 Verification of Tool Performance Specification

Performance specification is defined as the capability of the ILI system when travel in a specific pipeline to detect, locate, identify and size pipeline anomalies, components and features. The main objective of the validation is to evaluate essentially the inline inspection performance with respect to anomaly detection, identification, and sizing. The parameters considered in defining the ILI system performance are:

- Anomaly type (i.e., external metal loss, crack-field, long seam weld anomaly, dents, etc.) and its characteristics (i.e., associated with girth weld, surface breaking, etc.) covered by the performance specifications.
- Detection thresholds (i.e., crack detection threshold: (i.e., depth of 1 mm and 30 mm in length) and its probability of detection (POD) with a certainty of e.g., 80%.
- Adequacy of anomaly identification that is the probability of identification properly (POI) with a certainty of, e.g., 95%.
- Accuracy of sizing anomalies (e.g., ±10% wall thickness).
- Measurement accuracy of orientation and distance.
- Limitations (e.g., size metal loss anomaly in a dent, not able to report cracks on dents, etc.).

Example: Minimum characteristics and performance specification for an axial MFL ILI tool to be provided by ABC vendor

EXAMPLE OF AXIAL MFL ILI PERFORMANCE SPECIFICATION

Parameters	Specifications
Resolution	High resolution
Sensor types	Hall effect sensor
Maximum lost sensor channels on data	1%
ID/OD	Yes
Minimum bore restriction	20%
Flexibility	1,5D
Minimum MFL sampling rate	1 sample/5 mm
Maximum wall thickness	14 mm
Minimum tool velocity (m/s)	0,5 m/s
Maximum tool velocity (m/s)	5 m/s
Maximum pressure	1,600.00 psig
Maximum circumferential coverage per sensor	10 mm
Axial location accuracy	± 0.1 %, or ± 25 mm from reference girth weld
Circumferential location accuracy	± 15°
Maximum distance Inaccuracy	<0.50% Total Length
Minimum odometer resolution	<3% of the total length
Minimum metal loss orientation resolution	(±5°)
Temperature range	(15°C–50°C)
Battery life (hour)	24 hours
Number of primary sensors	120
Minimum probability of detection (POD)	90% wall thickness (w.t.)
Minimum probability of identification (POI)	Confidence level of 80%

| Probability of sizing | Longitudinal anomaly type | | | |
	General metal loss	Pin holes	Axial grooving	Axial Slotting[1]
POD=90% for depth detection (Fraction of w.t.)	0.2	0.2	0.2	0.2
Depth sizing accuracy at 80% confidence (Fraction of w.t.)	±0.15	±0.20	±0.15	±0.20
Width sizing accuracy at 80% confidence (mm)	±15	±15	±15	±10
Length sizing accuracy at 80% confidence (mm)	±15	±10	±15	±15

[1] Valid for axial slotting feature width ≥ 1 mm.

8.7.3 Determination of Validation Digs

Investigation of the existing anomalies detected by inline inspection runs is highly important for tool performance verification. The verification of tool performance is done by carrying out investigative digs for examination of either harmful anomalies or flaws that have not reached critical dimensions. This examination provides field information that is correlated again inline inspection data for validation of the inspection tool's capability to identify, characterize, and size anomalies adequacy. Field verification of the ILI reported features has three main aspects:

- Confirm the actual condition of the pipeline integrity through direct assessment of the reported features to operators, and help support any actions that may be taken
- Verify the actual tool performance to conduct feature dimension adjustments
- Accept the inline inspection, which is followed with the comparison of the vendor ILI tool tolerances

The ILI data provides a general view on the current state of the pipeline condition but the results are affected by measurement error. Initially, the current ILI data analysis comprises a correlation data between the most recent ILI tool run that identifies location, depth, length, and profiles of anomalies, and previous inline inspection data. The correlation is used to estimate the growth rates based on depth changes in matched anomaly taking into consideration new anomalies and the impact of measurement error needs to be considered. The measurement errors for both ILI tool techniques and inspection method used in the verification dig occur and it should be taken in consideration in the verification of ILI tool performance process. This requires trained field personnel to gather the data with the required accuracy and competency so that results can be relied upon. The ILI techniques and equipment used must be tested and certified in calibration, and the calibration and device tolerances must be taken into account when evaluating the results. The example below shows examples of direct inspection methods used when conducting anomaly measurements in the field for verification digs.

Example: Type of Non-Destructive Examination (NDE) methods for verifying ILI features during direct inspection and dig verification process.

EXAMPLE OF NON-DESTRUCTIVE EXAMINATION (NDE) METHODS FOR VERIFYING ILI FEATURES

Reported feature by ILI	Inspection methods[1] to be conducted in the field
External corrosion	Depth gauge micrometer, laser scanning or straight-beam UT probes
Internal corrosion	Straight-beam UT probes, Angle UT probes, UT (C) scanning, radiography
External crack	Magnetic particle, straight-beam UT probes, angle UT probes, phased-array UT probes, UT (C) scanning, ACFM (alternating current field measurement), EC methods
Internal crack	Angle UT probes, phased array, ToFD (time of flight diffraction)

[1]Best methods typically used in the field are underlined.

The advantages to determine the performance of the ILI inspection are to conduct the required preventive integrity plan (i.e., maintenance or repair) with the certainty operator's safety, and calculate the Factor of Safety (FOS) effectively using the actual ILI tolerances and certainty (not just the tolerance) matching those provided in the inspection report.

For validation, the level of confidence may be quantified by establishing minimum sample sizes [21]. These sample sizes represent the amount of feature that should be evaluated to achieve a given confidence level. An advanced statistical analysis are typically used to calculate the minimum statistically significant sample size from the inline tool dataset for anomaly type reported by inline inspection tools and estimates the number of the investigated or opportunistic digs to be performed. The purpose is ensuring the representativeness of the proposed excavation program based on ILI datasets. Calculation of the minimum sample sizes is based on the population of reported anomalies. Field Verification Procedure from Pipeline Operators Forum can be found on the internet public domain.

Several statistical approaches can be used in determining the sample size of anomalies to be investigated. One statistical method is the finite population proportion method, which can be found in several statistical textbooks.

8.7.4 Crack ILI Challenges and Confirmatory Hydrostatic Testing

Hydrostatic testing offers the binary response (Yes/No) to confirm the integrity of a pipeline associated with a given safety factor to the expected operating pressure and the leak tightness at one given time. Hydrostatic testing at higher than 100% SMYS levels during short periods (<2 hours) can also offer a "Spike" causing some manufacturing or environmentally assisted cracking to blunt (i.e., crack tip area plasticization) delaying their growth for some time [22]. However, the loss of validity of the in-line inspection data acquired until the hydrostatic testing, the growth of potential subcritical anomalies as well as the operational efforts and product quality challenges have posed the questions as to *when hydrostatic testing is essentially needed?*

Failures of pipelines inspected with crack ILI technologies have caused concerns about the ability for only one integrity assessment methodology to manage the integrity of seam weld anomalies [23].

Example: Identifying potential actions for minimizing effects of crack ILI challenges with seam weld cracking

Crack ILI: seam weld challenge	Potential actions for minimizing effects
Crack ILI detection, characterization, and sizing	• Multiple back-to-back ILI (i.e., CMFL and/or UTCD/Phased Array and/or EMAT): *complementing/improving data* • Multiple ILI-Repeatability Analysis: *commonalities and differences in reporting of anomalies* • 3rd Party Independent Reviewer of ILI raw data: *further reviews for field investigation of non-reported anomalies* • Build a crack ILI signal library correlated to field and laboratory findings: *increasing knowledge* • Work closely with ILI vendors for improving performance via hardware (e.g., sensors) and software (e.g., data analysis) changes

Crack ILI calibration affected by Non-Destructive Examination (NDE) human factors and technology selection	• Use of higher accuracy field NDE technologies (e.g., phased array, Inverse Wave Field Extrapolation—IWEX) • Pre-calibration of NDE technology and technicians using inspection coupons for laboratory verification • Define NDE qualification and inspection protocols for seam weld anomalies • Work closely with NDE vendors for improving performance via technology and qualification improvements
Estimation of the cracking predicted burst pressure and growth using ILI input	• Use of API 579 methodology • Use of actual material properties • Build maximum critical flaw size graphs tracking safety factor changes based on crack growth • Work closely with engineering consulting for improving assessment of anomalies

In addition, pipeline companies face the "time challenge" associated with the need for timely acting to prevent a seam weld-cracking incident while implementing integrity actions such as the ones mentioned in the example. Confirmatory hydrostatic testing may offer, if conducted minimizing the damage to or anomaly growth in the pipeline, a one-time verification of the integrity of the pipeline. Pipeline with high criticality (e.g., number of significant anomalies not detected by crack ILI) may be good candidates to apply confirmatory hydrostatic testing of short duration.

Ongoing Research and Development by special interest groups are also needed for tackling these historical challenges so they are overcome providing the safety and environmental protection required by the stakeholders.

8.8 CONTINUOUS IMPROVEMENT OF IN-LINE INSPECTION PROGRAMS

Some elements of continuous improvement include performing reassessment surveys, changing vendors for both better performance or a different look, using updated tool technologies, and overlaying additional information. The biggest potential for improvement is communication lines and responsibilities and agreed processes and procedures potentially based on industry standards establish a cost effective conduction of an in-line inspection program. Afterwards performance gaps can be identified and a repetition of low performance can be avoided. Subsequently improved technology can be developed to address these performance gaps.

8.8.1 Reassessment Surveys

A baseline assessment is the first time that a pipeline has a smart pig survey performed. The final step in the assessment is to determine the reassessment interval for the pipeline. The reassessment interval of regulated pipelines typically has a maximum interval. In the US, this maximum interval is 5 years for liquid pipelines and seven years for gas pipelines.

Operators can shorten the reassessment interval at their discretion. Accelerated corrosion mechanisms or a significant number of corrosion features that appear to have grown from one survey to the next would be examples of reasons to shorten the reassessment interval.

In the previous section, we discussed the usage of multiple survey techniques. Using the same survey technique with a different vendor can also be useful. ILI technology is rapidly changing and evolving. Using updated sensors can also improve tool performance.

The pipeline industry is very good at managing known threats. Failures still occur on threats that were not known to exist. Operators are encouraged to use different techniques for ILI surveys when the reassessment interval is over. For example, if an operator typically performs an MFL survey because the most likely threat is corrosion, then a crack detection survey may be considered later to be sure all threats are assessed. Operators could also elect to alternated AMFL and CMFL when the surveys are due. This would ensure that all types of corrosion are addressed.

8.8.2 Qualification of New Technologies

ILI vendors may have similar names and basic technique of tools but the functionality and the details of the analysis process can be very different. It is useful to qualify the same survey technology from a different vendor to understand the differences in performance. The comparison of the different runs will also benefit in getting a different perspective on the same feature. During qualification programs the operator gains an understanding of the performance the different technologies provide to the pipeline system. Therefore a false sense of security will be avoided.

8.8.3 Updated Sensors

Using the most up to date sensors on the ILI technique will provide the best results. As MFL and caliper surveys transitioned from low resolution to high resolution, operators were given the most accurate data possible. In MFL surveys, what could be first seen as a large area of corrosion could be regarded as an area of shallow wall loss with a few isolated deeper areas.

8.8.4 Importance of Data Alignment

ILI data is just one source of information for pipeline integrity. ILI data should be aligned with all other attributes of the pipeline for integrity management. For example, if there were an area of highly strained in a known landslide area then this would be an area to prioritize for remediation.

Operators are required to perform cathodic protection surveys and integrity assessments on their pipelines. An area reported with low cathodic protection potentials through an above ground survey technique and then shallow corrosion reported in an MFL survey could be a sign of active corrosion. If the cathodic protection in the area is improve, it could prevent further corrosion in the future.

8.9 ROBOTIC AND TETHERED TECHNOLOGY FOR NON-PIGGABLE PIPELINES

Robotics and tethered ILI tools can be used for a variety of applications where traditional free-swimming ILI tools cannot be used. These technologies typically require the complete shutdown and evacuation of the pipeline for assessment. Robotic and tethered tools typically have the same types of sensors that are offered on free-swimming ILI tools but they are put on chassis that allow for traversing the line on their own.

Robotic and tethered tools are useful for assessing:

- Tanks lines
- Loading lines
- Vent/flare lines
- Pump station piping
- Facility piping
- Wharf and jetty piping

Robotic crawler tools can be useful to obtain detailed assessment of features in the pipeline. Because line segments are isolated and evacuated of product in the assessment, repeated passes can be made with different technologies to get a full assessment. Figure 8.14 below is an example of a robotic crawler tool. Some robotic tools can be outfitted with EMAT ultrasonic, laser profiling, visual inspection, and girth weld scanning technologies.

Figure 8.15 below is the detailed scan of a robotic crawler with EMAT technology. The image on the left is the scan with EMAT while the image on the right is the associated visual image of the corrosion feature using the robotic tool.

Robotic crawler tools can be fitted with laser profiling tools. Laser profiling allows for very detailed study of internal anomalies (e.g., corrosion, dents, welds) so more advanced engineering analysis can be performed. With the detailed assessment, the proper course of action to mitigate the defect can be performed. Figure 8.16 shows an illustration of the tool on the left, the anomaly being reviewed in the center, and the results of the scan on the right.

Other vendors provide similar systems for crack detection with EMAT. In this case, a pulsed EMAT (PEMAT) system is used to avoid an unwanted blockage due to permanent magnets. Nowadays

FIG. 8.14 PHOTO OF A ROBOTIC INSPECTION TOOL (*Source:* Diakont)

also variants are available, which can travel inside a pressurized pipeline. Figure 8.17 depicts a robotic PEMAT tool. The equipment can be driven thru empty and pressurized pipelines as well. Also a data set obtained from an external crack colony is shown in Figure 8.18.

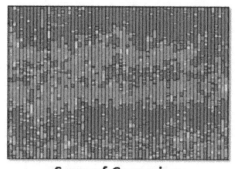

Scan of Corrosion   Associated Visual Image

FIG. 8.15 EMAT SCAN OF A ROBOTIC TOOL (LEFT) OF INTERNAL CORROSION (RIGHT) (*Source:* Diakont)

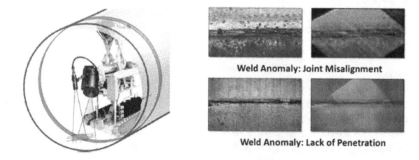

FIG. 8.16 LASER SCANNING AND SCAN RESULTS (*Source:* Diakont)

FIG. 8.17 ROBOTIC VEHICLE UTILIZING PEMAT SCANNER (*Source:* Rosen Group)

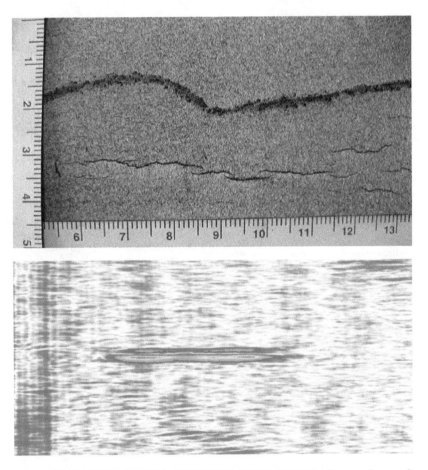

FIG. 8.18 CRACK AND SIGNAL OBTAINED FROM PEMAT (*Source:* Rosen Group)

Robotic and tethered tools are not without their own limiting factors. Tethered tools can be limited by the total radius of bends that are traversed. Beyond 350 degrees of bends, it can be difficult to pull the tethered tool to perform the inspection. Robotics and tethered tools can also be limited by bend radius in some cases. Tethered and robotic tools are advancing at a steady rate like ILI technologies. Limitations today may not be limitations tomorrow.

8.10   REFERENCES

1. Anon., 2013, *Recommended Practice 1160—Managing System Integrity for Hazardous Liquid Pipelines*, American Petroleum Institute, Washington DC, USA.

2. Anon., 2014, *Managing System Integrity of Gas Pipelines*, ASME B31.8S, American Society of Mechanical Engineers, New York, USA.

3. Anon., 2007, *Recommended Practice 579—Fitness for Service*, American Petroleum Institute, Washington DC, USA.

4. Anon., 2016, Recommended Practice 1176—*Assessment and Management of Cracking in Pipelines*, American Petroleum Institute, Washington DC, USA.

5. Interstate Natural Gas Association of America, 2012, "*Report to the National Transportation Safety Board on Historical and Future Development of Advanced In-Line Inspection Platforms for Use in Gas Transmission Pipelines*", USA.

6. Anon., 2013, *Recommended Practice 1163—In-Line Inspection Systems Qualification*, American Petroleum Institute, Washington DC, USA.

7. POF, 2009, Specifications *and Requirements for Intelligent Pig Inspection of Pipelines—Pipeline Oper*ators Forum (POF 2009), Europe http://www.pipelineoperators.org/.

8. Hall, R.J., McMahon, M.C., 2002, *Report on the Use of In-Line Inspection Tools for The Assessment of Pipeline Integrity*, U.S. Department of Transportation, Washington, DC USA.

9. Feng, Q., Sutherland, J. et al., 2011, "*Evolution of Triaxial Magnetic Flux Leakage Inspection for Mitigation of Spiral Weld Anomalies,*" 6th Pipeline Technology Conference.

10. Pipeline Research Council International, and Battelle Memorial Institute, 2013, "*Development of Dual Field Magnetic Flux Leakage (MFL) Inspection Technology to Detect Mechanical Damage*", Falls Church, VA, USA.

11. Hrncir, T., Turner, S., Polasik, S.J., Vieth, P., Allen, D., Lachtchouk, I., Senf, P., Foreman, G., *A case study of the crack sizing performance of the Ultrasonic Phased Array combined crack and wall loss inspection tool on the Centennial pipeline, the defect evaluation, including the defect evaluation, field feature verification and tool performance validation (performed by Marathon Oil, DNV and GE Oil & Gas)*, IPC2010-31079, ASME International Pipeline Conference, American Society of Mechanical Engineers, Calgary, AB, Canada.

12. Limon, S. et al., 2008, *A framework to manage the threat of SCC and other forms of cracking in pipelines using In-Line Inspection Tools*, International Pipeline Conference, IPC-200864090.

13. Michael Baker Jr., Inc, 2004, "Inspection Guidelines for Timely Response to Geometry Defects".

14. Russell, D.E., 2014, *What in-line technologies work best for condition assessment of pipelines and why*, Pipelines 2014 Conference, ASCE, Portland, OR, USA http://www.picacorp.com/Portals/1/documents/media/What%20technology%20works%20best%20for%20condition%20assessment%20of%20pipelines%20and%20why_ASCE%202014.pdf.

15. Georgiou, G.A., *Probability of Detection (PoD) Curves*, Jacob Consulting Ltd, UK Health & Safety Executive, Research Report 454, United Kingdom http://www.hse.gov.uk/research/rrpdf/rr454.pdf.

16. Beuker, T. et al., 2007, *Advantages of combining magnetic flux leakage and ultrasonic technologies in-line inspection solutions*, Rio Pipeline, IBP 1131_7.

17. Paeper, S. et al., 2006, *In-line inspection of dents and corrosion using "high-quality" multi-purpose smart pig inspection data*, International Pipeline Conference, Calgary, IPC 2006-10157.

18. Shie, T.M., Koch, G.H., Bubenik, T.A., 2010, *Unpiggable Pipelines*, Pipeline Pigging and Integrity Management—22nd Year, Houston, TX, USA.

19. Beller, M. et al., 2015, *Speciality Solutions for Challenging Pipelines*, Rio Pipeline, IBP1046_15.

20. Van der Graf, J., 2004, *Pipeline Data Logger—A new tool for the pipelines engineer*, pp. 49–51, 3R Vol 43.

21. Schaeffer, R.L., Mendenhall II, W., and Ott, R.L., 1996, *Elementary Survey Sampling*, 5th edition, Duxbury Press, USA.

22. Kiefner, J.F. and Maxey, W.A., 2000, *Hydrostatic Testing—Part 1—Pressure Ratios Key to Effectiveness; In-Line Inspection Complements*, Oil and Gas Journal, pp. 54–61 (July 31, 2000) and "Hydrostatic Testing—Conclusion—Model Helps Prevent Failures", Oil and Gas Journal, pp. 54–58 (August 7, 2000), USA.

23. Johnson, J., Nanney, S., 2014, *The Role, Limitation, and Value of Hydrostatic Testing vs In-Line Inspection in Pipeline Integrity Management*, IPC2014-33450, ASME International Pipeline Conference, American Society of Mechanical Engineers, Calgary, AB, Canada.

PRESSURE TESTING FOR PIPELINES

Engineering structures are tested before they are put into operation, to ensure they are strong enough and will perform safely.

Pipelines are made from lengths of line pipe, welded together in the field. Each section of line pipe is tested at the pipe mill by filling the pipe with water, and pressurizing the water to create a high hoop stress in the line pipe wall; this 'hydrotest' demonstrates strength and leak-tightness of each length of line pipe.

This type of 'pressure test' is also conducted on the entire pipeline before it enters service, using either a gas or water as the pressurizing medium. The test is performed at a pressure higher than the proposed pipeline operating pressure. A successful pressure test (i.e., no leakage of product or deformation in the pipeline) ensures:

- the pipeline is leak-tight;
- the pipeline does not contain large defects; and,
- the pipeline can withstand its maximum operating pressure.

These tests can be conducted on both new pipelines (all design standards require this type of test), and in-service pipelines. Testing of in-service lines is difficult and expensive, as it requires the pipeline to be taken out of service, but it does allow pipeline operators to assess the condition of their pipelines.

Pressure testing is one method of assessing the condition of an in-service pipeline: others include in-line inspection using 'smart pigs', and 'direct assessment'. An advantage of the pressure test is that it will detect (by failing) the largest defect in the pipeline that can fail due to internal pressure, whereas the other methods cannot offer this level of guarantee.

It is important to note that pressure testing does not guarantee a safe pipeline: subsequent growth of defects that have survived the test, or defects created after the test, can cause failure following the test.

This Chapter explains pressure testing, and the differing types used in the pipeline industry.

9.1 INTRODUCTION

All engineering structures receive some level of testing before they are put into service. These tests can range from simple quality control tests (for example, checking dimensions), to leak and strength tests; for example, a hot air balloon will be tested to ensure there are no leaks, and, ropes will be tested to ensure they can carry a specified load.

9.1.1 Pressure Testing Structures

Some structures can also 'pressure tested': any structure that is designed to contain pressure is a candidate for pressure testing. The test involves placing a fluid ('test medium') in the enclosed structure and raising the fluid's pressure (using a pump or compressor)

to test the strength and leak-tightness of the structure. A definition of 'pressure test' is [1]: "... *[a] means by which the integrity[1] of a piece of equipment (pipe) is assessed, in which the item is filled with a fluid, sealed, and subjected to pressure.*"

When a gas is used as the pressurizing medium the test is called a 'pneumatic' test. When water is used as the pressurizing medium, the test is called a 'hydrotest'. A definition of 'hydrostatic[2] test' or 'hydrotest' is [1]: "... *a pressure test using water as the test medium.*" therefore, hydrotesting means a pressure test using water, but the term 'hydrotesting' has been used to describe testing with any fluid (air, nitrogen, natural gas, etc.) [2].

9.1.2 Pressure Level and "Hold"

Operators will test to a pressure above the planned operating pressure of the structure. This pressure is held for a time to allow pressure to stabilize, and leaks detected, Figure 9.1. The higher the margin between the test pressure level and the operating pressure, the higher the confidence in the condition of the pipeline: "*Within limits, the greater the ratio of test pressure to operating pressure, the more effective the test*" [3].

9.1.3 Purpose of Pressure Test

9.1.3.1 General Pressure vessels and pressurized components (e.g., boilers and valves) have been pressure tested for many years, as it is easy to fill them with a gas or water and check for leaks, and also check the strength of the as-built structure. Obviously, it is far better to fail a structure during a pre-service pressure test, than during operation.

API 1110 [4] gives many reasons for pressure testing, including:

- to establish the operating pressure limit of a structure;
- to detect and eliminate defects in the structure[3];
- to show no evidence of leakage.

---

[1] 'Integrity' has many meanings. In this Chapter it means the ability of a pipeline to safely withstand all loads imposed on it, without failing (losing product) or becoming unserviceable.

[2] 'Hydro' means water, 'static' means stationery (not flowing). The word 'hydrostatic' has several meanings, but in the pipeline business it has two major meanings: it defines a 'stress state' in a structure where all the stresses are equal and are principal stresses (when a body is under hydrostatic stress, there is no shear stress, and the stresses on the body are equal in all directions); and, the pressure exerted by a liquid.

[3] The pressure test can fail defects in the pipeline; therefore, it is called a 'destructive' test. A 'non-destructive test' (for example, in-line inspection using 'smart' pigs) can inspect the pipeline but will not cause any failure.

Pressure or stress

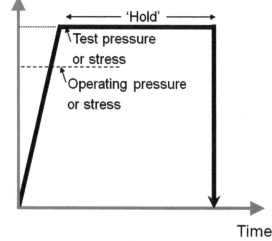

FIG. 9.1   PRESSURE TESTING OF PIPELINES

When the pressure test is at a pressure higher than the structure's subsequent operating pressure, the test 'proves' its strength prior to service [5].

9.1.3.2 Pipelines   Pipelines are pressure tested prior to being put into service: design standards require this test. This entails filling the pipeline with a fluid (usually water), and the fluid is pressurized to a pressure higher than the pipeline's design pressure.

The logic is... if a pipeline can withstand a test to a high pressure, it is logical that it can operate safely at a lower pressure.

It is important to note that this pressure test is checking the structure's fabrication and materials, rather than the design [6]: the design will have been checked by the designers via calculation. The designer has no control over the fabrication or the materials finally used in the fabrication; therefore, the pressure test is the final check on the fabrication and the materials used. The American Society of Mechanical Engineers is very clear on the purpose of a pressure test [1] "... *It is used to validate integrity and detect construction defects and defective materials.*"

Hence, a pressure test can demonstrate a pipeline's 'fitness for service'. It can be carried out:

- on new pipelines; or,
- to prove the integrity of existing lines; or,
- to prove the ability of existing lines for new operating conditions such as increased pressure, or a new fluid.

9.1.4   Passing a Pressure Test

A structure 'passes' a pressure test if the specified test parameters (pressure/stress and hold period) are achieved and verified/documented. Failures (e.g., a leaking flange) during a pressure test do not necessarily mean the test has failed. The failures can be rectified and the test restarted and repeated until the specified parameters are achieved. Obviously, if there are many of these 'test failures' or the test reveals a material or design fault, then the test would be considered a failure.

9.1.5   Stressing Caused by Pressure Testing

When a cylinder (pipe) is pressurized, stresses are induced in the pipe wall: the largest is the hoop stress, Figure 9.2. The hoop stress (σ_h) is a function of pipe outside diameter (D), wall thickness (t), and internal pressure (P). The hoop stress for straight pipe is calculated by the 'Barlow' equation: $\sigma_h = PD/2t$. The Barlow equation is applicable to 'thin shells' (pipe diameter to wall thickness ratios of >20). This hoop stress in a thin shell (high D/t pipe) is usually assumed to be a 'membrane stress' (a component of the structural stress that is uniformly distributed and equal to the average value of stress across the section thickness).

9.1.6   Energy in a Pressure Test

Energy is introduced into a vessel or a pipe when they are filled with a fluid and the fluid pressurized; for example, there is energy in the vessel wall due to straining, and within the fluid due to it being pressurized.

A release of this energy will create a hazard. The main hazard associated with pressure testing is this 'stored energy', E. This is the available energy during the test. It consists of [7]:

- strain energy, E_S, in the vessel caused by the stressing in the wall;
- fluid expansion energy, E_E (related to pressure × volume) which is high for a (compressible) gas, but very small when using an (incompressible) liquid; and,
- chemical reaction energy, E_C (e.g., caused by an ignition).

Hence, $E = E_E + E_C + E_S$ (Figure 9.3).

This latter E_c energy can be ignored if using non-flammable fluids (e.g., air or water). Therefore, if there is no chemical release, $E = E_E + E_S$.

After a failure, the conservation of energy requires the same amount of energy to be present, albeit in different forms [7]. After failure, the energy (E) is:

- kinetic energy of all fragments, E_K;
- energy of blast wave, E_B;

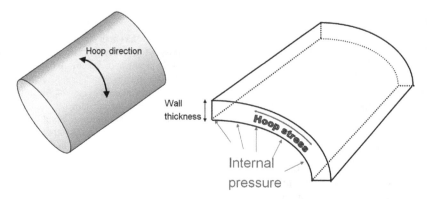

FIG. 9.2   HOOP STRESS CAUSED BY INTERNAL PRESSURE

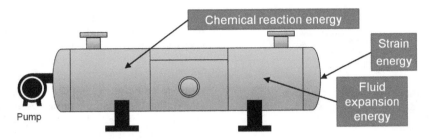

FIG. 9.3   ENERGY IN A PRESSURE TEST

TABLE 9.1   ENERGY IN A PRESSURE TEST [8]

| Test fluid | Internal energy in 1 m diameter vessel | |
	Energy	TNT equivalent
Water	0.59 MJ	0.13 kg
Nitrogen	39 MJ	8.7 kg

- energy of ground shock;
- thermal (heat) energy;
- radiant (light) energy;
- sound energy;
- chemical reaction energy products.

Assuming no chemical explosion, the energy is made up of missile and blast energy: $E = E_K + E_B$.

In a test using air, steam, or gas (a 'pneumatic' test), the stored energy creates a blast wave and missiles (broken parts of the vessel). In a test using a liquid, the blast wave is negligible (there will only be a sound wave (a loud 'bang')): it is missiles that are the major hazard.

The effect? The stored energy in a pneumatic test is huge compared to that in a liquid test, for the same pressure [8]. This difference is illustrated in Table 9.1 [8], for a small (3 m (118 inches) in length, 1 m (39 inches) diameter, and 100 barg (1450 psig) pressure at 57°C (135°F)) pressure vessel.

There are pressure testing safety procedures in the literature (for example, Reference 9).

9.1.7   The Role of Pressure Testing in Pipeline Integrity Management

Pressure testing is one method of assessing the condition of an in-service pipeline: others include internal inspection using using 'smart pigs', and 'direct assessment', Figure 9.4:

- smart pigs are tools that are inserted into a pipeline and can detect and measure defects in the pipe wall;
- direct assessment is a structured process where the operator integrates knowledge of the pipelines' physical characteristics and operating history with the results of inspections.

Other methods equivalent to these methods can also be used.

An advantage of a pressure test is that it will detect (by failing) the largest defect in the pipeline that can fail due to internal pressure, whereas the other assessment methods cannot offer this level of guarantee.

This Chapter will cover the testing of both new and in-service pipelines, but first the pressure testing of the line pipe that makes up the pipeline is covered.

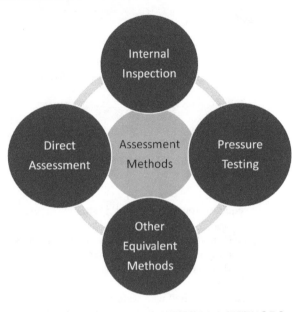

FIG. 9.4   CONDITION ASSESSMENT METHODS FOR PIPELINES

9.2   PRESSURE TESTING LINE PIPE

Pipelines that transport natural gas, and oil, are made from sections of high strength carbon steel tubes called 'line pipe', usually supplied in 12 m (40 feet) lengths, to engineering and metallurgical specifications (for example, API 5L [10]). The line pipe sections are welded together in the field with 'girth welds' to create the pipeline, Figure 9.5.

A line pipe manufacturer wants to produce a product that will meet the purchaser's requirements, particularly its internal pressure requirement; therefore, it is common-sense to test (pressurize) each section of line pipe before sending it to the purchaser.

Historically, line pipe manufacturers hydrostatically tested their line pipe sections in their pipe mill. This testing was not always a requirement, but USA line pipe manufacturers were hydrostatically testing each section of line pipe in their mills, to test the strength, in the late 1800s.

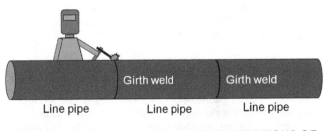

FIG. 9.5   PIPELINES ARE MADE FROM SECTIONS OF LINE PIPE

Section of line pipe (usually 12 m lengths)

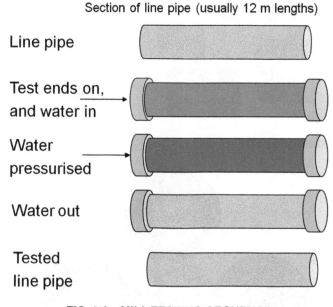

FIG. 9.6   MILL TESTING SEQUENCE

9.2.1   Mill Test

The testing sequence for this 'mill test' is simple (Figure 9.6):

- the line pipe is produced and quality checked (steel chemistry, strength and toughness, dimensional characteristics, etc.);
- test ends are then put on each end of each section of line pipe to provide a seal;
- the line pipe is then filled with water;
- the water is pressurized, using a pump, to a specified pressure (or hoop stress in the pipe wall);
- the specified pressure is maintained for a few seconds (e.g., 10 seconds);
- if there is no failure/leakage/deformation, then the pipe has passed the mill test.

The two key parameters for a pressure test are: the pressure level, P_{mt} (or hoop stress level); and, the time the pressure level is maintained for, Figure 9.7.

The mill test pressure is only maintained ('held') for a few seconds. This is 'custom and practice', and a practical limit, as the pipe mill will be producing hundreds of sections of line pipe: holding the pressure for many hours would create production delays. Also, most failures would be expected during the pressurization, and not during a hold period.

9.2.2   Pressure/Stress Level in Mill Test

Pipe mills use line pipe standards when choosing the required pressure level in the mill test; for example, API 5L [10]. The API 5L standard has gradually increased the required pressure level, to give a hoop stress of 90% of the specified minimum yield strength (SMYS[4]) in the line pipe wall [11], Table 9.2.

---

[4] Note that the yield point in line pipe steels is not clear; consequently, the yield strength is taken as the stress required to produce a small, specified amount, of plastic deformation in the tensile test specimen. The usual definition of yield strength is the 'offset' yield strength determined by the stress corresponding to the intersection of the stress-strain curve and a line parallel to the elastic part of the curve, offset by a specified strain. The offset is usually specified as a strain of 0.2 or 0.1.

Table 9.2 shows that line pipe was not always tested to high pressures/stresses, and it was not until the 1950s that API 5LX ('X' for high strength) established a mill test hoop stress level of 90% of the line pipe's specified minimum yield strength (SMYS) [12–14]. Hence, it is not certain how much line pipe was tested to 90% SMYS before 1949 [12], Table 9.3 [14]. Up to the 1950s and 1960s, line pipe manufacturers would hydrotest their line pipe to 60% to 90% of the SMYS [12–14], Table 9.4.

Additionally, the line pipe manufacturers would conduct some non-destructive testing: from the 1940s the manufacturers started to introduce non-destructive testing methods to evaluate the seam welds, but it was not until 1963 that API 5L had mandatory requirements for the non-destructive inspection [13, 14].

9.2.3   Perspective on Line Pipe in Older Pipelines

It is worth remembering that most pipelines 100 years ago did not operate at high pressures, because many of these pipelines were not welded together in the field (welding line pipe together did not start until the early 1900s), and the line pipe joints at that time (threaded, flanged, etc.) could easily leak. Also, 'Grade B' line pipe is a low strength line pipe today, but it was considered high strength many years ago: the high ('X') strength line pipe did not start to be produced until the mid-part of the 1900s.

The pipeline industry considers 'early' pipelines to be those constructed before 1940. After 1940, modern standards and specifications were available, and better materials and engineering allowed an improvement in pipelines. 'Vintage' pipelines are usually considered to be pipelines constructed from 1940 to 1970: the late 1960s and early 1970s was the start of pipeline regulations, which made minimum technical requirements mandatory. Again, better materials and engineering, coupled with the new regulations improved the quality of pipelines, and consequently 'modern' pipelines are considered those constructed from 1970.

An ageing asset is often described as 'old', and this word infers substandard, but there is a difference between 'old' and 'substandard':

- 'old' means a system that has not been designed and constructed using contemporary standards;
- 'substandard' refers to a system that does not comply with its original fabrication standard.

Therefore, 'old' does not mean 'substandard'.

9.2.4   Perspective on High Mill Test Pressures

Testing to higher than 90% SMYS may be a problem in the pipe mill [15]. If the hoop stress is greater than or equal to 95% SMYS there is a possibility the pipe will permanently expand. This is because the combination of the axial stress created by the pressure and the end loading (sealing pressures, Figure 9.6) is likely to produce an effective stress in the pipe wall of approximately 100% of SMYS [15]. Line pipe tested at less than 95% is less likely to expand during the mill test.

9.3   PRESSURE TESTING PIPELINES

Line pipe is made in a pipe mill. The line pipe is subjected to a pressure test at the mill, and should leave the mill in 'perfect'

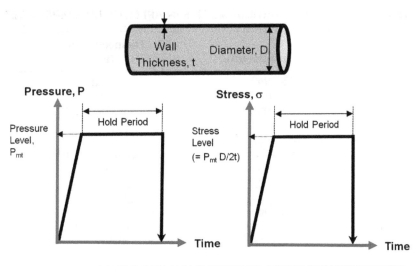

FIG. 9.7   PRESSURE AND HOLD PERIODS IN THE PRESSURE TEST

TABLE 9.2   TEST LEVEL REQUIREMENT IN API 5L [11]

Year	Standard	Hoop stress (% SMYS)
1928	API 5L	60
1942	API 5L	60 minimum, 80 maximum
1948	API 5LX (new standard for higher strength line pipe)	85
1956	API 5LX	90
1983	API 5L and API 5LX merged	90

condition. The line pipe then has to survive the transport to the construction site, and this transportation may expose the line pipe to damage and corrosion. The line pipe will also be subjected to threats during its storage (e.g., corrosion from the surrounding environment), and also during handling (e.g., denting), and fabrication (the line pipe is welded together, at the construction site ('in the field') to create the pipeline, Figure 9.5). Consequently, pipeline operators perform a pressure test on their completed pipelines to check the fabrication is satisfactory.

9.3.1   Pressure Testing Pipelines 'In the Field'
Pipelines are pressure tested 'in the field' to show the pipeline is leak tight, and can contain a specified pressure. This pressure test involves [4] (Figure 9.8):

- sealing the pipeline at both ends (using 'test ends');
- filling the pipeline with liquid (e.g., water) or gas (e.g., air);
- raising the pressure of the fluid (using pumps or compressors) inside the pipeline to a pressure greater than the normal operating pressure of the pipeline;
- holding the pressure for a number of hours to ensure there are no leaks in the pipeline;
- the pressure, and time at that pressure, are monitored during the test.

9.3.2   Purpose of the Pipeline Pressure Test
Pipeline standards require new pipelines to be pressure tested prior to service (e.g., [1]): this 'strength test' or 'proof test' is used to establish ("prove") the maximum operating pressure limit of a pipeline segment.

The pressure test demonstrates a new pipeline can safely withstand its design pressure on 'day one' of its design life, as the pressure of the test is set at a pressure higher than the design (or

TABLE 9.3   HISTORICAL TESTING REQUIREMENTS IN API 5L OF LINE PIPE PRIOR TO 1942 [14]

Type and grade of line pipe	Year of manufacture	Yield strength, lbf/in^2 (MPa)	Minimum mill test pressure (% SMYS)
Seamless, Grade A	1928, 1931	30,000 (207)	46.6
Seamless, Grade B	1928, 1929	40,000 (276)	45
Seamless, Grade C	1928, 1931	45,000 (310)	40
Lap-welded	1928–1941	25,000 (172)	56
Lap-welded	1928–1941	28,000 (193)	50
Lap-welded	1928–1941	30,000 (207)	46.6
Seamless, Grade B	1930, 1931	38,000 (262)	47.4
Seamless or Electric Welded, Grade A	1932–1941	30,000 (207)	46.6
Seamless or Electric Welded, Grade B	1932–1941	35,000 (241)	45.7
Seamless or Electric Welded, Grade C	1932–1941	45,000 (310)	40

TABLE 9.4   HISTORICAL TESTING REQUIREMENTS IN API 5L OF LINE PIPE AFTER 1941 [14]

Type and grade of line pipe	Year of manufacture	Diameter, inches (mm)	Standard mill test pressure (%SMYS)
Lap-welded steel pipe	1942–1962	All	60
Grades A and B	Before 1982	All	60
Grades A and B	1983–2007	≥2.375 (60.3)	60
Grade C (45,000 psi)	1942–1954	All	60
X Grades–all types	1949–1952	All	85
X Grades–all types	1953–1961	<8.625 (220.3)	75
X Grades–all types	1962–1969	4.5 (114.3)	60
X Grades–all types	1969–1982	≤4.5 (114.3)	60
X Grades–all types	1983–2007	≤5.56 (141.2)	60
X Grades–all types	1962–1999	6.625–8.625 (168.3–219.1)	75
X Grades–all types	1953–1955	≥10.75 (273.1)	85
X Grades–all types	1956–1999	10.75–18 (273.1–457.2)	85
X Grades–all types	2000–2007	8.625–18 (168.3–457.2)	85
X Grades–all types	1956–2007	≥20 (508)	90

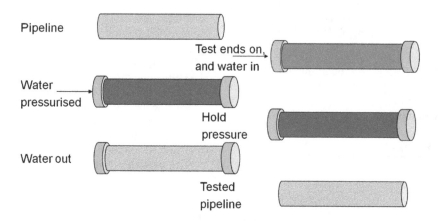

FIG. 9.8   PRESSURE TESTING PIPELINES IN THE FIELD

maximum operating pressure), Figure 9.9 (this Figure 9.9 assumes a test to 100% SMYS). This also means the pipeline has a higher failure pressure that the test pressure: this failure pressure will be between 125% SMYS and the ultimate tensile strength (UTS) of the line pipe, for a defect-free pipeline [16].

Pipeline standards [17–19] give all the main technical purposes of pressure testing:

• establish a maximum operating pressure limit for a pipeline (this means test to a pressure higher than the pipeline's maximum operating pressure);
• detect and eliminate defects in the pipeline (these defects could be in the line pipe material, welds, or could have been created during line pipe transportation, storage, or pipeline construction);
• check for leaks (for example, at welds in the line pipe, at flanges, etc.).

9.3.3   The Pipeline Pressure Test and Circumferential Defects

It should be noted that the pressure test imposes a high hoop stress (Figure 9.2) on the pipeline, but a much smaller axial stress (the axial stress is 30–50% of the hoop stress, depending on the restraint on the pipeline). This means that circumferential defects (such as defects in the pipeline's girth welds) will not be subjected to high stresses; therefore, the pressure test is of limited value in detecting circumferential defects (see Table 9.9 later).

Circumferential defects will usually be in the pipeline girth weld, Figure 9.5; therefore, the pressure test is of limited value for detecting defects in girth welds. This is not a major limitation: pipeline girth welds are usually subjected to low axial stresses during service, and are subjected to extensive inspection after fabrication, and these result in the welds being a low failure risk.

9.3.4   Requirements in Pipeline Standards

ASME B31.4 requires pipelines that will be operated above 20% SMYS to be tested using water (Figure 9.10):

• the test must not be less than 1.25 the pipeline's 'internal design pressure' (D_p); and,
• its duration must not be less than 4 hours.

ASME B31.4 also states that if sections of pipe can be visually inspected for leaks during the 'proof' test phase (the

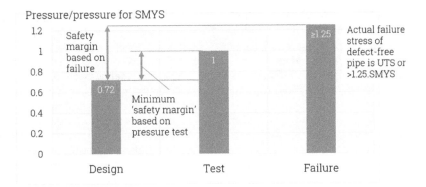

FIG. 9.9   SAFETY MARGIN FOLLOWING A SUCCESSFUL PRESSURE TEST

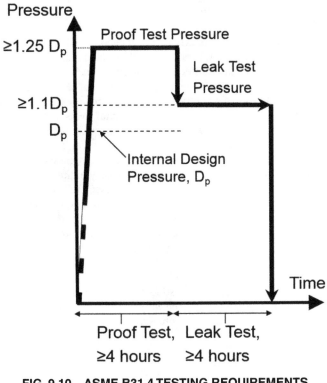

FIG. 9.10   ASME B31.4 TESTING REQUIREMENTS

highest pressure), no further testing is required. If sections cannot be inspected for leaks during pressure testing, then the proof test is followed by a leak test to:

- not less than 1.1 × internal design pressure;
- for not less than 4 hours.

ASME B31.8's [20] test requirements set the pipeline's maximum allowable operating pressure (MAOP), which must be less than the design pressure (D_P). The design pressure is obtained from a calculation using the pipeline's diameter, wall thickness, design factor, specified minimum yield strength, 'joint' factor, and 'temperature' factor. Hence, if the design pressure is 100 barg and the test pressure is 125 barg, then the MAOP will be 100 barg, but if the test pressure is 100 barg, then the MAOP is 100/1.25 = 80 barg.

The above information and requirements are clear, but it is useful to understand the reasons for these requirements and the history of their development.

9.3.5   Historical Background

9.3.5.1 Before 1955   The pressure vessel industry has tested its plant by over-pressurizing since the 1800s; the pipeline industry copied this good practice in the 1900s.

ASME published pipeline standards (ASME B31.4 [1] and B31.8 [20]) in 1955 requiring pressure testing of new constructions. There was no requirement before 1955; therefore, 1955 is a watershed in the pressure testing of pipelines.

TABLE 9.5   ASME B31.8 PRESSURE TEST REQUIREMENTS

Pipeline location classification	Test fluid	Test pressure (TP)		MAOP[2] (lesser of:)
		Minimum	Maximum	
Class 1 (Div 1)	Water	1.25 × MOP[1]	None	TP[3]/1.25 or D_P[4]
Class 1 (Div 2)	Water	1.25 × MOP	None	TP/1.25 or D_P
	Air or Gas	1.25 × MOP	1.25 × D_P	TP/1.25 or D_P
Class 2	Water	1.25 × MOP	None	TP/1.25 or D_P
	Air	1.25 × MOP	1.25 × D_P	TP/1.25 or D_P
Class 3	Water	1.5 × MOP	None	TP/1.5 or D_P
Class 4	Water	1.5 × MOP	None	TP/1.5 or D_P

[1] Maximum operating pressure (MOP) is the highest operating pressure during normal operation.
[2] MAOP is the maximum pressure allowed by ASME B31.8.
[3] TP is test pressure.
[4] D_P is the maximum pressure permitted by ASME B31.8 based on materials and location.

Prior to 1955, construction/operating practices only required a pipeline to withstand about 50 psig (3.5 barg) pressure higher than the pipeline's maximum pressure. Also, pressure testing, if performed, was usually performed using the commodity being transported as the test fluid [11], including gas or air. Operators were cautious of testing to high pressures, and wanted to limit commodity losses should the pipeline fail; therefore, test pressures ranged between 5 psig to 50 psig (0.3 to 3.5 barg), or 10%, higher than the operating pressure of the pipeline [15], Table 9.5.

There were many reasons why operators did not test to high pressures in the past, including:

- the standard used to construct pipelines in the USA (American Standards Association) ASA B31.1) did not specifically recommend pressure testing to establish the maximum operating pressures after the installation, prior to 1942 [11];
- operators did not have a requirement to test to high pressures (testing pipelines prior to operation was not mandatory (specified in regulations) in USA until 1970;
- the quality of the line pipe, and construction standards, were not as good as they are today (i.e., the pipe and its welds could easily fail on test);
- high pressure testing using a gas can lead to a long 'running fractures', should the pipeline fail.

There were additional reasons that prevented using water for testing:

- testing a small vessel is easy, but it is very difficult to test a long pipeline using water;
- large quantities of clean water were needed and obtaining the water for a hydrotest was both difficult and expensive in arid regions or remote regions;
- the need for recycling/discarding the water afterwards;
- gas pipeline operators did not want water remaining in their lines, as water left in a pipeline may lead to corrosion (drying methods were not developed in the early 1900s).

9.3.5.2 Testing to High Stresses The hydrotest is now recognized as an essential part of proving a pipeline's integrity. The improvement in line pipe quality and construction practices in the 1950s and 1960s gave operators confidence in the test, but failures still occurred.

Also, gas pipelines were being built in the 1950s, and liquid lines were being converted to gas use: gas pipelines operate at high pressure, and hence the hydrotest was of greater importance.

Testing pipelines to high stresses started in Texas in the 1950s [2], Table 9.6. Texas Eastern was converting products lines to gas, and incidents occurred in-service from manufacturing defects in the pipe. Texas Eastern's insurers would not support the pipelines unless the incidents stopped. Therefore, they hydrotested the lines to a high stress level (≥100% SMYS). They experienced "hundreds" of test breaks from the manufacturing defects, but after the testing there were no failures from manufacturing defects.

It is important to note that the stresses during the mill test and the field test are different [15]:

- The line pipe contracts axially (i.e., it becomes slightly shorter) in the mill hydrostatic test, due to 'Poisson's ratio', This contraction is the same effect that causes a tensile specimen to neck in as it is pulled. The hydraulic rams that seal the ends of the pipe follow that contraction, maintaining a seal (usually an O-ring seal). This means that the only axial stress in the pipe is due to the ram force, which needs only be more than the force needed to counteract the pressure in the pipe. The resulting axial stress in the line pipe is therefore compressive, which reduces the combined stress level to cause yielding.
- The field test creates a different axial stress: soil friction along a length of pipeline prevents the pipe from contracting. This results in a tensile axial stress which, when combined with the hoop stress, results in a greater combined stress required to achieve yielding. This has the effect of strengthening the pipeline.

Standards now encourage testing to high stresses; for example, Reference 18 states: "... *The highest degree of certainty... is achieved when the factor [test pressure to operating pressure] is*

TABLE 9.6   EARLY PRESSURE TESTING PRACTICES [2]

Time	Testing	Practice
1935–1952	Not required	A post-installation test on a pipeline was not required by standards. The operating pressure was based on pipe mill certificates, or calculation.
1940s	Low pressure tests	Pressure testing - if conducted - used the pipeline's product (e.g., natural gas or crude oil). Test pressures of 5 or 10 psig above operating pressure were used (but not more than 1.1 × operating pressure) due to fear of: • large losses of crude oil; or, • losses of large lengths of pipeline when using natural gas.
1950s	Hydrotesting	The Texas Eastern Transmission Corp (TETCO) acquired two War Emergency Pipelines. The pipelines were designed to carry crude oil, but TETCO converted them to natural gas, with many subsequent failures. The failures were due to manufacturing defects (which may have enlarged during the crude oil service period) and external corrosion (parts of the lines were originally installed uncoated to save construction time). TETCO decided to use water at high pressure (to give hoop stresses up to 109% SMYS [21]) to test the gas pipelines. The failures stopped: some parts of these pipelines were still in service in 2013.
1967	High stress tests	The concept of a 'high level' test (≥ 90% SMYS), or a 'test to yield' (100% SMYS) was introduced for gas pipelines in 1967 [21].

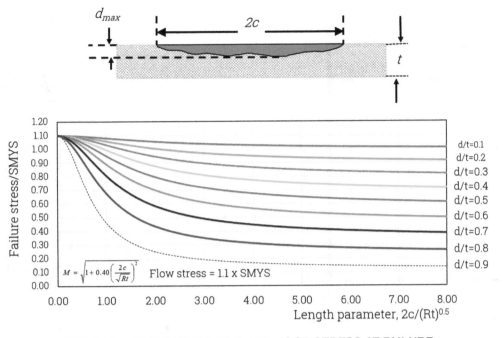

FIG. 9.11   ANOMALY PROFILE AND HOOP STRESS AT FAILURE

at least 1.25 and, additionally, the test pressure is such that the weaker pipes in the line yield a small amount in the test. This practice is called 'yield testing' or 'high level testing'... Records of the performance of pipelines subjected to high level testing have shown conclusively that the occurrence of ruptures in service from defects present at the time of the test has been almost eliminated."

Clearly the standards are encouraging high stress pressure tests, and the benefits are related to removing defects. The next Section gives an insight into the reasoning and technical basis for these high stress tests.

9.3.5.3 High Level Tests and 'Tests to Yield' Historically, pipelines have been tested to a pressure above the pipeline's operating pressure; for example, ASME B31.4 states that low stressed (≤20% SMYS) pipelines are tested to ≥ design pressure. This pressure may be above the operating pressure, but may give a low hoop stress in the pipe wall.

Stress is proportional to pressure: the higher the internal pressure, the higher the hoop stress in the pipeline. This is important, as increasing stress will fail more defects in the pipeline. Defects that can survive at high stress must be smaller than defects that can survive at a lower stress. Therefore, pipelines that have been tested to high stresses will contain small defects. These small defects (e.g., shallow fatigue cracks) will have a longer period to fail during service, as they will have to grow further to cause failure. Therefore, the high stress in a pressure test fails more defects, and will ensure the remaining defects take longer to fail the pipeline.

The stress levels that defects in pipelines can survive are illustrated in Figure 9.11. These failure curves are examples obtained from the following equations[5] [22, 23]:

$$\frac{\sigma_f}{\bar{\sigma}} = \frac{1 - \dfrac{d}{t}}{1 - \dfrac{d}{t}\dfrac{1}{M}} \qquad (1)$$

$$M = \sqrt{1 + 0.4\left(\frac{2c}{\sqrt{Rt}}\right)^2} \qquad (2)$$

$$\bar{\sigma} = 1.1\sigma_y \qquad (3)$$

σ_f	Hoop stress at failure.
$\bar{\sigma}$	Flow stress.
σ_y	Yield strength or SMYS.
R	Pipe outside radius.
t	Pipe wall thickness.
d	Defect depth.
$2c$	Defect length.

It was recognized that testing up to 80% or 90% SMYS may not reveal (fail) defects that may cause failure during operation at a lower stress (e.g., 72% SMYS). To ensure these defects failed, the concept of a 'high level' test (≥90% SMYS [17]), or a 'test to yield' (100% SMYS) was introduced for gas pipelines in 1967 [21, 24], Figure 9.12.

It is now generally recommended to test to the highest possible stress [24, 25], as high stresses should not damage the pipe or its coating:

- most line pipe in an order has a yield strength above the quoted SMYS, as (by definition) the SMYS is a lower bound value of yield strength;
- a buried pipeline yields at about 110% SMYS due to the biaxial stress in the pipeline; therefore, yielding is unlikely at 100% SMYS (see next Section);
- yielding at these stress levels (≤110% SMYS) does not harm pipe (most line pipe is 'cold expanded' to higher strains, in the pipe mill, and line pipe is strained to higher levels if formed into a field bend during pipeline construction);
- yielding at these stress levels does not harm external corrosion coating on the line pipe (the coating is designed to withstand higher strains during field bending).

---

[5] Other values of flow stress and the "Folias Factor" (M) are quoted in the literature [21, 22].

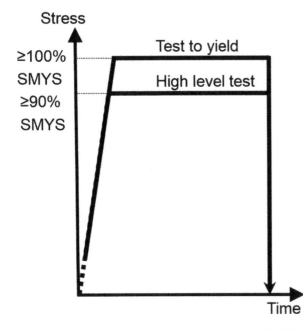

FIG. 9.12 THE 'HIGH LEVEL' TEST AND 'TEST TO YIELD'

Stress levels should not exceed 110% SMYS in the test [26]; for example, CSA Z662 [19, 27] states that the hydrotest can be up to 110% SMYS, depending on line pipe type and strength[6].

9.3.6 Stressing during the Test

Yielding of a metal is affected by the direction of stresses on the material. When a metal sample is tested to determine its strength, a simple 'stress-strain' test is created. The loading is in one direction: 'uniaxial' tension (loading along one axis). The metal yields at a certain stress, but if the same sample was loaded in two directions (two 'axes', or 'biaxial'), it would yield at a different stress.

When a buried pipe is pressurized, it is restrained by the soil from shortening. This restraint by the soil friction causes an axial tensile (positive) stress equal to Poisson's ratio (0.3 for steel) × hoop stress. The biaxial stress causes the buried pipeline to yield at about 109% SMYS [24, 25].

Figure 9.13 [15, 28] explains the high stress required to yield a pipeline during a field hydrotest. In the field test the longitudinal (axial) tensile stress is equal to Poisson's ratio times the hoop stress, but if a simple end-capped pressure vessel was pressurized, then the axial stress would reach 50% of the hoop stress (shown in Figure 9.13).

Figure 9.13 is a plot of the point at which the yield strength is reached, based on actual combined (biaxial) stresses:

- Line pipe undergoing a mill test is unrestrained and is free to move in the axial direction; therefore, it will attempt to shorten (contract) when pressurized due to Poisson's ratio[7]. The mill test uses a ram to seal the pipe ends: this hydraulic ram follows the pipe contraction, maintaining a seal. This results in a compressive (−) axial stress. This creates a combined stress (hoop and axial) that is less than the hoop stress (alone) to causes yielding.

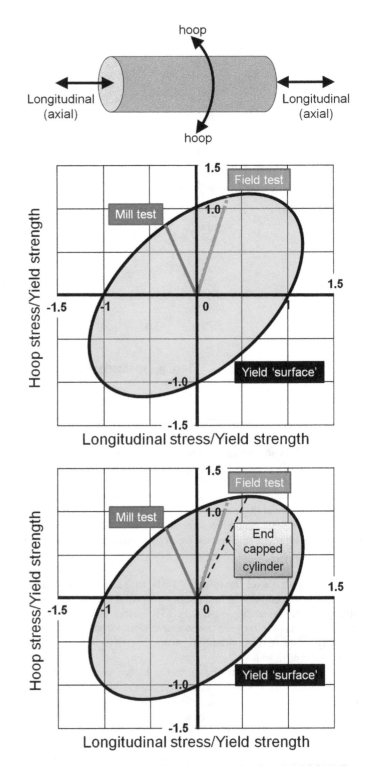

FIG. 9.13 BIAXIAL STRESSING DURING A PRESSURE TEST [15, 28]

- A pipeline is restrained and is not free to move: soil friction along the length of the pipeline axially restrains the pipe creating a tensile (+) axial stress, and prevents the pipeline from contracting. This creates a combined stress (hoop and axial) that is higher than the hoop stress alone) to causes yielding.
- (+) axial stress, and prevents the pipeline from contracting. This creates a combined stress (hoop and axial) that is higher than the hoop stress (alone) to causes yielding.

---

[6]There is some debate about testing to high stress levels in high strength steels, and steels with FBE coatings. This is because the FBE coating process requires heating the line pipe to about 250°C, and this can cause 'strain aging', which can increase strength and decrease yield strength to ultimate tensile tetsrength ratio ('Y/T').

[7] This is the same phenomenon that causes a tensile specimen to neck in as it is pulled.

Examples of the magnitude of these stresses are [15]:

- a mill test conducted at a hoop stress of 95% SMYS, with a compressive axial stress provided by a ram, creates a combined stress of approximately 100% SMYS.
- a field test conducted at approximately 108% SMYS, with the pipeline restrained by soil friction, creates a combined stress of approximately 100% SMYS.

This example shows that the pressure to cause through thickness yielding in a buried pipe is about 13% higher than for a pipe under hydrostatic test in the mill.

9.4   TYPES OF PRESSURE TESTS

Pressure testing is a recognized method of establishing or revalidating the integrity of a pipeline [3]. There are three basic types of pressure tests [4, 17, 18]:

- a 'strength test': this test establishes the operating pressure limit of a pipeline segment.
- a 'leak test': this test is used to determine that a pipeline segment does not show evidence of leakage.
- a 'spike test': this test is used to verify the integrity of a pipeline containing 'time dependent' defects.

This Section will cover these three types of test.

9.4.1   Strength Test

A 'strength test' or 'proof test' is used to establish ('prove') the maximum operating pressure limit of a pipeline segment. Typically (Figure 9.14) [4]:

- the strength test pressure ratio (test pressure/maximum operating pressure (MOP)) is 1.25, and,
- the duration is 4 hours or longer.

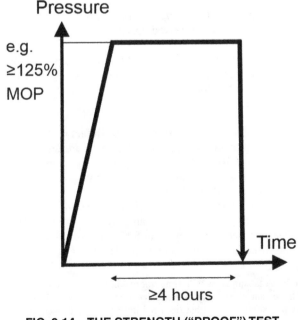

FIG. 9.14   THE STRENGTH ("PROOF") TEST

Strength tests are considered successful if no pipe ruptures or leaks occur, according to established acceptance criteria [4].

The hold period commences when the test pressure or stress level has been reached, and the pressure is stable. A period of temperature[8] stabilization may be required before the start of the test as [4]:

- pressure transients can occur during the pressurization process; and,
- residual air may go into solution in water (see Section 9.7.2 later).

The time for this temperature stabilization depends on [4]:

- the temperature of the test medium at the time of filling;
- heat capacity of the test medium;
- pipe diameter;
- depth of pipe burial; and,
- the ground temperature.

9.4.1.1 Basis of the Pressure Levels Used in Strength Testing   The basis of the effectiveness of a pressure test is the fact that there is a relationship between the stress applied to the structure and the size of defect that will cause the structure to fail at that stress level [29]: the higher the stress level, the smaller the defect required to fail the structure, Figure 9.11. Therefore, the higher the test pressure to maximum operating pressure ratio, the more effective the test [29].

Historically, pipelines, pressure vessels, and piping have been tested from 1.1 to 1.5 times their design pressure. 1.25 times the design pressure is often quoted in standards (Figure 9.10), with a minimum level of 1.1 times design pressure (this 10% above design pressure is due to most standards allowing a surge pressure equal to 10% the design pressure (e.g., Reference [1])).

The 1.25 times design pressure appears to be based on both technical and operational experience [18], but, certainly, the higher the ratio of test pressure to design pressure, the higher the certainty that any defect present at the time of the test will not fail at the subsequent design pressure [4, 18].

9.4.1.2 Basis of the Hold Times Used in Strength Testing   Test pressure needs to be 'held' for a time period to allow temperatures, etc., to settle: as temperature drops, pipeline pressure drops. Leaks are detected by pressure drop; hence, temperature changes must not be confused with leaks. Pressure/temperature charts can help differentiate between changes in pressure due to temperature variation rather than a leak.

Long (8 to 24 hours) hold periods allow the water in the pipe to stabilize with the surrounding soil. Historically, the pressure was held for 24 hours:

- to allow leak detection in long pipelines; and,
- due to failures being observed during the hold period.

Many pipeline standards do not now require a 24 hour hold period; for example, ASME B31.4 requires a minimum 4 hour hold period, at a minimum of 1.25 times the maximum operating pressure, Figure 9.10.

---

[8] Hydrotesting in cold climates or cold periods will require the use of anti-freezing agents in the water. This can lead to both temperature stabilization problems, and water disposal problems.

The value of hold time is in establishing that the test segment is free of leaks. It does not contribute to the value of the test in terms of the margin of safety [29–31], and the hold period duration is considered 'arbitrary' [21, 32].

The strength test pressure needs only to be held for 2 hours. Longer periods are often quoted for tests using gases [19] as the compressive nature of the gases mean they are not as sensitive as a liquid medium at identifying a leak, and longer periods are sometimes used for lines with stress corrosion cracking. Longer than 8 hours may not add value for verifying the strength of the pipe [18]. Extensive load hold periods and repeated cycling during the test are not beneficial [33, 34], particularly at high stresses [27].

9.4.2 The Leak Test

A 'leak test' is used to determine that a pipeline segment does not show evidence of leakage. Typically (Figure 9.15) [4]:

- the test pressure ratio (test pressure/maximum operating pressure (MOP)) is less than 1.25; and,
- the duration is 2 hours or longer.

The length of hold time to detect leaks should be based on the volume of water in the test section. The larger the volume, the longer holding at constant pressure is required to detect a leak of a given size.

Leak tests are determined to be successful if all pressure variations can be explained by the established acceptance criteria [4].

Locating leaks during a hydrotest can be a difficult and time-consuming process. Detecting and repairing leaks in an uncovered pipeline, in a trench, is easy; however, if the pipeline is buried or under water, leak detection is difficult. Various methods are available to detect leaks [4]:

- sectioning or segmenting the pipeline (e.g., by valve closure, or freeze plugs) and monitoring the pressure of each section.
- test water can use dyes to help locate a leak;
- acoustic monitoring;
- odorants or tracers introduced into the test medium during the filling process.

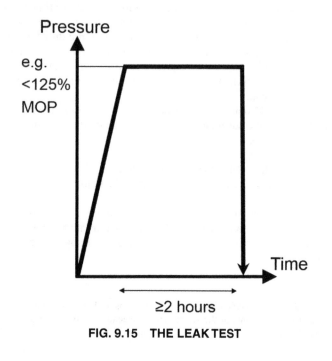

FIG. 9.15 THE LEAK TEST

9.4.3 Testing Pipelines Containing Defects

9.4.3.1 Defects and the Pressure Test Pressure testing is a recognized method of demonstrating the integrity of a new pipeline. The high stresses created by the pressure test will fail defects of a certain size in the pipeline; consequently, the test is also of use for proving the integrity of a pipeline containing defects.

Unfortunately, pipelines can contain many defects, as they age, including:

- stress corrosion cracks, caused by a combination of high stresses, and a corrosive environment;
- fatigue cracks caused by high, repeated stressing;
- corrosion.

The pressure test can be used to detect (fail) these defects, and then the pipeline can be returned to service.

API 1110 [4] lists the variety of purposes of the pressure test, and highlights its use to fail defects:

- detect and eliminate 'time dependent'[9] anomalies in a pipeline segment (the greater the ratio between the test pressure and operating pressure, the more effective the test);
- detect and eliminate stable[10] anomalies and verify the structural integrity of a pipeline segment (by testing the pipeline segment to a pressure higher than its operating pressure);
- verify the integrity of a pipeline before returning it to service after it has been idle or inactive;
- verifying the integrity of a pipeline when changing its service.

The test has some limitations when assessing defects [3]:

- the only defects 'identified' by a test are those that fail during the test;
- defects with failure pressures above the test pressure will not be discovered (i.e., short, deep defects that have high failure pressures, Figure 9.11);
- testing is not usually used for assessing metal loss caused by corrosion, as short, deep defects will not fail during the test, Figure 9.11 (in-line inspection tools may be more effective for these type of metal loss defects);
- the test gives no knowledge of the numbers and locations of anomalies that have survived the test.

Despite these limitations, the pressure test offers a good method for proving the integrity of in-service pipelines containing defects [20], and the next Sections will cover the use of the pressure test on pipelines with known defects.

9.4.3.2 Limitations with Short, Deep Defects The previous section has highlighted the problem with pressure testing, and its ability to detect (fail) short, deep defects. Figure 9.16 uses Equation (1) to plot the defect sizes that would fail a high pressure test (giving a hoop stress of 100% SMYS).

It can be seen that short, deep defects can survive the pressure test: defects of length (2c) shorter than $2c/(Rt)^{0.5} = 0.5$ would not fail providing their depths are less than 80% pipe wall thickness. This 0.5 value of $2c/(Rt)^{0.5}$ corresponds to a length of 1.5 in. (38 mm)

---

[9] Defects that can grow with time; for example, fatigue cracks.
[10] Stable- and time independent defects do not grow with time; for example, some defects that occurred during the manufacture of the line pipe.

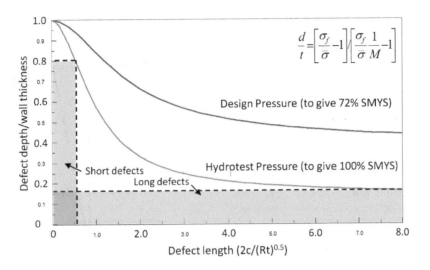

FIG. 9.16   PROBLEM WITH SHORT, DEEP DEFECTS

in a pipe of radius (R) 18 inch (457 mm) and wall thickness of 0.5 inch (12.7 mm). This means that deep corrosion pits will not fail during a pressure test.

Figure 9.16 also shows that long, shallow (about <15% wall thickness) defects cannot fail during a high stress pressure test. This means that long, shallow corrosion defects will not fail during a pressure test.

It is concluded from Figure 9.16 that the pressure test cannot fail all defects, even at high stresses. This means that hydrotested pipelines can still fail, if the defects that survive are allowed to grow during subsequent service. Section 9.4.3.10 covers this growth. This limitation of detecting short, deep defects, and long, shallow defects also applies to inspection methods such as in-line methods using smart pigs.

9.4.3.3 Defect Growth during a Pressure Test   A strength test conducted in-service subjects a pipeline to a high stress and this will fail defects of a certain size: more defects will fail as the test pressure is raised, Figure 9.11.

The defects fail by growing in size, and then failing; however, some defects can grow in size but not fail before the test is completed: the defect has grown during the pressure test. This means that it is possible for a defect to increase in size during the pressure test, and remain in the pipeline when it is returned to service, Figure 9.17. This enlarged surviving defect can fail during a subsequent pressure test, at a pressure lower than the pressure test level it had previously survived.

This is called a 'pressure reversal' [31, 35–39].

9.4.3.4 Pressure Reversals   A pressure reversal is a phenomenon where a pipeline fails at progressively lower pressures during subsequent tests [4], Figure 9.18.

The reversal is caused by a defect in the pipeline (e.g., a crack). This defect has grown during the pressurization and/or the hold period: it is known that most of a defect's growth occurs during the first 2 hours of the hold period [35].

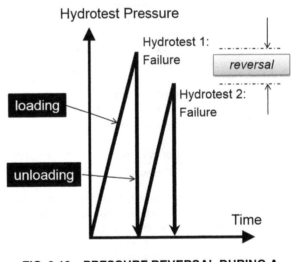

FIG. 9.18   PRESSURE REVERSAL DURING A
PRESSURE TEST

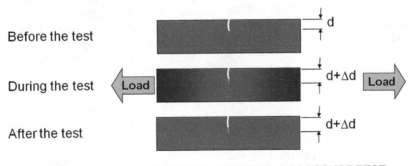

FIG. 9.17   DEFECT GROWTH (Δd) DURING A PRESSURE TEST

Consequently, a pressure reversal can occur when:

- a previous hydrostatic test causes a defect to grow nearly to failure; and,
- subsequently, additional defect extension occurs during pressure unloading.

The defect has grown due to:

- the growth of plasticity around the defect; and,
- crack growth at the defect tip.

This plasticity and growth requires time, and a pressure level close to the failure pressure of the defect; consequently, a pressure reversal requires [3]:

- a defect to be present;
- time for growth to occur; and,
- ductile line pipe (it will not occur in brittle pipe, as the defect needs to be in ductile materials to allow defect growth).

The growth of a defect allows it to increase in depth, and become sharp, and the pipeline can fail at a pressure lower than the previous hydrostatic test pressure.

Pressure reversals are defined by Equation (4), and:

- are not common in pipelines that have been tested to a pressure of 1.25 × maximum operating pressure or higher [18, 40];
- are usually between 1 to 5% of the previous pressure level [14];
- large and frequent reversals are usually associated with longitudinal seam welds in older electric resistance welded (ERW) line pipe [25], but experience and analysis have shown that the possibility of a pressure reversal causing a failure in service in this type of ERW seam welds is remote providing the test-pressure to operating pressure ratio is equal to or greater than 1.25 [40];
- are not usually a major problem, unless there are many defects in a pipeline [41]; accordingly, many pressure reversals will indicate a pipeline with many defects.

$$Pressure\ Reversal = \frac{Original\ Pressure - Failure\ Pressure}{Original\ Pressure \times 100} \quad (4)$$

A special pressure test can be used to both demonstrate the integrity of a pipeline containing defects, and also reduce the incidence of pressure reversals. This is the 'spike' test.

9.4.3.5 The Spike Test
When a pipeline with defects needed to be tested there are two conflicting requirements [41]:

- to avoid failures (but this would mean avoiding high stresses (as they may grow and cause a failure or reversal), and limiting the hold period (but this would not allow time to detect leaks).
- to satisfy pressure level and hold requirements in standards and regulations.

A spike test can satisfy these requirements: the spike test is a special form of the strength pressure test. Higher test pressures are used, and these are typically:

- at high hoop stresses (100% and 110% SMYS); or,
- in terms of maximum operating pressure (MOP) 1.39 and 1.53 times MOP [29, 38].

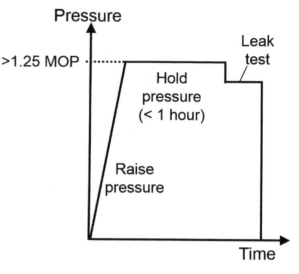

FIG. 9.19 THE SPIKE TEST

These high values of stress and pressure assume the pipeline is designed to operate at a high hoop stress (=72% SMYS). A spike test to below 100% SMYS (or below 1.39 or 1.53 × MOP [29]) is still of use if a pipeline is not operating at such high hoop stresses (≤72%SMYS), but it may not necessarily result in the best assessment of pipeline integrity.

The initial (high) stress/pressure is held for a short time (e.g., less than 1 hour), Figure 9.19. The high pressure fails defects, but the short duration reduces defect growth. The most important consideration is attaining the highest possible test pressure, if only for a few minutes [29].

The spike test has been used for many years in the pipeline business [36], and was first used on pipelines containing SCC [42].

Spike tests are determined to be successful if no pipe ruptures occur according to pre-determined acceptance criteria [4].

9.4.3.6 Stress Level to be Used in the Spike Test
A spike test is similar to a normal pressure test, but spike tests are conducted at high stresses, for a very short duration; for example, ASME B31.8S [43] recommends the following spike test to help control SCC in gas pipelines:

- test pressure equivalent to (give a hoop stress of) a minimum of 100% SMYS;
- test pressure to be maintained for a minimum period of 10 minutes.

A maximum test pressure to give a hoop stress of between 100 and 110% SMYS provides a balance between removing large flaws that could cause failures in service, against producing growth only in a relatively few near-critical flaws [31].

Guidance on setting the spike test pressure (*TP*), based on a pipeline's maximum operating pressure (*MOP*) is given in the literature, for example [29]:

$$\frac{TP}{MOP} = -0.00736(\%SMYS\ at\ MOP) + 1.919 \quad (5)$$

$$\frac{TP}{MOP} = -0.02136(\%SMYS\ at\ MOP) + 3.068 \quad (6)$$

TABLE 9.7 MANAGEMENT OF THREATS USING A SPIKE TEST [4]

Threat	Management	Assessment interval
Internal and external corrosion threats.	Spike test, strength test, and/or a leak test.	Depends on test pressure.
SCC threats.	Spike test and/or a strength test.	Depends on test pressure.
Manufacturing threats (such as fatigue susceptible seam weld flaws that grow due to pressure cycling).	Spike test and/or a strength test.	Depends on test pressure.
Materials and construction threats.	Spike test and/or a strength test.	These 'time independent' or 'stable' threats do not require reassessment to establish their integrity, providing no in-service conditions affect them.

TABLE 9.8 ASSESSMENT INTERVALS FROM ASME B31.8S-2014 [43]

Maximum assessment interval (years) for pressure testing	Operating hoop stress >50% SMYS	Operating hoop stress >30% SMYS, but ≤50% SMYS	Operating hoop stress ≤30% SMYS
5	Test Pressure (TP) to 1.25 times maximum allowable operating pressure (MAOP)	TP to 1.39 × MAOP	TP to 1.65 × MAOP
10	TP to 1.39 × MAOP	TP to 1.65 × MAOP	TP to 2.20 × MAOP
15	Not allowed	TP to 2.00 × MAOP	TP to 2.75 × MAOP
20	Not allowed	Not allowed	TP to 3.33 × MAOP

Note that conditions apply to these equations, and the reader should refer to the original reference before using them on a specific pipeline.

Equation (5) is for use when fatigue cracking from pressure cycling is the problem defect, and Equation (6) is for use when selective seam weld corrosion or stress corrosion cracking are the problem defects.

For example, a pipeline operating at an MOP to give a stress of 72% SMYS would require a *TP/MOP* ratio of 1.39 (*TP* = 100% SMYS) using Equation (5), and 1.53 (*TP* = 110% SMYS) using Equation (6) [29].

9.4.3.7 Hold Period for a Spike Test A high pressure/stress in a pressure test is recommended to fail defects in the pipeline; however, a long hold period may not be beneficial as defects can grow during this hold period [4].

Most defect growth occurs within the first 2 hours of a hold period [35], and, also, reducing the pressure on a defect to 90% to 95% of its failure pressure will limit/stop its growth. Consequently, spike tests are held for less than 2 hours[11] (typically longer than 5 minutes but shorter than 1 hour [3, 4, 38, 39]) to minimize the growth of defects that are too small to fail during the test. This hold period is followed by a lower pressure hold period, for leak testing. Note that the duration of the hold period has no significance—it is to give confidence that the peak pressure has been reached [39].

9.4.3.8 Leak Test Following Spike Test The spike test pressure is lowered (e.g., to 1.25 × MOP [4]) after the hold period to allow for leak testing. This lower pressure should be:

- at least 5% below the spike-test level, but preferably 10%;
- equal to, or greater than, 110% of the maximum operating pressure [39];
- held for a longer period (as specified in standards), to allow leak detection and meet requirements in standards.

9.4.3.9 Purpose of a Spike Test A spike test finds (fails) defects [3, 4], but minimizes the growth of defects [35, 38, 39]; consequently, it is intended to verify the structural integrity of a pipeline with time dependent anomalies such as:

- stress corrosion cracks (SCC);
- fatigue;
- internal corrosion; or,
- external corrosion.

The purpose of a spike test is to demonstrate a pipeline's integrity, or allow an interval until the next assessment of its integrity [3].

9.4.3.10 Defects 'Managed' by the Spike Test The spike test can be used to manage 'threats' to a pipeline [39], such as fatigue. Table 9.7 [4] gives examples of defect threats that can be managed by the spike test. The pressure level of the test will dictate the next integrity assessment; for example, the higher the test pressure, the longer the period to the next spike test.

This assessment interval can be calculated using corrosion growth rates, crack growth rates, etc. (see next Section), although some standards prescribe maximum intervals, Table 9.8, [43].

---

[11] The test duration should be sufficient to allow any pressure and temperature transients in the test medium caused by the pressurization process to stabilize.

9.5 USING THE PRESSURE TEST TO DEFINE ASSESSMENT INTERVALS

The pressure test has a key role in pipeline integrity management. The test removes defects by failing them: the higher the stresses created by the test, the smaller the size of defect removed. The defects that fail at the hoop stresses created by the pressure test can be calculated using various methods, for example, Equation (1), Figure 9.20.

There will be a difference in size between the defects surviving the test, and those (larger) defects that can subsequently fail the pipeline at the maximum operating pressure (MOP). The difference in size is the amount of growth available for the surviving defects (Figure 9.20): the surviving defects must grow by this amount to fail the pipeline during service. Growth calculations can be performed to predict the time required for this growth to occur, using published methods (for example, [5, 44] and these calculations allow an assessment interval (time between tests) to be set.

It is important to emphasize the role of pipeline defects on pipeline failure; for example, there are no known cases of fatigue failures due to 'pressure cycle effects' (fatigue) in gas or liquid pipelines in the absence of some sort of significant defect, damage, or incorrect design feature that concentrates stresses locally. Therefore, the elimination of larger defects is important, and hence the role of the hydrotest in eliminating these defects is important [45]. Indeed, it has been reported [46] that natural gas pipelines subjected to pressure cycling typical for a gas pipeline would not be expected to experience a fatigue failure in less than 100 to 500 years from any defects surviving a hydrostatic test of 1.1 times its maximum allowable operating pressure. Liquid pipelines are subjected to more aggressive pressure cycling, and would have a much shorter fatigue life.

9.6 PRESSURE TEST FAILURES

Modern pipeline materials and construction practices mean that failures during pressure testing are rare. The absence of failures means that failure data are no longer collected; however, some historical data is available. Failure data are available in the literature,

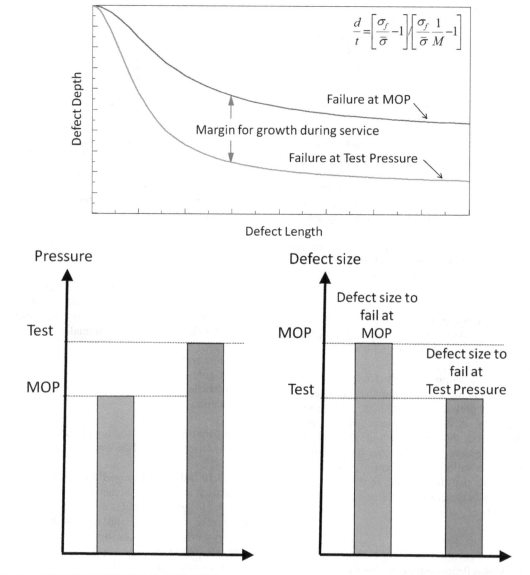

$$\frac{d}{t} = \left[\frac{\sigma_f}{\bar{\sigma}} - 1\right] \Big/ \left[\frac{\sigma_f}{\bar{\sigma}}\frac{1}{M} - 1\right]$$

FIG. 9.20   DEFECTS SURVIVING THE PRESSURE TEST AND MAXIMUM OPERATING PRESSURE (MOP)

for both onshore and subsea pipelines [47]. This Section looks back on historical pressure test failures; therefore, its relevance is not with new pipeline constructions, but with testing older pipelines.

9.6.1 Historical Data

From 1950 to 1965, pressure testing in the USA failed many defects in pipelines, but most of the failures were in the line pipes' longitudinal welds, Table 9.9 [48]. Most of the defects failed on pressurization, or within the first two hours of the test, Figure 9.21.

These pressure test failures occurring within 2 hours, or at 2 hours are probably due to the common practice of re-pressurizing at this 2 hour point to accommodate small pressure drops arising from relaxation of the pipeline [49].

In the USA, earlier experience of hydrostatic testing indicated that 50% of failures occurred during initial pressurization, 18% in the next 4 hours and 16% in the final 19 hours [21]. The experience of British Gas, UK, was similar (Figure 9.22): out of 165 failures that occurred during testing to the nominal yield strength of the line pipe, 100 occurred during the 24 hour hold [50, 51]. Some hydrotest failures occurred after a period of approximately 15 hours [21]:

- 40% failed during initial pressurization;
- 21% during first two hours of hold period;
- 29% in the final 22 hours;
- 10% during the re-pressurization to the original maximum pressure.

9.6.2 Why Do Defects Fail during the Hold Period?

Defects such as a gouge can fail when held at a constant load below the 'straight-off to failure' load. This is because ductile materials, such as line pipe, exhibit 'time dependent behavior' caused by 'stress-activated creep' [52]. This creep can lead to local plastic flow, followed by crack initiation and/or crack extension.

TABLE 9.9 LOCATION OF PRESSURE TEST FAILURES

Failures location	Percentage of failures on test
Longitudinal welds	>80%
Girth welds	≤5%
In the plate	≤10%

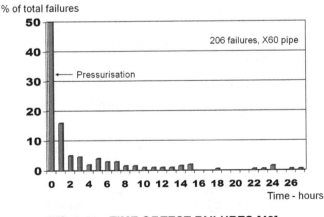

FIG. 9.21 TIME OF TEST FAILURES [48]

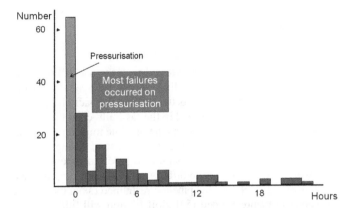

FIG. 9.22   UK EXPERIENCE OF PRESSURE TEST FAILURES IN NATURAL GAS PIPELINES (UP TO 1974) [21]

Available data indicates that this 'creep' is exhausted in approximately 28 hours, with 90% of this process completed within about 2 or 3 hours. These values are supported by [52]:

- observations that very few failures occur in the final hours of a 24 hour hydrotest;
- the longest time to failure recorded in tests of part wall defects in pipe held at constant pressure is 23 hours;
- tests using standard fracture mechanics specimens failed after 72 hours at constant load.

9.7 OTHER CONSIDERATIONS

9.7.1 Test Lengths and Elevation

Pipelines are pressure tested in segments. The 'test length' can be as long as practicable (e.g., 15 to 20 miles (24 to 32 km)), but testing in short sections both increases the sensitivity to detecting small leaks, and reduces the amount of water used.

The pipeline configuration and elevation changes will affect this length: changes in pipeline elevation will affect the pressure in the pipeline during hydrotest. Accordingly, elevation changes (e.g., over 60 m) are limited by standards when liquid is used. The permissible elevation range is the range which allows the minimum test pressure to be achieved at the highest elevation without exceeding the maximum test pressure at the lowest elevation [53].

Changes in pressure caused by this 'hydrostatic head' are accommodated by dividing the pipeline into appropriate test segments when the elevation changes are significant [54].

9.7.2 Effect of Air

Air in the water used to perform a hydrotest can corrupt pressure readings: air trapped in the test section can mask pressure changes caused by a volume loss either because the air expands to fill the space, or because of temperature effects on the bubbles of air [53]. Trapped air will partly dissolve as the pressure is raised, and throughout the test: this can cause a drop in pressure readings, falsely indicating a leak. Consequently, the air content in the test water must be small (e.g., less than 0.2% of the fill volume of a pipeline under test).

Fluid volume, pressure and temperature are measured during the pressure test by a variety of equipment (pressure gauges, dead weight testers, pressure recorders, thermocouples, etc.). These allow a graphical plot of pressure against volume during the pressurization period. This plot represents a stress/strain graph for the pipeline. Air content is determined from the pressure/volume plot (see, for example, reference 53).

9.7.3 Effect of Backfill and 'Restraint'

An onshore pipeline is usually tested with the backfill over the pipeline: this is because the pipeline can be damaged during the backfilling process. Hence, it is better to have the backfill in place, as the pipeline is then being tested in the 'as-built' condition.

Additionally, the backfill protects the pipeline immediately from damage, and also helps temperatures to be constant during the test.

The backfill provides some restraint to pipe movement during the pressure test. There is little difference between a pressure test of a pipeline segment with the pipeline restrained (axial movement is prevented) or unrestrained [53]. Soil friction will fully restrain a long section of pipe, but few other situations give this level of restraint; therefore, buried test sections longer than 2 km can be considered to be fully restrained [53].

9.7.4 Test Records

The minimum records (signed and dated by the test engineer) following completion of a hydrotest on a section are [25]:

- date and time of test;
- description of pipe in test section;
- any permits needed for the test;
- permits for discharge of test water;
- test medium used;
- temperature of test medium;
- calculations supporting the basis of the test pressure (including account of elevation);
- elevation profile for test section;
- test instrument calibration;
- explanation of any temperature or pressure discontinuity, or failure;
- any records required by other organizations with jurisdiction over the testing operation.

9.7.5 Highway and Rail Crossings

Sections of an onshore pipeline at road, rail, and water crossings usually have thicker wall (thicker than the rest of the pipeline). Depending on their length, these crossings can be pre-tested to the appropriate maximum test pressure, before being included in a test section that may have different wall thickness or steel properties [18].

9.7.6 Testing Subsea Pipelines

Subsea pipelines are typically tested to 1.25 × maximum allowable operating pressure, or 90% SMYS. Subsea pipeline design standards usually require water to be used: gas cannot be used in many standards.

The hold period is often quoted as 24 hour (for example, [55, 56]), but other standards require shorter hold periods (for example, [27]).

Hydrotesting an offshore line can be time-consuming:

- filling with water;
- cleaning and gauging the line;
- temperature stabilization;
- dissolving residual air;
- pressurization to the required test level;
- hydrostatic strength test (and leak test);
- emptying the pipeline (after a successful hydrotest);

- disposal of the test water; and,
- drying the pipeline.

The British Standards Institute pipeline document PD 8010-2 [55] says a 24 hour hold period is needed:

- to ensure the detection of small leaks in large pipelines; and,
- to cover the possibility of creep mechanism causing failures after a number of hours.

9.7.7 Replacing the Pressure Test

The pre-commissioning pressure test should only be replaced when:

- the pressure test is either a safety hazard, environmental hazard, not cost beneficial, or does not achieve its stated aim of proving the integrity of the system prior to service; and,
- alternative methods and procedures will achieve the stated objectives of the pressure test.

In-line inspection is another method of proving the integrity of a pipeline and is often proposed as an alternative method to the pressure test [47]. An inspection alone will not replace the pressure test, as smart pigs are not yet able to detect the range and combination of defects (types, sizes, and locations) that are detected by a pressure test. However, it can be used in conjunction with other procedures to prove the integrity of a pipeline. Indeed, some operators now use a smart pig to 'fingerprint' a pipeline before it goes into service, or shortly after it goes into service, to complement the pressure test. This fingerprint (or 'baseline') inspection can detect a range of defects that the pressure test cannot detect, and gives the operator a through appraisal of the starting condition of the pipeline.

9.7.8 In-Service Pressure Tests

A pre-service pressure test is conducted on new pipelines, but standards (e.g., [20]) state that the test can also be used to determine the (hoop stress) integrity of an in-service pipeline.

This can be an attractive alternative to other assessment methods (for example, in-line inspection), but in-service hydrotesting is a problem due to [11]:

- the high cost of taking the pipeline out of service (on average, an in-service hydrotest takes about 18 days, and alternative supplies may be needed);
- obtaining permits to acquire, treat, and dispose of the test water (the water may be contaminated from debris in the pipeline, or from the cleaning, and require treatment before an environmentally-friendly (and possibly expensive) disposal);
- water storage may be needed to minimize test time;
- failure during the test may cause contaminated water to go into the surrounding environment;
- the pipeline may be close to population;
- old pipelines may have multiple diameters, wall thicknesses, and grades that complicate pipeline cleaning and specifying the design parameters for the test

The costs (2002) of in-service testing can be very high [57]:

- preparation for testing: $1,250–$5,000/mile for a 16–24" pipeline, plus $2,000/mile to conduct the test;
- loss of throughput is the biggest cost component for hydrostatic testing; for example, $7,000/mile for a 24" gas line, to $90,000 for a 16" petroleum line.

Another obvious problem with in-service testing is the possibility of a failure during the test, as often the main reason for the in-service test is a suspicion of large defects being present. The high pressures used during the test can lead to leaks and ruptures which can threaten both public safety and the environment.

Finally, an operator must take care not to introduce new threats to the pipeline's integrity through pressure testing; for example, poor quality water can contaminate a pipeline, and inadequate drying after a hydrotest can leave water in the line and lead to corrosion during service.

9.7.9 Other Benefits from Pressure Testing

Pressure testing has many other benefits for the pipeline's integrity:

- Defects that survive the pressure test can have their tips 'blunted'. Both fracture and fatigue are affected by the defect tip sharpness: blunting can improve both fracture resistance and the subsequent fatigue life of a defect [49];
- This blunting, and the creation of residual stresses around the defect tip, can also improve low temperature toughness (this is called 'warm pre-stressing' (WPS)). BSI 7910 [5] gives information on warm pre-stressing stating, *"The benefits of a WPS... have been attributed... mainly to the establishment of a compressive residual stress zone ahead of the crack tip. However, the effects of crack tip blunting and strain hardening have also been claimed as significant by a number of authors."*
- Dents in a pipeline will move outwards during a hydrostatic test (their depth will decrease). This can increase their fatigue life.

- The hydrostatic test also decreases stresses in weldments. These 'residual' stresses are a major concern in thick-walled pressure vessels (they can lead to failure), but are of little significance today in modern thin-walled line pipe.

9.8 MANAGEMENT SYSTEM APPROACH FOR PRESSURE TESTING OF PIPELINES

Figure 9.23 shows a plan-do-check-act summary for this Chapter:

- PLAN: a pressure test has to be planned, and this will include satisfying the requirements of relevant regulations and standards, specifying the pressure test type (pressure level, medium, and hold period for the pressure), purpose (for example, is the test being used to detect defects in the pipeline by failing the pipeline?), any impact on the operation of the pipeline (the test will suspend pipeline operation), and the environment (water is usually used for testing at high pressures, and this water will need to be acquired, and then returned to its source or disposed of).
- DO: the test requires close working with the pipeline's control room, as the pipeline will be taken out of service, and operations suspended. The test procedure will have been written before the test, and the required pressures and hold periods specified. Monitoring must ensure that any leak is detected, and that sections of the pipeline are not subjected to overloads that may cause deformation.

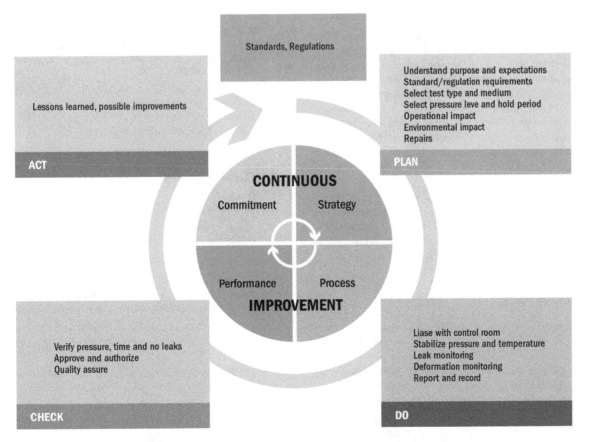

FIG. 9.23 PLAN-DO-CHECK-ACT FOR PRESSURE TESTING

- CHECK: all parameters set for the test must be verified, including checks for leaks. A full record of the test needs to be approved and authorized.
- ACT: the test should have satisfied all aspects of the Plan, but any areas of improvement should be specified and implemented in later tests.

9.9   CONCLUSIONS

Pipelines are pressure tested before entering service to ensure: the pipeline is leak-tight; the pipeline does not contain large defects; and, the pipeline can withstand its maximum operating pressure.

These pre-service tests are performed at pressures above the pipeline's maximum operating pressure, They can also be conducted on an in-service pipelines to allow pipeline operators to assess the condition of their pipelines; although, testing of in-service lines is difficult and expensive, as it requires the pipeline to be taken out of service.

There are three types of pressure test:

- a 'strength test': this test establishes the operating pressure limit of a pipeline segment.
- a 'leak test': this test is used to determine that a pipeline segment does not show evidence of leakage.
- a 'spike test': this test is used to verify the integrity of a pipeline containing 'time dependent' defects, such as fatigue cracks.

These tests can be conducted on both new pipelines (all design standards require this type of test), and in-service pipelines.

The pressure test has a key role in pipeline integrity management. The test removes defects by failing them: the higher the stresses created by the test, the smaller the size of defect removed. This means that there will be a difference in size between the defects surviving the test, and those (larger) defects that can subsequently fail the pipeline at the maximum operating pressure. The difference in size is the amount of growth available for the surviving defects: the surviving defects must grow by this amount to fail the pipeline during service. Growth calculations can be performed to predict the time required for this growth to occur, and these calculations allow an assessment interval (time between tests) to be set.

It is important to note that pressure testing does not guarantee a safe pipeline: subsequent growth of defects that have survived the test can cause failure following the test. Following a management system approach would provide assurance of adequacy, implementation, and effectiveness of a pressure testing program.

9.10   REFERENCES

1. Anon., 'Pipeline Transportation Systems for Liquid Hydrocarbons and Other Liquids', American Society of Mechanical Engineers. ASME B31.4-2012. ASME International. 2014.

2. M J Rosenfeld, R W Gailing, 'Pressure testing and recordkeeping: Reconciling historic pipeline practices with new requirements', Journal of Pipeline Engineering. 1st Quarter 2013. p. 15.

3. Anon., 'Managing System Integrity for Hazardous Liquid Pipelines', American Petroleum Institute. API Recommended Practice 1160. Second Edition. September 2013.

4. Anon., 'Recommended Practice for the Pressure Testing of Steel Pipelines for the Transportation of Gas, Petroleum Gas, Hazardous Liquids, Highly Volatile Liquids, or Carbon Dioxide', American Petroleum Institute, USA. API Recommended Practice 1110. 6th Edition, February 2013.

5. Anon., 'Guide to methods for assessing the acceptability of flaws in metallic structures', BS 7910:2013. The British Standards Institution. UK. 3rd Edition. 2013.

6. J P Ellenberger, 'Piping Systems and Pipelines'. ASME Code Simplified', McGraw-Hill. 2005.

7. G Saville, S M Richardson, B J Skillerne de Bristowe, 'Pressure Test Safety', HSE Contract Research Report 168/1998. HMSO. UK. 1998, and G Saville, 'Pressure Test Safety', HSE Guidance Note GS4. Health and Safety Executive. 1998.

8. G Saville, 'Pressure Test Safety', HSE Guidance Note GS4. Health and Safety Executive. 1998.

9. Anon., 'Pressure Testing (Hydrostatic/Pneumatic) Safety Guidelines', Construction Safety Consensus Guidelines. The INGAA Foundation, Inc., USA. September, 2012.

10. Anon., 'Specification for Line Pipe', American Petroleum Institute. API 5L-2012. Edition 45. 2012.

11. Anon., 'Technical, Operational, Practical, and Safety Considerations of Hydrostatic Pressure Testing Existing Pipelines'. INGAA Foundation Report. Number 2013/03. Prepared by Jacobs Consultancy. December, 2013.

12. T Shires et al., 'Development of the B31.8 Code and Federal Pipeline Safety Regulations: Implications for Today's Natural Gas Pipeline System', Gas Research Institute Report, GRI-98/0367.1. 1998.

13. C J Trench, J F Kiefner, 'Oil Pipeline Characteristics and Risk Factors: Illustrations from the Decade of Construction'. American Petroleum Institute. December 2001.

14. J F Kiefner, 'Evaluating the Stability of Manufacturing and Construction Defects in Natural Gas Pipelines', Final Report for Department of Transportation, Office of Pipeline Safety, Report 05-12R. April 2007.

15. Anon., 'Identification of Pipe with Low and Variable Mechanical Properties in High Strength, Low Alloy Steels Energy Pipeline Industry Pipe Quality Action Plan September, 2009'. White Paper. The INGAA Foundation. USA. 2009.

16. B N Leis, T Thomas, 'Linepipe Property Issues in Pipeline Design and Re-establishing MAOP', International Congress on Pipelines, PEMEX, Merida, Mexico, Paper ARC-17, November 2001.

17. Anon., 'Code of practice for Pipelines-Part 1: Steel pipelines on land', BSI PD 8010-1. BSI. UK. 2004.

18. Anon., 'Pipelines-Gas and liquid petroleum. Part 1: Design and Construction', AS 2885.1-2007. Australian Standard. 2007.

19. Anon., 'Commentary on CSA Z662-11, Oil and gas pipeline systems'. Canadian Standards Association. Z662.1-11. 2011.

20. Anon., 'Gas Transmission and Distribution Piping Systems', ASME B31.8-2014. American Society of Mechanical Engineers. USA. 2014.

21. D G Jones, 'Notes on the Philosophy and History of Pressure Testing', IMechE Seminar on Developments in Pressure Vessel Technology, London, UK.

22. J F Kiefner et al., 'Failure Stress Levels of Flaws in Pressurized Cylinders', American Society for Testing and Materials. USA. ASTM STP 536. 1973. pp. 461–481.

23. J Kiefner, K Leewis, 'Pipeline Defect Assessment: A Review and Comparison of Commonly-used Methods', Pipeline Research Council International. PRCI. USA. May 2011. Report No. L52314.

24. A R Duffy, M McClure, W A Maxey, T J Atterbury, 'Study of Feasibility of Basing Natural Gas Pipeline Operating Pressure on Hydrostatic Test Pressure', Pipeline Research Committee, American Gas Association, No. L30050, February 1968.

25. J F Kiefner, W A Maxey, 'The Benefits and Limitations of Hydrostatic Testing', Oil and Gas Journal, July–August 2000.

26. E W McAllister, 'Pipeline rules of thumb handbook', 6th Edition. Elsevier. 2005. p. 143.

27. M Law et al., 'Understanding the Hydrostatic Strength Test', 14th EPRG/PRCI Meeting. Berlin. 2003. Paper 28.

28. Anon., 'Oil & Gas Pipeline Systems', CSA Z662. Canadian Standards Association. Canada. 2011.

29. Anon., 'Spike Hydrostatic Test Evaluation', Final Report. Michael Baker Jr. Inc. For Department of Transportation Research and Special Programs Administration, Office of Pipeline Safety. TTO Number 6. July 2004.

30. J F Kiefner, W A Maxey, R J Eiber, 'A Study of the Causes of Failure of Defects That Have Survived a Prior Hydrostatic Test', Pipeline Research Committee, American Gas Association, NG-18 Report No. 111, November 3, 1980.

31. B N Leis, F W Brust, 'Hydrotest Strategies for Gas Transmission Pipelines Based on Ductile-Flaw Growth Considerations', American Gas Association, Pipeline Research Committee, NG-18 Report No. 194, July 27, 1992.

32. H Haines, J Kiefner. M Rosenfeld, 'Study questions specified hydrotest hold time's value', Oil and Gas Journal, March 2012.

33. S J Garwood et al., 'The Application of CTOD Methods for Safety Assessment in Ductile Pipeline Steels', Fitness-for-Purpose Validation of Welded Structures Conference, IMechE, London, November 1981.

34. L E Brooks, 'Hydrostatic Testing Procedures and Results', Pipeline Engineer. September 1967. p.27.

35. W Maxey, 'Hydrostatic Test Considerations', OPS Public Meeting on Enhanced Protection in High Consequence Areas. Hernden, Virginia. November 1999. www.kiefner.com.

36. J Kiefner, W A Maxey, 'Hydrostatic Testing-1: Pressure ratios key to effectiveness; in-line inspection complements'. Oil and Gas Journal. July 31 2000.

37. Anon., 'Natural Gas Pipeline Rupture Transcanada Pipeline Limited 762-millimetre-diameter Pipeline Line 100-1—Main Line Valve 111a-1 from Kilometres 11.12 to 11.16 Near Marten River, Ontario 26 September 2009', Pipeline Investigation Report P09H0083. Transportation Safety Board of Canada. 2009.

38. Anon., 'NTSB Safety Recommendation'. D A P Hersman Letter to California Public Utilities Commission, San Francisco, USA. January 3, 2011.

39. M Rosenfeld, 'Hydrostatic Pressure Spike Testing of Pipelines: Why and When?' The Journal of Pipeline Engineering. Vol. 13. No. 4. December 2014.

40. B N Leis, B A Young, J F Kiefner, J B Nestleroth, J A Beavers, G T Quickel, C S Brossia, 'Final Summary Report and Recommendations for the Comprehensive Study to Understand Longitudinal ERW Seam Failures—Phase One', Final Report—Task 4.5. U.S. Department of Transportation Pipeline and Hazardous Materials Safety Administration 1200 New Jersey Ave., SE Washington DC 20590. October 23, 2013.

41. B Leis, R Galliher, 'Hydrotest Protocol for Applications Involving Lower-Toughness Steels'. ASME IPC2004. Calgary 2004. IPC04-0665.

42. R R Fessler, 'Overview of Solutions to the Stress-Corrosion Cracking Problem', 6th Symposium on Line Pipe Research, A.G.A. Catalogue No. L30175, 1979.

43. Anon., 'Managing System Integrity of Gas Pipelines', ASME B31.8S-2014. American Society of Mechanical Engineers. New York, USA. 2014.

44. Anon., 'Fitness-For-Service', API 579-1/ASME FFS-1, Second Edition, American Petroleum Institute, 2007.

45. J F Kiefner, W A Maxey, 'Model helps prevent failures from pressure-induced fatigue' Oil and Gas Journal. 31st July, 2000, and 7th August, 2000.

46. M. J. Rosenfeld, J. F. Kiefner, 'Basics of Metal Fatigue in Natural Gas Pipeline Systems - A Primer for Gas Pipeline Operators'. Pipeline Research Council International, Inc.. Catalog No. L52270. June, 2006.

47. M Kirkwood, A Cosham, 'Can the pre-service hydrotest be eliminated?', Pipes and Pipelines International, July–August 2000. pp. 5–19.

48. Anon., 'Safety of Interstate Natural Gas Pipelines', US Federal Commission. April 1966.

49. G D Fearnehough, D G Jones, 'An Approach to Defect Tolerance in Pipelines', IMechE, Conference on Defect tolerance of Pressure Vessels, May 1978. London, UK.

50. G D Fearnehough, P, Hopkins, D G Jones, 'The Effect of the Hydrostatic Test Hold Period on Defect Behaviour and Surviving Defect Population', 8th EPRG/AGA Joint Technical Meeting on Line Pipe Research, Paris. 1991.

51. A Cosham, P Hopkins, 'A Review of the Time Dependent Behaviour of Line Pipe Steel', Proceedings of IPC 2004: International Pipeline Conference October 4–8, 2004; Calgary, Alberta, Canada. IPC04-0084. 2004.

52. B Leis et al., 'The assessment of time-delayed failure under constant pressure', 17th EPRG/PRCI Joint Technical Meeting, Milan 2009.

53. Anon., 'Pipelines - Gas and liquid petroleum', Part 5. Field pressure testing. Australian/New Zealand Standard, AS/NZS 2885:2012.

54. J F Kiefner, 'Overview of Hydrostatic Testing', Hydrostatic Testing Symposium. California Public Utilities Commission. May 6, 2011.

55. Anon., 'Code of practice for pipelines-Part 2: Subsea pipelines', BSI PD 8010-2:2004. British Standards Institute. UK. 2004.

56. Anon., 'Submarine Pipeline Systems', Offshore Standard. Det Norske Veritas. DNV-OS-F101. Norway. October 2010.

57. P H Vieth et al., 'Integrity Verification methods support US efforts in pipeline safety', Oil & Gas Journal. December 16, 2002. pp. 52–59.

CHAPTER 10

DIRECT ASSESSMENT: PROCESS, TECHNOLOGIES AND MODELING

10.1 INTRODUCTION

Direct assessment methodologies are only applicable for time dependent threats (i.e., internal corrosion, external corrosion, and stress corrosion cracking). Time-independent threats (i.e., operator error, weather related/outside force, and third party/excavation damage) and stable threats (i.e., equipment failure, manufacturing defects, and construction defects) do not have industry recognized direct assessment methodologies. Prudent operators should assess time-independent and stable threats by other means than direct assessment. Operator training audits, monitoring of third party activity, inspection of equipment, pressure testing, and In-Line Inspection (ILI) are options for assessing these threats.

NACE International (NACE) has a series of standard practice documents on direct assessment for pipeline integrity management. Unpiggable pipelines are the most likely candidates for integrity assessment by direct assessment. ILI surveys and the data integration to perform direct assessment are complementary methodologies. As with any integrity assessment, a risk assessment will determine the proper direct assessment methodology.

There are six direct assessment methodologies currently published by NACE and a seven is under development as of the writing of this book. The authors of this book recommend that the reader examine these documents for more details. In this chapter, we will give a high-level overview of the methodologies and provide information on the things the reader needs to consider when performing these assessments so they can be as effective as possible. The Direct Assessment methodologies are as follows:

- NACE SP0502—External Corrosion Direct Assessment (ECDA) [1]
- NACE SP0210—External Corrosion Confirmatory Direct Assessment (ECCDA) [2]
  - Used for reassessment interval when ECDA has already been performed
- NACE SP0206—Dry Gas Internal Corrosion Direct Assessment (DG-ICDA) [3]
- NACE SP0208—Liquid Petroleum Internal Corrosion Direct Assessment (LP-ICDA) [4]
- NACE SP0110—Wet Gas Internal Corrosion Direct Assessment (WG-ICDA) [5]
- Under Development—Multiphase Flow Internal Corrosion Direct Assessment (MF–ICDA) [6]
- NACE SP0204—SCC Direct Assessment (SCCDA) [7]

For corrosion to occur internally or externally, an anode, cathode, metallic path, and an electrolyte all need to be present to complete

the circuit. ECDA focuses on monitoring the performance of the anode in protecting the cathode and looking for areas where coating is no longer preventing the electrolyte from touching the cathode. ICDA focuses on looking for where the electrolyte could settle in the line and form a corrosion cell. SCCDA relies more on a data integration process to look for area with the highest likelihood of the threat. Figure 10.1 below shows an illustration of how the Direct Assessment methodologies can be applied to pipeline by threat.

10.2 TERMINOLOGY

Anode—The location in a corrosion cell where oxidation occurs. The anode is used to protect the cathode.

Cathode—The location in a corrosion cell where reduction occurs. The pipeline is desired to be the cathode.

Disbonded—Coating that is not adhered to the pipe surface.

Electrolyte—A liquid medium that allows the transfer of ions in the corrosion cell.

Holiday—A small location where pipe steel is exposed through the coating.

Shielding—Prevention of cathodic protection current from reaching the pipe.

10.3 DIRECT ASSESSMENT OVERVIEW: FOUR STEP PROCESS

There are four steps in every integrity assessment performed by a direct assessment methodology. Each step is explained in later sections for the overall process and specific considerations within each direct assessment methodology.

1. Preassessment
2. Indirect Inspection
3. Direct or Detailed (ICDA) Examinations
4. Post Assessment

Direct Assessment also follows the PLAN-DO-CHECK-ACT principles of a management system. Figure 10.2 Illustrates the four integrity management principles with their corresponding step in the Direct Assessment methodology. The bullets in each box are actions taken in that step of the Direct Assessment methodology.

Direct Assessment is an approved integrity assessment methodology in industry regulations [8] and industry codes [9]. Direct Assessment has been challenged recently on whether it is as

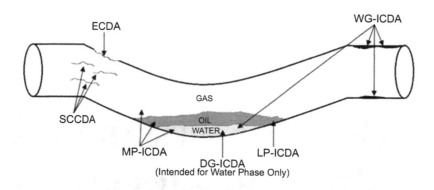

FIG. 10.1   GRAPHICAL ILLUSTRATION OF THE NACE DIRECT ASSESSMENT METHODOLOGY (*Source:* NACE International)

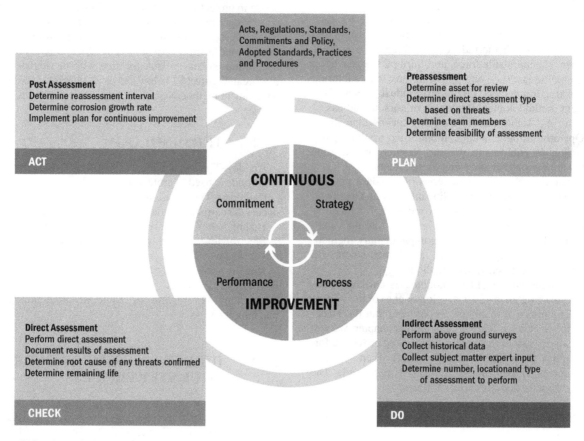

FIG. 10.2   PIPELINE INTEGRITY MANAGEMENT SYSTEM DIAGRAM FOR DIRECT ASSESSMENT

effective as other integrity assessment techniques in addressing integrity concerns. The National Transportation Safety Board (NTSB) in the United States issued a report [10] that recommended, "increasing the use of ILI and eliminating the use of direct assessment as the sole integrity assessment method." This report came following three high profile natural gas pipeline failures in high consequence areas. In one of these failures [11] Direct Assessment was the sole means of integrity assessment and this allowed for a gap in integrity management. In another case, there was no integrity management performed because it was not a regulated pipeline.

Integrity management needs to cover all nine threats to pipelines in accordance with ASME B31.8S [12]. A fundamental weakness of Direct Assessment is that the methodology followed only covers

one threat and not all nine. To properly manage pipeline risks, work needs to be performed above and beyond direct assessment to ensure adequate pipeline integrity management.

As will be shown in this chapter, the user of Direct Assessment methodology has to be careful when performing the four steps in the methodology. Each step needs to be carefully considered so that when the data collected is integrated, that the most likely locations for the threat to exist are investigated and mitigated. Without adequate care in performing Direct Assessment, practitioners may be given a false sense of security in the risk level of the pipeline being assessed. A diagram of an ECDA process is shown in Figure 10.3 below. Similar workflow patterns should be followed for other Direct Assessment methods.

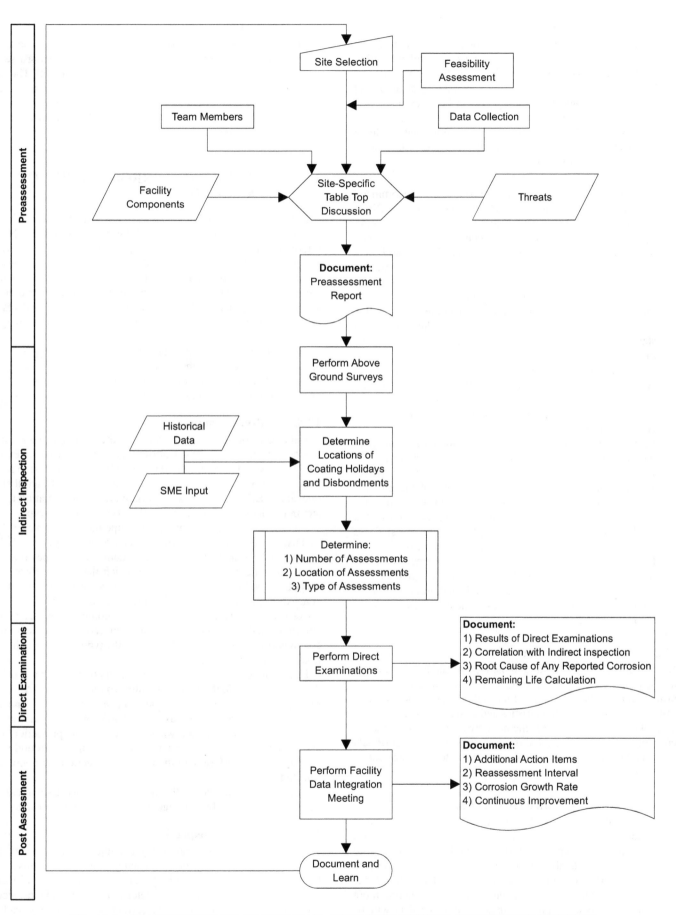

FIG. 10.3   ILLUSTRATION OF AN EXTERNAL CORROSION DIRECT ASSESSMENT PROCESS

10.3.1 Preassessment

Proper performance of this critical first step will advise the direct assessment practitioner if the process will be successful at locating the threat being addressed. This step is a research step that will collect relevant data and analyze it for the applicability of the direct assessment methodology. The primary goal of Preassessment is to determine if Direct Assessment is feasible on the pipeline. Once the feasibility has been confirmed, the most appropriate Indirect Inspection techniques for the pipeline are selected. There are a number of subsections to the Preassessment.

1. Collect Pipeline Data—The NACE documents recommend the types of information that should be collected for Direct Assessments. Current and historic information are both important in the process. Threats could have been initiated due to a change in the operation or the threat could have existed in the past and then been remediated. It is also important to collect asset knowledge from long-term company employees. These individuals may have information that is relevant to the process that cannot be documented on paper. Some operators elect to include key individuals at review stages to ensure their thoughts are included in getting the most out of the process.

2. Determine the Feasibility—Not all pipelines can have Direct Assessment performed. The factors that affect feasibility are unique to the threat that is being addressed. Please see subsequent sections for feasibility considerations for each Direct Assessment type.

3. Select Indirect Inspections—These techniques are unique to the threat being assessed and will be discussed below.

4. Establish Preliminary Direct Assessment Regions—Direct Assessment aims to group similar regions of the pipeline and then to excavate in representative locations in each region. A preliminary assessment of regions is performed in the Preassessment step for future planning purposes. These regions are then further refined as more is learned.

5. Document Preassessment Results—Documentation is important during any pipeline integrity management process. Operators should have an established process to document all stages of the Direct Assessment process and store the results for future use.

10.3.2 Indirect Inspection

Indirect Inspections is unique to the type of threat that is being addressed. Indirect Inspection for ECDA focuses on above ground measurements of characteristics of the cathodic protection system. ICDA focuses on flow characteristics and design elements of pipelines looking for areas where internal corrosion is most likely to occur. For SCCDA, this step focuses both utilizes a number of data source that are design, operational, and related to cathodic protection performance.

10.3.3 Direct or Detailed (ICDA) Examination

In this step of the process, the pipeline will be excavated based on the prioritized list formulated from the results of the previous two steps. Company excavation forms should be examined before performing direct or detailed examinations to ensure that the data suggested in NACE documents can be collected and stored for future reference. Direct Examination excavations require more information that standard pipeline digs because the data will be applied to the broader integrity management of the pipeline.

10.3.4 Post Assessment

The Post Assessment determines the effectiveness of the Direct Assessment process. This step reprioritizes the threat being assessed on the pipeline and determines the reassessment interval. The reassessment interval is determined based on the severity of the features that are found in the Direct Examination step and the remaining life of the pipeline. There are a number of methods for determining the remaining life of features found during excavations.

10.4 EXTERNAL CORROSION DIRECT ASSESSMENT

External Corrosion Direct Assessment (ECDA) was the first of the Direct Assessment methodologies produced by NACE International. External corrosion is a potential threat on most if not all pipelines. Many of the methodologies used in the ECDA process had been used on pipelines prior to a formalized integrity assessment process. ECDA brings together techniques that are used for assessment of pipelines for external corrosion and formalizes the process for how the information is examined and used for integrity management.

ECDA is typically used in pipelines that are not piggable. It is also possible to use this methodology to confirm the results of ILI and to further assess the condition of a pipeline beyond what is provided through an ILI assessment.

10.4.1 Preassessment

Data collection is the first stage of the ECDA Preassessment process. Both current and historical data are relevant to ECDA. Table 1 of NACE SP0502-2010 Pipeline External Corrosion Direct Assessment Methodology lists a number of types of data that are relevant to ECDA. The most important type of information to collect in the data collection phase is the external coating type.

The type of external coating on a pipeline plays a key role in ECDA. If the coating is known to cause shielding, then ECDA might not be practical. A shielding coating (e.g., tape) does not allow cathodic protection current to reach the steel surface when it is disbonded (i.e., not adhered to the pipe surface) and not broken. Aboveground survey techniques are not able to function as effectively as possible when there is a shielding coating. The return current from the pipeline would give an inaccurate statement on the coatings condition. Bare pipelines limit the types of indirect inspections can be performed.

Feasibility of ECDA is determined at this stage. We have already discussed the difficulties that shielding coatings in ECDA. ECDA may be limited or not feasible on right-of-ways that are inaccessible or are limited on the season of excavation. Rocky terrain can also limit the electronic pathways for the cathodic protection to be assessed. Congested corridors and significant below ground metal structures can interfere with the indirect inspection methodologies selected.

The results of the data collection and feasibility assessment will be used to select the best indirect inspection for performance.

10.4.2 Indirect Inspection

Appropriate above ground survey techniques are selected in order to locate areas of coating holidays and disbondments. Subsections below detail above ground survey techniques that can be utilized to locate coating issues. It is best to determine the cathodic protection levels by interrupting impressed currents and obtaining readings outlined on the ground level.

Above ground survey data should be correlated utilizing GPS coordinates and integrated to identify areas with the highest susceptibility for external corrosion metal loss. Overlaying piping drawings with the integrated survey data can aid in locating initial areas for excavation. NACE recommends that two or more Indirect Inspection techniques be performed in this step of ECDA. Table 2 of NACE 0502 provides a tool selection matrix for indirect survey techniques based on the condition of the pipeline.

10.4.2.1 Annual test station surveys Annual test station surveys are required by most industry codes and governmental regulations. Test stations are placed at prescribed intervals and unique locations (e.g. casings, foreign crossings, etc.) along the pipeline. Each year, a qualified cathodic protection technician uses a reference cell electrode and a voltmeter to take the potential at that location. Testing these stations ensures that the pipeline is meeting the cathodic protection design criteria. If the desired criteria are not achieved, then the potential for corrosion is higher in that location and requires mitigation. Mitigation could be achieved by either turning up the output voltage of the rectifiers on either side of the location or adding a new rectifier with associated ground bed near that location.

10.4.2.2 Alternating Current Voltage Gradient (ACVG) ACVG is an above ground survey technique that measures the current change from an induced low voltage AC signal over a given length and measuring the return. This method can be used to pin point coating anomalies. This is typically performed at targeted areas following a close interval survey and is the preferred method of secondary above ground indirect inspection. NACE TM0109 [13] provides guidance on the performance of ACVG as an Indirect Inspection.

10.4.2.3 Close-Interval Surveys (CIS) CIS is an above ground survey technique that measures the effectiveness of the cathodic protection system over a length of the pipeline. A rectifier interrupter can be used to establish both an on and an off potential of the pipeline. This is effective in making sure the pipeline is meeting the specified criteria. An additional technique such as ACVG or DCVG should be performed to further evaluate condition of the coating. NACE SP0207 [14] provides guidance on the performance of CIS as an Indirect Inspection. Figure 10.4 below shows a crew

performing ACVG (left) and CIS (right) on a pipeline. Figure 10.5 shows the results of the CIS, ACVG, and depth of cover surveys of a pipeline. Locations in the Figure that are shown to be below the −850 mV criteria and have an ACVG indication have the potential to be locations for Direct Examination.

10.4.2.4 Direct Current Voltage Gradient (DCVG) DCVG is an above ground survey technique that measures the current change from the rectifier output over a given length. This method can be used to identify areas of coating holidays (i.e. small locations of exposed metal). This is typically performed at targeted areas following a close interval survey. NACE SP0207 and NACE TM0109 provide guidance on the performance of DCVG as an Indirect Inspection. Figure 10.6 below shows a surveyor performing DCVG on a pipeline.

10.4.2.5 Ground Penetrating Radar Ground Penetrating Radar: This above ground survey technique used pulses of electromagnetic radiation to locate below ground structures. This can be used in congested areas to confirm location of the target asset and other assets in the area. Ground penetrating radar is best used to locate structures but does not provide any assessment of condition of the asset identified.

10.4.2.6 Buried Coupons Buried coupons are useful tools in determining the current electrical state of the pipeline. There are a number of different types of coupons available for pipeline applications and the user should select coupons based on the needs of the assessment.

10.4.2.6.1 AC Coupons. AC current can be detrimental to pipeline integrity if it is not properly managed. AC coupons are placed in areas with a high potential for AC current to be induced on the pipeline from foreign sources. The most likely sources of induced AC are electrical transmission lines and other cathodic protection systems. When the AC levels indicated on the coupon are above the operator's defined threshold, mitigation should be performed.

10.4.2.6.2 Corrosion Coupons. In this section we are specifically referring to external corrosion coupons. Corrosion coupons are buried next to the subject pipeline and are connected electrically to the cathodic protection system through the test station.

FIG. 10.4   ACVG (LEFT) AND CIS (RIGHT) BEING PERFORMED (*Source:* JW's Pipeline Integrity Services)

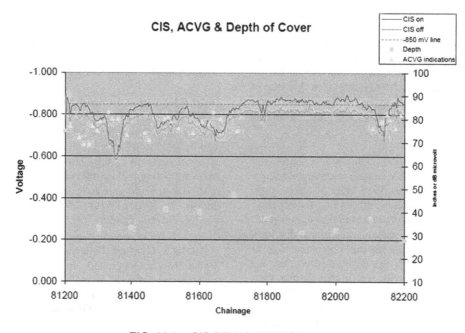

FIG. 10.5   CIS BEING PERFORMED
(*Source:* JW's Pipeline Integrity Services)

FIG. 10.6   DCVG BEING PERFORMED
(*Source:* JW's Pipeline Integrity Services)

Overtime, the coupon mimics a coating holiday on the pipeline by polarizing itself from with the cathodic protection. These coupons are helpful in determining if the 100 mV shift criteria for cathodic protection are achieved.

10.4.2.7 Soil Resistivity   This is the soils ability to resist the flow of electricity. Soil resistivity is a factor in both how cathodic protection migrates through the ground and how the electrochemical involved with a corrosion cell occurs on a pipes external surface.

10.4.3   Direct Examination

The number of excavations that are required to be performed in the ECDA process are guided by NACE SP0210. NACE recommends that one excavation be performed in each unique ECDA region that was identified in the Indirect Inspection phase. NACE also provides guidance on how to prioritize Indirect Inspection indications based on:

1. Immediate action required,
2. Scheduled action required, and
3. Suitable for monitoring

The value in performing multiple above ground surveys is the integration of the data collected. When there are indications from multiple above ground surveys, these should be prioritized for excavation.

Example
   If there is an ACVG or DCVG reading in the same location as a low close-interval-survey potential reading, this is a solid indicator that there is a coating holiday with inadequate cathodic protection. This location would have a high potential for active corrosion.

Direct Examination should follow the existing procedures for excavation performance that operators follow. External corrosion Direct Assessments serves to confirm or refute the presence of external corrosion on the pipeline segment. During the excavation, the pipe and coating condition will be evaluated in order to determine if the indirect survey data is correlating with suspected findings.

Indirect data validation with direct examination findings will dictate the path forward for each site being evaluated. Guided wave technologies can be utilized where applicable in excavations to extend the limits of the corrosion metal loss investigation. Guided wave uses ultrasonics to send a sound wave longitudinally along the pipeline. The guided wave signal can extend for up to 100 feet in both direction from the coupler in the right circumstances. It is recommended to follow with detailed ultrasonic scans at locations of guided wave indications.

Corrosion damage should be documented and the root cause identified which will be utilized for any subsequent excavations. During these excavations, it is possible to identify other threats than external corrosion (e.g. stress corrosion cracking, mechanical damage, etc.). The user is encouraged to be prepared to evaluate these types of threats before they are actually uncovered in the field.

If corrosion defects are discovered during excavation, remaining strength calculations shall be performed to ensure that the pipeline is still safe to operate. Acceptable limits of remaining strength should be determined before the excavation occurs. In the event the remaining strength is below the acceptable limits, a pressure reduction shall be taken to be sure that the pipeline is safe to operate. Remaining strength calculations can be performed with B31G, Modified B31G, CorLAS™, or other fracture mechanics software.

Root cause analysis of significant external corrosion features can be beneficial to overall improvement of an integrity management program. ECDA may identify uniform corrosion or pitting corrosion. The causes of these forms of external corrosion commonly occur when there is disbonded coating in areas of low cathodic protection potentials. If there are external corrosion features discovered during excavations that are unique or uncommon, these would warrant further investigation. Some examples of features that required additional investigation would be: AC corrosion, preferential seam corrosion, microbiologically induced corrosion, and others.

10.4.4   Post Assessment

The Post Assessment step has a number of deliverables. Determination of subsequent external corrosion reassessment intervals is one of the goals of this step. The reassessment interval is determined based on the corrosion rate calculated from the external corrosion that was discovered with the ECDA process.

The Post Assessment step can also serve as a look back on the ECDA process as a whole to determine if applying or enhancing mitigation measures can be implemented or improved upon. The ultimate goal of any integrity management process is continuous improvement. The Post Assessment sets the stage for improving the next assessment of the line.

Industry Context [15]

On August 24, 1996 an 8-in. diameter liquid Butane pipeline ruptured due to external corrosion. The rupture resulted in the deaths of two residents as they drove through the vapor cloud and the evacuation of about 25 families. The pipeline in this case had a MFL survey performed. It would have been difficult to perform ECDA on this pipeline because it had tape coating which would have shielded the cathodic protection surveys performed. This incident led to a number of industry improvements

- Increased public awareness related to the hazards of pipelines
- Increased assessment of cathodic protection systems (49 CFR 195.416)
- Additional requirements to examine all exposed pipelines for external corrosion (49 CFR 195.416)
- Additional requirements to replace or repair pipelines joints with significant corrosion (49 CFR 195.416)
- Development of microbiological testing procedures by NACE

10.4.5   Advantages

ECDA can be advantageous in maintaining pipeline integrity. When the performance of a cathodic protection system is monitored and assessed in detail, it is possible to find defects before they become significant features. For example, an excavation could be performed at locations where an ACVG feature and low CIS potentials overlap. In this case, when the excavation is performed there could be relatively shallow corrosion found under poorly coated pipe. The location had the elements of a potentially severe corrosion feature, but the excavation was performed and the condition remediated. The results of ILI are lagging indicators for pipeline integrity. By closely monitoring the cathodic protection system, the leading indicators from the survey can be followed and improve the overall integrity of the pipeline.

10.4.6   Limitations

There are a number of limitations to ECDA. When the pipeline contains shielding coatings, there are a limited number of Indirect Inspection techniques. Even when the pipeline does not have shielding coatings, the ECDA practitioner may not always choose the right locations to perform excavations. ECDA excavations occur at sampled locations along the pipeline. Not all sampled locations will be perfect. It is up to the ECDA practitioner to use engineering judgement as to whether a sufficient amount of information has been gained in the process.

Direct Assessment may not be the best methodology to perform on pipelines with a high number of external corrosion features. In this case, a large number of excavations would be required to meet the intention of the NACE ECDA standard. ILI would give a holistic review of the pipeline in this case and allow for prioritized remediation of the deepest features. ILI also has a higher probability of detection for significant features. ECDA has the potential to miss features if the data is not perfectly aligned. ILI gives an assessment of the entire line but as mentioned before is a lagging indicator.

10.5   INTERNAL CORROSION DIRECT ASSESSMENT

There are always four elements required for corrosion to occur: (1) anode, (2) cathode, (3) metallic path, and (4) electrolyte. Minimizing the exposure to the electrolyte or neutralizing the effect of the electrolyte on the steel is the goal of the internal corrosion mitigation process. NACE currently has three internal corrosion direct assessment (ICDA) methodologies available with a fourth on the way.

- NACE SP0206—Dry Gas Internal Corrosion Direct Assessment
- NACE SP0208—Liquid Petroleum Internal Corrosion Direct Assessment
- NACE SP0110—Wet Gas Internal Corrosion Direct Assessment
- Under Development—Multiphase Flow Internal Corrosion Direct Assessment

Each of the documents follows the four step direct assessment methodology and have unique elements to each based on the product type. The user is encouraged to review these documents in detail for more guidance.

10.5.1 Preassessment

Determining the feasibility of ICDA is an important Preassessment phase. There are limitations on the applicability of ICDA based on the product type in the pipeline. Limitations are discussed later in this chapter by product type. ICDA proceeds by collecting operational and product information for use in flow modeling simulations. The flow modeling occurs in the Indirect Inspection step. The three NACE ICDA standards provide guidance on what data is the most important for ICDA work.

10.5.2 Indirect Inspection

Modeling of the pipeline occurs in the Indirect Inspection step. Software for modeling the corrosion rate prediction is available from a number of different software vendors. As with any computer program, the results that it provides are dependent upon the quality of the data that is input into the model.

The profile of the pipeline plays an important role in the accuracy of the ICDA process. In all three methodologies, it is important to know where liquid is expected to drop out of the flow and settle on the pipe steel. This location is where the internal corrosion cell could occur. Internal corrosion software models take into account product type, flow regime, protection from water wetting, down time, product quality and other factors.

The results of the flow modeling aid in the selection of locations for excavation. The results of corrosion coupons and other corrosion measurement devices aid in prioritization of which lines should be investigated first.

10.5.2.1 Linear Polarization Resistance (LPR) Probe

LPR probe require a conductive environment to attain readings because it functions by taking a reading between two electrodes. The probes can be used to provide readings on corrosion rates at any given moment in time. This can be useful in batched operations where one product stream could be more corrosive then another. These probes can also be helpful in determining when the inhibitor has either run out or is no longer effective. These probes come in a wide range of pressure and temperature options. Figure 10.7 shows some examples of LPR probes.

10.5.2.2 Electrical Resistance (ER) Probes

ER probes change the signal that is produced based on the volume of corrosion that is occurring. Data from the probe can then be analyzed to determine the rate of corrosion that is occurring. These probes can be used in both conductive and non-conductive environments and come in a wide range of pressure and temperature options. Figure 10.8 shows some examples of ER probes.

10.5.2.3 Internal Corrosion Coupons

Internal corrosion coupons are installed on pipelines to get actual corrosion measurements from the product stream. Coupons have a defined material, shape, size, and surface area that aid in the determination of corrosion rates. Internal corrosion coupons are typically installed for six months to a year and then removed from the system. Once the coupon is removed, it is cleaned weighed and examined for corrosion. The weight loss from the coupon can then be extrapolated into a corrosion growth rate for the period of time. It can be beneficial to have a setup where one corrosion coupon is changed every six months and one is changed every year. It is best to install coupons in locations where there is expected to be liquid because if the product is always dry then no corrosion cell would occur on the coupon and could give a false sense of security.

Internal corrosion coupons can be installed on a vessel called a side stream analyzer. An example of a side stream analyzer and the schematic of how the product flows through the pipeline is shown in Figure 10.9. Side stream analyzers can be advantageous because multiple probes can be used in the same location and water drops can be allowed to accumulate. When side stream analyzers are not used, retractable corrosion coupons are useful. It is essential that the coupon from the retractable coupon, see Figure 10.10, be inserted into the liquid phase of the product stream.

10.5.3 Detailed Examination

Because internal corrosion is not visible from the external surface, detailed examination requires ultrasonics to obtain a profile of the internal surface of the pipe. Measurement techniques range from ultrasonic range from single point-wall thickness measurement to sophisticated full scans of the pipeline. With ultrasonic inspection, as resolution increases cost also increases.

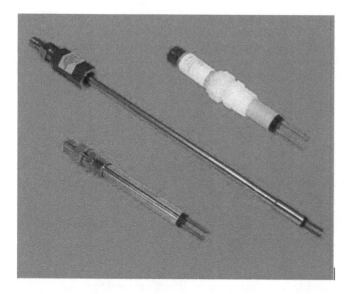

FIG. 10.7 EXAMPLES OF LINEAR POLARIZATION RESISTANCE (LPR) PROBES (*Source:* EnhanceCo)

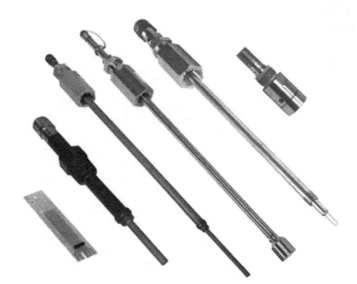

FIG. 10.8 EXAMPLES OF ELECTRICAL RESISTANCE (ER) PROBES (*Source:* EnhanceCo)

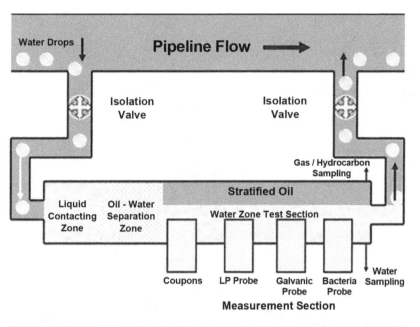

FIG. 10.9 SETUP OF A SIDE STREAM INTERNAL CORROSION ANALYZER AND PHOTO OF COMPLETE SYSTEM (*Source: EnhanceCo*)

When full scans of the pipeline are performed, there are a number of options. Non-destructive testing vendors can provide the following types of scans:

- Automatic ultrasonic scans
- Digital radiography
- Eddy current testing
- Electromagnetic surveys
- Guided wave testing
  - Can extent up to 100 feet in both longitudinal directions from the coupler
  - Requires detailed scans at severe anomalies identified with this method
- Phased array

When selecting the assessment methodology to assess internal corrosion features, it is important to consider the type of feature that is expected. Single point ultrasonic wall thickness measurement would not be the best choice if microbiologically influenced corrosion or pitting corrosion is suspected to exist on a pipeline. It would be the best choice if the user desired to create a river bottom profile for remaining strength calculations.

10.5.4   Post Assessment

Post Assessment is used to establish the reassessment interval for the ICDA process. Comparisons should be made to determine if the model predicted what was actually found in the field. As more investigations are performed the information can be fed back into the model to determine better and better predicted corrosion rates. If significant

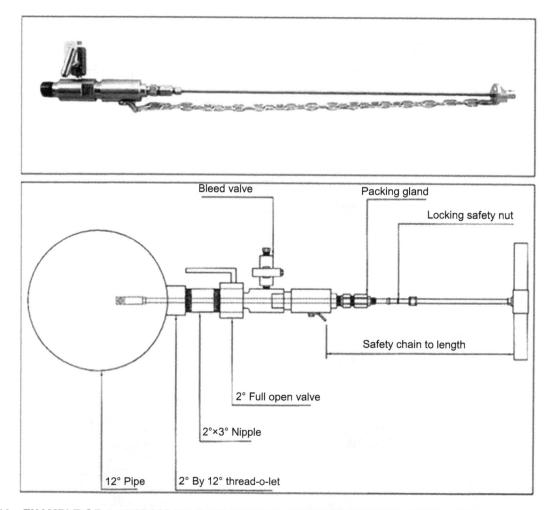

FIG. 10.10 EXAMPLE OF A RETRACTABLE COUPON HOLDER (INSERT FROM COPYRIGHT) (*Source:* EnhanceCo)

or unexpected internal corrosion is found then a root cause analysis should be performed and mitigation methods identified.

Determining if additional coupons or monitoring probes are required in the system is a beneficial step in the ICDA process. As the program matures, better results can be obtained.

> *Industry Context* [16]
>
> On August 19, 2000 a natural gas pipeline near Carlsbad, New Mexico ruptured. The rupture and resulting fire killed 12 people who were camping near the location of the failure. The cause of the rupture was determined to be internal corrosion. This failure led to changes in the United Stated Code of Federal Regulations. This incident also led NACE International and other organizations [17] to create and improve and number of internal corrosion related documents including the Direct Assessment methodologies.

10.5.5 Dry Gas Considerations and Limitations

Pipelines that accumulate solids are not ideal candidates for dry gas ICDA because it is difficult to predict how the solid will affect the local corrosion rate. It is possible for condensation to occur on normally dry gas pipelines. Dry gas ICDA is not helpful in predicting this type of corrosion because the most models focus on water

settling in the pipeline. Some models predict condensed water corrosion, but this is not always available.

Both maintenance pigging and the use of corrosion inhibitors are good barriers in prevention of internal corrosion however they limit the accuracy of dry gas ICDA. Maintenance pigs push liquid from low spots of the line and reduce the likelihood of a corrosion cell developing in the low spot. At this liquid moves along the line, it can wet the surface of the pipe wall and allow for internal corrosion to occur in an area not predicted by dry gas ICDA. Similarly, corrosion inhibitors prevent liquid that settles in the line from forming an internal corrosion environment. There could be a case where the inhibitor prevents corrosion in one section of the line and not in the other and the ICDA process would not distinguish this phenomenon.

Dry gas ICDA cannot be performed on pipelines with a continuous internal coating. It is difficult to predict where the coating has broken down and allowed a corrosion cell to form. Any known locations of breaks in the internal coating should be prioritized for Indirect Inspection.

In general, the dry gas ICDA methodology is best for standard pipelines systems that do not have unique operation of flow characterizations. It is best used for lines that only have occasional upsets in the saturation level of the gas stream. It can aid prediction of the location of liquid fallout if it were to occur in an operational upset. Therefore, ICDA cannot be used for dry gas pipelines that contain liquids or glycols.

10.5.6 Wet Gas Considerations and Limitations

Direct assessment methodologies can cast a wide net when collecting data for assessment. Wet gas ICDA is dependent upon the accuracy of the information that is collected. Operating characteristics such as moisture content, flow rates, contaminants, and the like can be ever changing. Without being able to accurately identify these values, it is difficult to accurately apply the methodology.

10.5.7 Liquid Petroleum Considerations and Limitations

Similarly for ECDA, if a pipeline has a large number of internal corrosion features, it could be cost prohibitive to perform an ICDA process. There could be a cost advantage in this case to making necessary modifications to the pipeline to make the line piggable. Making the line piggable would give a complete assessment of the pipeline and not point locations. One reason why there would be a large number of internal corrosion features would be because there is a large water separated layer in the pipe. In this case it would be difficult to prioritize where corrosion is most likely to occur.

10.6 STRESS CORROSION CRACKING

As illustrated in Figure 10.11, Stress Corrosion Cracking (SCC) is susceptible to occur in pipelines when

1. Stress,
2. Susceptible material, and
3. Susceptible environment

SCCDA focuses on finding locations of susceptible environments. High pH and Near-neutral pH SCC are the two forms of SCC that are common to pipelines. There are unique attributes to each type that can be investigated in the SCCDA process.

10.6.1 Preassessment

There are a wide variety of conditions and ranges of conditions that SCC can occur. Table 1 of NACE SP0208 gives a listing of the types of data that can be collected in the SCCDA process. What is

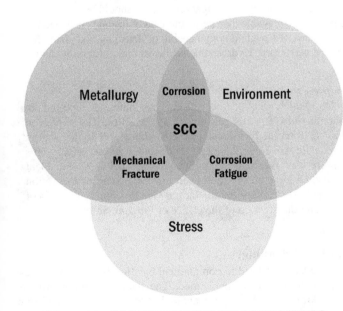

FIG. 10.11 SCC SUSCEPTIBILITY COMPONENTS

helpful about this table is it provides (1) a prioritization of what data to collect, (2) how the information is useful to the process, and (3) the use of the results.

The list of data collected in this step can be daunting. Prioritizing effort to the collecting the data that is identified as having the most value to the process is what the user is encouraged to perform. Collecting current and historical data, when possible, can be important to SCC. SCC is a unique phenomenon in that it can grow, arrest and then grow again. If SCC is discovered on the pipeline through this process, knowing whether the conditions that led to its existence are active or mitigated aid the response to what is found.

The Preassessment step should prioritize pipeline based on their highest susceptibility to SCC. Segments can be prioritized based on history of SCC, operating stress, operating temperature, density of deformation features, and, soil type, cathodic protection performance, and others. If a pipeline has previously had SCC, the focus for investigation should be to find similar locations on the pipeline.

10.6.1.1 Near-Neutral pH SCC Considerations Below are some factors that are commonly associated with Near-Neutral pH SCC. Not all factors will be present and some are more significant than others are, but they can all be used to select locations for excavations or determine which type of SCC has been discovered.

- Location—Within 20 miles of pump/compressor station
- Temperature—No correlation with temperature
- Electrolyte—Diluted bicarbonate
- pH—5.5 to 7
- Cathodic Protection Potentials—−760 to −790 mV (CuSO$_4$)
- Cathodic Protection Effectiveness—Does not reach pipe surface (typically due to shielding qualities of coating)
- Metallographic Observations—Wide, transgranular cracks with minimal branching and typically has corrosion within crack
- Operating stress—>60% SMYS
  - Note: Can exist at lower operating stress due to local residual stress
- Presence of Corrosion on Outer Surface—Light localized corrosion
- Presence of Stress Risers—Typically in the presence of dents, gouges and other stress risers
- Vintage—>10 years
- Predominate Coating—Tape
- Location—Top of the pipe
- Topography—Localized depression with ground water, or alternating wet and dry conditions

10.6.1.2 High pH SCC Considerations Below are some factors that are commonly associated with High pH SCC. Not all factors will be present and some are more significant than others are, but they can all be used to select locations for excavations or determine which type of SCC has been discovered.

- Location—Within 10 miles of pump/compressor station
- Temperature—Increased likelihood greater than 100°F
- Electrolyte—Concentrated bicarbonate-carbonate
- pH—7 to 11
- Cathodic Protection Potentials—−850 to −2000 mV (CuSO$_4$)
  - Typically close to −850 mV
- Cathodic Protection Effectiveness—CP can get to pipe surface and contributes to raising pH

- Metallographic Observations—Narrow, intergranular cracks typically associated with branching and no corrosion within crack
- Operating stress—>60% SMYS
- Presence of Corrosion on Outer Surface—Not typically
- Presence of Stress Risers—Not typically
- Vintage—>10 years
- Predominate Coating—Coal tar and Asphalt
- Location—Bottom of the pipe
- Topography—No correlation

10.6.2   Indirect Inspection

The Indirect Inspection techniques discussed in the ECDA section are relevant to SCCDA as well. IF deformation and metal loss ILI survey results are available they should be used for excavation site selection. The data collected in the first two steps of SCCDA should be overlay and integrated so that the most likely locations for the existence of SCC can be identified.

Example: Prioritization of Near-Neutral pH SCC Pipeline Segment

The most significant factors in predicting the existence of Near-Neutral pH SCC are coating type, locations of stress risers, and existence of corrosion. In the situations below, which segment should be prioritized? Hint: Use resources in Chapter 4.

Pipeline A
- Tape wrapped coating
- Cathodic protection levels mostly –850 mV to –1.2V with some areas below –800 mV
- ILI data shows deformations with some corrosion nearby

Pipeline B
- Tape wrapped coating
- Cathodic protection levels are –850 mV to –1.2V
- ILI data shows deformations without corrosion nearby

Answer: Pipeline A
Pipeline A would be prioritized for assessment because it has deformations with corrosion in the area and some locations with low cathodic protection potentials. If a pipe joint showed a deformation with corrosion elsewhere on the joint in an area of low cathodic protection potentials, then this would be a candidate for assessment.

Example: Prioritization of High pH SCC Pipeline Segment

The most significant factors in predicting the existence of High pH SCC are coating type, operating stress, and operating temperature. In the situations below, which segment should be prioritized?

Pipeline A
- Non-Fusion bonded epoxy coating
- Operating stress greater than 60%
- Operating temperature less than 100°F

Pipeline B
- Non-Fusion bonded epoxy coating
- Operating stress greater than 60%
- Operating temperature greater than 100°F

Answer: Pipeline B
All things being equal, the pipeline with the higher operating temperature would be prioritized for investigation. This is particularly true for lines that operate above 100°F.

10.6.3   Direct Examination

There is no industry guidance on how many direct examinations should be performed to confirm or refute the presence of SCC. The user should determine how many excavations are prudent based on the data that is collected using engineering judgment. One methodology is to select more sites that are planned to excavate. For example, if it is determined that four excavations should be performed then 8 could be selected. If no SCC is found in the first four locations, then the excavation program can stop. If SCC is found, then the extent and severity of what is found would determine if more excavations are required.

Table 2 of NACE SP0208 provide the types of data to collect in SCC investigations, the importance of the data, and how the data is used in the process.

Magnetic particle inspection (MPI) should be performed at all SCCDA excavations. At SCC excavations, the entire exposed section of pipe should be inspected with MPI. It is good practice to perform MPI at all dent excavations. The blast media to remove the coating is important for SCC investigations. A less aggressive blast media like walnut shells should be used. When more aggressive blast media is used, it has the potential to mask SCC features by folding over the crack opening and giving a false negative reading.

When SCC is discovered during excavations, data should be collected to determine which type of SCC is present. The best way to determine the type of SCC that is present is to perform destructive metallography of the feature. This is not possible in all cases and investigation and judgment is required to determine the type. Some operators have had luck using in-situ metallography to determine if the cracking is transgranular or intergranular and thus determining the type of SCC.

10.6.4   Post Assessment

As with the other direct assessment methodologies, the post assessment determines the reassessment intervals and the mitigation strategy that is necessary. If significant numbers and significant severity SCC features are found, then a mitigation strategy should be developed. If the pipeline is piggable then performing an ultrasonic crack detection tool or other ILI methodology capable of detecting and sizing SCC could be performed. Hydrotesting is also an option for mitigation.

Industry Context

On March 4, 1965, a 32-in. gas pipeline ruptured in Natchitoches, LA and resulted in 17 fatalities. The explosion was so forceful that local residents thought that the nearby Barksdale Air Force Base had come under atomic attack. This rupture was eventually classified as High pH SCC. President Johnson addressed this failure in his 1967 State of the Union Address. This failure led to Natural Gas Pipeline Safety Act of 1968 and the creation of the Office of Pipeline Safety.

10.6.5   Advantages

SCCDA can be used to complement existing ILI data if a crack detection tool was not performed. The data integration process required for SCCDA is significant and gaps in pipeline knowledge can be identified and mitigated.

10.6.6 Limitations

There are broad factors that contribute to the presence of SCC on pipelines. A very experience subject matter expert may not always select the correct locations for assessment. The results of the SCCDA process are subjective and the repeatability of the process is not always possible.

10.7 PREDICTIVE MODELING TECHNIQUES

Methodologies are being developed that use statistics to analyze pipeline data to predict where threats are most likely to occur. Information collected in the direct assessment methodologies mentioned previously in this chapter is also used in these statistical methodologies. Predictive models also use local knowledge and subject matter expertise to determine a probability of failure along the pipeline for a given threat. When the location of a high probability of failure is combined with a high consequence are, a true risk picture is developed.

A limitation of any direct assessment methodology is that information is location specific, time specific, and has the chance for error in measurement. Information about pipelines changes over time. The information can change because more is learned about the pipeline's condition through further assessment. The condition of the pipeline can also degrade over time simply because it is aging and this may not be reflected in the information. The predictive modeling techniques take all of these changing variables into account when determining the probability of failure.

In predictive models, all of the data used to calculate the probability of failure has a distribution applied to it to account for inaccurate data, and changing data. Similar to when a subset of data from a larger set is examined, there is a mean, standard deviation, and confidence interval that can be calculated. Through statistical analysis, we can be confident that we have determined the true mean of the whole set of data through analysis of the subset of data. This is essentially how the data collected is used in the predictive model.

Predictive modeling using statistics and subject matter input requires a significant amount of data to get the most out of the methodology. The model is only as good as the data that goes into it and the rigor around the justification for the model structure and value given to individual data. As the saying goes…"there are three kinds of lies: lies, damned lies, and statistics." Statistics can give a false sense of security because there is implied accuracy in the significant figures that the calculations can generate. The user is cautioned that these methodologies are useful but are not to replace sound engineering judgment and diligence.

10.8 USING DIRECT ASSESSMENT TO JUSTIFY MAKING A LINE PIGGABLE

The data collected in a Direct Assessment could aid in making a justification to make a pipeline piggable. Below are some example of reasons that could justify making a pipeline piggable

- External corrosion, internal corrosion, or SCC were found to be severe or widespread during Direct Assessments
- The location of direct assessments are in areas that are difficult to access or are in environmentally sensitive areas that could make the dig prohibitive from a cost or environmental perspective

- Multiple threats are determined to be active on the pipeline and performance of multiple direct assessments are time prohibitive
- The consequence of failure of the pipeline is above the threshold of tolerance for the operator
- Failures of the pipeline from external corrosion, internal corrosion, or SCC following performance of direct assessment
- Direct assessment was determined to not be feasible or the results were inconclusive to maintain integrity

10.9 CASE STUDY 1—FACILITY PIPING

Context

- You are performing an ECDA program at a terminal
- The facility has a congested underground network of piping
- Good records for pipeline location are not available
- Performing large excavations are not desirable to the terminal manager
- Excavations must minimize impact to below ground piping that are not in the scope of the ECDA program
- The pipeline to be assessed has coal tar enamel coating

Assignment

1. Determine which Assessment options are most appropriate in determining locations for excavations.
2. Explain why these are the most appropriate options.

Result

ACVG—ACVG is used to locate coating holidays by measuring the voltage gradient that exists when a holiday is present. ACVG has a number of advantages in this application. ACVG inputs its own voltage onto the pipe and measures any change in that signal. This serves two purposes. First, it helps ensure the pipe location in question because the user is looking for a known signal. Secondly, it locates coating holidays for further investigation. ACVG works well when there is the potential for AC interference and gives the user confidence in the results.

Ground Penetrating Radar—When a coating holiday is located with ACVG, ground penetrating radar can be used to ensure where other below ground structures are located so excavations can be as minimally invasive as possible. Hand digging in congest areas is appropriate.

10.10 CASE STUDY 2—UNPIGGABLE PIPELINE

Context

- 5-mile long dry gas pipeline
- Internal corrosion coupons historically have shown minimal to no weight loss
- The dehydration units are considered to be very reliable
- Recent maintenance pigging has not yielded any free liquid
- The pipeline is less than ten years old
- The pipeline is coated with fusion bonded epoxy
- The pipeline operated below 100°F
- The pipeline operates below 50% of SMYS

Assignment

1. Select the most appropriate Direct Assessment methodology for this pipeline.
2. Explain why this methodology is the most appropriate.

Result

External Corrosion Direct Assessment is the most appropriate methodology for this pipeline.

Dry-Gas ICDA may be appropriate for the pipeline but here were clues to why it would not be the first choice. When dry gas pipelines have reliable dehydration units before the gas flows into the pipeline or the transport gas comes from a cryogenic process, the likelihood of internal corrosion is low. Without moisture to complete the corrosion cell, then corrosion is not likely. This low likelihood is confirmed by the results of the corrosion coupons and the lack of liquid following maintenance pigging.

SCCDA would not be the most appropriate because the pipeline does not have any of the traditional risk factors associate with SCC. The line is lower temperature and pressure with new fusion bonded epoxy coating. If the opposite was true of all these factors, then the line would be considered to have higher susceptibility to high-pH SCC.

10.11 REFERENCES

1. Anon., 2010, *Pipeline External Corrosion Direct Assessment, NACE SP0502*, Houston, USA.

2. Anon., 2010, *Pipeline External Corrosion Confirmatory Direct Assessment, NACE SP0210*, Houston, USA.

3. Anon., 2006, *Internal Corrosion Direct Assessment Methodology for Pipelines Carrying Normally Dry Natural Gas, NACE SP0206*, Houston, USA.

4. Anon., 2008, *Internal Corrosion Direct Assessment Methodology for Liquid Petroleum Pipelines, NACE SP0208*, Houston, USA.

5. Anon., 2010, *Wet Gas Internal Corrosion Direct Assessment Methodology for Pipelines, NACE SP0110*, Houston, USA.

6. Anon., *Multiphase Flow Internal Corrosion Direct Assessment, NACE TG-426*, Houston, USA.

7. Anon., 2008, *SCC Direct Assessment Methodology, NACE SP0204*, Houston, USA.

8. Z662, 192 cfr, 195 cfr.

9. ASME B31.8S Others.

10. NTSB/SS-15/01 PB2015-102735 Integrity Management of Gas Transmission Pipelines in High Consequence Areas, April 2015.

11. NTSB/PAR-11/01 Accident Report, Pacific Gas and Electric Company, Natural Gas Transmission Pipeline Rupture and Fire, San Bruno, CA, September 9, 2010.

12. Anon., 2014, Managing System Integrity of Gas Pipelines, ASME B31.8S, American Society of Mechanical Engineers, New York, USA.

13. Anon., 2009, *Aboveground Survey Techniques for Evaluation of Underground Pipeline Coating Condition, NACE TM0109*, Houston, USA.

14. Anon., 2007, *Performing Close Interval Potential Surveys and DC Surface Potential Gradient Surveys on Buried or Submerged Metallic Pipelines, NACE SP0207*, Houston, USA.

15. NTSB/PAR-98/02/SUM Pipeline Accident Summary Report—Liquid Butane Release, and Fire Lively, Texas; August 24, 1996.

16. NTSB/PAR-03/01 Pipeline Accident Report—Natural Gas Pipeline Rupture and Fire Near Carlsbad, New Mexico August 19, 2000.

17. Anon., 2002, *Internal Corrosion Direct Assessment of Gas Transmission Lines—Methodology*, Gas Technology Institute, GRI-02-0057, Des Plaines, IL, USA.

PREVENTION, MITIGATION, AND MONITORING MEASURES FOR PIPELINES

In this chapter, we examine how to prevent, mitigate and monitor threats that can cause unplanned events on pipelines. The names of the integrity threats have been selected from the ASME B31.8S-2014 industry-recognized standard. The American Society of Mechanical Engineers (ASME) publishes *Managing System Integrity of Gas Pipeline*. This is document is commonly referred to as ASME B31.8S. ASME B31.S paragraph 2.2 defines nine (9) threats to gas pipelines. These nine threats are divided into three main categories: (1) Time Dependent, (2) Stable, and (3) Time Independent. Some of the nine threats have subcategories of things to consider in a robust integrity management program. While ASME B31.8S focuses on gas pipelines, the discussion of threats is applicable to liquid pipelines as well.

Furthermore, Table 11.1 provides a correlation of the integrity threats mentioned in the ASME B31.8S, CSA-Z662 (Canada), API RP 1160 and UKOPA (United Kingdom) that may help the reader with following the chapter within their own jurisdiction.

As illustrated in Figure 11.1, a Management System approach provides companies with the guidance for planning (PLAN), implementing (DO), verifying (CHECK) and reviewing (ACT) the prevention, mitigation and monitoring measures to continuously improving the integrity of a pipeline. This chapter focuses on identifying and describing those measures with a practical approach.

11.1   TERMINOLOGY

Consequence—Any unplanned effect on the environment, people, or pipeline operation

Likelihood—The probability that an event occurs

Injurious—A feature that could lead to failure of the pipeline

Inspection Test Plan (ITP)—Document that covers all of the steps for a paint or coating application to a pipeline

Mitigation—The act of removing a threat or reducing the likelihood of a threat causing a consequence on the pipeline

Monitoring—The act of ensuring that known threats do not increase in likelihood of failure

Prevention—The act of stopping a threat from occurring on a pipeline during any phase of a pipeline's life cycle

11.2   INTRODUCTION: CONCEPTS, RELATIONSHIPS, AND DIFFERENCES

11.2.1   Integrity Prevention of Integrity Threat and Consequences

As initially discussed in Chapter 3, prevention is stopping a threat from occurring. Preventions of threats through proper design, operation, maintenance, and integrity management plans are essential for cost effective and safe operation of pipelines. The design phase of any pipeline project should examine all potential threats to the pipeline and prevent these threats from occurring. Not all threats can be prevented.

When threats cannot be prevented, it is necessary to either mitigate or monitor the threats to ensure pipeline integrity. The preference is to mitigate the threats identified instead of only monitor them. Even though mitigating consequence is challenging, those measures should also be considered.

Tables 11.2 through 11.4 provide some examples of prevention of integrity threats and consequences as well as an exercise.

11.2.2   Integrity Mitigation of Integrity Threat and Consequences

As initially discussed in Chapter 3, when new or existing threats are identified to exist on a pipeline, it is necessary to mitigate them. Mitigation can take a number of forms. Ideally, a physical mitigation method can be used to remove the threat from the system permanently. In this case, a feature could be removed by a repair method. Following mitigation, it would be prudent to investigate why the threat existed and put preventative barriers in place to keep it from happening again. Features that cannot be permanently removed may require barriers to prevent them from growing or to monitor the status of the threat. Mitigation can also take the form of reducing the likelihood of the threat from reforming or occurring again. The discussion in this chapter will focus on mitigation measures as they reduce the likelihood of reoccurrence per integrity threat and the consequences.

Tables 11.5 and 11.6 provide some examples of mitigation of integrity threats and consequences.

11.2.3   Integrity Monitoring of Integrity Threat and Consequences

Following the initial discussion from Chapter 3, monitoring can serve two (2) purposes in integrity management. The first purpose is that it can determine when a threat or consequence has occurred (e.g., new corrosion or dwelling) or is likely to occur (e.g., new pressure cycling or new facility). The second purpose is to identify whether the threats or consequence have changed (e.g., crack or population growth). Early detection of a threat (e.g., ILI) or consequence (e.g., request for municipality approval) occurring or changing over time can lead to intervention at a more cost effective stage of an asset's lifecycle.

One limitation is that some threats may not be able to be monitored all of the time. A typical case is aerial or satellite monitoring of the right-of-way may not be available exactly when the excavation activity could be taking place over the pipeline.

Tables 11.7 and 11.8 provide some examples of mitigation of integrity threats and consequences.

TABLE 11.1 CORRELATION OF PIPELINE INTEGRITY THREAT NAMES FROM INDUSTRY STANDARDS

ASME B31.8S Gas transmission pipelines	CSA-Z662 (Canada) Gas, liquid hydrocarbons, oilfield water and steam, liquid or dense phase carbon dioxide pipeline systems	API 1160 Hazardous liquid pipelines	UKOPA (United Kingdom) Major accident hazard pipelines
External corrosion	Metal loss	External corrosion	External corrosion
Internal corrosion	Metal loss	Internal corrosion	Internal corrosion
Stress corrosion cracking—SCC (i.e., environmentally-assisted cracking)	Cracking	Crack and crack-like	Other
Manufacturing-related (e.g., seam weld and pipe body)	Material or manufacturing	Design and material	Seam weld defect
Welding/fabrication related (e.g., girth weld, wrinkle bend, threads)	Construction	Construction	Girth weld and pipe defect, construction/material
Mechanical damage (e.g., immediate/delayed failure, vandalism/theft)	External interference	Third party damage	External interference
Incorrect operations (e.g., procedure)		Operation errors	Operator error
Weather-related and outside force (e.g., hydro-geotechnical, cold weather, lightning, heavy rains or floods, Earth movements)	Weather		
	Geotechnical failures	Ground movement	Ground movement
Pipeline equipment (e.g., valves, seals)	Ancillary equipment	System operations	N/A

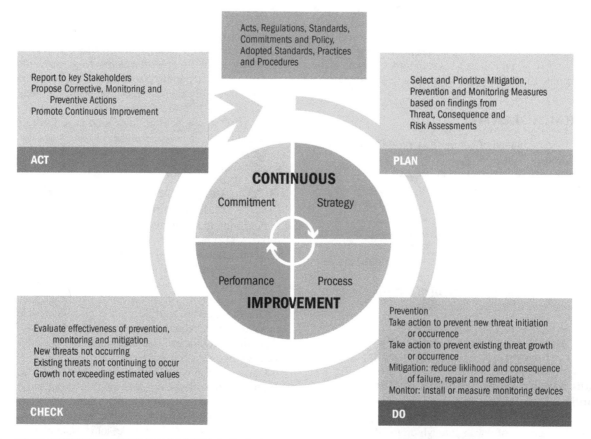

FIG. 11.1 PIPELINE INTEGRITY MANAGEMENT SYSTEM APPROACH TO PREVENTION, MITIGATION, AND MONITORING

Example: Prevention measures for each of the integrity threats:

TABLE 11.2 PIPELINE INTEGRITY THREAT PREVENTION EXAMPLES

Integrity threat	Examples of prevention measures *Effectiveness changes depending on the pipe characteristics and conditions*
External corrosion	• Coating selection • Coating application during construction • Wall thickness design
Internal corrosion	• Flow corrosivity reduction at source • Biocide • Inhibitor • Internal lining • Internal cleaning
Stress corrosion cracking—SCC (i.e., environmentally assisted cracking)	• SCC-resistant coating • Pipe residual stress treatment at mill
Manufacturing-related (e.g., seam weld and pipe body)	• Mill testing quality control • Hydrostatic testing at 100%SMYS (e.g., defect removal)
Welding/fabrication related (e.g., girth weld, wrinkle bend, threads)	• Welding specifications and procedures • Construction quality control
Mechanical damage (e.g., immediate/delayed failure, vandalism/theft)	• RoW activity-driven depth of cover design • Public awareness • Damage prevention
Incorrect operations (e.g., operational, workplace and human factors)	• Controller training (e.g., knowledge and stress) • Pipeline operations workplace design for controller function (e.g., displays, automatic systems: valves, overpressure) • Effective and simplified operational procedures (e.g., shutdown, alarm management)
Weather-related and outside force (e.g., hydro-geotechnical, cold weather, lightning, Heavy rains or floods, Earth movements)	• Pipeline route design minimizing effects from hydrogeotechnically-sensitive areas • Depth of Cover at water crossings
Pipeline equipment (e.g., valves, seals)	• Valve design for optimal maintenance

11.2.4 Relationship and Differences among Prevention, Mitigation, and Monitoring

Prevention, mitigation, and monitoring for both integrity threats and consequences require working together to achieve an adequate,

Example: Prevention measures for each of the type of consequences

TABLE 11.3 PIPELINE INTEGRITY CONSEQUENCE PREVENTION EXAMPLES

Integrity-related consequence	Examples of consequence prevention measures *Effectiveness changes depending on the pipe characteristics and conditions*
Safety and health (e.g., employees, landowners in proximity, and public in general)	• Distance to the pipeline minimizing release impact to people • Working with municipality planning departments to coordinate dwelling expansions near to the pipeline
Environmental (e.g., air, land, and water including habitat—flora and fauna)	• Reduction of pipeline release volume by system modifications (e.g., auto shutoff and remotely-operated valves, valve separation) • Improved detection and isolation time response in the event of a release
Regulatory (e.g., penalties, fines and enforcement)	• Commitment to follow the law and regulations

implemented, and effective Pipeline Integrity Management System (PIMS).

Regarding integrity threats, the best approach is to prevent them through the entire pipeline cycle design, manufacturing, operation, deactivation/reactivation, and abandonment. Even though the best efforts, it is always necessary to mitigate threats when they are identified through monitoring. If mitigation is not performed, then monitoring can be used to ensure the integrity condition is not changing or worsening. Monitoring can also be used to determine when an integrity threat has emerged on a pipeline.

Regarding consequences of a pipeline, multiple stakeholders influence the increase or decrease of consequences related to safety and health, environmental and regulatory. Consequence monitoring should enable cost effective strategies to prevent population getting close to the pipeline area of impact; whereas, consequence mitigation would be minimized if all involved stakeholders are consulted leading to proactive collaboration and agreement for everyone's benefit. Preferably, consequence mitigation should be minimized in light of greater efforts in prevention and monitoring to avoid significant effort in terms of time and resources from different stakeholders.

Regarding the pipeline life cycle, prevention, mitigation, and monitoring of threats and consequences may occur at the same or different stages. Threat and consequence prevention occurs before the asset begins to operate and once a threat or consequence has been identified, to keep it from occurring again or from growing and becoming a larger undertaking. Mitigation removes the threat or consequence from the system and monitoring is oversight of the initiation or growth defined within PIMS.

Exercise: Determine integrity prevention strategies during the design of a new NPS 30 API 5L X100 pipeline going through hydro-geotechnically-sensitive areas. The pipe body will be coated with fusion bonded epoxy (FBE) in the mill and shrink sleeves applied on each girth weld. The pipeline will transport hydrocarbon liquids exposed to pressure cycling.

TABLE 11.4 PIPELINE INTEGRITY THREAT PREVENTION EXERCISE

Hazard concern	Integrity threat (Pipeline life stage)	Expected issue	Proposed prevention measure
API 5L X100 lower weldability	Girth weld cracking (construction)	Combined construction-girth weld cracking exposed to axial/bending stresses	Increased Construction quality assurance and control of girth welds in hydro-geotechnical areas
Hydro-geotechnical longitudinal or transverse stresses on the pipe	Outside force (in-service)		
Shrink sleeve disbondment causing cathodic protection losses	Corrosion and cracking	Corrosion initiation/growth that may lead to SCC	Replace shrink sleeves with a more compatible coating (e.g., field-applied)
Pressure cycling inducing fatigue on pipe seam/girth welds	Cracking	Cracking growth	Minimize pressure cycling via system modifications (e.g., Variable Frequency Drive—VFD)

Example: Mitigation measures for each of the integrity threats:

TABLE 11.5 PIPELINE INTEGRITY THREAT MITIGATION EXAMPLES

Pipeline integrity threat	Examples of threat mitigation measures [1] *Effectiveness changes depending on the pipe characteristics and conditions*
External corrosion	• Pipe cleaning and recoating • Cathodic protection (e.g., sacrificial anodes, impressed current) effective for certain coatings • Coating reinforcement according to pipeline loading (e.g., landslide, water crossings, subsidence) • Proper training of coating installers, including field coating applications at joints
Internal corrosion	• Biocide • Inhibitor • Internal cleaning (e.g., pigs, fluid flushing, chemical)
Stress corrosion cracking—SCC (i.e., environmentally assisted cracking)	• Spike testing (e.g., crack blunting) • Stress reduction (e.g., growth reduction)
Manufacturing-related (e.g., seam weld and pipe body)	• Operational pressure cycling reduction • Pressure testing with higher stress levels (e.g., longer remaining life)
Welding/fabrication related (e.g., girth weld, wrinkle bend, threads)	• Axial and bending stress reduction (e.g., landslide mitigation) • Ultrasonic NDE (e.g., shear-wave, phased array) specified for both volumetric and planar anomalies (e.g., reduce/repairs defects)
Mechanical damage (e.g., immediate/delayed failure, vandalism/theft)	• Concrete slabs or steel plates (e.g., protect pipe from external impact) • Depth of cover greater than area activity depths (e.g., farming, construction) • Fines for recurrent damage or unauthorized excavation activity • Actions according to the law for RoW unauthorized activities, vandalism, or theft
Incorrect operations (e.g., operational, workplace and human factors)	• Reduction of procedure task complexity balancing workload: *operational factor* • Adequate consoles and SCADA (e.g., information content, self-explanatory presentation and access): *workplace factor* • Controller's stress management: *human factor*
Weather-related and outside force (e.g., hydro-geotechnical, cold weather, lightning, Heavy rains or floods, Earth movements)	• Casing or bridging in-service pipelines (e.g., vehicular) • Pipeline stress relief (e.g., excavation/load removal) • French and horizontal drainage (e.g., landslides) • Wall thickness increase (e.g., subsidence) • Pipeline re-route (e.g., water crossing migration)
Pipeline equipment (e.g., valves, seals)	• Valve maintenance (e.g., grease, seals) • Tubing repairs (e.g., melting or freezing)

Note: Pressure reduction and specific repair methods such as recoating, buffing (shallow anomalies) and grinding, pipe replacement and sleeves (e.g., composite, fiberglass, steel) varies depending on the actual conditions. Please refer to CSA-Z662, ASME B31.8S and API 1160.

Example: Mitigation measures for each of the type of consequences:

TABLE 11.6 PIPELINE INTEGRITY CONSEQUENCE MITIGATION EXAMPLES

Integrity-related consequence	Examples of consequence mitigation measures *Effectiveness changes depending on the pipe characteristics and conditions*
Safety and health (e.g., employees, landowners in proximity, and public in general)	• Timely and effective emergency response ensuring synergy and coordination from all responders • Landowners' awareness of emergency procedures
Environmental (e.g., air, land and water including habitat—flora and fauna)	• Determining spill pathways and employing mitigation measures (spill response) • Adequate clean-up equipment and qualified personnel availability • Cooperation agreements for timely response and regular drills including equipment and materials
Regulatory (e.g., penalties, fines and enforcement)	• Proactive and collaborative attitude towards stakeholders (e.g., regulators, responders, community) • Transparency (e.g., self-disclosure compliance)
Financial (e.g., repair and clean-up cost, insurance premium or direct cost)	• Service rate agreements with regional contractors • Clean-up insurance (e.g., self or external)

11.3 PREVENTION MEASURES PER INTEGRITY THREAT

There are two distinct periods in a pipeline's life cycle where prevention of unplanned events can take place. The most cost effective area to prevent unplanned events is in the design phase of a pipeline. The second period is during the operation of the pipeline. Prevention can take many forms during operation. Prevention includes how the pipeline is operated and maintained. It can also include how the pipe is assessed to ensure integrity and reliability.

As discussed in the introduction—threats to pipelines are grouped as (1) time dependent, (2) stable, and (3) time independent. For time dependent threats, preventative measures can be put in place to prevent initiation and growth over time. Stable threats are introduced during the manufacturing and construction phases of the pipeline life cycle and do not typically have a predictable time to failure. Prevention in of stable threats takes place in the initiation and degradation of the threat. Time independent threats have no time scale and can occur at any time. The goal of prevention of time independent threats is to have sufficient barriers to keep them from happening.

Section 10 of API Recommended Practice 1160 Managing System Integrity for Hazardous Liquid Pipelines (API 1160) is Preventive and Mitigative Measures to Assure Pipeline Integrity. This document provides the reader with a resource for things to consider in their specific application.

Example: Monitoring measures for each of the integrity threats

TABLE 11.7 PIPELINE INTEGRITY THREAT MONITORING EXAMPLES

Integrity threat	Examples of monitoring measures *Effectiveness changes depending on the pipe characteristics and conditions*
External corrosion	• Cathodic protection coupons (e.g., monitoring the environment not corrosion on the pipe) • ILI multi-run comparison (i.e., actual monitoring of corrosion via detection and sizing of corrosion)
Internal corrosion	• Corrosion (environment) growth coupons (e.g., weight) • Cleaning pig residue (e.g., chemical analysis)
Stress corrosion cracking—SCC (i.e., environmentally assisted cracking)	• In-Line Inspection (e.g., new or larger SCC) • Cathodic protection effectiveness • Stress level changes (i.e., decrease)
Manufacturing-related (e.g., seam weld and pipe body)	• Crack size changes (e.g., field monitoring) • In-Line Inspection multi-run comparison (e.g., growth)
Welding/fabrication related (e.g., girth weld, wrinkle bend, threads)	• NDE field investigation • Customized ILI (e.g., transverse Ultrasonic)
Mechanical damage (e.g., immediate/delayed failure, vandalism/theft)	• Right-of-Way surveillance (e.g., unauthorized activities) • New geometry anomalies reported by ILI
Incorrect operations (e.g., operational, workplace and human factors)	• Upset conditions (e.g., pressure, product) • Situation awareness indicators (e.g., missed events)
Weather-related and outside force (e.g., hydro-geotechnical, cold weather, lightning, Heavy rains or floods, Earth movements)	• Flooding effects (e.g., scour and bank erosion) • ILI strain and pipe geometry/damage
Pipeline equipment (e.g., valves, seals)	• Maintenance frequency (e.g., valve) • Hours of pump seal usage (e.g., failure frequency)

Example: Monitoring measures for each of the consequences

TABLE 11.8 PIPELINE INTEGRITY CONSEQUENCE MONITORING EXAMPLES

Integrity-related consequence	Examples of consequence monitoring
Safety and Health (e.g., employees, landowners in proximity, and public in general)	• Population growth • Right-of-Way encroachment
Environmental (e.g., air, land and water including habitat—flora and fauna)	• Air and water quality • Species inventory
Regulatory (e.g., penalties, fines, and enforcement)	• Integrity regulatory changes

11.3.1 External Corrosion

11.3.1.1 Preventing Initiation of External Corrosion   There are three main barriers to the occurrence of external corrosion: (1) paint/coating, (2) cathodic protection, and (3) wall thickness.

Paint or coating that is suitable to the environment and operating conditions of the pipeline is a robust barrier to the threat of external corrosion. An excellent coating system can be weakened by poor application. Quality application of the coating system can be assured by having a well-defined inspection test plan (ITP) prior to application. ITPs take the applicator step by step through the process of applying coatings and the acceptance criteria of that step. The most important elements of the ITP are the blast media, chloride content before blasting and before applying a coat, anchor profile, dry film thickness, and inspection of the final product. You only get what you inspect in a coating program. Hiring a skilled coating inspector that has experience with the coating that is being applied can help ensure a proper coating application. Figure 11.2 shows a completed coating system during construction.

Cathodic protection is the second most important barrier in preventing external corrosion. For corrosion to occur there needs to be an anode, a cathode, an electrolyte, and a metallic path. Cathodic protection ensures that the pipe is the cathode and is protected by the sacrificial anode through electrical continuity. Cathodic protection prevents corrosion from initiating when the coating system has a holiday or a disbonded area. Sometimes it is possible to remove the electrolyte from the equation to prevent corrosion but this can be unpredictable.

Cathodic protection and the coating system of the pipeline are intertwined. No coating system is perfect. Sometimes the coating system is excellent, but the coating system was poorly applied or it degraded over time. Cathodic protection protects inadequacies of the coating system. When selecting a coating system, it is important to specify a system that allows the cathodic protection to effectively reach the pipe surface when it fails (e.g., disbondment or holiday).

Example: Dielectric tape coatings can disbond and not allow cathodic protection to be effective. It is also difficult to monitor the cathodic protection system with indirect assessment methods in pipelines that have dielectric tape.

If a pipeline is designed for performance in a corrosive environment, a thicker wall thickness could be specified. The thicker wall thickness would allow some external corrosion to occur and not adversely affect the operation of the pipeline. Thicker wall thickness pipe is more expensive and still requires quality coating and coating application with effective cathodic protection to prevent the initiation of external corrosion.

When excavations are performed on a pipeline and the coating is removed to perform examination of the external or internal surface of the pipe, it is important to perform the reapplication of the coating to prevent new initiation of corrosion at the same location.

FIG. 11.2   COMPLETED FIELD JOINT COATING FOLLOWING COMPLETION OF A GIRTH WELD
(*Source:* Gus Gonzalez-Franchi)

NACE RP0105 provides guidance on how to apply liquid epoxy for repair of existing pipelines. The surface prep of the steel surface is the same as any other coating application. Where the repair coating meets the existing coating, NACE recommends that the edges of the existing coating be roughened, but not removed, and an overlap of at least four inches of repair coating be applied.

11.31.2 Preventing Growth of External Corrosion Once external corrosion has initiated, there are only a few options for preventing growth. Effective cathodic protection is the best method to prevent growth of external corrosion. Prevention in this case works hand in hand with monitoring. Above ground cathodic protection surveys can be performed to locate areas of poor performing coating and the cathodic protection can be adjusted to prevent further growth.

11.3.2 Internal Corrosion

11.3.2.1 Preventing Initiation of Internal Corrosion Prevention of internal corrosion in the design phase takes a number of forms. If a corrosive transportation product is expected, then a corrosion allowance can be added to the pipe to have thicker pipe to gain a longer life. It is also possible to install protective internal coatings and liners. The drawbacks to internal coatings and liners are that it limits the assessment options of the pipeline in operation. In the event the coating or liner fails, there may not be much that an operator can do to assess the situation.

When negotiating the transportation contract with the shippers on a pipeline system, it may be possible to have rejection limits on the quality of the product that can be shipped in the pipeline. By controlling the quality and corrosivity of the product being transported, the likelihood of internal corrosion can be decreased.

> *Example:* Gas transportation pipelines can have contract limits on the moisture content of the gas stream that enters the pipeline. If the gas is sufficiently dry and is not expected to condense in the pipeline, then the electrolyte of the corrosion cell has been removed.

11.3.2.2 Preventing Growth of Internal Corrosion Biocides, inhibitors and maintenance pigging have roles in both prevention of initiation and growth of internal corrosion.

Biocides target microbiological activity. Biocides can target the specific species of bacteria that is present and can kill the bacteria in a number of ways. Biocides can treat product before it enters the pipeline or once bacteria has entered the pipeline. It is always best to prevent the bacteria from entering the pipeline. Testing of the product to ensure the effectiveness of the biocide is important. NACE Standard TM0194 *Field Monitoring of Bacterial Growth in Oil and Gas Systems* is a good resource for guidance.

Inhibitors function by creating a protective barrier between the internal steel surface of the pipeline and the product being transported. Inhibitors are specific to the type of corrosion that is being prevented and the environment that is occurring inside the pipeline. If the inhibitor is not selected appropriately, then the inhibitor can actually make the problem worse. It is important to test inhibitors both in the field and in the laboratory for their effectiveness.

Maintenance pigging helps prevent initiation and growth of internal corrosion in a number of ways. Maintenance pigging ensures that the corrosive elements of the product stream are not allowed to settle in low areas of the pipeline and are swept along

to the end of the line. When corrosion cells are occurring, aggressive brush cleaning pigs can be used to remove the environment from pits and allow the inhibitor or biocide to reach the pipe surface. Maintenance pigging can insure biocides and inhibitors are as effective as possible. It can be detrimental in some cases because it can move the environment to a location where it did not exist previously and make ICDA models inaccurate.

11.3.3 Stress Corrosion Cracking (SCC)

11.3.3.1 Preventing Initiation of SCC SCC requires a susceptible material, an environment, and stress to form. Pipeline steels are susceptible materials so SCC prevention is performed through controlling the environmental and operating pressure exposure to the pipeline. Proper selection and application of coatings can prevent an environment susceptible to SCC from forming. Controlling operating pressure and pressure cycling can perform to prevent stress from contributing.

The Canadian Energy Pipeline Association (CEPA) published its second edition of the Stress Corrosion Cracking Recommended Practices in 2007. In this document, SCC prevention is discussed through coating and cathodic protection performance. CEPA's SCC document focuses on the Near-Neutral pH form of SCC but the coating discussion is applicable to both mechanisms. CEPA recommends that coatings should have the following characteristics:

- Adhesion/Resistance to Disbonding
- Low Water Permeability
- Effective Electrical Insulator
- Abrasion and Impact Resistance
- Temperature Effects/Sufficiently Ductile
- Resistance to Degradation
- Retention of Mechanical/Physical Properties
- Non-Shielding to Cathodic Protection if Disbonded

These characteristics combine to form CEPA's three requirements for a coating to prevent SCC initiation.

1. Prevent the environment from contacting the steel surface
2. Allow cathodic protection to pass through the coating
3. Surface prep prior to coating should render it less susceptible to SCC

The third requirement is not a property of the coating but is essential for prevention of SCC. SCC can best be prevented through the proper surface preparation of the pipe before applying a coating or coating system that is generally considered to be non-susceptible to SCC. Fusion bonded epoxy (FBE) is not considered susceptible to SCC. During surface preparation before applying the FBE, a slight residual compressive stress is introduced to the exterior surface of the pipe and this prevents SCC initiation.

> *Note on Coatings:* A proper application of a good coating system can prevent two of the three time dependent threats to pipelines. Good surface prep and application reduces the likelihood of coating failure leading to external corrosion and can prevent SCC initiation. Diligence in coating selection and application can have long-term cost and integrity benefits for pipelines.

Coatings work together with cathodic protection to prevent SCC and external corrosion. CEPA states that Near-Neutral pH SCC occurs at cathodic protection potentials that are below criteria (e.g.,

−850 mV or 100 mV shift). This occurs due to poor cathodic protection system performance or the external coating preventing the current from reaching the surface of the pipe (i.e., shielding). Cathodic protection is not as significant of a factor for prevention of High pH SCC.

Pipeline Research Council International, Inc. (PRCI) prepared a document for the United States Department of Transportation called *Development of Guidelines for Identification of SCC Sites and Estimation of Re-inspection Intervals for SCC Direct Assessment.* This document goes into significant detail on the prevention of initiation and growth of both forms of SCC (i.e., Near-Neutral pH SCC and High pH SCC).

PRCI's work showed that tensile stress plays a key factor in the initiation of both Near-Neutral and high pH SCC. For High pH SCC, high operating pressures and pressure cycling are the most important factors in combination with elevated temperature for cracks to initiate. Near-neutral pH SCC typically requires a stress riser to elevate the localized tensile effect of the operating pressure. Near-Neutral pH features are more likely to initiate in the presence of high frequency and large amplitude pressure cycles. Controlling the operation of the pipeline to reduce cycling can prevent initiation of both forms of SCC.

For High pH SCC, temperature is a significant factor in the initiation of SCC features. Some products require elevated temperatures for flowing. For pipelines that do not require elevated temperatures for operation, product can be run through coolers to ensure that the temperature remains below the 100°F (38°C) threshold to prevent High pH SCC.

11.3.3.2 Preventing Growth of SCC Once an SCC feature has initiated, it can continue to grow by an SCC mechanism or it can begin to grow by a fatigue or corrosion fatigue mechanism [1]. In any of these three growth cases, an important preventative measure to crack growth is to reduce the cyclic stress on the feature due to pipeline operation. It is not always possible to reduce a pipeline's pressure cycling. SCC is sometimes best managed with mitigation and monitoring.

11.3.4 Manufacturing-Related: Seam Weld and Pipe

11.3.4.1 Preventing Initiation of Seam Weld and Pipe Defects Manufacturing defects are initiated when pipe is made. These features are typically a function of poor quality control and quality assurance during the pipe making process. Manufacturing threats affect both the longitudinal seam weld and the pipe body. Longitudinal seam weld defects include lack of fusion, hook cracks, inclusions, and others. Pipe body defects include laminations, inclusions, scabs, slivers, etc. These features can be prevented by having well defined quality plans during the manufacturing process.

The American Petroleum Institute (API) publishes a variety of white papers and recommended practices for the oil and gas industry. API Specification 5L is a Specification for Line Pipe manufacturing. This document provides some guidelines for quality control during pipe manufacture.

Transport fatigues is a phenomena where pipe joints can be manufactured and leave a pipe mill without any defects and then develop fatigue cracks while they are being transported from the mill to the construction site. Transport fatigue is most commonly associated with railcar transportation but can also occur during truck shipment. API publishes the *Recommended Practice for Railroad Transportation of Line Pipe* (API 5L1) to prevent this type of fatigue from occurring. According to API 5L1, pipe with a diameter to wall thickness ration (*D/t*) greater than 50 is susceptible

to transport fatigue. Pipe that meet this criteria should follow the instructions in API 5L1 to prevent transport fatigue from occurring. There are similar standards for barge transportation (API 5LW), and truck transportation (API 5LT) of manufactured line pipe.

11.3.4.2 Preventing Growth of Seam Weld and Pipe Defects Not all manufacturing defects are injurious to pipeline integrity. Scabs and slivers on the pipe body are blemishes that can affect the quality of the coating application but are generally not the root cause of a failure. Features like laminations, inclusions, and lack of fusion are considered to be non-injurious if they are not surface breaking. If they are surface breaking, then they should be assessed and treated like a crack. Surface breaking crack-like defects grow through the cyclical operation of pipelines. If the features cannot be mitigated, then care should be taken to ensure that the pipeline is operated in such a way as to not allow them to grow in service.

11.3.5 Welding/Fabrication Related

11.3.5.1 Preventing Initiation of Welding/Fabrication Defects Similar to manufacturing defects, welding/fabrication defects are initiated when the welding process is performed. These welds are typically field tie-in welds and girth welds performed during pipeline construction. These features can be prevented through proper weld procedures and welder qualifications. API Recommended Practice 1104 *Welding of Pipelines and Related Facilities* (API 1104) provides extensive information on what constitutes a proper weld procedure and how welders need to be qualified. The ASME B31 code series also provides welding guidance for specific types of oil and gas assets. Figure 11.3 shows a girth weld being performed during pipeline construction.

Wrinkles and buckles can be introduced into the pipeline when bends are made in the field. Wrinkles and buckles are both integrity threats and can make it difficult to inspect the pipeline using ILI surveys. Crews that are performing field bends should be properly trained to a qualified procedure before they are allowed to construct a bend. Proper procedures and training can prevent these threats.

11.3.5.2 Preventing Growth of Weld/Fabrication Defects Weld/fabrication defects would not be expected to grow in service unless another force acted upon the weld. The hoop

FIG. 11.3 WELDING OF A GIRTH WELD DURING PIPELINE CONSTRUCTION (*Source:* Gus Gonzalez-Franchi)

stress of the pipeline operation does not affect a girth weld as if it were a longitudinal seam weld. Stress can be applied to girth welds through subsidence of the weld due to large scale (e.g. whole hillside movement) or small scale (e.g., washout of creek bank) ground movement. Small scale ground movements can be prevented by ensuring that the pipeline is supported properly when laid in the ditch and sufficient compaction is provided to the surrounding soil. Large scale ground movements will be covered later in this chapter when landslides are discussed.

11.3.6 Preventing Equipment Failure

Equipment failures can be prevented in the design phase and through maintenance of the equipment during the piece of equipment's lifecycle. Equipment should be designed to withstand the operating ranges that it is expected to survive during the lifecycle. Operating limits include, but are not limited to, pressure ranges, temperature ranges, corrosivity of the operating environment (e.g., ambient temperature and corrosivity). When a well designed and built piece of equipment is installed, then a maintenance plan needs to be developed to ensure that it is reliably maintained. Maintenance of equipment is not typically in the purview of an integrity group, but it is essential that it be performed to ensure the pipeline operates to the stated capacity.

11.3.7 Mechanical Damage: First/Second/Third Parties, Delayed Failure, Vandalism/Theft

11.3.7.1 Preventing Initiation of Mechanical Damage Mechanical damage caused by first, second and third parties can be prevented by methods such as:

1. Damage prevention program,
2. Public awareness program, and
3. Pipeline mechanical design (e.g., wall thickness, material toughness)
4. Right-of-way routing
5. Depth of Cover

For complete damage prevention, detected party activity must have an intervention to ensure that the activity is necessary and that it is performed to prevent damage to the pipeline. Damage prevention programs for first parties (i.e., company) and second parties (i.e., company contractors) start with training, communicating, and implementing company procedures. Early detection techniques support the damage prevention program, which is discussed in the Monitoring section.

Public awareness involves communication with external parties to prevent second and third party damage. Communication with the public and company contractors about the location of below ground assets (e.g., line markers) and one-call programs (e.g., 811 in the US) for locating assets before excavations can go a long way in preventing second and third party damage. Communication with the public is not limited to people that are not in the oil and gas industry. Communication should be performed with all landowners in the right-of-way and any equipment owner or operator that has the potential to harm the pipe. Mailings, phone calls, and face to face conversations all play a role in public awareness.

Design of pipelines with sufficient depth of cover to reduce the likelihood of contact with small machines is an effective preventative barrier. Monitoring such as aerial patrol and fiber optics with signal to the control center are effective barriers.

One-call systems can identify potential areas of damage before they happen. One-call systems rely on the diligence of the person performing the excavation to be an effective barrier to damage. Furthermore, API 1160 provides guidance with respect to the following preventative strategies:

• Pipeline Mapping
• One-Call systems
• Locating and Marking
• Communication with an Excavator and Monitoring of Excavation
• Public Awareness
• Right-of-Way Maintenance and Surveillance
• Permanent Markers, Warning Techniques, and Physical Barriers

API Standard 1166 on *Excavation Monitoring and Observations* can also be a good source of information to the user during excavation.

Concrete slabs over pipelines can serve as a physical barrier deters first/second/third party damage. In locations where there is activity above the pipeline and depth of cover cannot be increased, a concrete slab can be placed over the pipeline to ensure that the load of overhead traffic is spread over a larger area. Warning tape with distinctive colors can also be barrier on top of pipelines to give identification that a below ground structure is present.

Preventing the installation of illegal taps or accessories for unauthorized extraction of pipeline fluids may require further investigation of the socio-economic and legal framework of the regions along the pipeline. Even though the immediate motivation may be clear, minimizing the root-causes for the theft act as well as communicating and enforcing the legal consequence may help preventing the installation of illegal taps.

11.3.7.2 Preventing Growth of Mechanical Damage Some third party damage fails as soon as the contact is made with the pipe. In other cases third party damage can develop cracks over time and grow to failure through crack growth mechanisms like SCC and fatigue. Identification, mitigation, and monitoring are important factors in preventing failures of mechanical damage that do not fail at the time of contact and then fail in service.

11.3.8 Preventing Incorrect Operation

Pipelines and pipeline networks are operated by pipeline personnel called controllers that operate supervisory control and data acquisition (SCADA) systems. SCADA systems are typically in controlled environments called control centers or control rooms. Incorrect operation can occur both in the control center and in the field. Typically, incorrect operation has the same root cause of poor training of the individual that performed an incorrect task. Control center staff needs to be trained on how the actions they take related to increasing and decreasing set points in the system affect the integrity of the pipelines. Field staff needs to be trained on the details of the facilities that they service. Understanding the mechanisms of how pressure control valves work on the pipeline network is essential for all staff that could use these valves.

In some cases, control room employees are limited by the display that they see and how valves are controlled. Automated systems aid in smooth operation of pipelines, but are not a substitute for well-trained employees. API Recommended Practice 1168 Pipeline Control Room Management provides guidance on:

• Personnel roles, authorities, and responsibilities
• Guidelines for shift turnover

- Provide adequate information
- Fatigue mitigation
- Change management
- Training
- Operating experience
- Workload of pipeline controllers

Incorrect operation is preventable through correct design and maintenance of the operational system of the pipeline and training of control center and field staff. Proper identification and training procedures for staff that have access to pressure control devices is essential for preventing incorrect operation.

11.3.9 Preventing Weather-Related and Outside Force Effects

There are not many preventative measures for the threat of weather-related and outside forces. Pipeline design to reduce the likelihood of weather-related and outside forces can be done but the focus of this threat is on mitigation and monitoring. Pipelines can be designed for sufficient depth of cover to ensure that they are protected from freezing environments and washout at river crossings. Erosion of the bank at the edges of waterways can be prevented with sufficient depth of cover and monitoring of that depth of cover. During operation rivers can freeze on the surface and have accelerated flow underneath the ice. This increased flow can wash away the cover of the pipeline and prevent access in the event of a release. Proper depth in the middle of a river can prevent this phenomenon.

Design can also identify areas of active or potential mining activity in the pipeline right-of-way. As mines are depleted and abandoned, there is the potential of subsidence and cave-in of the mine. By avoiding these areas when selecting right-of-way, this threat can be prevented.

Lightning can strike operating rectifiers for cathodic protection systems. Proper rectifier design would have a grounding rod and a lightning arrestor. These two barriers prevent lightning strikes from having larger consequences when they happen but do not prevent lightning strikes.

API 1160 provides guidance for considerations in preventing and mitigating weather and outside force threats.

11.4 MITIGATION MEASURES PER INTEGRITY THREAT

In the introduction to this chapter we discussed that mitigation could either (1) remove a threat from a pipeline, or (2) reduce the likelihood of reoccurrence. Our discussion in this section will focus on how to mitigate as it related to risk (i.e., likelihood of failure and consequence of failure).

There are many industry resources that provide guidance on mitigation techniques that remove threats or repair the threats. Some sources of additional information on repair include:

- Local regulatory body (i.e., PHMSA, CSA)
- Company policies and procedures
- API 1104—*Welding of Pipelines and Related Facilities*
- API 2200—*Repairing of Hazardous Liquid Pipelines*
- ASME B31.4—*Pipeline Transportation Systems for Liquids and Slurries (Chapter VII)*
- ASME B31G—*Manual for Determining the Remaining Strength of Corroded Pipelines*

- ASME PCC-2—*Repair of Pressure Equipment and Piping*
- ASME Section IX—*Welding and Brazing Qualifications*
- PRCI—*Updated Pipeline Repair Manual*

11.4.1 Mitigating External Corrosion

External corrosion is easiest to prevent through proper coatings and designs of a cathodic protection system. Mitigation of external corrosion could involve coating rehabilitation projects if the coating system that was originally installed was significantly deteriorated in isolated sections. Rehabilitation of external coating will help get better performance out of the cathodic protection system as well as remove the threat of external corrosion in that area assuming the new coating was properly applied.

Coating rehabilitation could include changing of the type of coating that is reinstalled. When a river crossing shows signs of significant external corrosion, it may not be practical to perform an excavation. A directional drill could be used to install a new line at the crossing to avoid the safety and environmental risks of performing excavations at water crossings. When the coating is selected for the directional drill, it should be a two layer coating system. The first layer would be the primary barrier to long-term corrosion protection and the second layer would be an abrasion resistant outer coating to protect the primary layer as the pipe is pulled through during installation.

Indirect assessments discussed in the Direct Assessment chapter can be used to determine the need for mitigation. These above ground survey methodologies focus on the review of the performance of the cathodic protection system. Mitigation of external corrosion involving the cathodic protection could include installing a new rectifier and anode ground bed in a location with low cathodic protection potentials. Mitigation could also include turning up the output of the rectifier in a low potential section to ensure that the cathodic protection potentials are meeting the criteria.

Example: MFL and CPCM ILI surveys can be performed to assess where external corrosion is forming on a pipeline and what the cathodic protection potentials are in that area. If there was a density of external metal loss features reported in a joint of pipe and it corresponded to a high current density reported by the CPCM tool, this would be a prime location for coating rehabilitation. Even if the metal loss features are not significant, the location is consuming cathodic protection current and could eventually cause a problem over a wider area.

Alternating Current (AC) can cause corrosion on pipelines in the form of interference. Cathodic protection systems operated on Direct Current (DC) and any AC current is from foreign sources (e.g., electric power lines and other cathodic protection systems.). NACE SP0177 *Mitigation of Alternating Current and Lightning Effects on Metallic Structures and Corrosion Control Systems* provides guidance on how to mitigate AC induced corrosion. Some of the most common ways to manage the risk of AC induced corrosion are:

- Bonding to existing structures to act as a current drain
- Anodes installed specifically for AC current drain
- Zinc ribbon installed with the affected pipeline as a sacrificial anode for the AC current
- Installation of isolating flanges
- Installations of decouplers where continuous AC current is expected (See Figure 11.4 below)

FIG. 11.4 EXAMPLE OF AC DE-COUPLER INSTALLATION (*Source:* TRC Solutions)

11.4.2 Mitigating Internal Corrosion

Mitigation and prevention work together to reduce the likelihood of internal corrosion. The methodologies of using inhibitors, biocides, and maintenance pigging can prevent internal corrosion from occurring and they can also be used to mitigate internal corrosion from growing. To effectively mitigate internal corrosion, the user must know something about the type of internal corrosion that is occurring. Internal corrosion that is bacterial in nature can be mitigated with biocides. Inhibitors can be used to protect the pipe from the corrosive environment. Maintenance pigging allows biocides and inhibitors to function as effectively as possible. Inhibitors are very specific, care should be taken that the correct inhibitor for the application is used.

Example: During some refining process, there is a cycling of feedstocks. When feedstocks change, the contents of the export product can change (e.g., H_2S or CO_2) and create a more corrosive environment. If the pipeline is not designed to take on the changing stream, it may be necessary for the operator to reject the stream if it cannot be otherwise treated.

11.4.3 Mitigating Stress Corrosion Cracking

Mitigation by removal of SCC is the best method for SCC management. If SCC is widespread on a pipeline, or it cannot be managed with an excavation program, it may be necessary to perform other mitigation methods. One option for mitigation includes a spike hydrotest. This type of test is discussed in the hydrotesting chapter. The object of a spike hydrotest is to arrest SCC crack growth by blunting the tip of the SCC cracks. Care should be taken in designing a spike test because there could be unintended consequences. These consequences could include growth of existing crack-like features or yielding of joints of pipe. CEPA provides additional considerations in their SCC document for hydrotesting to mitigate SCC.

PRCI's work showed that both forms of SCC that are present in pipelines grow through pressure cycling during line operation. If possible, operating procedures could be used to reduce the number of cycles that occur on the line and reduce the likelihood of further growth and new initiation of the SCC features. Severity of pressure cycling can also be reduced through a pressure reduction. If the maximum pressure on the lien is lower than the worst case pressure cycle on the pipeline is lower. A lower operating pressure can also provide a safety factor for pipelines with a large number of SCC features.

A number of industry documents provide guidance on appropriate repair methodologies for SCC. CEPA provides guidance on how to perform sequential grinding on the pipeline to mitigate SCC features in Section 9.4.2. Local regulatory practices should also be consulted when performing grinding on pipelines. The deeper the grinding performed, the lower the pressure that is allowed in the pipe.

11.4.4 Mitigating Manufacturing-Related: Seam Weld and Pipe Defects

Manufacturing related defects can be mitigated with quality control practices when the pipe is manufactured. Quality control at this stage would also be considered prevention. Specific tests to mitigate the risk of manufacturing related defects is hydrotesting the pipe sections to higher pressures both in the mill and during the post construction hydrotest. By testing the pipe to higher pressures, the size of a potential remaining flow following the test is lower. Similar to the hydrotest for SCC, care should be taken at the higher pressure to not yield the pipe. Growth of manufacturing-related defects can be mitigated through controlling the pressure cycling of the pipeline.

Grinding has proven to be effective in mitigating crack-like defects from pipelines. Grinding is an approved method of repair in various industry codes and government regulations. When grinding is to be performed, users are encouraged to have a plan for grinding prior to performing the action. In some regulatory jurisdictions, having a plan for grinding is a mandatory requirement. The PRCI Updated Pipeline Repair Manual allows for cracks to be ground out up to 40% without any additional reinforcement as long as a safe operating pressure is maintained; however, the CSA-Z662-15 standard allows for grinding provided that an engineering assessment is conducted ensuring safe operating pressures are maintained at all times.

The user shall exercise caution when using grinding as a mitigation method as some coincidental anomalies may not be detected by the Non-Destruction Examination causing to interconnect and fail putting at risk the safety of the people and the environment. Some longitudinal seam welds (i.e., low toughness ERW, and flash welded pipe) have additional limitations for grinding.

11.4.5 Mitigating Welding/Fabrication Related Defects

Prevention and monitoring is the most productive way to manage welding/fabrication related features. Removal or strengthening these locations are the only mitigation methods that is applicable to this threat. Preventing the initiation of these features during construction and ensuring that outside forces do not weaken the weld should be the goal of the integrity management program for these features.

Older vintage pipelines used older standards considered acceptable at the time of construction for assessing girth welds, but failures have still occurred. The use of X-Ray techniques on girth weld have allowed for evaluating volumetric anomalies and establishing repair/mitigative actions during construction and in-service pipeline cutouts. However, the presence of planar anomalies such as cracking from either construction or in-service growth has required revisiting the mitigation of girth weld anomalies to prevent in-service failures.

The development of inspection protocols for assessing girth welds during pipeline service (i.e., Non-Destructive Examination— NDE-) may contribute in mitigating (i.e., repair) anomalies

becoming defects. They may be needed due to historical evidence or susceptibility to:

1. Cracking on girth weld (e.g., construction quality),
2. Rapid or frequent hydraulic transients (e.g., Operational closure of valves, abnormal/upset conditions), and/or
3. Geohazards affecting the pipeline (e.g., Right-of-Way lateral or rotational displacement)

Example: Recommended Elements for a Girth Weld Assessment Protocol of In-Service Pipelines

- Pipeline characteristics and girth weld material properties
- NDE anomaly information (e.g., depth, length, orientation and coincidental anomalies)
- Depth criterion for leak-dependent (e.g., maximum depth allowed)
- Length criterion and anomy type for girth weld acceptability (e.g., ASME B31.3-2014)
- Rationale for whether girth weld coincidental anomalies are acceptable (e.g., ability to size and assess)
- Differentiation between brittle fracture and plastic failure assessment cases
- Evaluation of potential for girth weld anomaly to grow over time
- Acceptance criteria: safety factor, remaining life and combined effective stress (e.g., internal pressure + lateral displacement)
- Acceptable repair method (i.e., recoating, grinding, type B sleeve and cut-out)

Girth weld defects cannot be permanently repaired by type A or reinforcement sleeves due to their limitation of not containing a potential release or defect rupture.

11.4.6 Mitigating Equipment Failure

All equipment requires proper maintenance throughout its lifecycle. Maintenance mitigates failure by lowering the likelihood of equipment failure. Valve functionality is reliable when they are maintained through functionality tests. Operating conditions is an important factor for equipment.

In locations where the pipe could potentially freeze, smaller tubing is most susceptible to the weather conditions. It may be possible to place a building over pipeline facilities that operate in cold weather environments. This can reduce exposure of the piping and components to the severe weather elements. Insulation could also be installed to prevent freezing. When insulation is used, a maintenance program should be in place to ensure that corrosion does not adversely affect the operation of the component.

11.4.7 Mitigating Mechanical Damage

Mitigation measures for mechanical damage (e.g., dents) start with remediation techniques to recover and improve the integrity condition of a pipe. Damage already occurred to the pipeline need to be timely identified and assessed as they may suddenly fail some time after the time of the damage (e.g., delayed failure).

The presence of stress concentrators (e.g., scratches, gouges) and cold working (i.e., strain hardening) caused by excavator teeth would reduce the time-to-failure of a dent. Unless of a pipe cut-out, removal of stress risers and minimizing the effects of rebounding and re-rounding of the dents (i.e., change of dent diameter contour) become the following priority in remediating dents. The use of compression and epoxy-filled sleeves has provided some alternatives for in-service repairs if API 579 level 2 fatigue life estimation is acceptable under the expected pipeline pressure cycling. However, estimating the time of the dent initiation is always challenging.

Other mitigation measures such as concrete slabs or steel grating in high traffic construction areas or highly congested locations provide a barrier to any excavation in the area. Hence, excavations would hit the secondary structure first and thus protect the pipeline.

Furthermore, Depth of Cover (DoC) surveys and follow up remediation can reduce the likelihood of mechanical damage. There have been cases in industry where repeated plowing in the same area over time has removed the depth of cover. When the soil cover got thin enough, contact was made with the pipeline. It cannot be assumed that the depth of cover at the time of install is still intact. Surveys must be performed to confirm that the depth of cover is maintained.

Local law and other forms of legal action can reduce the likelihood of mechanical damage as well as theft. For mechanical damage, enforced *Call before you Dig* and *Click before you Dig* programs and fines for repeat offenders can be effective deterrents for excavating companies. For theft, penalties for offenders are typically included in the law.

11.4.8   Mitigating Incorrect Operation

Mitigation and prevention are complementary activities to reduce the likelihood of incorrect operation. When controllers have reasonable expectations and workloads, then they are less likely to perform incorrect actions. When the task to be performed is clear, operators are also less likely to make mistakes. Likelihood can be reduced with proper training, procedures, and clear user interface with the pipeline system.

11.4.9   Mitigating Weather-Related and Outside Force Effects

Depending on the type of outside force, there are a number of things that can be done to mitigate the threat if it had not resulted in instantaneous pipeline failure. When a pipeline has been in the ground for a long time, population can encroach upon the right-of-way. Casings can be installed when transportation is required to occur over the pipeline. If casings are installed, care should be taken to ensure that sufficient separation is maintained between the casing and the pipe such that it does not become shorted. When the pipeline and casing are shorted together, the casing receives cathodic protection and the effectiveness of the system is reduced. End seals, or wax infill can prevent this from happening.

Additional drainage around the pipeline can reduce the strain put on the pipeline in the event of flooding. Examples of additional drainage around the pipe are French and horizontal drains. Similarly, removal of the load that is causing the strain on the pipe due to ground movement also mitigates the likelihood of failure due to ground movement. Rerouting of pipelines away from geological movement (e.g., land or waterway) can be costly but it is also effective in mitigating outside force threats.

11.5   MONITORING MEASURES PER INTEGRITY THREAT

All but two, incorrect operation and equipment failure, of the threats we have discussed in this chapter can be detected and monitored using one of the integrity assessment options that have been discussed elsewhere in this book. The design phase of the pipeline is crucial for preventing threats from being introduced. Threats can still be monitored if they are introduced during operation. Monitoring can also detect or predict when a threat has or will occur.

11.5.1   Monitoring of External and Internal Corrosion

MFL and CMFL surveys are the best ILI technologies to monitor external and internal corrosion. Performing run-to-run (i.e., comparing one survey to another) comparisons can give an estimated growth rate to features that are reported. If a high growth rate is determined to exist, that location can be prioritized for remediation. High growth rates may also lead to shortening of the time between integrity assessment methods. Pipelines with coatings that shield cathodic protection either at the girth welds or for the entire pipe length are good candidates for monitoring in this fashion.

In the Direct Assessment chapter we discussed a number of indirect assessment methods for both ECDA and ICDA. For pipelines that are piggable, these assessments serve as monitoring for external and internal corrosion. External corrosion can be monitored with cathodic protection surveys and corrosion coupons. Internal corrosion can be monitored by corrosion coupons and analysis of any product expelled during maintenance pigging.

Example: Internal corrosion coupons must be placed in the water phase of the product to be effective. If the coupon is not in the water phase then the testing is not going to be valid and will give a false sense of security. When designing a corrosion coupon be sure that the rod that is extending the coupon into the pipeline is sufficiently long to have the coupon in contact with the water phase. Using side stream analyzers can reduce this risk. Side stream analyzers take an intermittent flow of product and it sits stagnant for a period of time to allow for saturation of the coupon.

11.5.2   Monitoring of Stress Corrosion Cracking and Manufacturing Related Defects

Crack detection ILI surveys are the best method for monitoring of cracks (i.e., SCC or manufacturing related). When monitoring cracks, the report from the vendor should be reviewed for locations of newly reported cracks and for crack features that increased in size. Most vendors that have crack detection tools report features in ranges of depth. If a feature moves up a range from one survey to the next, this could be an indication that the feature has grown.

Unlike MFL and CMFL surveys, it is difficult to accurately perform a signal comparison to determine if an individual crack has grown. Development of a unity plot to correlate reported feature size with field reported size could be useful in management of cracks.

Monitoring of the cathodic protection system can be helpful in the management of SCC. SCC occurs in specific ranges of cathodic protection levels. Care should be taken to ensure that the cathodic protection levels on the pipeline are more negative than the −850 mV criteria that most pipeline operators aim to achieve.

11.5.3   Monitoring of Manufacturing Related: Seam Weld and Pipe Anomalies

API 1176 Recommended Practice for Assessment and Management of Cracking in Pipelines provides guidance on how to use pressure cycle analysis when monitoring crack-like defects.

API recommends that pressure cycle analysis be performed on a regular bases (i.e., every one, three, or 5 years) to see how pressure cycles have changed over time. If pressure cycles become more aggressive, then the remaining life would be expected to shorten. Reassessment intervals should be determined based on the most conservative analysis performed.

Monitoring pressure data also requires reviewing the pressure changes including shutdowns, spikes, abnormal operating conditions, and exceeding the maximum operating pressure of the pipeline. The operational effects in the integrity of pipeline should be monitored and tracked with key performance indicators (KPI).

11.5.4   Monitoring Welding/Fabrication Related Anomalies

Welding defects related to the girth weld are difficult to monitor other than by direct examination and inspection. If girth welds meet vintage criteria or similar welds have failed in service, it may be necessary to install pumpkin reinforcement sleeves to maintain the integrity of the weld. Traditional ILI methods cannot see defects in girth welds. Locations of high strain reported by strain measurement tools should be cross-referenced with locations of girth welds that may not have had sufficient quality assurance and quality control during construction. These locations should be monitored and mitigated as necessary.

Acceptable "imperfections" from construction may experience growth and failure during in-service in locations where thermal (e.g., temperature) or external force (e.g., displacement) changes may occur. It is important to monitor these locations with nondestructive methods and repair as necessary.

Wrinkles and buckles that were introduced into the pipe that still allow passage of an ILI tool can be monitored with deformation detection surveys. When wrinkles or buckles are identified in a pipeline, they should be mitigated promptly as they are a threat to pipeline integrity.

11.5.5   Monitoring Equipment Failure

Equipment used in the oil and gas industry have design lives. Trending of the equipment used in the pipeline system versus the common performance life of equivalent equipment can be used for monitoring. As equipment nears the end of the expected life, the equipment can be budgeted for replacement if the equipment can no longer be effectively maintained.

11.5.6   Monitoring Mechanical Damage

Deformation surveys can be used to monitor for new excavation damage. New deformations reporting on the top half of the pipe should be prioritized for assessment even if they do not meet any of the regulatory required assessment criteria. Third party damage can fail immediately or they can have delayed failure from cracks that grow over time. It is best to remediate these features as soon as they are reported by the deformation survey.

Monitoring of right-of-way activity can anticipate first/second/third party damage. Early detection can be performed by

- aerial patrol,
- walking/driving the right-of-way,
- satellite monitoring, or other forms of remote monitoring such as
  - Video surveillance
  - Monitoring of acoustic frequencies of activity above the pipeline
  - Fiber optic cable that sends an alert when it is disturbed

Aerial patrols and walking of the pipeline can identify activity over the line that is not included in the one-call. Density of right-of-way activity can be used to identify areas for additional monitoring between deformation ILI surveys.

Figure 11.5 below shows an example of a monitoring activity study over a pipeline. The image on the left is a combined hot spot analysis of third party activity over the pipeline. This shows the density of potentially injurious excavation activity. The image on the right is just the instances of third party activity for the same pipeline.

Fiber optic lines can be used to identify ground movement and excavation activity. One methodology measures vibration around the pipeline. When the pipeline has sufficient vibration activity it will send an alarm and location of the alarm for investigation. The operator can then dispatch a right-of-way patrol to determine if the pipeline has been damaged. The second methodology is a continuous fiber that runs the length of the pipeline. When the fiber is broken, it sends an alarm and location. These are evolving technologies and can be useful for locations with high construction activity.

Monitoring also includes attending excavations that occur close to the operator's pipeline but are not on the pipeline. Representatives of the potentially affected pipeline can watch and supervise excavations close to their pipeline to ensure they are not infringed upon. This individual can also enforce hand digging as they get close to the pipeline to prevent damage.

Monitoring of the installation of illegal taps or accessories for unauthorized extraction of pipeline fluids can be conducted using the following inspection, surveying, operational monitoring, and surveillance technologies.

- Axial or Circumferential MFL and Wall Measurement Ultrasonic in-line inspections have identified some type of illegal taps as a "metal loss" circumference resulting from the perforation or "round and isolated anomalies" located at periodic or aligned distances from each other as well as metal object in close proximity to the pipeline.
- Direct Current Voltage Gradient (DCVG) surveys are able to detect coating fault and buried metal object signals triggered by the installation of illegal taps.
- Operational monitoring with Leak detection systems and volume inventory checks have been able to identify variances in the transportation of the fluids (e.g., diesel, gasoline, jet fuel); however, gaseous fluids are more challenging to be verified.
- Right-Of-Way (ROW) surveillance via ground, aerial and Unmanned Aerial Vehicles (UAV) for detecting ground disturbance or the presence of unauthorized activities. Technology such as thermal infrared (i.e., electromagnetic energy) can be added for surveillance at night.

11.5.7   Monitoring Incorrect Operation

Pressure excursions or unplanned rises in operating pressures can be examples of incorrect operation. These instances can be monitored with the SCADA system. In some regulatory jurisdictions, exceeding the MOP, or a percentage of the MOP, is a reportable event. Pressure excursions can be leading indicators as to whether other integrity threats have become or could become a more significant threat to the pipeline. A number of the threats that we have discussed in this chapter are dependent upon consistent operation of the pipeline from a pressure perspective to avoid initiation and growth.

Incorrect operation can also lead to upset conditions from a product quality perspective. Product quality can lead to the rejection of refined product. Upset conditions can also introduce moisture into dry pipeline systems. Moisture in the pipeline that is not dried

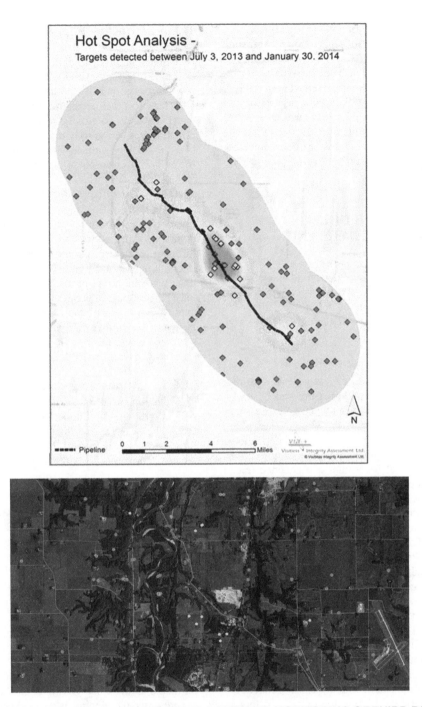

FIG. 11.5   EXAMPLE "HOT SPOT ANALYSIS" THAT COMBINES MONITORING OF THIRD PARTY ACTIVITY
(*Source:* Visitless Integrity Assessment Ltd)

sufficiently quickly can lead to internal corrosion of the pipeline. Monitoring for upset conditions can protect the pipeline's integrity and ensure maximum profit from shipping product.

Example: When refined product such as gasoline and diesel are batched in the same pipeline, there is a volume of mixed product between the two batches. If this mixed area becomes too large, the receiving station has the contractual authority to reject delivery. This can be a significant cost to the shipping refinery.

11.5.8   Monitoring Weather-Related and Outside Forces

Landslides and ground movement can be monitored with some of the same methodologies as third party damage. Aerial patrols are able to see locations where the soil surrounding the pipeline either has moved or shows signs that it could move imminently. Strain detection ILI tools can also show where the pipe has moved from its original position.

InSAR data reviews can monitor for location and magnitude of ground movement. InSAR has the capability to monitor for landsides as well as subsidence (e.g., from mine collapse). LiDAR can be used for details scanning of locations of ground movement.

Cold weather is best to be designed for and not to be monitored. Extreme cold can pose a health and safety risk so ensuring a pipeline and its components can operate in extreme temperatures should be considered in the design phase. Following major storms, an aerial patrol or right-of way survey can identify any locations of damage. Slope inclinometers are used to measure the rate of movement at a given location. Piezometers are able to determine the effect that ground water movement has on the overall pipe movement process.

11.6 SPECIAL MEASURES DRIVEN BY INTEGRITY THREATS AND CONSEQUENCES

Some time special measures are required beyond the regular prevention, mitigation and monitoring. These special measures are intended to minimize both likelihood of failure (i.e., integrity threat) and potential consequences intervening as either temporary or permanent measures.

11.6.1 Pressure Reduction

Pipeline threats that have a failure mode (e.g., large leak, rupture-dependent) that is dependent on operating pressure can be mitigated through a pressure reduction. However, reducing the pressure typically decreases the fluid flow reducing the volume (i.e., consequence extent) of a potential release even for leak-dependent anomalies. Pressure reduction strategies are preferable mitigative options in pipelines with potential for a high consequence (e.g., safety, environment, supply, community, reputation)

Pressure reductions provide a factor of safety between a known pressure where the pipeline did not fail and a future failure pressure after growth. For example, a 20% reduction of the referenced pressure (100%) provides a 1.25 factor of safety (i.e., 1/0.80) over the reduced pressure (80%). Pressure reductions can be temporary or permanent. Temporary reductions could be used either when repairing a pipeline or while gathering more information allowing for an interim safe operation.

The local regulatory context for pressure reduction reporting and duration should be considered when planning a pressure reduction to ensure compliance accounting for the effectiveness of the pressure reduction level [17]. Both forms of corrosion, SCC, manufacturing related defects and cracks that could grow from pressure cycling can be mitigated to some extent with a pressure reduction. Pressure reductions do not remove the pipeline threat, it offers a factor of safety while pursuing a more permanent solution. API 1160 provides guidance on pressure reduction in response to ILI and by for relevant pipeline threats.

11.6.2 Higher Stress Level Hydrostatic Testing

Cracking integrity threats may benefit from a hydrostatic testing at a higher stress level for both removing near-critical anomalies and extending the remaining life via crack tip blunting. Higher stress level hydrostatic testing also reduces the chance of an unintended release or consequence resulting from ILI-non-reported cracking. This temporary measure would increase confidence in the operation and stakeholders, but requiring other monitoring techniques (e.g., ILI re-inspection, pressure cycling and crack growth) to become effective over time.

11.6.3 Pipeline Replacement

Review of the effectiveness of repeated mitigation driven by accelerated anomaly initiation and growth (e.g., SCC in telescopic-designed pipelines coated with tape, insulated pipeline with microbial corrosion or MIC, combined SCC, and seam weld cracking) may lead to pipeline replacement. Other factors such as maintenance capacity to timely repair several pipeline locations, cost effectiveness of multiple repairs over several years and likelihood of experiencing a failure with significant consequences from which the authorities may not allow for operating for a while or anymore.

11.6.4 Decommissioning or Abandonment

When mitigation of pipeline threats is to occur over a large portion of the pipeline or may be too costly, decommissioning or abandonment of the asset is an option for mitigation. The decision to mitigate by decommissioning or abandonment is typically made in combination of the risk of the asset and the business need. If an asset has a high risk profile and a short remaining business life, it may be prudent to decommission the asset. Sometimes even with extensive mitigation, the remaining risk may be too high for the business to continue to operate the asset. There are a number of scenarios where this option may be used and should be considered on a case-by-case basis.

11.6.5 Minimizing Consequence of Releases

Reduction of consequence can be just as valuable in lowering risk as reducing likelihood. If threats occur in high consequence areas or next to known dwellings, it may be prudent to mitigate these areas first to ensure the protection of people, property, and the environment.

Example: ILI surveys are performed to detect strain locations in areas of suspected ground movement. When the report from the ILI vendor is received, the locations of strains should be plotted along the right of way. The same strain value has a different risk if it occurs in the middle of an unused field or if it occurs near a cabin or waterway. It would be prudent to mitigate the higher consequence location before the other location if they both required mitigation.

The consequence of a pipeline failure can be reduced in a number of ways. Early detection and shutdown of the pipeline system upon detection of a release reduces the spill volume. By reducing the spill volume, the effect to health, safety, and environment is mitigated to an extent. Giving control center workers the authority to shut down the pipeline if deemed necessary in addition to all the protective measures is essential to safe pipeline operation. Once a pipeline is shutdown, a detailed review of the pipeline status needs to be performed before the line can be restarted. Industry failures have been exacerbated by a control center not diligently checking why the pipeline was shut down in the first place.

Early detection from means other than the control center can also reduce the consequence of a leak. If a landowner or other third party can identify a leak and know whom to call before the SCADA system recognizes a problem then the pipeline can be shutdown sooner and mitigate the consequence. Ideally, the control center would identify any release first but early detection is key regardless of the source.

Pipeline threats can sometime warrant shutting down of pipelines because they present an imminent safety hazard. Pipeline shutdowns can be costly to the business if they are not necessary. If a threat to a pipeline is considered a temporary condition, then a reduction in flow may reduce the cost of restarting the pipeline once the upset condition has ended. The decision to reduce flow versus shutting down should be weighed carefully and be customized to the threat that is present. High profile industry failures have been escalated due to a failure to fully shutdown the pipeline when it was prudent. The user is cautioned to consult their company's guidelines when weighing a reduced flow or shutdown.

API 1160 Section 10.6 is "Detecting and Minimizing the Consequence of Intended Releases." To reduce the consequence of a leak or a rupture, API states that operator integrity management plans should include:

- Methods to minimize the time required to detect the release
- Methods to minimize the time required to locate a release
- Methods to minimize the volume that is released
- Methods to minimizing the emergency response time
- Methods for protecting the public and limiting effects on the environment

All of the elements in the API document, if fully implemented in a robust integrity management plan would aid in the reduction of the consequence of a release.

11.7   REFERENCES

1. Maier, C.J., Shie, T.M., Beavers, J.A., Vieth, P.H., "Interpretation of External Cracking on Underground Pipelines," Paper No. IPC2006-10176, ASME International, New York, International Pipeline Conference (IPC 2006), Calgary, Alberta, September 2006.

2. Anon., 2012, *Pipeline Transportation Systems for Liquids and Slurries*, ASME B31.4, American Society of Mechanical Engineers, New York, USA.

3. Anon., 2014, Managing System Integrity of Gas Pipelines, ASME B31.8S, American Society of Mechanical Engineers, New York, USA.

4. Anon., 2015, Repair of Pressure Equipment and Piping, ASME PCC-2, American Society of Mechanical Engineers, New York, USA.

5. Anon., 2013, Welding and Brazing Qualifications, ASME Section IX, American Society of Mechanical Engineers, New York, USA.

6. Jaske, C.E., Hart, B.O., Bruce, W.A., 2006, Updated Pipeline Repair Manual, Pipeline Research Council International, Inc., Washington, D.C., USA.

7. King, Fraser, 2010, Development of Guidelines for Identification of SCC Sites and Estimation of Re-inspection Intervals for SCC Direct Assessment, Pipeline Research Council International, Inc., Washington, D.C., USA.

8. Anon., 2007, Stress Corrosion Cracking Recommended Practices—Second Edition, Canadian Energy Pipeline Associate, Calgary, Alberta, Canada.

9. Anon., 2013, Welding of Pipelines and Related Facilities, API Standard 1104, American Petroleum Institute, Washington D.C., USA.

10. Anon., 2009, Railroad Transportation of Line Pipe, API Recommended Practice 5L1, American Petroleum Institute, Washington D.C., USA.

11. Anon., 2013, Truck Transportation of Line Pipe, API Recommended Practice 5LT, American Petroleum Institute, Washington D.C., USA.

12. Anon., 2009, Transportation of Line Pipe on Barges and Marine Vessels, API Recommended Practice 5LW, American Petroleum Institute, Washington D.C., USA.

13. Anon., 2013, Specification for Line Pipe, API 5L, American Petroleum Institute, Washington D.C., USA.

14. Anon., 2013, Managing System Integrity for Hazardous Liquid Pipelines, API Recommended Practice 1160, American Petroleum Institute, Washington D.C., USA.

15. Anon., 2015, Excavation Monitoring and Observations, API Technical Report 1166, American Petroleum Institute, Washington D.C., USA.

16. Anon., 2014, Mitigation of Alternating Current and Lightening Effects on Metallic Structures and Corrosion Control Systems, NACE SP0177, NACE International, Houston, USA.

17. Anon., 2014, Field Monitoring of Bacterial Growth in Oil and Gas Systems, NACE TM0194, NACE International, Houston, USA.

18. Hall, R.J., McMahon, M.C., 1997, Evaluation of the Effectiveness of a 20% Pressure Reduction After a Pipeline Failure, Report No. DTRS56-96-C-0002-001, US Department of Transportation, Washington, D.C., USA.

CHAPTER 12

FITNESS-FOR-SERVICE ASSESSMENTS FOR PIPELINES

12.1 INTRODUCTION

Pipelines are a safe form of transportation, due to the high standards used in their design, construction and operation [1–12]. Inevitably, pipelines will contain some defects, and these defects will require assessment.

A Fitness-For-Service assessment provides the short and long-term integrity mitigation, prevention and monitoring measures as well as recommendation for system modifications to ensure a safe, environmentally-responsible and reliable service during a defined period until the next pipeline integrity assessment. FFS provides a timeframe for which the integrity condition of a pipeline is acceptable provided the FFS proposed measures are found to be effective over time.

For developing a FFS, this chapter has been started with a summary of fracture mechanics methods and theories, and a section on fatigue is also included. Fracture mechanics forms the basis of most of the fitness-for-service methods used in the pipeline industry, and an understanding is essential. The chapter then progresses onto specific methods for assessing specific defects, including corrosion, dents, and cracks. This chapter also covers defect assessment methods factoring the ILI performance. An example of a typical Table of Contents for a FFS assessment report is also included.

As illustrated in Figure 12.1, a management system approach (i.e., plan-do-check-act) is recommended for developing a FFS assessment. The following are some considerations to be made in the preparation of a FFS assessment:

PLANNING (Plan)

- Identify the goals and objectives of the FFS assessment ensuring alignment with the stakeholder expectations (e.g., Society, Regulators, Communities, Business)
- Understand the consequences of 'getting it wrong';
- Define sources of data understanding validity and limitations;
- Select the best assessment methods (i.e., applicability, accuracy) and software tools (do not use "black boxes");

IMPLEMENTATION (Do)

- Use qualified, experienced and competent staff from multiple disciplines;
- Understand and challenge methodology and software results (i.e., model versus reality check) - perform sensitivity analysis;
- Review all input data and assumptions, and document clearly;
- Seek advice when in doubt;

VERIFICATION (Check)

- Use a cold eye review approach (e.g., experienced reviewer not involved in the analysis)
- Perform a 'sense' check on answers ("*does the answer make sense, is it safe, can it be justified to all stakeholders?*");
- Validate the results (e.g., within the project or similar projects)
- Check, check, and check again;

MANAGEMENT REVIEW (Act)

- Were the FFS goals and objectives achieved? Why not? Disclaimer needed?
- Were the assumptions within reason? What would go wrong if they are incorrect?
- What is the impact of the FFS results? Reasonable? Credible?
- Learn lessons, and continuously improve;
- Never stop learning!

12.1.1 Line Pipe

Pipelines are made from steel tubes called 'line pipe'. The line pipe is ordered (usually in lengths of 12 m) according to its mechanical properties:

- strength (yield and ultimate);
- ductility (the ability to deform);
- toughness (the ability to resist the presence of defects such as cracks);

These mechanical properties govern how the line pipe reacts to stress (strength), strain (ductility), and defects (toughness).

Modern line pipe is strong and ductile, Figure 12.2a. Its toughness is measured using a small (cross-section is 10 mm [0.394"] × 10 mm [0.394"]) notched impact specimen called a Charpy specimen. The toughness undergoes a 'transition' from high toughness to low toughness, as the test temperature of the Charpy specimen is reduced, Figure 12.2b. Additionally, at high temperatures, the specimens have 100% ductility (often called 'shear area') on their fracture surfaces, but at low temperatures the specimens have little/no ductility: they exhibit a brittle fracture, Figure 12.2c.

Line pipe must be bought to operate at temperatures where it will be high toughness, and ductile. Line pipe purchased to specifications such as API 5L [13] will meet these toughness and ductility requirements.

12.1.2 What is a 'Defect'?

The word 'defect' implies substandard. Substandard refers to a system that does not comply with its original fabrication standard.

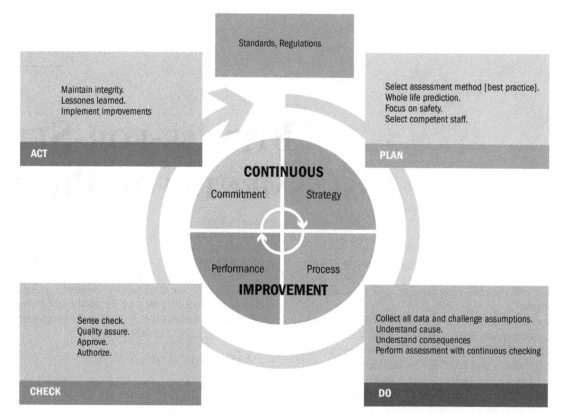

FIG. 12.1   PLAN-DO-CHECK-ACT

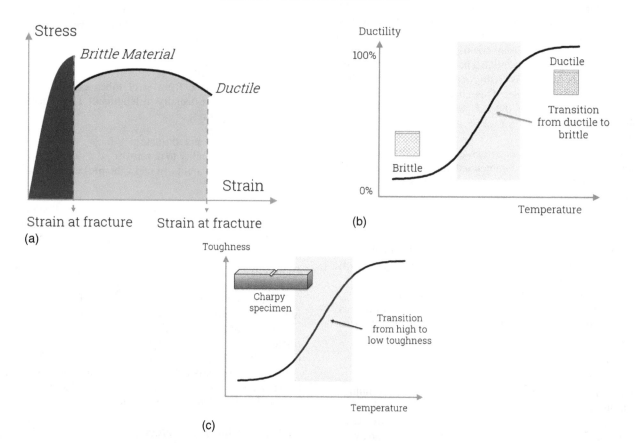

FIG. 12.2   LINE PIPE DUCTILITY AND TOUGHNESS. (A) TYPICAL STRESS AND STRAIN CURVES FOR BRITTLE AND DUCTILE MATERIALS, (B) TYPICAL DUCTILITY CURVE AS A FUNCTION OF TEMPERATURE DEPICTING BRITTLE, TRANSITION AND DUCTILE REGIONS, AND (C) TYPICAL TOUGHNESS CURVE AS A FUNCTION OF TEMPERATURE

Hence, 'defect' does not mean 'failure': it means it is not acceptable to a standard.

Table 12.1 depicts the definition of defect, imperfection, flaw, discontinuity and anomaly from reference [14] gives definitions of defects, and related items.

Pipelines are structures that are usually buried underground, or laid underwater. These environments expose the pipeline to corrosion, and the pipeline can be damaged by earth moving equipment (onshore), or anchors or dropped objects (offshore). Consequently, pipelines will contain defects during their service life.

These defects can be assessed using 'fitness-for-service' methods.

12.1.3 'Fitness-for-Service'

Most structures contain defects, but these defects are limited by quality control levels. These levels can be both arbitrary and conservative, but are essential for the monitoring and maintenance of quality during production [15]. Defects acceptable to quality control levels are allowed to remain in the structure, but defects not acceptable to these levels are not necessarily a threat to the structure's integrity [15].

The 'fitness-for-service' principle is that a structure is considered to be adequate for its purpose, provided the conditions to cause failure are not reached [15]. It involves conducting a quantitative engineering evaluation to demonstrate the structural integrity of a component that may contain a flaw or damage [16]. This evaluation is often called an 'engineering critical assessment' (ECA).

It should be note that the term 'fitness-for-service' when used in pipeline defect assessment, is related to the definitions in engineering standards [e.g., 15, 16]: it does not relate to any legal definition that may imply a warranty.

12.1.4 Pipelines and Fitness-for-Service

The failure of defects in pipelines has been the subject of considerable study over the past 50 years with many publications (e.g., [17–69]), based on full scale tests, theories, and analyses.

These defect assessments can be grouped into two categories: 'generic' methods; and, 'pipeline specific' methods.

12.1.4.1 Generic There are various technical procedures available for assessing the significance of defects in a range of structures; for example, References 15 and 16. These methods use fracture mechanics principles and can be applied to most defects in pipelines.

12.1.4.2 Pipeline-Specific The above generic standards can be overly-conservative when applied to specific structures such as pipelines. Therefore, the pipeline industry has developed its own fitness-for-service methods over the past 50 years; however, it should be noted that these pipeline-specific methods are usually based on experiments, with limited theoretical validation (i.e., they are 'semi-empirical'). This means that the methods may become invalid or unreliable if they are applied outside these empirical limits.

The pipeline industry developed pipeline-specific methods for assessing the most common defects in operating pipelines, for example corrosion [59], and these methods have been widely-used over the past 50 years.

Example: Table of Contents of a Fitness-For-Service Assessment

1. Executive Summary
2. Pipeline System Description
3. Scope and Objectives
4. Regulatory, Industry and Company Framework
5. FFS Assessment Methodology for Pipelines

6. Understanding their Design, Manufacturing, Construction and Commissioning
7. Pipeline In-Service: Past, Present and Future
8. Integrity Criticality (Threat) Assessment
8.1. Threat Inclusion/Exclusion (Identified, Susceptible and No Threats)
8.2. Threat 1 (e.g., Current, estimated future and proposed improvements: integrity Status/Inspections & Testing and Growth, mitigation, prevention and monitoring)
8.3. Threat 2
9. Consequences of Applicable Threats (e.g., Safety, Environmental, Public and Supply)
9.1. Leak Effect
9.2. Rupture Effect
10. Risk Analysis, Evaluation and Ranking
11. Fitness-For-Service Assessment Results
11.1. Short-Term Mitigation and Field Validation
11.2. Long-Term Mitigation and Prevention
11.3. In-Service Monitoring and Program Enhancements
12. Recommended Path Forward
Appendix A: Records
Appendix B: Regulatory and Industry Standards Requirements

TABLE 12.1   DEFINITIONS [14]

Term	Definition
Defect	*"Imperfection of a type or magnitude exceeding acceptable criteria"*
Imperfection	*"Flaw or other discontinuity noted during inspection that may be subject to acceptance criteria during an engineering and inspection analysis"*
Flaw	*"Imperfection that is smaller than the maximum allowable size"*
Discontinuity	*"Interruption of the typical structure of a material, such as a lack of homogeneity in its mechanical, metallurgical, or physical characteristics"*
Anomaly	*"A possible deviation from sound pipe material or weld"*

12.2   FRACTURE MECHANICS

Fitness-for-service methods are based on fracture mechanics. Fracture mechanics provides the scientific understanding of the behavior of defects in structures. It is easy to demonstrate and visualize the deleterious effect of defects on structures [70]; for example, a rubber balloon can be inflated until it bursts. It will not burst at low inflation (low stress), but when the rubber reaches its yield point it will burst. However, the balloon can burst at much lower levels of inflation if it is pricked with a tiny pin. The pin introduces a 'defect', and this tiny defect causes the balloon to burst at a stress level well below its yield strength.

A defect such as a crack, creates two problems in a structure (Figure 12.3):

- it reduces the load-bearing capacity of the structure; and,
- it locally increases stresses (the stresses which the structure would have supported with no crack present, will concentrate and gather around the tip of the crack, if a crack is present).

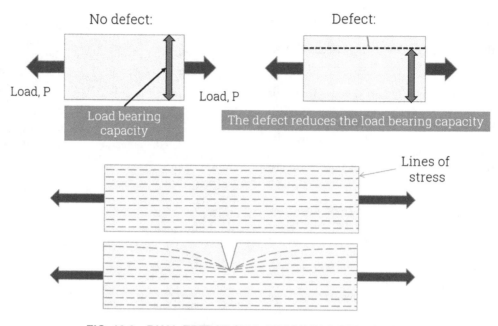

FIG. 12.3   DUAL EFFECT OF A CRACK IN A STRUCTURE

This loss of load-bearing capacity, and stress intensification can lead to failure at stress levels well below the yield strength of the material. Fracture mechanics explains and predicts these effects.

This Section briefly covers basic fracture mechanics principles, as these principles are essential to the understanding of how defects in any structure are assessed.

12.2.1   The Early Days

The effect of defects on structures was studied as long ago as the 15th century by Leonardo da Vinci (1452–1519). He had sketched in his notebooks a test of the tensile strength of a wire, and investigated metal failure [71]. He noted, as expected, that the strength of the wire was constant (for a specific length of wire tested).

Da Vinci failed a long length of wire, then tested half of the remaining length, and noted that this half-length failed at a higher load, Figure 12.4. He then tested a half of the remaining length

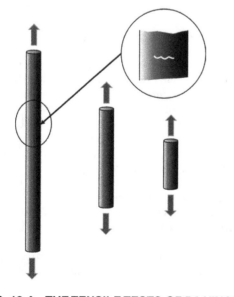

FIG. 12.4   THE TENSILE TESTS OF DA VINCI

(i.e., a quarter of the original length) and noted this quarter length had a higher failure load than the full and half lengths. The fracture strength was inversely proportional to the length of wire, which was not expected.

Drawn wire in the 15–16th centuries will not have been good nor have consistent quality; i.e., the longer the length, the more likely it was to contain a defect, Figure 12.4. Da Vinci concluded that the longest wire failed at the lowest load because it was more likely to contain the largest defect. The defects were small and not easily seen, but their effect was significant. He had observed a fact: the failure strength of a material was dependent on the presence of defects.

12.2.2   Inglis

The theoretical basis of fracture mechanics started in the early part of the 20th century. In 1913, C E Inglis [72] showed that the maximum stress (σ_{max}) at the edge of a hole of length b, and width a, subjected to a stress of σ is (Figure 12.5):

$$\sigma_{max} = \sigma(1 + (2b/a)) \qquad (1)$$

$$\sigma_{max} = \sigma(1 + 2(a/r)^{0.5}) \qquad (2)$$

where r is the radius of the notch.

A circular hole is when $a = b$, and this gives $\sigma_{max} = 3\sigma$.

Three features of these equations are:

- if b is reduced, the maximum stress reduces and eventually becomes equal to σ;
- if a is reduced, the maximum stress becomes very large (as a approaches zero, the stress become infinite);
- if r is reduced (the notch resembles a crack), the maximum stress becomes very large (as a approaches zero, the stress become infinite).

These high stresses are higher than the material's strength; therefore, the notch - theoretically - will fail at very low applied stresses. This is not the case, as modern materials resists these high stresses through a combination of 'ductility' and 'toughness'; however,

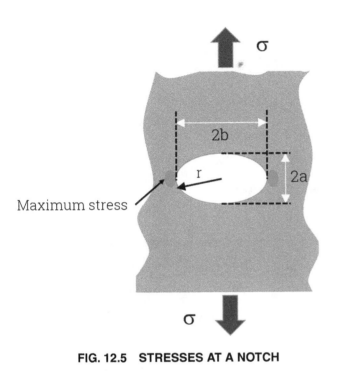

σ

2b

r

2a

Maximum stress

σ

FIG. 12.5   STRESSES AT A NOTCH

Inglis, showed the elevation of stresses caused by defects such as notches, and predicted very high stresses at the tip of a crack.

12.2.3   Griffith

A few years after Inglis'' work, A A Griffith [73] considered the failure of elastic, brittle materials (in which no plastic deformation[1] took place). Griffith investigated the fracture of glass - which is brittle - as he wanted to explain the huge ($\times 1000$) difference between the theoretical failure stress of glass compared to the measured failure stress. He observed that prior to glass failure, micro-cracks grow in length (propagated). Griffith discovered that there were many microscopic cracks in every material, which were present at all times.

Griffith hypothesized that these small cracks actually lowered the overall strength of the material, because when a load is applied to these cracks, a stress concentration is created, Figure 12.6 and Figure 12.7 (as predicted by Inglis). This stress concentration magnifies the stresses at the crack tip, and these cracks will grow much more quickly, thus causing the material to fracture, long before it ever reaches its theoretical strength. Griffith developed a fracture mechanics theory for cracks in brittle materials, such as glass.

12.2.4   Irwin

In the first half of the 20th century, failure reports of engineering structures did not usually consider the presence of cracks. Cracks were considered unacceptable in terms of quality, and there seemed little purpose in emphasizing this. Additionally, the effect of cracks could not be quantified, as the early fracture mechanics work of Griffith could not be applied to engineering materials, since it was only applicable to brittle materials; i.e., it was not directly applicable to engineering materials which exhibit ductility and deformation (plasticity) before they fail.

Griffith's work was significant; however, it did not include ductile materials in its consideration. In the 1950s there was major

interest in fracture in the aircraft industry in the USA, particularly in aluminum, and G R Irwin [74, 75], began to see how Griffith's theory would apply to ductile materials.

Irwin and his co-workers found that the stress fields around a crack tip could be represented by a stress 'intensity', K ('K' after Irwin's co-worker Kies), Figure 12.7. The maximum 'K' a material could withstand, K_c, was its fracture toughness. Irwin's work was the start of 'liner elastic fracture mechanics', where fracture mechanics models were developed for materials that exhibited some, but limited, plasticity at the crack tip.

Irwin discovered that K is a function of the applied stress, σ, the crack size, a, and other geometrical and crack features, Y. 'Y' is a parameter that accounts for differing shapes of cracks and structures, and these Y values are obtained from handbooks.

$$K = Y\sigma(\pi a)^{0.5} \qquad (3)$$

The fracture toughness (K_c) is measured in the laboratory, by creating a crack in a specimen, then failing the specimen in bending. The stress (load) at failure, s_c, and the depth of the crack at failure, a_c, gives K_c:

$$K_c = Y\sigma_c(\pi a_c)^{0.5} \qquad (4)$$

K is an excellent measure of stresses around a crack tip, but it becomes invalid when there is extensive plasticity (deformation, called a 'plastic zone') at the tip, Figure 12.7. This occurs in very ductile/tough materials - the toughness resists fracture, and allows extensive deformation rather than fracture, Figure 12.7.

This 'plasticity' limit on K means that it is not valid past the 'linear elastic' (small deformation/ductility) region of a transition curve, Figure 12.8. Some other measures of the stress intensity in the 'transition' and 'upper shelf' regions are needed.

12.2.5   Wells and Rice

Other fracture mechanics models and toughness parameters need to be used in the more ductile steels. In the 1960s two workers, A A Wells, and J R Rice, developed fracture mechanics parameters that could accommodate ductility:

- 'crack tip opening displacement' (δ)—this is a measure of the strain at the crack tip [76, 77]; and,
- 'J-integral' (J) - this is a measure of the energy around the crack tip [78].

These parameters are more appropriate for ductile materials such as line pipe steel, Figure 12.9.

K, δ, and J are all measures of the crack tip stresses/strains/energies, and toughness can be measured in terms of these parameters (K_c, d_c, J_c), Figure 12.9. Failure occurs when these toughnesses are reached. The parameters are linked when conditions are elastic:

$$(K^2/E') = J = m\sigma_y\delta \qquad (5)$$

where:

m is a constant (in the range 1 to 2) that depends on specimen geometry and work hardening behavior of the material.

σ_y is the yield strength.

E' = Young's modulus, E, for 'plane stress' (usually thin materials).

$E' = E/(1 - v^2)$ for 'plane strain' (usually thick materials), where v is Poisson's ratio.

---

[1] Deformation of a structure can be 'elastic' (meaning the deformation is recoverable, and none remains when the load is removed), or 'plastic' (meaning some of the deformation remains when the load is removed).

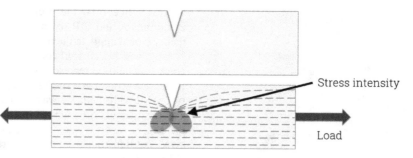

FIG. 12.6   STRESS INTENSITY AT CRACK TIP

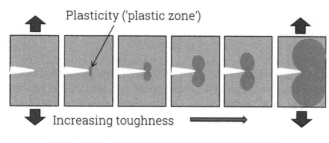

FIG. 12.7   PLASTICITY AT A CRACK TIP

12.2.6   The Failure Assessment Diagram

The fracture mechanics parameters K, J, and δ all allow the failure stress of a defective structure to be calculated, providing the material toughness is known. But this fracture calculation can be misleading, as the calculation can predict failure stresses above the material's ultimate tensile strength, if the toughness is high, or the defect is small, Equation (4). Obviously, the material cannot survive beyond its ultimate tensile strength, σ_u: the remaining ligament below any crack would 'collapse' at this strength, Figure 12.10. Therefore, two calculations are needed to determine if a defect such as a crack will fail a structure: a 'fracture' calculation, using fracture toughness, to determine the stress at fracture, and a 'collapse' calculation, to determine if the remaining ligament below the crack will collapse by reaching its ultimate (or yield) strength, Figure 12.11. Failure occurs at the lower of the two calculated values.

BSI 7910 [15] and API 579 [16] use a 'failure assessment diagram' (FAD) to assess defects in structures, which accommodates these two calculations. The FAD was developed in the mid-1970s, in the UK by the Central Electricity Generating Board (CEGB) to cover all possible failures: brittle, transitional, and upper shelf (where the remaining cross-section around a defect collapses due to excessive plasticity - 'plastic collapse'), Figure 12.11. All these

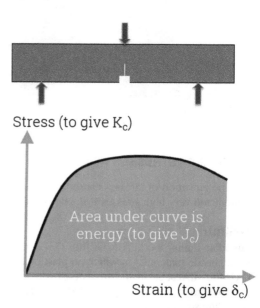

FIG. 12.9   LINKING K, J, AND δ

possible failures are assessed on the single diagram - the FAD - and a failure line ("locus") connects the two extreme types of failure (brittle [no plasticity at the crack tip] and collapse [extensive plasticity at the crack tip]), Figure 12.11.

The FAD is ensuring: the crack does not fail the structure; and, the remaining cross-section is sufficient to carry all the loads. The FAD is based on the principles of fracture mechanics:

- the vertical axis is a ratio of the applied stress to the stress to cause fracture (at the toughness, K_c) at the crack tip;
- the horizontal axis is the ratio of the applied load to that required to cause collapse (at σ_y or a function of σ_y) in the remaining ligament around the defect.

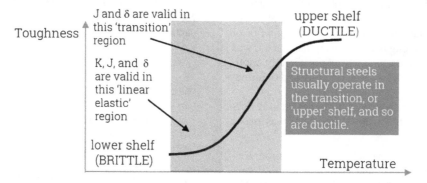

FIG. 12.8   VALIDITY OF IRWIN'S K IS RESTRICTED TO THE LINEAR ELASTIC REGION

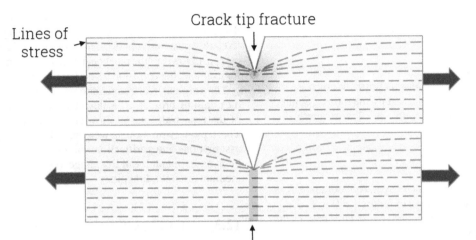

FIG. 12.10 FRACTURE (AT CRACK TIP) AND COLLAPSE (IN REMAINING LIGAMENT) OF A STRUCTURE

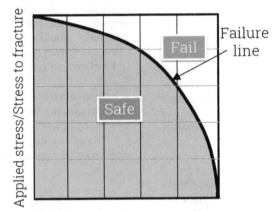

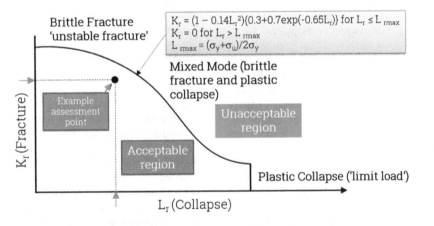

FIG. 12.11 THE FAILURE ASSESSMENT DIAGRAM ("FAD")

The assessment line is plotted on the diagram to predict the failure stress (σ_f) for both fracture and collapse, Figure 12.11, and has the general format:

$$K_c = \sigma_y \left\{ \frac{8a}{\pi} \ln \sec \left[\frac{\pi}{2} \frac{\sigma_f}{\sigma_y} \right] \right\}^{0.5} \qquad (6)$$

where:

K_c fracture toughness
a defect size
σ_y yield strength
σ_f failure stress

In a FAD, the proximity to fracture is given on the vertical axis as the ratio of applied stress intensity, K, to fracture toughness, K_c:

$$K_r = K/K_c \tag{7}$$

i.e., if $K_r = 1$, failure is predicted to occur by brittle fracture.

The proximity to plastic collapse is given by the ratio of applied net section stress, σ_n, to a 'flow strength', σ_f (flow strength is a stress between yield and UTS):

$$S_r = \sigma_n/\sigma_f \tag{8}$$

i.e., if $S_r = 1$, failure is predicted to occur by plastic collapse. S_r is often written as L_r (ratio of applied load to failure load) in FADs.

These values can be placed into Equation (6), and this creates the failure assessment line (Figure 12.11):

$$K_r = S_r \left\{ \frac{8}{\pi^2} \ln \sec \left(\frac{\pi}{2} S_r \right) \right\}^{-0.5} \tag{9}$$

12.2.7 Pipelines

The 1950s and 1960s was a period where the safety of transmission pipelines was of interest, primarily in the USA, where most of the long distance pipelines were operating. Early workers on pipeline defects were faced with problems: pipelines were thin walled, increasingly made of tough materials, and exhibited extensive plasticity before failure. Consequently, the fracture mechanics methods at that time could not reliably be applied to the failure of defective pipelines. To have been able to accurately predict the failure stress of a defect in a pipeline the workers would have needed:

- quantitative fracture toughness data, including measures of crack initiation and tearing (the workers did not have these measures, they only had the drop weight tear test [DWTT] [63] and Charpy V-notch impact energy);
- a predictive model for the fracture of a defect in a thin-walled pipe (they had models developed from the nuclear pressure vessel industry, but none for thin-walled line pipe); and,
- a model to predict the collapse of a defective cylinder.

Workers at the Battelle Memorial Institute in Columbus, Ohio, USA, decided to develop methods based on existing (in the 1960s) fracture mechanics models, but they overcame the above deficiencies in fracture mechanics' knowledge by a combination of expert engineering assumptions and calibrating their methods against the results of full scale tests. The resulting methods were therefore 'semi-empirical', but have proven to be accurate and prescient.

The workers noted from full scale tests, that line pipe with defects tended to fail in a ductile manner with plasticity, but that two basic distinctions could be made:

- 'Toughness dependent' - these tests failed at lower stresses (pressures). To predict the failure stress of these tests a measure of the material toughness was required (e.g., the stress intensity factor, K_c, or an empirical correlation based on upper shelf Charpy impact energy).
- Strength (or 'flow stress'[2]) dependent - these tests failed at higher stresses. To predict the failure stress of these tests only a measure of the material strength (or flow stress) was required.

---

[2] A function of the material's yield and ultimate tensile strengths.

The work at Battelle led to the development of the flow stress dependent and the toughness dependent, through-wall and part-wall 'NG-18' equations [17–19]. This work formed the basis of the development of many of the pipeline defect assessment methods used today; for example, the corrosion assessment methods in ASME B31G [59].

The original work and models accommodated the very complex failure process of a defect in a pipeline, involving bulging of the pipe wall, plastic flow, crack initiation and ductile tearing. These pioneering models were safe due to inherently conservative assumptions and verification via testing, but they were limited by their experimental validity range (generally, thin walled, low to medium strength, line pipe).

12.3 FATIGUE

Very few structures are subjected to a single ('static') stress during service. Most structures are subjected to varying, or 'cyclic', stresses during operation. A structure subjected to a cyclic (repeated) stress is said to have been 'fatigued'.

Fatigue has been known and understood for many years: railway engineers in the 19th century would say that railroad car axles and rail tracks would fail after a certain length of time, as they were 'fatigued' (tired) by the long service.

Pipelines are subjected to cyclic stresses due to:

- internal pressure variations;
- changes in temperatures;
- external loads (e.g., traffic loading, or movement on a seabed from currents);
- repeated pressure testing;
- etc..

These cyclic stresses range from a minimum stress (σ_{min}) to a maximum stress (σ_{max}) to give a stress range of $\Delta\sigma$. These cycles in a pipelines will not be regular and equal - they will be variable, but the number of stress cycles a structure can withstand, at a specific stress range, is called the 'fatigue life' of the structure; for example "10,000 cycles at a stress range of 100 N/mm²," Figure 12.12.

12.3.1 S-N Curves

The simplest way to design against fatigue is to use 'S-N' cures (stress range [S] versus number of cycles [N] to failure). The curves are derived from experimental testing of materials or components under a known cyclic stress range. These curves are obtained from standards/publications, Figure 12.12. It should be noted that the fatigue life of metals decreases when they are exposed to a corrosive environment: this 'corrosion fatigue' is 'environmentally assisted cracking' (EAC), and is caused by the combined actions of cyclic loading and a corrosive environment.

S-N curves are limited to design: they cannot be used if a defect is present (such as a crack). They are used on new structures, which are notionally 'defect-free'.

12.3.2 Fatigue Fracture Mechanics

Fracture mechanics is used to predict the fatigue life of a defect. API 579 [16] or BSI 7910 [15] give guidance on the fracture mechanics assessment of an existing defect under cyclic loading.

The calculation process is:

- an 'initial' defect, a_i, (e.g., a crack) is present in the structure; and...

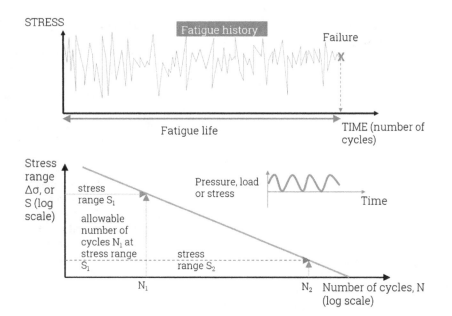

FIG. 12.12   FATIGUE AND THE S-N CURVE

- it will grow (Δa) under cyclic loading (e.g., pressure cycles in a pipeline); until...
- it reaches a 'final' defect size, a_f, which is large enough to...
- cause failure (fracture or collapse) of the structure.

The defect grows with every cycle, providing the cyclic stress ($\Delta\sigma$) is large enough to cause it to grow: large stresses grow it quickly; and, small stresses grow it slowly. Also, the bigger the defect becomes, the faster it grows.

This defect growth is predicted using fracture mechanics, as the stresses (both static and cyclic) create a stress intensity factor, K, Figure 12.13. Defect growth is related to the cyclic stress intensity factor, ΔK. ΔK is a function of the applied cyclic stress ($\Delta\sigma$), defect size (a), and the defect and structure shape (Y):

$$\Delta K = Y\Delta\sigma(\pi a)^{0.5} \quad (10)$$

where $\Delta\sigma = \sigma_{max} - \sigma_{min}$ and $\Delta K = K_{max} - K_{min}$. ΔK describes the cyclic stresses around the tip of the crack and will govern the speed of crack growth from the crack tip.

The crack growth rate is da/dN. The 'instantaneous' rate of crack growth (the slope, da/dN), is related to the stress intensity range, ΔK, Figure 12.13. The stress intensity range is the main contributor to growth rate, hence, the rate of crack growth (da/dN) can be plotted against ΔK, Figure 12.12.

A plot of log (da/dN) versus log (ΔK) will create a crack growth curve. The growth rate and the cyclic stress intensity have a linear relationship in the central portion of the curve. A straight line on a log (y) – log (x) plot gives:

$$\log(y) = m.\log(x) + \log C \quad (11)$$

where m is the gradient. Therefore, when $y = da/dN$, and $x = \Delta K$:

$$da/dN = C(\Delta K)^m \quad (12)$$

This linear relationship in the centre of this sigmoidal curve is called the 'Paris Law' where the parameters 'C' and 'm' are obtained from standards (e.g., [14–16]). For most practical cases

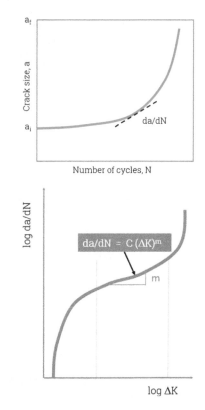

FIG. 12.13   CRACK GROWTH AND THE PARIS LAW

the region of the curve before the Paris Law can be ignored, as it is many stress cycles, but little crack growth. Similarly, the region after the Paris Law can be ignored as it is rapid growth, but in few cycles. Therefore, the Paris Law is assumed to represent most of the crack growth. This gives an equation to calculate the number of cycles (N_f) to grow a crack from an initial size (a_i), to a final size (a_f), for a specific stress range ($\Delta\sigma$).

The number of cycles to grow a crack to failure can be obtained by integrating the Paris Law. This integration assumes Y is a constant. Y is a constant if the crack is small and the structure is much larger, but in pipelines this is not the case; therefore, this simple integration is for illustration purposes only:

$$\frac{da}{dN} = C\left(Y\Delta\sigma\sqrt{\pi a}\right)^m \qquad (13)$$

$$\int_0^{N_f} dN = \int_{a_i}^{a_f} \frac{da}{C(\Delta\sigma)^m \pi^{\frac{m}{2}} Y^m a^{\frac{m}{2}}} \qquad (14)$$

$$N_f = \frac{2\left(a_f^{\frac{2-m}{2}} - a_i^{\frac{2-m}{2}}\right)}{(2-m)C(\Delta\sigma Y\sqrt{\pi})^m} \qquad (15)$$

Here is an example of a crack growth calculation using this equation. Assume an initial crack size of 1.0 mm, and a final crack size of 5.0 mm. Now, calculate N (cycles) to grow the crack from 1.0 to 5.0 mm, assuming $C = 1.36 \times 10^{-13}$ (units are MPam$^{1/4}$ for K, and m/cycle for da/dN), $m = 4$, Y is constant at 1.1, and stress range = 150 MPa.

Growth rate/cycle is... $da/dN = C(\Delta K)^m$... therefore, $\Delta K = 1.1 \times 150 \times (\pi a)^{0.5}$... resulting in cycles to failure = 80,346, Table 12.2. Note how the crack growth accelerates as the crack becomes deeper.

A corrosive environment can accelerate crack growth as the Paris constants (C, m) change with changing environments. Corrosive or hydrogen-enriched environments can increase crack growth rates by an order of magnitude compared with rates for non-aggressive environments [14].

When the parameter Y varies, the above integration is invalid, and a stepwise calculation is necessary, Table 12.3.

TABLE 12.2   CRACK GROWTH CALCULATION

Growth, mm	Cycles	% Life
1.0–1.5	33478	42
1.5–2.0	16739	21
2.0–2.5	10043	12
2.5–3.0	6696	8
3.0–3.5	4783	6
3.5–4.0	3587	4
4.0–4.5	2790	3
4.5–5.0	2232	3
TOTAL	80,346*	100*

*Rounding effects.

TABLE 12.3   CRACK GROWTH CALCULATIONS WITH Y VARYING

Cyclic stress	Cycle number	Initial defect size	Y value	Cyclic stress intensity	Crack growth
$\Delta\sigma$	1	a	Y	ΔK	Δa
$\Delta\sigma$	2	$a + \Delta a$	Y_1	ΔK_1	Δa_1
$\Delta\sigma$	3	$a + \Delta a + \Delta a_1$	Y_2	ΔK_2	Δa_2
$\Delta\sigma$	4	$a + \Delta a + \Delta a_1 + \Delta a_2$	Y_3	ΔK_3	Δa_3
$\Delta\sigma$	etc.	etc.	etc.	etc.	etc.

12.4   CORROSION ASSESSMENT

Corrosion is a major cause of failures in pipelines. Corrosion is the deterioration (loss) of a material resulting from a reaction with its environment. Therefore, corrosion is a time dependent, environmentally-assisted mechanism that causes a metal to deteriorate.

Corrosion is an electro-chemical process, and therefore requires an electro-chemical cell containing a cathode, anode, a connection between the anode and cathode, and an electrolyte.

12.4.1   Corrosion in Pipelines

Pipelines have external coatings and cathodic protection to protect them against external corrosion. Modern line pipe has external coatings applied in a factory, which are superior to field applied coatings. These factory coatings are excellent, but they will never by perfect; hence, the need for cathodic protection.

Pipelines can also be affected by corrosion on their internal surfaces. There must be liquid water present for corrosion to occur, and the water must wet the wall of the pipe: internal corrosion generally cannot occur in a pipeline unless there is an electrolyte to complete the corrosion cell.

Water or other aqueous materials (such as glycols from dehydration processes) are needed to form the electrolyte. Also, other chemicals usually must be present; for example, carbon dioxide (CO_2) for the formation of diluted organic and inorganic acids, or, sulfur for the formation of acid or growth of bacteria. Once introduced, the corrosive materials may continue to damage the pipeline until they are removed, or until they are consumed in corrosion reactions.

12.4.2   Corrosion and Cracking

Corrosion can result in:

- metal loss (the corrosion defect can have a smooth or irregular profile, and possibly contain blunt or sharp features); and/or,
- cracking

Cracks caused by a corrosive environment are called 'environmental cracks'. Environmentally-assisted cracking includes stress corrosion cracking, sulfide stress corrosion cracking, and hydrogen induced cracking. This Section will focus on blunt corrosion: cracks will be dealt with in a later Section.

12.4.3   Assessment of Corrosion

Corrosion assessment is important as inspection methods, such as 'smart' pigs, now easily detect its presence and size, so there is an increasing need to determine its severity, rather than continuously excavate and repair. This Section will cover the major methods for assessing corrosion.

12.4.3.1 Background to Current Corrosion Assessment Methods   In the 1960s and 1970s, workers [17–19] at the Battelle Memorial Institute in Columbus, Ohio decided to develop defect assessment methods based on existing fracture mechanics models. Over a 12-year period, up to 1973 [19], over 300 full-scale tests were completed, but the main focus was on (Figure 12.14):

- 92 tests on axially-orientated artificial through-wall defects; and,
- 48 tests on axially-orientated artificial part-wall defects (machined V-shaped notches).

These defects modeled part-wall corrosion in pipelines (Figure 12.14). The workers noted that line pipe containing defects tended

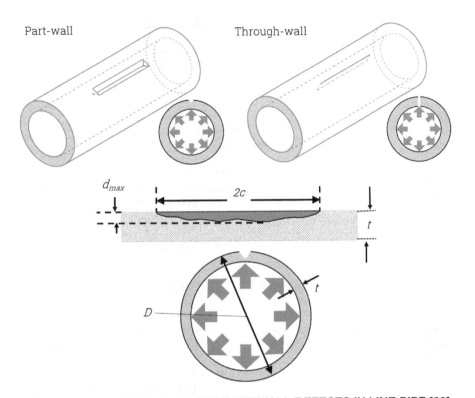

FIG. 12.14   THROUGH-WALL AND PART-WALL DEFECTS IN LINE PIPE [68]

to fail in a ductile manner, and final failure was by collapse, although very low toughness line pipe could fail in a brittle manner. The Battelle workers concluded that two basic distinctions could be made:

- 'toughness dependent' - these tests failed at lower stresses (pressures) and required a measure of the material toughness (e.g., the upper shelf Charpy impact energy) to predict the failure stress;
- 'strength dependent' - these tests failed at higher stresses but only needed a measure of the material's tensile properties to predict the failure stress.

12.4.3.2 The Basic Equations   The work at Battelle led to the development of strength ('flow stress' [19]) dependent, and toughness dependent, through-wall and part-wall equations.

Flow stress was a concept introduced by Battelle to help model the complex plastic flow and work hardening associated with structural collapse. Flow strength is a notional material property with a value between yield strength and ultimate tensile strength [19].

12.4.3.3 Through-Wall Defect   The Battelle workers produced equations [19] that could predict when a through-wall defect (Figure 12.15) would extend in length ('rupture'):

$$\frac{K_c^2 \pi}{8c\bar{\sigma}^2} = \frac{C_v \dfrac{12}{A} E\pi}{8c\bar{\sigma}^2} = \ln \sec\left(\frac{\pi M \sigma_\theta}{2\bar{\sigma}}\right) \quad \text{toughness dependent} \tag{16}$$

$$M = \sqrt{1 + 0.314\left(\frac{2c}{\sqrt{Rt}}\right)^2 - 0.00084\left(\frac{2c}{\sqrt{Rt}}\right)^4} \tag{17}$$

$$\sigma_\theta = M^{-1}\bar{\sigma} \quad \text{strength dependent} \tag{18}$$

12.4.3.4 Part-Wall Defect   In parallel, the Battelle workers produced an equation [19] that could predict the pipeline hoop stress when a part-wall defect (Figure 12.15) failed:

$$\frac{K_c^2 \pi}{8c\bar{\sigma}^2} = \frac{C_v \dfrac{12}{A} E\pi}{8c\bar{\sigma}^2} = \ln \sec\left(\frac{\pi M_p \sigma_\theta}{2\bar{\sigma}}\right) \quad \text{toughness dependent} \tag{19}$$

$$M_P = \left[\frac{1 - \dfrac{d}{t}\left(\dfrac{1}{M}\right)}{1 - \dfrac{d}{t}}\right] \tag{20}$$

$$\sigma_\theta = \bar{\sigma}\left[\frac{1 - \dfrac{d}{t}}{1 - \dfrac{d}{t}\left(\dfrac{1}{M}\right)}\right] \quad \text{strength dependent} \tag{21}$$

D	outside diameter of pipe (R=D/2=radius)
t	pipe wall thickness
E	elastic modulus
M	bulging ("Folias") factor (examples are given in the literature (for example [19, 59]))
R	radius of pipe
d	part-wall defect depth
σ_θ	hoop (circumferential) stress at failure (or σ_f)
$2c$	defect axial length
C_v	upper shelf Charpy V-notch impact energy
A	area of Charpy specimen fracture surface
σ	flow stress (function of σ_u [ultimate tensile strength] and σ_y [yield strength])

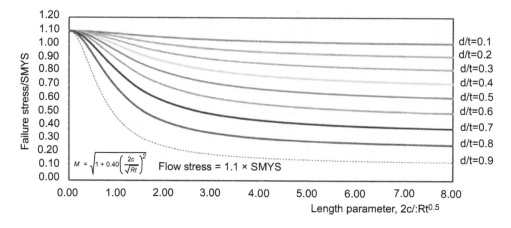

FIG. 12.15   STRENGTH DEPENDENT EQUATION FOR PART-WALL DEFECTS IN PIPELINES

The strength dependent equation can be put into graphical format by assuming a flow stress, and a bulging factor (*M*), Figure 12.15.

12.4.3.5 Assessing Corrosion in Line Pipe The corrosion assessment methods in use today require the line pipe to be 'ductile'. 'Ductile' means the material:

- has passed a 'drop weight tear test' (DWTT) criteria; and,
- is on the Charpy toughness 'upper shelf', i.e., 100% shear area.

Line pipe that meets contemporary specifications [13] would satisfy these criteria, but because corrosion is blunt, the methods are applicable to lower Charpy toughness line pipe [58].

Corrosion can be in any location in a pipeline, and in any orientation, Figure 12.16, but the first corrosion type considered in this Chapter will be corrosion that is primarily in the axial direction, and subjected to the full hoop stress in the pipeline. Corrosion in other orientations can be projected onto the longitudinal axis, and the projected length used in the calculation (see Reference 59 for further details). Corrosion in the circumferential direction will be covered later.

12.4.3.6 ASME B31G-2012 The popular methods used for assessing corrosion orientated in the axial direction are based on the research at Battelle Memorial Institute in the 1960s and 1970s. The first recognized method was published in the 1980s, as the pipeline industry identified a need for standardized guidelines for the assessment of corrosion in pipelines.

In 1984, the American Society of Mechanical Engineers (ASME) produced ASME B31G (now [59]) for the assessment of corrosion defects: ASME B31G considers corrosion in pipelines under internal pressure loading only: it does not cover external loads.

FIG. 12.16   ORIENTATION OF CORROSION

ASME B31G is applicable to pipelines and bends containing:

- metal loss due to corrosion or grinding;
- metal loss that affects longitudinal or helical electric seam welds or circumferential electric welds.

Note that the welds must be of 'sound quality' [59].

ASME B31G gives the user a choice of four assessment levels, each with decreasing conservatism, Table 12.4. There are also choices of methods within the levels. The ASME B31G standard in now the benchmark standard for the assessment of corrosion in line pipe. This standard can allow large areas of corrosion to safely remain in an operational pipeline.

ASME B31G gives the user choices on the level of assessment, and on input data; for example, it gives a choice of flow stress, Table 12.5. Also, it should be noted that ASME B31G does not calculate the failure stress of a corrosion defect: it gives acceptance levels. 'Acceptance' is defined in ASME B31G as: "*A flaw or anomaly is considered acceptable where the computed failure stress is equal to*

TABLE 12.4   METAL LOSS ASSESSMENT LEVELS IN ASME B31G [59]

Level	Method
0	'Original' B31G from 1984, in tabular form.
1	'Original' B31G, but in equation form. 'Modified' B31G [58] API 579 (Level 1) [16]
2	'Effective area method' [19] API 579 (Level 2) [16]
3	'Detailed' (e.g., finite element stress) analysis

TABLE 12.5   CHOICE OF FLOW STRESS IN ASME B31G

Flow stress	Comments
1.1 × SMYS	For plain carbon steel operating at temperatures below 250°F (120°C)
SMYS + 10 ksi (69 N/mm²)	For line pipe of SMYS ≤X70 (483 MPa)
(SMYS + SMUTS)/2	For line pipe of SMYS ≤X80 (551 MPa)

or greater than the hoop stress at the operating pressure multiplied by a suitable safety factor."

The minimum safety factor recommended in ASME B31G equals the ratio of the minimum hydrotest pressure required for the given type of pipeline construction to the maximum operating pressure (MOP), or maximum allowable operating pressure (MAOP), but usually not less than 1.25.

It is not intended to cover these Levels in detail as they have been detailed before (e.g., [58]), but a brief summary, with some worked examples, are presented below.

12.4.3.6.1 Level 0.

Level 0 presents the user with metal loss acceptance levels in the form of tables. The equations used to develop the Level 0 tables are from the original (1984) version of ASME B31G, and corrosion defects are assumed to have a parabolic shape (hence the '2/3' ratio in the equations):

$$\sigma_f = \bar{\sigma}\left[\frac{1-\dfrac{2}{3}\dfrac{d}{t}}{1-\dfrac{2}{3}\dfrac{d}{t}\dfrac{1}{M}}\right] \qquad (22)$$

$$M = \sqrt{1+0.8\left(\frac{l}{\sqrt{Dt}}\right)^2} = \sqrt{1+0.4\left(\frac{2c}{\sqrt{Rt}}\right)^2} \qquad (23)$$

when $M \geq 4.12$, then:

$$\sigma_f = \bar{\sigma}\left[1-\frac{d}{t}\right] \qquad (24)$$

$$\bar{\sigma} = 1.1 \times SMYS \qquad (25)$$

$$\sigma_f = 100\% SMYS \qquad (26)$$

12.4.3.6.2 Level 1 (Original Equations)

Level 0 Uses the original ASME B31G equations in tabular form. The first method in Level 1 uses these equations to calculate acceptable defect sizes. These simple calculations rely on single measurements of the maximum depth and axial extent of metal loss, and are intended to be conducted in the field.

The historical equations given above can be represented on a single acceptance plot, Figure 12.17. Figure 12.17 also shows these equations plotted using an actual pipeline geometry, and showing the acceptable and unacceptable areas of the plot.

Table 12.6 gives a worked example of this level.

$$\left(\frac{l}{\sqrt{Dt}}\right)^2 = 1.146$$

$$\frac{d}{t} = 0.45$$

$$M = \sqrt{1+0.8\left(\frac{1}{\sqrt{Dt}}\right)^2} = \sqrt{1+0.8 \times 1.15} = 1.385$$

$$\sigma_f = \bar{\sigma}\left[\frac{1-\dfrac{2}{3}\dfrac{d}{t}}{1-\dfrac{2}{3}\dfrac{d}{t}\dfrac{1}{M}}\right] = 1.1 \times 358 \times \left[\frac{1-\dfrac{2}{3}0.45}{1-\dfrac{2}{3}0.45\dfrac{1}{1.385}}\right]$$

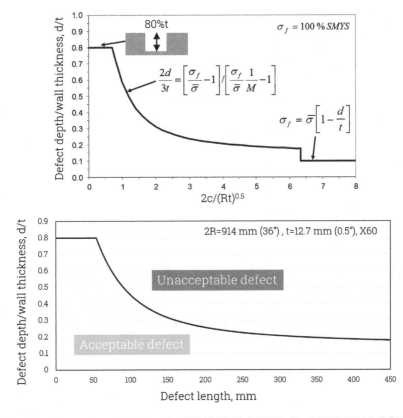

FIG. 12.17 LEVEL 1 OF ASME B31G USING THE ORIGINAL ASME B31G EQUATIONS

TABLE 12.6 INPUT FOR WORKED EXAMPLE (LEVEL 1, ASME B31G)

Input	Value	Input	Value
Defect length, l	100 mm (3.94")	Maximum defect depth, d	6.4 mm (0.25")
D	610 mm (24")	t	14.3 mm (0.562")
Grade	358 N/mm² (X52)	Pre-service hydrotest	100% SMYS
Design factor	0.72		

Failure stress = 352 N/mm² = 51,118 lbf/in² (= 98% SMYS). This defect will not fail the pipeline at its design stress (= 72% SMYS) but... the acceptance criterion for B31G Level 1 for this defect is that a defect must survive a stress of 100% SMYS; i.e., a safety factor of 1.39 on a design stress of 72% SMYS (although smaller safety factors can be used). Hence, using this safety factor, this defect is unacceptable.

12.4.3.6.3 Level 1 ('Modified B31G'). The first method in Level 1 used the historical ASME B31G equations. The original B31G criterion from 1984 was perceived to be overly-conservative. Most of the over-conservatism in the original B31G arose from the length limitation (at M ≥ 4.12); consequently, a second method was developed known as 'modified B31G' [58]. This is now the second method in Level 1, and is less conservative than Level 0, and Level 1 (using the original B31G equations).

The modified B31G method uses similar equations to the original B31G criterion from 1984, but:

- the flow stress was changed from 1.1 × SMYS to SMYS + 69 N/mm² (SMYS + 10,000 lbf/in²).
- the shape factor was changed from the parabolic shape ('2/3') to 0.85.
- the bulging factor was changed.

$$\sigma_f = \bar{\sigma}\left[\frac{1-0.85\dfrac{d}{t}}{1-0.85\dfrac{d}{t}\dfrac{1}{M}}\right] \quad (27)$$

for $\left(\dfrac{l}{\sqrt{Dt}}\right)^2 \le 50.0$ use:

$$M = \sqrt{1+0.6275\left(\frac{l}{\sqrt{Dt}}\right)^2 - 0.003375\left(\frac{l}{\sqrt{Dt}}\right)^4} \quad (28)$$

for $\left(\dfrac{l}{\sqrt{Dt}}\right)^2 > 50.0$ use:

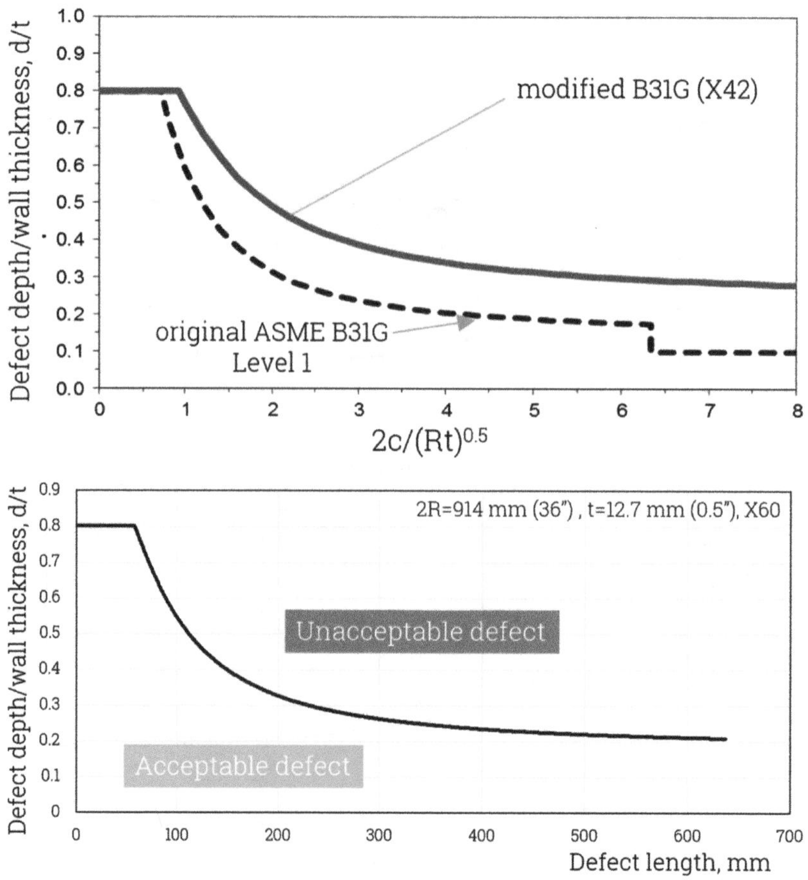

FIG. 12.18   LEVEL 1 ('MODIFIED B31G') OF ASME B31G

$$M = 0.032\left(\frac{l}{\sqrt{Dt}}\right)^2 + 3.3 \qquad (29)$$

$$\bar{\sigma} = \text{SMYS} + 10\,\text{ksi} = \text{SMYS} + 69\,\text{N/mm}^2 \qquad (30)$$

Figure 12.18 plots these equations. A single curve for all pipeline geometries and grade cannot be plotted, due to the '+10,000 lbf/in²' in the flow stress equation. Figure 12.18 also gives an example of these equations, for a specific pipeline geometry.

12.4.3.6.4 Level 1 (API 579). The third method in Level 1 (API 579), Table 12.4, is very similar to the modified B31G method; therefore, it is not covered here.

12.4.3.6.5 Level 2 (Effective Area Method). Corrosion has an irregular profile, and the Level 1 methods in ASME B31G simplify the calculations by assuming a parabolic shape, or a shape factor of 0.85 on the corrosion's maximum depth. These are conservative assumptions, but the shape of the corrosion can be used in calculations. The Battelle equation can use defect area (A), rather than maximum depth:

$$\sigma_f = \bar{\sigma}\left[\frac{1 - \dfrac{A}{A_O}}{1 - \dfrac{A}{A_O}\dfrac{1}{M}}\right] \qquad (31)$$

where A_o is the cross-sectional area of wall thickness (t) occupied by the defect of length 2c (=2ct), Figure 12.19.

The corrosion will be three-dimensional; therefore, a method to determine where to take depth measurements is needed. The 'river bottom' method is used to select the location of the depth measurements [59]: it can be visualized as water flowing down a valley where the valley is the corrosion, and the water flows into the base of the corrosion. This creates the longitudinal path of minimum thickness, through the deepest areas of metal loss, Figure 12.20.

The effective area method will give less conservative predictions of the failure stress of a metal loss defect, but a user needs to be careful when measuring defect area when there are large deep areas of corrosion within a shallow area. This is because the effective area method can 'mask' a deep area of corrosion that will fail at a lower stress than the overall defect area.

This error can be avoided by systematically measuring areas and continuously calculating failure stresses for all combinations of the defect area. This process will calculate the lowest failure stress for any combination of depths within the corroded area, Figure 12.21.

12.4.3.6.6 Level 3. Level 3 allows more detailed assessment methods such as finite element analysis.

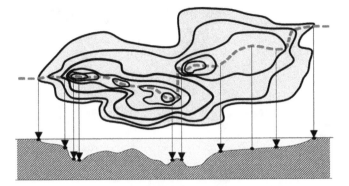

FIG. 12.20 'RIVER BOTTOM' METHOD OF SELECTING METAL LOSS DEPTHS IN A CORRODED AREA [87]

12.4.3.7 DNV RP-F101 The Norwegian organization DNV has published guidance on assessing corrosion in line pipe [55]. DNV RP-F101 is based on full scale tests and numerical analyses of corrosion defects [79, 80], and gives guidance on the assessment of:

- single defects and interacting defects;
- complex shaped defects (i.e., assessing the actual profile of the defect); and,
- combined loading.

It is not applicable to line pipe grades above X80, or to cracks, and corrosion defect depth must be ≤ 85% wall thickness.

The hoop stress to cause failure is given by:

$$\sigma_\theta = \sigma_u\left(\frac{1 - \dfrac{d}{t}}{1 - \dfrac{d}{t}\dfrac{1}{Q}}\right) \qquad (32)$$

$$Q = \sqrt{1 + 0.31\left(\frac{2c}{\sqrt{Dt}}\right)^2} \qquad (33)$$

These equations are similar to the original Battelle equations. DNV-RP-F101 incorporates safety factors by calculating a 'safe working pressure':

$$P_{sw} = F \times P_f \qquad (34)$$

where:

P_f is the failure pressure obtained from Equation (32), and F = 'total usage factor' = $F_1 F_2$.

F_1 = 'modeling factor' = 0.9.

F_2 = operational 'usage factor' (normally taken as equal to the 'design factor' taken from the pipeline's design standard).

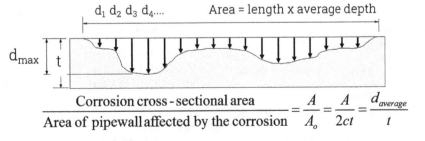

$$\frac{\text{Corrosion cross-sectional area}}{\text{Area of pipewall affected by the corrosion}} = \frac{A}{A_o} = \frac{A}{2ct} = \frac{d_{average}}{t}$$

FIG. 12.19 USING THE CORROSION SHAPE IN ASME B31G

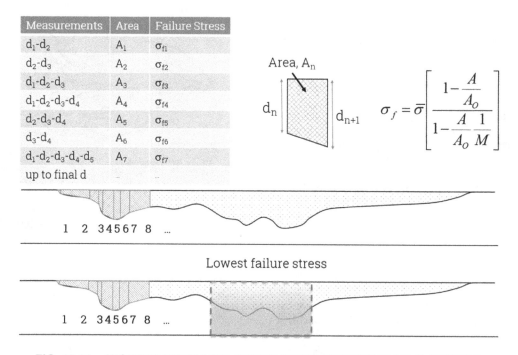

Measurements	Area	Failure Stress
d_1-d_2	A_1	σ_{f1}
d_2-d_3	A_2	σ_{f2}
d_1-d_2-d_3	A_3	σ_{f3}
d_1-d_2-d_3-d_4	A_4	σ_{f4}
d_2-d_3-d_4	A_5	σ_{f5}
d_3-d_4	A_6	σ_{f6}
d_1-d_2-d_3-d_4-d_5	A_7	σ_{f7}
up to final d		

$$\sigma_f = \overline{\sigma}\left[\frac{1-\dfrac{A}{A_o}}{1-\dfrac{A}{A_o}\dfrac{1}{M}}\right]$$

Lowest failure stress

FIG. 12.21 AVOIDING MISCALCULATIONS IN THE EFFECTIVE AREA METHOD

DNV has said [80]: "*For old pipelines, or pipelines where the material might not have sufficient ductility, the... DNV... criteria should not be used. Modern pipeline steel materials normally have sufficient toughness to expect plastic collapse failure.*" 'Plastic collapse' means the remaining ligament below the corrosion defect can tolerate ultimate tensile strength. Line pipe will collapse if the toughness is very 'high', but... what is 'high', and will older line pipe 'collapse'?

Modern line pipe is 'very high' toughness, and should fail by 'plastic collapse', but older steels do not have 'high' toughness, Table 12.7 [81]. The DNV standard itself states that its methods should not be applied to line pipe steels materials with Charpy values less than 27 J (20 ft lb f). For the weld, a minimum full size Charpy value of 30 J is recommended. Reference 82 supports this 27 J (20 ft lb) limit for plastic collapse: line pipe toughnesses less than this value may not be able to support plastic collapse [21].

Therefore, a lower bound toughness to support plastic collapse of corrosion defects in line pipe material is ≥27 J (≥20 ft lb). There are higher estimates in the literature, and these will be more appropriate: plastic collapse can be expected with a minimum toughness of 82 to 102 J (60 to 75 ft lb) [83], which is similar to another estimate of 90 J (68 ft lb) [81]. These values indicate that line pipe steels fabricated after 1980 are more likely to 'collapse' than line pipe fabricated before 1980, Table 12.7.

12.4.3.8 Other Corrosion Assessment Methods There are many other corrosion assessment methods available. Two popular methods

(software) are detailed in References 26 and 43. These methods use and extend the methods developed by the Battelle workers. The methods are reviewed in Reference 58, and therefore are not reviewed here.

12.4.3.9 Comparison of Assessment Methods Reference 58 compared the assessment methods detailed above. The methods were assessed against a large body of full scale test data. The predicted to actual failure pressures (P_a/P_f) are presented in Table 12.8. 'RSTRENG' (effective area methods in ASME B31G) gives the most accurate predictions of P_a/P_f.

12.4.4 Corrosion on Line Pipe Welds

It is now generally considered that longitudinal corrosion across seam welds in line pipe (other than 'autogenous' welds [a welding procedure that does not use filler metal, such as electric resistance welded line pipe]) can be treated as corrosion in 'parent plate' (i.e., as though the corrosion is in the line pipe), and this is supported by test data (e.g., [86]).

TABLE 12.7 TYPICAL CHARPY (CVN) TOUGHNESS IN LINE PIPE OVER SEVEN DECADES

Decade	1950s	1960s	1970s	1980s	1990s
Grade	X42/52	X52/60	X60/65	X65/70	X75
Typical CVN, J (ft lb)	27 (20)	41 (30)	54 (40)	88 (65)	109 (80)

TABLE 12.8 COMPARISON OF CORROSION ASSESSMENT METHODS [58]

Assessment method	P_a/P_f		P_a/P_f (All data except early grade B tests)	
	Mean	Standard deviation	Mean	Standard deviation
ASME B31G (Level 1)	1.330	0.468	1.347	0.479
Modified B31G [58]	1.184	0.285	1.194	0.289
'RSTRENG' [84, 85]	1.170	0.177	1.188	0.168
DNV RP F101 [55]	1.178	0.318	1.205	0.309
PCORR [43]	1.191	0.310	1.220	0.301
API 579 [16]	1.436	0.407	1.465	0.403

Accordingly, standards [e.g., 15, 55, and 59] allow the assessment of corrosion on welds using methods such as those in ASME B31G provided the weld mechanical properties are similar or superior to the line pipe, and the weld must be free from other defects.

12.4.5   Assessing Corrosion in the Circumferential Direction

Corrosion can sometimes be in the circumferential direction. This orientation subjects the corrosion to both the axial stress due to the internal pressure (this will be between 30% and 50% of the hoop stress, depending on the pipeline end restraints) and any other axial stress (for example, due to thermal loads, ground or pipe movement, loss of support [e.g., spanning], bends, supports, etc.). Hence, it is the axial stress, not the hoop stress, which may be the major stress acting on corrosion defect. Therefore, an assessment method is needed to assess corrosion in a pipeline subjected to high axial loads when the corrosion is primarily in the circumferential direction, or, the corrosion has extensive width.

Kastner et al. [22] published a failure criterion for a circumferential part-wall defect subject to internal pressure, axial and/or bending loads. This is now the most popular method for assessing circumferential corrosion. The axial stress at failure (σ_f) is given by:

$$\frac{\sigma_f}{\overline{\sigma}} = \frac{\eta(\pi - \beta[1-\eta])}{\eta\pi + 2[1-\eta]\sin(\beta)} \qquad (35)$$

where $\eta = 1 - (d/t)$ and $\beta = c/R$ (Figure 12.22), and the flow stress is the average of the yield strength and ultimate tensile strength.

Figure 12.22 shows the size of circumferential corrosion that would fail at axial stress levels of 100% SMYS, for two grades of line pipe steel.

Note that safety factors must be applied to this predicted axial stress at failure, and metal loss having a significant circumferential extent and acted on by high longitudinal stresses in compression could be susceptible to wrinkling or buckling.

Corrosion with both axial extent and circumferential extent can fail due to hoop stress and axial stress. Under pressure loading only, the axial dimension is the critical dimension, and unless the circumferential length > axial length, the circumferential failure need not be considered [85]. However, if there are very high external loads (e.g., mining subsidence or spanning) then two calculations must be conducted: failure under pressure loading (e.g., ASME B31G); and, failure due to the axial loads (e.g., the Kastner equation).

12.4.6   Assessing Groups of Corrosion Defects

Corrosion often occurs in groups ('colonies'), Figure 12.23. The failure stress of a corrosion defect can be reduced by the presence of another corrosion defect. When the failure stress of an individual defect is reduced by the presence of a neighboring defect, the defects are said to 'interact'.

ASME B31G [59] considers 'interaction': "*Corrosion may occur such that multiple areas of metal loss are closely spaced longitudinally or transversely. If spaced sufficiently closely, the metal loss areas may interact so as to result in failure at a lower pressure than would be expected based on an analysis of the separate flaws.*"

If the defects interact, the defects are assessed as a single defect of length and width equal to the total dimensions, and a depth equal to the maximum depth of the group. The new total length, width and maximum depth has to be input into the failure equations.

Various standards give interaction 'rules', usually related to the line pipe's wall thickness, t, and outside diameter D, Table 12.9.

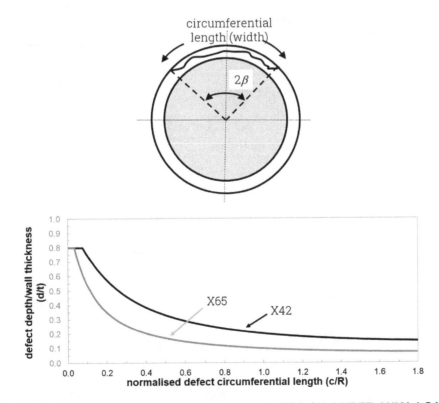

FIG. 12.22   FAILURE OF CIRCUMFERENTIAL CORROSION, UNDER AXIAL LOADS

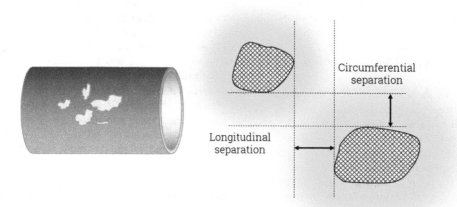

FIG. 12.23   INTERACTION OF CORROSION COLONIES

TABLE 12.9   INTERACTION CRITERIA FOR CORROSION DEFECTS

Reference	Circumferential	Longitudinal
ASME B31G [59]	3 × wall thickness (t)	3t
ASME B31.4 [8]	6t	1" (25.4 mm)
Pipeline Operators Forum [88]	6t	6t
CSA Z662 [89]	Less than the longitudinal length of the smallest area	
DNV-RP-F101 [55]	360 $(t/D)^{0.5}$ (degrees) (circumferential angular spacing)	2.0 $(Dt)^{0.5}$

12.4.7   Corrosion Growth

Corrosion is time dependent: this means it increases in size with time, and this corrosion growth must be included in any assessment. Figure 12.24 illustrates this growth on an assessment diagram.

API 1160 [61] gives guidance on external surface corrosion growth rates, Tables 12.10 and 12.11 are typical corrosion growth rates based on soil resistivity and corrosion type; however, it is recommended to assess specific pipeline and environmental conditions.

TABLE 12.10   CORROSION GROWTH RATES IN SOILS

Soil resistivity (ohm-cm)	Corrosion rate, 0.001 in./year (mm/year)
>15,000 and no active corrosion	3 (0.08)
1,000 to 15,000 and/or active corrosion	6 (0.15)
<1,000 (worst case for soil)	12 (0.30)

TABLE 12.11   GROWTH RATES FOR DIFFERING TYPES OF CORROSION

Corrosion type	Rate, 0.001 in./year (mm/year)
Stray current or bacteria	16 (0.41)
Corrosion along the seam weld of ERW or EF line pipe.	Up to ×4 the rates in soil
Stress corrosion cracking	24 (0.61) (worst case)

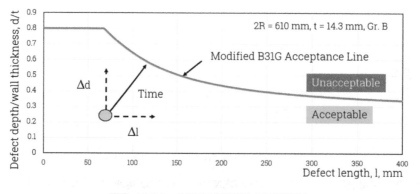

FIG. 12.24   CORROSION GROWTH

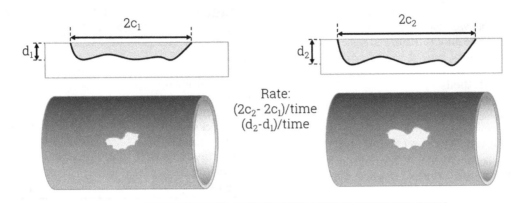

FIG. 12.25   DETERMINING CORROSION RATES FROM PIG RUNS

Corrosion growth (Figure 12.25) can also be estimated from two or more smart pig runs, although care needs to be taken when considering smart pigs with differing accuracies and reliabilities [90].

12.5   CRACKS

12.5.1   Introduction

A crack is a very sharp defect: it is a 'planar' (two-dimensional) defect. ASME B31.8S-2012 [67] states that a crack is a... *"very narrow, elongated defect caused by mechanical splitting into two parts."*

Cracking in a pipeline is usually:

- in or around welds (e.g., cracks caused by poor welding);
- environmentally-created (e.g., stress corrosion cracking) in the welds or line pipe, or,
- in fittings.

The occurrence of cracking is generally an indication of:

- a difference between the conditions expected at the design stage and the actual operating conditions (e.g., fatigue);
- poor fabrication, manufacturing, or construction control procedures; or,
- some other poor practice, or an unexpected event.

BSI 7910 [15] states: *"Cracks can be indicative of poor design features, workmanship or inspection, or stress corrosion, fatigue or corrosion fatigue, and therefore the cause should be ascertained before proceeding with an assessment."*

Accordingly, a crack is an indication that something has gone wrong, or something is going wrong; therefore, standards do not allow them unless a rigorous assessment is conducted. ASME B31.4 [8] states: *"Verified cracks except shallow crater cracks or star cracks in girth welds shall be considered defects and removed or repaired unless an engineering evaluation shows that they pose no risk to pipeline integrity...".*

There are assessment methods for cracks in pipelines (e.g., Table 12.12), but cracks are both difficult to detect and difficult to size, leading to uncertainties in their assessment. Cracks also grow quickly if the environmental conditions are favorable (e.g., if corrosion is present, or the structure is fatigued). This growth can be difficult to predict and calculate, again adding to uncertainties in crack assessment. Therefore, cracks are treated with extreme caution: it is better to prevent their occurrence than assess them.

TABLE 12.12   ASSESSMENT METHODS FOR CRACKS IN PIPELINES

Method	Summary
API 579 [16]	General fracture mechanics method.
BSI 7910 [15]	General fracture mechanics method.
Battelle's 'ln sec' formula [58]	Pipeline-specific, developed at Battelle.
'PAFFC' [26]	Pipeline-specific (software) developed at Battelle in the USA.
'CorLAS' [43]	Pipeline-specific (software) developed at DNV in the USA.

12.5.2   Effect of toughness

The models used to assess corrosion in the previous Section are applicable to blunt defects, where the blunt defects fail by 'net section collapse' net section collapse can be considered failure when the average stress in the remaining ligament below the defect reaches the flow stress of the material (Figure 12.10). Net section collapse is the same as plastic collapse, if the flow stress equals the ultimate tensile stress.

Generally, in net section collapse, failure is not preceded by the development of a slowly tearing crack at the base of the blunt defect. Even if some tearing does occur, the strain hardening of the material and/or the redistribution of stress during the yielding will ensure the ligament does not fail until the flow stress level is reached [58]. Blunt defects, such as corrosion, do not easily tear through the remaining ligament (as they are blunt). This allows the average stress in the ligament to reach the flow stress.

A crack can cause tearing at its tip, and this tearing can continue through the ligament. This tearing crack can prevent the stress in the ligament reaching the flow stress, and this can lead to a failure stress lower than predicted by blunt defect models: the toughness of the line pipe now plays a key role [58]. One reference [90] indicates that a sharp defect will behave like a blunt defect if the line pipe has an upper shelf Charpy toughness of 100 ft lb (135 J).

Net section collapse cannot occur if the toughness is low, as low toughness material will allow the crack to tear quickly and easily; therefore, low toughness material can lower the failure stress of a defect in a pipeline.

Predicting the failure pressure of a crack in a pipeline uses:

- 'linear elastic fracture mechanics' for very brittle (low toughness) material; and, for a very tough material,
- the equations used for blunt defects such as corrosion.

However, the task of predicting failure pressures for cracks in materials with toughness between these two extreme values (e.g., some line pipe materials made prior to 1980, and some made after that time) is more difficult [58].

12.5.3 Assessment Methods

There are several accepted methods for assessing cracks in pipelines, Table 12.12. These will now be summarized.

12.5.3.1 API 579 and BSI 7910 API 579 and BSI 7910 use the 'failure assessment diagram' (FAD) to determine if a crack is acceptable. Two calculations are performed to assess a crack:

- a fracture (using G, K, J, CTOD) calculation; and,
- a plastic collapse (using material strength) calculation.

The structure's failure stress is obtained from these two calculations, and compared to a failure line (Figure 12.11):

$$K_r = (1 - 0.14L_r^2)\{0.3 + 0.7 \exp(-0.65L_r)\} \text{ for } L_r \le L_{rmax} \quad (36)$$

$$K_r = 0 \text{ for } L_0 > L_{rmax} \quad (37)$$

$$L_{rmax} = (\sigma_y + \sigma_u)/2\sigma_y \quad (38)$$

The FAD is ensuring: the crack does not fail the structure; and, the remaining cross-section is sufficient to carry all the loads.

The FADs in BSI 7910 and API 579 (Figure 12.11) use a fracture toughness measured in 'stress intensity factor' (K), or 'J integral' (J), or 'crack tip opening displacement' (δ). Line pipe and its welds will rarely have these measures: usually only a Charpy value of toughness is available. Hence, BSI 7910 and API 579 use simple correlations between Charpy (CVN) and K, J, δ: these correlations will introduce inaccuracies in the calculations, and make comparisons between the standards difficult, Figure 12.26 [60].

The FAD can be used to assess cracks in line pipe steels: Table 12.13 gives the input data for an assessment using Figure 12.27. The predicted failure pressure using BSI 7910 is 697 psig (48 barg). This is above the MAOP, Figure 12.27, and therefore the defect is not predicted to fail.

TABLE 12.13 EXAMPLE OF A CRACK ASSESSMENT USING A FAD (INPUTS)

Input	Value
Crack of depth (d)	0.157 inches (4 mm)
Length (2c)	6 inches (152.4 mm)
Pipe diameter (D)	48 inches (1219.2 mm)
Wall thickness (t)	0.5 inch (12.7 mm)
Grade (strength)	X52 (358 N/mm²)
MAOP (maximum allowable operating pressure)	623 psig (43 barg)
Toughness	15 ft lbs (20J)

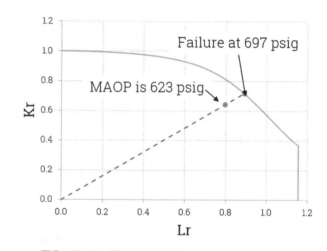

FIG. 12.27 EXAMPLE OF AN ASSESSMENT OF A CRACK USING THE FAD

12.5.3.2 Battelle Formula Battelle in the USA developed an equation (called the 'ln secant' or 'NG 18' equation) that can be used to assess cracks in line pipe under pressure. It is a complex equation [19]:

$$\sigma_\theta = \frac{2\overline{\sigma}}{\pi M_P} \cos^{-1} \exp - \left[\frac{C_v \dfrac{12}{A_c} E\pi}{8c\overline{\sigma}^2} \right] \quad (39)$$

$$M_P = \left[\frac{1 - \dfrac{A}{A_o}\left(\dfrac{1}{M_T}\right)}{1 - \dfrac{A}{A_o}} \right] \quad (40)$$

$$M_T = \left[1 + 1.255\frac{c^2}{Rt} - 0.0135\frac{c^4}{R^2t^2} \right]^{0.5} \quad (41)$$

where:

t	pipe wall thickness
E	elastic modulus
R	radius of pipe
σ_θ	hoop stress at failure
$2c$	defect axial length
C_v	upper shelf Charpy V-notch impact energy
A_c	area of Charpy specimen fracture surface (0.124 in²)
σ	flow stress

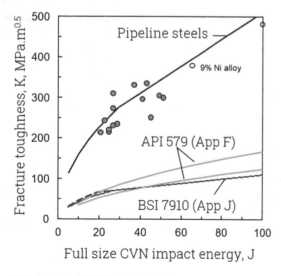

FIG. 12.26 TOUGHNESS CORRELATIONS [60]

Note that this equation uses an upper shelf Charpy toughness: if the Charpy toughness of the material around the crack is not on the Charpy upper shelf, then API 579 or BSI 7910 should be used, with appropriate toughness.

The ln sec formula was developed in the 1960s/70s, is partly empirical, and will have limitations [58]. It can be very conservative with long, shallow defects, and if the defect depth is set to zero, the model predicts decreasing failure pressure, with decreasing defect length: this is clearly incorrect. Accordingly, a modified ln sec model has been proposed [58]:

$$\sigma_\theta = \frac{\overline{\sigma}}{M_P} \cos^{-1}(e^{-x})/(\cos^{-1}(e^{-y})) \tag{42}$$

$$x = \left[\frac{C_v \dfrac{12}{A_c} E\pi}{8c\overline{\sigma}^2} \right] \tag{43}$$

$$y = \left[\frac{C_v \dfrac{12}{A_c} E\pi}{8c\overline{\sigma}^2} \right] \left(1 - (d/t)^{0.8}\right)^{-1} \tag{44}$$

This form of the equation uses imperial units, and the flow stress is SMYS + 10,000 lbf/in^2.

The ln sec equation can be used to assess cracks in line pipe or welds with Charpy toughness on the upper shelf. For example, assume a crack in a longitudinal weld of pipe of diameter 24 inch (610 mm), wall thickness 0.5 inch (12.7 mm), Grade X52 (358 N/mm^2). Design pressure is 1560 psig (108 barg). The weld has a toughness (upper shelf) of 20 ft lbs (27 J). The pipeline has previously been hydrotested to 1950 psig (134 barg), and the crack has been in the pipeline since this hydrotest. In this example it is assumed the crack has a semi-elliptical shape: (area = ($\pi \times$ d $\times$ 2c/4) = 0.31 inches^2).

The calculation is as follows:

$$\sigma_\theta = \frac{2(62000)}{\pi(1.048)} \cos^{-1} \exp - \left[\frac{20\dfrac{12}{0.124}30000000\pi}{8(2)(62000)^2} \right] \tag{45}$$

$$M_P = \left[\frac{1 - \dfrac{0.31}{2}\left(\dfrac{1}{1.353}\right)}{1 - \dfrac{0.31}{2}} \right] = 1.049 \tag{46}$$

$$M_T = \left[1 + 1.255\frac{2^2}{12(0.5)} - 0.0135\frac{2^4}{12^2(0.5^2)} \right]^{0.5} = 1.353 \tag{47}$$

$\sigma_\theta = 37640.3 \times \cos^{-1}\exp - (2.966) = 37640.3 \times 1.5193 = 57185$ lbf/in^2

Failure pressure = 57,185 × 2 × 0.5/24 = 2383 psig.

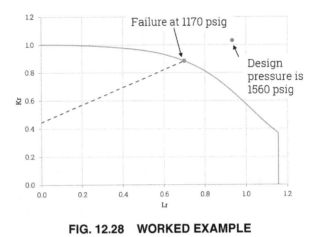

FIG. 12.28 WORKED EXAMPLE

The crack is predicted to fail at 2383 psig (164 barg) using the ln sec equation, but using BSI 7910, the crack is assessed to fail at 1170 psig (81 barg), Figure 12.28. This lower prediction is due to assumptions about welding residual stress, Charpy conversions, etc., in BSI 7910.

The ln sec equation can be shown to be a failure assessment diagram [60], by replacing the Charpy toughness with a stress intensity fracture toughness:

$$\frac{K_c^2 \pi}{8c\overline{\sigma}^2} = \ln\left[\sec\left(\frac{\pi M \sigma_\theta}{2\overline{\sigma}} \right) \right] \tag{48}$$

$$K_c^2 = c\overline{\sigma}^2 \frac{8}{\pi} \ln\left[\sec\left(\frac{\pi}{2}\left\{ \frac{M\sigma_\theta}{\overline{\sigma}} \right\} \right) \right] \tag{49}$$

$$\frac{1}{K_c^2} = \frac{1}{c\overline{\sigma}^2}\left\{ \frac{8}{\pi} \ln\left[\sec\left(\frac{\pi}{2}\left\{ \frac{M\sigma_\theta}{\overline{\sigma}} \right\} \right) \right] \right\}^{-1} \tag{50}$$

$$\frac{M^2\sigma_\theta^2 \pi c}{K_c^2} = \frac{M^2\sigma_\theta^2 \pi c}{c\overline{\sigma}^2}\left\{ \frac{8}{\pi} \ln\left[\sec\left(\frac{\pi}{2}\left\{ \frac{M\sigma_\theta}{\overline{\sigma}} \right\} \right) \right] \right\}^{-1} \tag{51}$$

$$\left\{ \frac{M\sigma_\theta\sqrt{\pi c}}{K_c} \right\} = \left\{ \frac{M\sigma_\theta}{\overline{\sigma}} \right\}\left\{ \frac{8}{\pi^2} \ln\left[\sec\left(\frac{\pi}{2}\left\{ \frac{M\sigma_\theta}{\overline{\sigma}} \right\} \right) \right] \right\}^{-\frac{1}{2}} \tag{52}$$

$$K_r = S_r\left\{ \frac{8}{\pi^2} \ln\left[\sec\left(\frac{\pi}{2}S_r \right) \right] \right\}^{-\frac{1}{2}} \tag{53}$$

This is the same equation as in an FAD (Figure 12.11, and Equation (9)).

12.5.4 Comparison of Crack Assessment Methods

The assessment methods in Table 12.12 have been compared; for example, a 2009 [92] publication compared the predicted failure stresses using the 'ln sec' formula, PAFFC, CorLAS, and BSI 7910 (see Table 12.12) with the failure stresses of real cracks. The methods generally gave conservative results, but difficulties in modeling the irregular crack shapes found in the field, and the differing toughness correlations/assumptions in the methods ensured scatter in the predictions. CorLAS showed superior agreement between its predictions and the observed failure stresses.

A 2010 [93] publication concluded that cracks failing by ductile tearing in line pipe can be assessed using a variety of methods: API 579 showed very good agreement with experimental results; BSI 7910 was the most conservative method; and, the 'ln sec' formula provided conservative collapse pressure predictions.

It is again emphasized that the toughness needed to use documents such as BSI 7910 and API 579 is measured in 'stress intensity factor', or '*J* integral', or 'crack tip opening displacement' (CTOD), but line pipe and its welds have toughness measured by Charpy. Hence, BSI 7910 and API 579 use simple correlations between Charpy and J, CTOD, etc. and these correlations introduce inaccuracies in the calculations [60].

12.5.5 Effect of Fatigue and a Corrosive Environment

A crack in a pipeline subjected to cyclic stresses (fatigue), or continued exposure to a corrosive environment, can grow rapidly. API 579 and BSI 7910 gives further guidance on this 'time dependent' growth, and fatigue has been covered earlier in this Chapter.

12.5.6 Cracks in Welds

12.5.6.1 General Great care should be exercised when a crack is in a weld in a pipeline, as:

- the strength and toughness of the weld are rarely known;
- the shape of the weld will introduce stress concentrations;
- residual stresses could lower the failure stress;
- the shape of the crack will be irregular; and,
- the crack will be difficult to size and detect.

Large safety factors will be needed to accommodate these uncertainties.

12.5.6.2 ERW Line Pipe Line pipe produced using electric resistance welding (ERW) method involves shaping a plate of steel into a tube then welding the ends together by a combination of pressure and temperature. Older ERW line pipe (often called 'low frequency') has a history of problems at or around the longitudinal weld [36, 40, 42, 91].

Older ERW line pipe was manufactured using low frequency (60–360 Hz or less) AC, or DC, welding methods. Also, some ERW line pipe was made using 'dirty' (high percentages of non-metallic inclusions). The weld produced using the older low frequency welding method (up to the 1960s) was narrow, with excess material ground flat.

The older ERW line pipe can have problems; for example, the bond line/heat affected zone was often subjected to a post weld heat treatment after manufacture, which improved quality (e.g., reduced hardness), but this did not occur in all ERW pipe mills, and if the heat treatment was poorly conducted, the weld could crack. Prior to 1960, many sizes and grades of ERW pipe were mill tested to about 75% of SMYS, so large defects could survive this test. Consequently, some older ERW line pipe is poor quality, resulting in the creation of a variety of defects, such as [94]:

- 'cold welds' (lack of fusion between the weld and the line pipe);
- 'hook cracks' (metal separation around non-metallic inclusions in the steel plate);

TABLE 12.14   FAILURES IN ERW LINE PIPE

Defect	% of failures in liquid lines
Lack of fusion	23%
Selective corrosion	23%
Fatigue/corrosion fatigue	32%
Hook cracks	15%

- 'selective' corrosion in the weld (both internal and external surfaces), as some of the line pipe used for this pipe contained high sulfur materials that have not been heat treated [95, 96].

These issues resulted in the ERW line pipe having weld problems, with 98% of in-service failures associated with the longitudinal weld [97], Table 12.14.

ERW line pipe manufactured before 1970 poses the biggest failure risk, but gas pipelines do not have the fatigue problem highlighted in Table 12.14.

Defects in ERW line pipe, in, or near, the weld are difficult to assess, due to uncertainties in the weld and heat affected zone toughness and strength [36, 98–103]. The defects are also difficult to size leading to even more uncertainty. Methods such as the 'ln sec' equation are not recommended as these types of methods require the material toughness to be on the 'upper shelf', and many of these welds are not on the upper shelf. More appropriate models are BSI 7910 and API 579, providing the weld toughness is known and fatigue/corrosion rates are known,

Many operators have the older ERW line pipe, but not all older ERW line pipe is poor quality: most has been operating satisfactorily for decades. Criteria exist to assess if a pipeline made of older ERW line pipe will have these problems [103].

12.5.6.3 Girth Welds Line pipe in joined in the field by girth welding to a standard; for example, API 1104 [62], Figure 12.29. All welds contain defects; consequently, all welding standards allow some defects to remain in the weld: American, Canadian, Australian and European standards for girth welds allow similar levels of defects. These levels are 'workmanship' levels—defects that a good worker, using good processes and materials will leave in the weld. Typical workmanship limits are surface-breaking defects of length 25 mm, and embedded defects of 50 mm length. Note that most welding standards still give defect acceptance levels based only on length, as radiography is used to detect the defects. Radiography gives a two-dimensional image, and therefore cannot determine defect depth. Cracks are not usually allowed, as cracks can indicate bad welding, poor material, poor quality control, etc. Cracks are usually repaired.

Older girth welds did not have the rigorous welding standards or inspection of today's welds. This means they can contain defects outside current standards [104], Figure 12.29.

Most of the defects in a girth weld will be in the circumferential direction. This means the major stress the defects will be subjected to will be the axial stress, not the hoop stress. Fortunately, most pipelines are not subjected to high axial stresses; therefore, girth welds are generally under low loads.

Girth weld failures are rare: only 2% of all failures in USA pipelines are due to girth welds; and the majority gives leaks not ruptures. The consequences are also low: there has been no recorded casualty/fatality caused by them in the USA [57].

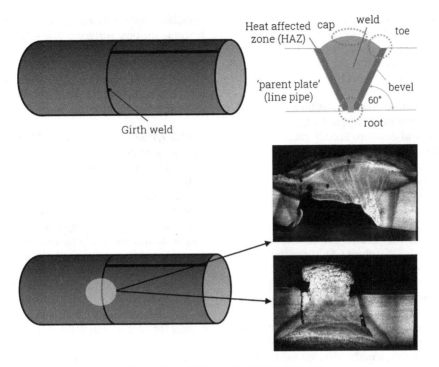

FIG. 12.29   PIPELINE GIRTH WELDS

Defects in girth welds larger than the workmanship levels must either be repaired or assessed using fracture mechanics. Welding standards usually allow the use of fracture mechanics methods such as API 579. Indeed, API 1104 has a fracture mechanics appendix [62], and the European Pipeline Research Group has published fracture mechanics-based guidelines [27, 69]. Note that all these methods require the toughness of the weld to be known.

12.5.7   Stress Corrosion Cracking

Pipelines are highly stressed, due to the high internal pressures, and they are surrounded by a hostile environment in terms of corrosion (for example, soil contains water which will contain carbon dioxide (CO_2) from the decay of organic matter, and seawater is another aggressive electrolyte). This environment can lead cause both corrosion and cracking. The cracking is called 'environmental cracking'.

12.5.7.1 Stress Corrosion Cracking in Pipelines   Stress corrosion cracking (SCC) is environmental cracking, and occurs in many industries, including nuclear and shipping. ASME B31.8S [67] defines SCC as a: "… *environmental attack of the metal involving an interaction of a local corrosive environment and tensile stresses in the metal, resulting in formation and growth of cracks.*" External SCC in pipelines has been well known for many years. The first documented case of SCC causing a pipeline failure was in 1965 in Natchitoches, Louisiana [105]. This rupture caused a gas release and fire resulting in 17 fatalities.

SCC can be in liquid and gas lines, but most incidents are on natural gas pipelines. The SCC is usually in high hoop stress (at or above 60% SMYS) pipelines, occurring in clusters ('colonies'), growing perpendicular to the maximum stress, with the fracture faces covered in magnetite (Fe_3O_4) and carbonate films.

SCC can occur at any location in the pipeline, although most failures are within 20 miles (32 km) downstream of a gas compressor station (where temperatures and pressure will be high). Pressure fluctuations (fatigue) assists/accelerates this cracking.

12.5.7.2 Cause of Stress Corrosion Cracking   Pipeline coatings can deteriorate and disbond. The line pipe steel will then come into contact with ground water in places where the coating is damaged, disbonded, or porous. This can cause corrosion.

The steel then corrodes and cracks due to the presence of hydrogen (from a corrosion process), but often without any obvious deformation or material deterioration (although pitting corrosion is associated with the cracking). The outside surface of the pipeline is cracked by the SCC.

12.5.7.3 Types   There are two types of external SCC in the pipeline business[3] [105]:

- 'high pH SCC' (occurs in a high pH environment of 9, and was first reported in 1965 in the USA);
- 'near-neutral pH SCC' (occurs in a pH environment of between 5 and 7.5, and was first reported in 1985 in Canada).

High pH SCC is associated with high temperatures, and occurs more frequently at higher temperature locations (>100°F [38°C]), whereas temperature is not implicated in near-neutral pH SCC, but high temperature can damage coating.

12.5.7.4 High pH or Near-Neutral pH SCC Creation   The electrolyte (wet soil) around an onshore pipeline contains carbon dioxide (CO_2) from the decay of organic matter. The CO_2 can dissolve in water and form carbonic acid (H_2CO_3), which then can dissociate to form hydrogen (H^+), bicarbonate (HCO_3^-), and carbonate (CO_3^{2-}) ions.

The pipeline's cathodic protection (CP) plays a key role in creating the pH value at the pipe's surface. API 1176 [14] states: "*The balance between the level of cathodic protection reaching the*

---

[3] It is the pH value at the tip of the SCC that is important, not the pH of the soil around the pipe.

surface and the CO_2 partial pressure in the ground water is critical in determining the pH at the steel surface, and hence, whether high-pH stress corrosion cracking, near-neutral-pH stress corrosion cracking, or neither of these, occurs."

The CP will attempt to stop the corrosion [106–108]:

- If the surface under a disbonded coating is still receiving CP current (i.e., the coating is not shielding the CP), a carbonate-bicarbonate environment will develop. The CP causes the pH of the electrolyte to increase, and the CO_2 dissolves in the elevated pH electrolyte, resulting in a concentrated carbonate-bicarbonate (CO_3-HCO_3) electrolyte.
- If the disbonded coating is being shielded from the CP, there will be some corrosion.

The cracking environment for near-neutral pH SCC appears to be a dilute groundwater containing dissolved CO_2 (from the decay of organic matter and geochemical reactions in the soil). There is little, if any, CP current reaching the pipe surface, either because of the presence of a shielding coating, a high resistivity soil, or inadequate CP design.

The environment for high pH SCC, also involves CO_2. Cathodic protection causes the pH of the electrolyte beneath disbonded coatings to increase, and the CO_2 dissolves in the elevated-pH electrolyte. This generates a concentrated CO_3-HCO_3 electrolyte.

12.5.7.5 Cracking Types High pH cracking is usually 'intergranular' (the crack passes between the grains in the steel microstructure), and narrow, whereas low pH cracking is 'transgranular' (the crack travels through the grains) and wider, with less branches.

12.5.7.6 Primary Causes of High pH and Near-Neutral pH SCC [105–112] The principal factor contributing to high pH SCC is coating condition. A groundwater/carbonate/bicarbonate solution (high pH) needs to be in contact with the pipe through a defect in the pipeline coating. The surface under the disbonded coating is still receiving CP current, which allows a carbonate-bicarbonate environment to develop. The CP causes the pH of the electrolyte to increase, and the CO_2 dissolves in the elevated pH electrolyte, resulting in a concentrated carbonate-bicarbonate (CO_3-HCO_3) electrolyte. It is not associated with surface corrosion, due to the CP being in the partially-protected range (CP is –600 to –700 mV [Cu/CuSO$_4$]).

The principal factor contributing to near-neutral pH SCC is a breakdown in the pipe coating and the CP. A coating defect must exist which is then exposed to a groundwater solution containing dissolved CO_2. The surface under the disbonded coating is being shielded from CP, resulting in little, if any, CP current reaching the pipe surface (e.g., the pipe coating is shielding the CP). It is

associated with light surface corrosion due to the absence of CP reaching the pipe surface (CP is –760 to –790 mV [Cu/CuSO$_4$]).

SCC in oil and gas pipelines has never been reported in subsea pipelines, or line pipe with factory (as opposed to field-applied) coatings [111], and there have been no reports of in-service or hydrotest failures in pipelines installed after 1981 (presumably due to these 'modern' pipelines being factory coated).

12.5.7.7 SCC in Canadian Pipelines [110, 112, 114] There have been many SCC failures in Canada: 17% of Canadian pipelines failures have been by SCC, compared to 2% in the USA. Most of the Canadian failures have been at/near the longitudinal weld, and are near-neutral pH SCC.

Most of the Canadian failures have been on tape wrapped pipelines: tape wrap is usually put on the pipeline in the field. The longitudinal seam weld contains residual stresses, microstructural changes, and causes 'tenting' of the coating, around the 'peaking' at the longitudinal weld (longitudinally welded line pipe cannot be a perfect circular shape at this weld), Figure 12.30. 'Tenting' at the seam weld can allow moisture to penetrate, and if the coating is polyethylene (PE) tape, there will be shielding from the CP. This is perfect for near-neutral pH SCC to occur.

12.5.7.8 SCC Assessment SCC can be assessed using the crack assessment methods covered earlier in this chapter: API 579; BSI 7910; PAFFC; 'ln sec'; and, CorLAS [112]. A review [113] of SCC assessment noted that CorLAS usually resulted in the best overall performance, but all the published methods show scatter in their predictions of failure stresses of SCC [106].

SCC grows with time: this growth must be quantified if SCC remains in the pipeline. After initiation, stress corrosion cracks usually show an approximately uniform velocity, provided that the stresses are maintained and the local environment remains unchanged (see BSI 7910, Figure 12.31). SCC failure is a very complex process that can involve the formation, initiation, coalescence, interlinking and interaction of numerous cracks. This leads to highly varied growth rates for both high pH and near-neutral pH SCC. Growth rates from field measurements and laboratory testing range from 0.06 to 0.88 mm/year [112]. API 1160 [61] notes a worst-case rate of growth of 0.61 mm/year, Table 12.11.

12.5.8   Laminations
Laminations are a plane of non-fusion in the interior of a steel plate created during the steel manufacturing process. A 'delamination' is the separation of a lamination, under stress.

Laminations are normally in the centre of the plate, parallel to the plate's surface. They were originally inclusions or blow holes in the steel ingot that become elongated during the rolling process.

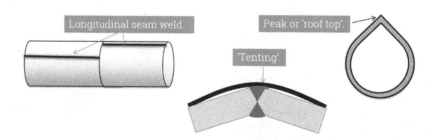

FIG. 12.30   'TENTING' OF TAPE WRAP COATING AT A LONGITUDINAL WELD

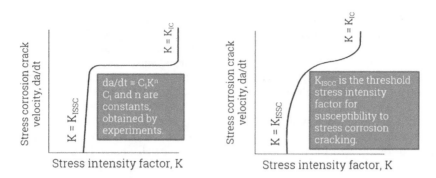

FIG. 12.31 GROWTH OF SCC [15]

Laminations are not structurally significant if:

- they have survived the pipe mill test and field hydrotest;
- they are not in a hydrogen-charging environment; and,
- they are parallel to the surface of the plate.

These laminations rarely cause failures [102] unless they can cause a leak path due to:

- being inclined to the surface of the plate; or,
- many being present (multiple laminations) through the wall thickness.

There have been failures due to laminations when they have been orientated at an angle to the pipe wall: they reduce the effective wall thickness and have caused failures. Usually the failures occur during the pre-service test.

API 579 [16] says: "*A remaining life evaluation [for laminations] is typically not required for shell components under internal pressure loads (i.e., membrane tensile loads) because there is no loss of strength and no associated effect on the internal inspection interval except for the special inspection requirements for in-service monitoring. A remaining life may be required for laminations subject to compressive stresses or bending stresses, or for laminations in components subject to cyclic loads.*"

12.6 MECHANICAL DAMAGE

12.6.1 Introduction

Pipelines can be impacted by excavating machines (onshore), anchors (subsea), etc. This impact is called 'external interference'. External interference causes 'mechanical damage' in the line pipe, such as a gouge or a dent. This damage is sometimes called 'third party damage' as most (but not all) of the damage is caused by third parties (the 'first party' is the pipeline owner, the 'second party' are those authorized to work on the pipeline, the 'third party' are those not authorized to work on the pipeline).

ASME B31.8S [67] defines:

- a gouge as a... "*mechanically induced metal-loss, which causes localized elongated grooves or cavities in a metal pipeline.*"
- a dent as... "*permanent deformation of the circular cross-section of the pipe that produces a decrease in the diameter and is concave inward.*"

Most (>80%) onshore pipelines contain many dents, but they do not contain any other defect in the dent (e.g., a gouge) [115].

TABLE 12.15 PIPELINE FAILURES AND CASUALTIES (USA, 1993–2012) [116]

Cause	Natural gas		Hazardous liquids	
	Fatalities	Casualties	Fatalities	Casualties
Corrosion	13	4	1	18
Excavation damage	15	51	12	38
Incorrect operation	0	9	9	20
Material/weld/ equipment failure	8	71	4	12
Natural force	0	2	0	1
Other outside force	0	13	3	5
All other causes	6	38	10	43

Pipeline failures due to these 'plain' dents (dents with no surface damage) are not a significant portion of reported failures [45]. This indicates that plain dents are not a major threat to pipelines, but other types of mechanical damage (e.g., dents with gouges) are a major threat, and can lead to serious consequences, Table 12.15.

Mechanical damage will be exposed to all the stresses on the pipeline. This means it will have to resist 'static' stresses (e.g., the maximum hoop stress), and cyclic stresses (e.g., from internal pressure variations). Therefore, damage needs to be assessed against static stresses ('burst strength'), and cyclic stresses ('fatigue life').

12.6.2 Delayed and Immediate Failures

Mechanical damage usually fails a pipeline immediately [41, 117], but the damage can fail some time after it has been introduced ('delayed failure') [61]. The damage can gradually grow by, for example, fatigue.

Most mechanical damage incidents occur immediately; therefore, inspection methods ('reactive' methods) will only reduce the failures/consequences by a small percentage [118]. It is better to protect pipelines against damage (be 'proactive'), by using methods such are protective barriers, 'awareness' methods (such as markers), 'one-call-systems', etc. [118].

12.6.3 Gouges

A gouge is surface damage to a pipeline. It is usually caused by foreign objects removing part of the pipe wall. Gouges can be assessed as a part-wall defect in line pipe, but this assessment requires caution as:

- the gouge could be associated with a dent;
- the gouge can be associated with a 'hard layer' and cracking;
- the gouge can subsequently crack due to hydrogen cracking;
- the gouge can be susceptible to fatigue cracking.

12.6.3.1 Associated Denting Gouges may be associated with denting, as after impact and removal of the impactor, the dent will attempt to [23]:

- 'springback', due to the elasticity in the pipe wall; and then,
- 'reround', due to the internal pressure in the pipeline.

Cracks can form at the base of a gouge in the dent, during this springback and rerounding of the dent, Figure 12.32.

A combined dent and gouge is the most severe form of mechanical damage, and can fail at stresses well below that of a gouge, or a dent, alone.

12.6.3.2 Hard Layer An impact on the line pipe will crush the steel, leading to 'work hardening' or 'cold work' (plastic flow). This reduces the line pipe's ductility. An impact on a pipeline will also create high energy. Some of this energy is heat from friction between the line pipe and the impacting object, and this heat can be very intense, and can change the microstructure of the line pipe. As the metal subsequently cools, it becomes brittle (low toughness). All these factors lead to a 'hard layer' (low ductility and low toughness) around the gouge [23, 47, 117, 119].

The gouge is surrounded by the hard layer, and the hard layer cracks easily. Additionally, any 'springback' or 'rerounding' of the pipe will crack the layer. The crack can sometimes be hidden by folded metal, Figure 12.33. These folds of metal are sometimes described as 'spalling' or 'pancaking'.

The presence of the hardened layer and possible cracking below the gouge needs to be taken into account when assessing a gouge in line pipe. The 'Pipeline Defect Assessment Manual' recommends that the measured depth of the gouge should be increased by 0.5 mm to account for these factors [23]. Additionally, the line pipe should be shown to have a toughness in excess of 20 J (15 ft lb) [21, 120–122].

12.6.3.3 Hydrogen Cracking Risk Impacts on a pipeline will damage/remove coating on the pipeline, allowing the environment to contact the line pipe. This will lead to corrosion, but the cathodic protection should reduce the effect. This is due to the corrosion being part of an electrochemical cell (which consists of a cathode, anode, connection [to allow the flow of electrons from the anode to the cathode] between the anode and cathode, and an electrolyte [to allow the flow of ions surrounding the anode and cathode). The anodic reaction and a cathodic reactions are:

- at the anode electrons are lost and there is corrosion: $Fe \rightarrow Fe^{2+} + 2e^-$;
- at the cathode electrons leaving the metal enter the electrolyte (e.g., water) reacting with hydrogen ions and forming hydrogen gas: $2H^+ + 2e^- \rightarrow H_2$.

Unfortunately, the hard layer around a gouge can cause a problem, as it can crack due to 'hydrogen ('stress') cracking'. This cracking needs:

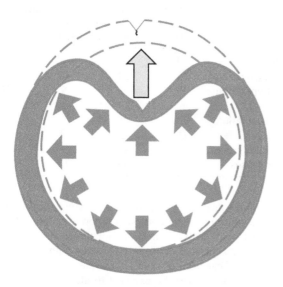

FIG. 12.32   DENT SPRINGING BACK AND REROUNDING DUE TO INTERNAL PRESSURE [47]

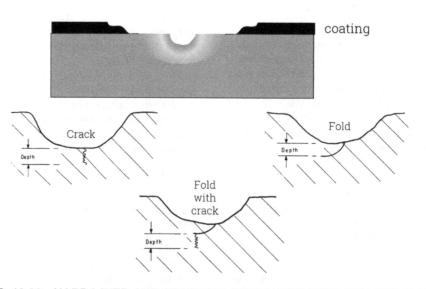

FIG. 12.33   HARD LAYER AND CRACKING ASSOCIATED WITH GOUGING [119]

- a steel of high hardness and high strength;
- sufficient sustained tensile stresses (e.g., >50% SMYS); and,
- a source of atomic hydrogen.

Clearly, the cathode reaction around a gouge creates a suitable environment for hydrogen cracking.

12.6.3.4 Fatigue Cracking A gouge can have a sharp base and cracks in its surface. Under cyclic stresses these gouges/cracks may grow and cause failure due to fatigue [123].

12.6.3.5 Caution with Gouges Standards treat gouges with great caution; for example [9], "*Gouges, grooves, and notches have been found to be an important cause of pipeline failures, and... must be prevented, eliminated, or repaired.*"
This cautious approach is in recognition of [38]:

- the known severity of mechanical damage; and,
- the fact that a residual dent may have been overlooked around a gouge.

12.6.3.6 Assessment Axially-oriented gouges in line pipe can be assessed as part-wall defects [124, 125], but some allowance must be made for the possibility of a hard layer/crack. The equation in Section 12.4.3.4 can be used, with the gouge depth increased (d_c) to allow for this cracking/hard layer:

$$\frac{\sigma_f}{\bar{\sigma}} = \frac{1 - \dfrac{d + d_c}{t}}{1 - \dfrac{d + d_c}{t}\dfrac{1}{M}} \tag{54}$$

$$M = \sqrt{1 + 0.26\left(\frac{2c}{\sqrt{Rt}}\right)^2} \tag{55}$$

with the flow stress taken as the average of the yield and ultimate tensile strength, and an assurance that the line pipe has a minimum toughness of 20 J (15 ft lb).
Mechanical damage such as gouges and dents have caused ruptures on pipelines operating below 30% SMYS, but none below 25% SMYS [126, 127]. Ruptures at these low stresses are unlikely because of the combination of the following factors that need to be present [127]:

- defects must be long;
- defects must lay axially along the pipe;
- defects must be very deep (about 80%–90% of the wall thickness); and,
- line pipe must have low toughness.

Circumferentially-oriented gouges in line pipe can be assessed using the equation in Section 12.4.5, again with an allowance on the gouge depth for the cracking/hard layer, and a line pipe toughness in excess of 20 J (15 ft lb).
Note that in all assessments of gouges, the effect of fatigue must always being considered and, where necessary, included, and the risk of hydrogen cracking must be eliminated.

12.6.4 Dents

Dents are often detected in pipelines. Their severity is usually measured by their depth, H. This depth is usually measured as a

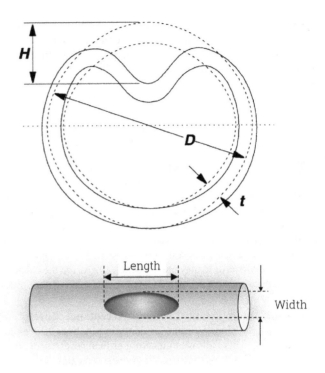

FIG. 12.34 DENT NOMENCLATURE [47]

percentage of the pipeline's outside diameter, Figure 12.34. The length and width of the dent affects:

- the stress and strain distribution in the dent; and,
- its geometric stiffness,

but these effects are secondary compared to the dent depth when considering the dent's 'static strength' and 'fatigue life' (see later).
Dents are common in pipelines; therefore, standards give acceptance levels. ASME B31.4 [8] requires the following dents detected during operation to be removed or repaired, or be the subject of an engineering evaluation if they are to remain in the pipeline:

- dents with a depth greater than 6% of the nominal pipe diameter (0.250 in. (6.4 mm) in depth for a pipe diameter NPS 4 and smaller);
- dents containing gouges, grooving, scratches, cracking or other stress riser;
- dents containing metal loss resulting from corrosion or grinding where less than 87.5% of the nominal wall thickness remains;
- dents that affect pipe curvature at a girth weld or a longitudinal seam weld.

12.6.4.1 Dents (Static Loading) A 'plain' dent is damage to the pipewall which causes a 'smooth' change in curvature of the pipe wall, without a reduction in pipe wall thickness (i.e., it contains no defects or imperfections). Many experiments have shown that plain dents do not affect the strength of pipelines when the pipeline is subjected to static pressure. The plain dent introduces high localized stresses (causing yielding and wall thinning) in the dented area, but these high stresses and strains are accommodated by the ductility of the line pipe [47].
On pressurization, the dent attempts to move outward, allowing the pipe to regain its original circular shape, through 'spring back' and 'rerounding' (Figure 12.32). The dent will not reduce the static

strength of the pipe provided nothing restricts the movement or acts as a stress concentration (e.g., a gouge or a kink in the dent) [47].

Experiments have shown dents of depth ≤ 7% D are acceptable under 'static' (no cycling) pressure [47]. This ≤ 7% D is a technical limit, but deep dents:

- may restrict product flow;
- can prevent the passage of pigs along the pipeline; and,
- create high strains that may disbond coatings.

These operational considerations lead to dent acceptance limits being usually set at 6% D in standards [38, 41].

ASME B31.8 [9] details a dent assessment based on strains, which can allow plain dents of ≤ 6% pipe diameter to be accepted in the line pipe providing strains in the dent are ≤ 6% (≤4% if the dent affects a weld) [37, 38]. The various strains are calculated using an effective strain calculated using the von Mises Theory [45]. The equations are listed in the standard, so they are not detailed here, but the calculations will need accurate three-dimensional measurements of dent profile.

12.6.4.2 Dents (Static Loading) Acceptance Levels Many references give simple limits for dents in pipelines [for example, 61 and 67], and also allow engineering critical assessments if limits are exceeded. Table 12.16 [61, 67] gives some examples.

12.6.4.3 Dents (Fatigue Loading) Pipelines can be heavily pressure cycled; for example, a gas line could experience 60 cycles per year of a pressure differential of 200 psig (13.8 barg), whereas, the same pressure differential can occur more than 1800 times in a liquid line in a year [54]. Therefore, liquid pipeline operators are more concerned about fatigue.

Under cyclic pressure loading there will be large cyclic stress and strains localized in the dent. The pressure cycles will make the dent move in and out, bending the line pipe. Cyclic pressure fatigue tests on plain dents in pipes indicate that plain dents reduce the fatigue life compared to plain circular pipe [47].

The fatigue life of a dent in a pipeline that is pressure cycled can be calculated using two models:

- the 'SES' model (from the American Gas Association [AGA]) [128];
- the European Pipeline Research Group (EPRG) model [129].

These models use S-N curves, modified for the stress concentration due to the dent, Figure 12.35. Table 12.17 shows the formulae used in these models.

Fatigue models rely on empirical data (e.g., S-N curves), and this leads to uncertainties in determining fatigue life. Safety factors are applied to fatigue life calculations: a factor of 10 is often used. Accordingly, the models in Table 12.17 will require safety factors; for example, a safety factor of 13.3 is recommended when using the EPRG model [23].

12.6.4.4 Dents on Welds A dent may be on, or coincide, with a weld. Pipeline welds are usually stronger ('over-matched') than their 'parent' material (line pipe), and can withstand high stresses, but other properties, such as ductility and toughness, may be inferior, and this could lead to cracking in the weld if it is subjected to high stresses. Additionally, the fatigue life of a weld is always lower than the parent material; therefore, a dent with a weld will have a lower fatigue life than a dent in parent material.

TABLE 12.16   DENT ACCEPTANCE LEVELS

API 1160		ASME B31.8S	
Requiring immediate response in all pipeline segments requiring 365-day response in 'critical' areas		**Requiring condition assessment within a period not to exceed 5 days**	**Requiring a 'scheduled' (Time to become critical) response**
Any dent containing cracking.		Dents with gouges.	Dents with cracks.
Any dent above the 4 and 8 o'clock position that contains stress raisers (gouges, notches, scratches), or corrosion.	A dent below the 4 and 8 o'clock position that contains stress raisers (gouges, notches, scratches), or corrosion.		
Any dent above the 4 and 8 o'clock position with a depth greater than 6% of the nominal pipe diameter.	A dent above the 4 and 8 o'clock position with a depth greater than 2% of the nominal pipe diameter (0.250 in. (6.4 mm) in depth for a pipe diameter less than NPS 12).		A plain dent that exceeds 6% of the nominal pipe diameter.
	A dent below the 4 and 8 o'clock position with a depth greater than 6% the pipe diameter where critical strain levels have been exceeded.		
	A dent with a depth greater than 2% of the nominal pipe diameter [0.250 in. (6.4 mm) in depth for a pipe diameter less than NPS 12] that affects a girth or longitudinal weld exceeded.		Dents that affect ductile girth or seam welds if the depth is in excess of 2% of the nominal pipe diameter.
			Dents of any depth that affect non-ductile welds.

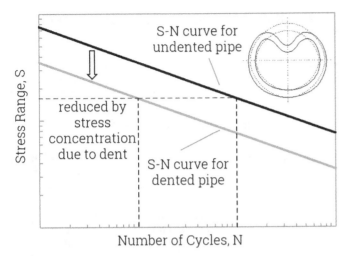

FIG. 12.35 MODIFYING AN S-N CURVE FOR A DENT

TABLE 12.17   DENT FATIGUE MODELS

SES (Imperial units)		EPRG (Metric units)	
$N = 2.0 \times 10^{6} \left(\left[\dfrac{\Delta\sigma'}{\Delta p'} \right] \Delta p \, \dfrac{1}{11400} \right)^{-3.74}$		$N = 1000 \left[\dfrac{(\sigma_U - 50)}{2\sigma_A K_S} \right]^{4.292}$	
N	Predicted fatigue life of plain dent	σ_a	Function of stress range and σ_u
t/D	Wall thickness(mm)/ diameter(mm)	H_0	Dent depth (at zero pressure)
Δp	Cyclic pressure, psi	σ_u	Ultimate tensile strength
$\left[\dfrac{\Delta\sigma'}{\Delta p'} \right]$	'Stress intensification factor'	K_s	$2.871.(H_0(t/D))^{0.5}$

Experiments on dented seam welds in line pipe have shown very low static failure stresses (as low as 7% SMYS) and very short fatigue lives [23].

Dents on good quality, ductile, welds, free from major defects, can have high failure stresses and fatigue lives; however, in some impacts, the large stress and strains associated with the denting process cause cracking in the welds, Figure 12.36, leading to low failure stresses and short fatigue lives.

API 1160 [61] and ASME B31.8S [67] give guidance on a dent affecting longitudinal and girth welds, Table 12.16, and both indicate that dents of depth ≤2% pipe outside diameter (API 1160 has a depth limit of 0.250 inch [6.4 mm] in pipe diameters less than NPS 12) affecting a good quality weld, with its coating intact, are insignificant, but note these standards have other conditions attached to these depths, and the reader is directed to the standards for these details.

When a pipeline is pressure cycled, the fatigue life of the dent affecting the weld must be assessed using a suitable S-N curve for the weld, and the increased stresses in the dent.

12.6.5   Restrained Dents

Dents created by impact can move outwards; therefore, these are 'unconstrained' (or unrestrained) dents Pipelines can be dented by rocks that rest under them: these dents are 'constrained' (they cannot move outwards due to the presence of the rock), but the overburden on a pipeline can cause the rock to puncture the pipeline.

The constraint in a dent reduces their threat [58]:

- A publication [130] concludes that constrained dents less than 6 percent of the outside pipe diameter in depth are not a problem, unless they involve a sharp object which might cause a puncture, or they interfere with operations such as pigging.
- Another publication [14] notes that plain dents that have been excavated and that have had the indenting object removed (allowing the pipe circular cross section to partially re-round under internal pressure) are susceptible to fatigue from pressure cycles acting on the unrestrained residual dent.

Dents created by rocks can pose long term problems [37]:

- coating damage;
- corrosion;
- shielding from cathodic protection;
- stress corrosion cracking (SCC) or hydrogen cracking; and,
- punctures due to continued settlement (if rock is sharp and overburden load is high).

12.6.6   Dents with Corrosion

Tests on corrosion in rock dents have shown the dent and corrosion to have high static failure strength, due to the dent preventing the corrosion 'bulging' [37]. ASME B31.8 [9] allows the assessment of corrosion in dents using ASME B31G [59].

Corrosion protection will still be needed around the dent to prevent further corrosion and possible SCC.

12.6.7   Dents with Gouges

Most onshore pipelines contain many dents [115], but pipeline failures due to plain dents are not a significant portion of reported failures [45]. However, dents containing gouges do fail pipelines [131].

An impact to a pipe can cause both denting and gouging. The line pipe's ductility will accommodate the associated high strains and stress concentrations, but the large strains created as a dent moves outwards (under the action of the internal pressure) can cause tearing of any defect within the dent. The tearing goes through the remaining ligament of the pipewall, and is resisted by the line pipe's toughness; therefore, good line pipe ductility and toughness is needed to resist a gouge in a dent [54].

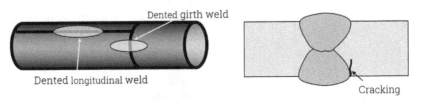

FIG. 12.36   DENTS ON WELDS

TABLE 12.18 EFFECT OF DENTS CONTAINING DEFECTS [54]

Dent depth (% Pipe diameter)	Notch depth (% Wall)	Burst stress (% SMYS)
5	5	141
5	10	120
10	10	96
15	15	53
12	10	99
15	5	51

Dents containing defects can record low failure stresses, Table 12.18, as the line pipe cannot withstand the high strains around the gouge [54], and are usually not allowed in pipelines, Table 12.16.

There is a model to predict the failure stress of a combined dent and gouge [16, 132], but it is prone to scatter, and needs suitable safety factors applied to its predictions [23]. The combined dent and gouge also can record very low fatigue lives: the fatigue life can be as low as one or ten percent the fatigue life of an equivalent plain dent [54].

12.6.8 Ripples in Bends

Pipelines can contain ripples ('wrinkles'). These ripples are usually at the intrados of a field bend in a pipeline, Figure 12.37. This is accidental damage, not caused by impact—it is caused by poor control of the field bending. These ripples should not be confused with large wrinkles created deliberately by the bending process during pipeline construction to produce a 'hot wrinkle bend' [10, 52]. These wrinkle bends are a very old construction practice, not used today.

Many pipeline standards contain limits on ripples in bends [133]. Allowable limits on the ripples are not set by static failure pressure, as ripples have little effect on static failure strength, but they can be a weakness in the line pipe when the pipe is under compressive loading, and it has been noted that longitudinal cyclic loading may cause a fatigue crack to grow at the ripple [39, 46].

ASME B31.4 [8] and ASME B31.8 [9] both contain simple guidelines on the depth of ripples (h) as a percentage of the pipeline's outside diameter (D); for example, Table 12.19.

ASME B31.8 allows ripples having a dimension measured from peak to valley $\leq$ 1% pipe outside diameter in all gas lines, and deeper ripples ($\leq$2%) in low stressed pipelines.

12.7 CONCLUSIONS

Pipelines are a very safe form of transportation, and they have a very good safety record, but defects such as corrosion and denting will inevitably occur during service, and a pipeline operator must continually check a pipeline for these defects.

Many of these defects can be both detected and assessed. These assessments ('fitness-for-service') can allow many defects to

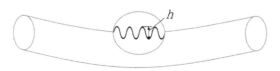

Ripples in a bend

FIG. 12.37 RIPPLES IN FIELD BENDS

TABLE 12.19 RIPPLE ACCEPTANCE CRITERIA IN ASME B31.4

Maximum operating hoop stress, S	(h/D) × 100 cannot exceed*:
$\leq$20,000 psi (138 MPa)	2
>20,000 psi (138 MPa) $\leq$30,000 psi (207 MPa)	[((30000 − S)/10000) +1]
$\leq$47,000 psi (324 MPa)	0.5[((47000 − S)/17000) +1]
>47,000 psi (324 MPa)	0.5

remain safely in the pipeline. Fracture mechanics forms the basis of fitness-for-service assessments. This science has developed over centuries, but the major developments have occurred since the 1950s. A problem with the methods for use on pipelines, particularly when assessing cracks, is that they require a detailed knowledge of material properties such as fracture toughness. Unfortunately, the toughness of line pipe is measured by the Charpy test, which is only an indication of toughness, and will lead to uncertainties in any fitness-for-service assessment.

The methods used for assessing corrosion, and for assessing plain dents, do not require this detailed knowledge of material properties, and are mainly dependent on accurate measurements of size.

12.7.1 Competence

Defect assessment requires competent staff. 'Competence' is a mix of practical and thinking skills, experience, and knowledge, and also depends on the individual's values, and this competence must be 'demonstrable', through a qualification or a certification.

Standards have always emphasized the need for competent staff (e.g., [8, 9]), but the USA's NTSB has recently stated [134], *"The NTSB concludes that professional qualification criteria for pipeline operator personnel performing IM [integrity management] functions are inadequate."*

Developing and maintaining competency involves education, training, mentoring, etc., and all companies supporting staff in defect assessment must ensure these staff are demonstrably competent.

12.7.2 The Five Rules of Defect Assessment

Any person assessing a defect should check that he/she has followed these rules:

- always understand the cause of any defect being assessed;
- understand and use the best assessment practices;
- use all relevant data (e.g., inspection data, operations records, maps, etc.);
- check calculations, inputs, outputs, and assumptions; and,
- always appreciate the consequences of any failure of the defect.

The cause of the defect may indicate a more general problem in the pipeline, and will certainly affect the choice of assessment method. The consequences of any possible failure will affect the safety factor applied to the calculations.

12.7.3 Corrosion

Pipelines can tolerate corrosion, as most corrosion is blunt, and line pipe usually has sufficient ductility and toughness to tolerate large areas. Corrosion can be assessed using a variety of methods, but ASME B31G [59] is the most recognized.

The assessment of corrosion should follow a process:

- corrosion is suspected/detected by inspection;
- assess if corrosion is credible in the location identified (check corrosion monitoring records, coating surveys, low points in the pipeline, changes in pipeline elevation, changes in product composition, changes in temperatures, etc.);
- explain the reason for the corrosion, and any implications for the rest of the pipeline;
- question if the reported size of the corrosion is credible given the pipeline's service life and operating conditions;
- ensure the reported defect is corrosion and not mechanical damage (for example, corrosion is often located around the 6 o'clock position of the pipeline, whereas mechanical damage can be at the top of the pipeline);
- check for neighboring corrosion defects, and check for 'interaction';
- check that associated cracking is unlikely (stress corrosion cracking and hydrogen induced cracking can be associated with corrosion);
- check probabilities of detection and detection accuracies for the corrosion, and include allowances in calculations;
- consider location of corrosion along the pipeline route in terms of impact on people safety and environmental damage should the corrosion fail at a future date;
- assemble all necessary data for the assessment process (e.g., material yield strength);
- perform calculations, and include any future growth of the corrosion;
- report and record.

12.7.4   Cracks

Cracks are sharp defects which indicate something is going/gone wrong in a pipeline; consequently, cracks are treated with great caution. A crack assessment requires two calculations, Figure 12.38:

- a calculation to determine its failure stress; and,
- a calculation to predict any future growth (by fatigue or environmentally-assistance).

There are a number of crack assessment methodologies (e.g., [15, 16]). A specific cracking problem in pipelines is stress corrosion cracking (SCC). SCC is caused by high stress and a corrosion environment. It can be assessed, but the predictions have scatter, and growth is difficult to predict.

A crack assessment is critically dependent on input data, in particular, the crack size, location (weld or line pipe material) and the material fracture toughness. Uncertainties or inaccuracies in these key parameters will have an adverse effect on the calculation.

12.7.5   Defects in Welds

All welds contain defects, but older welds may contain defects much bigger than current acceptance levels. A major problem with assessing weld defects is the absence of material properties (strength and toughness), unknown stresses (due to stress concentrations and welding residual stress), and assuming values will lead to uncertainties in the predicted failure stress. Additionally, defects in longitudinal welds are sensitive to fatigue loading; therefore, longitudinal defects need to be assessed for both fracture and fatigue.

Girth welds can contain defects larger than acceptance standards. These girth weld defects can be assessed, but again toughness properties are needed which are not usually available.

Older electric resistance welded (ERW) line pipe can contain defects in the weld and weld region. These ERW weld defects can cause failures, and need to be managed. Poor and/or unknown toughness properties in and around the weld mean the assessment of these defects is difficult.

12.7.6   Mechanical Damage

Mechanical damage (e.g., gouges, dents) is a major cause of pipeline failures, and casualties; therefore, pipelines need to be protected against this damage. Gouges can be assessed, but the possibility of associated denting, surface cracking, hydrogen cracking, and susceptibility to fatigue means that standards are cautious and require their prevention, elimination, or repair.

Dents containing no surface damage (such as gouging) fail at high stresses, and standards allow many of these dents to remain in service, but they can be a fatigue problem and require fatigue assessment. Dents containing welds can have low failure stresses and fatigue lives, and dents containing gouges are a severe form of damage and are usually repaired.

12.7.7   Damage Assessment Process

The assessment of mechanical damage should follow a process:

- damage is suspected/detected by inspection;
- assess if damage is credible in the location identified (check excavation activities, shipping lanes, surveillance records, one-call records, rocky terrains, nearby building/constructions, coating surveys, highway/road/river crossings, etc.);

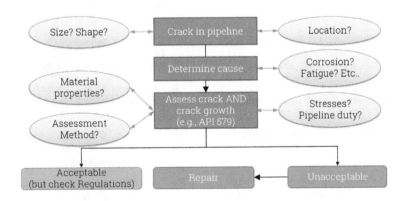

FIG. 12.38   CRACK ASSESSMENT METHODOLOGY

- check if damage is coincident with a weld (the damage can crack a weld);
- explain the reason for the damage, and any implications for the rest of the pipeline;
- question if the reported size of the damage is credible given the pipeline's operating conditions and design (for example, a pipeline with a deep depth of cover is unlikely to be damaged by surface excavation equipment);
- ensure the defect is accurately reported, and not another defect, such as corrosion damage (for example, corrosion is often located around the 6 o'clock position of the pipeline, whereas mechanical damage can be at the top of the pipeline and denting due to rocks can be at the bottom of the pipeline);
- ensure that other damage coincident or close to the reported damage has been correctly identified and reported (for example, a reported dent may contain a gouge, or a reported gouge may have associated denting);
- check that associated cracking is unlikely (stress corrosion cracking and hydrogen induced cracking can be associated with damage, and damage can extend due to fatigue cracking);
- check probabilities of detection and detection accuracies for the damage, and include allowances in calculations;
- consider location of damage along the pipeline route in terms of impact on people safety and environmental damage should the damage fail at a future date;
- assemble all necessary data for the assessment process (e.g., material yield strength, pressure cycling data, and material toughness);
- perform calculations, and include any possible future growth (fatigue and corrosion);
- report and record.

12.8   REFERENCES

1. Anon., US Office of Pipeline Safety Incident Database. https://hip.phmsa.dot.gov/analyticsSOAP/saw.dll?Portalpages.

2. http://www.pipeline101.com/Are-Pipelines-Safe/What-Is-The-Safety-Record.

3. Anon., 'Gas Pipeline Incidents'. 9th Report of the European Gas Pipeline Incident Data Group. Doc. Number EGIG 14.R.0403. February 2015. www.egig.nl.

4. P M Davis et al., 'Performance of European cross-country oil pipelines. Statistical summary of reported spillages in 2013 and since 1971', CONCAWE (conservation of clean air and water in Europe). Report 4/15. May, 2015. www.concawe.be.

5. Anon., 'Pocket Guide to Transportation. 2014', US Department of Transportation. RITA Bureau of Transportation Statistics.

6. http://www.ntsb.gov/investigations/data/Pages/Data_Stats.aspx.

7. Anon., 'Onshore Pipelines. The Road to Success'. International Pipe Line & Offshore Contractors Association. 3rd edition. September 2013. http://wiki.iploca.com/pages/viewpage.action?pageId=1803629.

8. Anon., 'Pipeline Transportation Systems for Liquids and Slurries', ASME B31.4—2012, American Society of Mechanical Engineers. New York, USA, 2012.

9. Anon., 'Gas Transmission and Distribution Piping Systems'. ASME B31.8—2014, American Society of Mechanical Engineers. New York, USA, 2014.

10. J F Kiefner, M J Rosenfeld, 'The Role Of Pipeline Age In Pipeline Safety', Prepared for the INGAA Foundation, Inc. INGAA. November 8, 2012.

11. http://primis.phmsa.dot.gov/comm/reports/safety/SigPSIDet_1994_2013_US.html#_pipepliquid.

12. http://primis.phmsa.dot.gov/comm/reports/safety/SigPSIDet_1994_2013_US.html#_ngtrans.

13. Anon., 'Specification for Line Pipe', American Petroleum Institute. API 5L-2012. Edition 45. 2012.

14. Anon., 'Recommended Practice for Assessment and Management of Cracking in Pipelines', API RP 1176. American Petroleum Institute. 2015.

15. Anon., "Guide to Methods for Assessing the Acceptability of Flaws in Metallic Structures", BS 7910: 2013, British Standards Institution, London, UK, 2013.

16. Anon., 'Fitness-For-Service, API Recommended Practice 579-1/ASME FFS-1 2007', Second Edition, The American Society of Mechanical Engineers, New York, USA, June 2007.

17. A R Duffy, 'Full-Scale Studies', Third Symposium on Line Pipe Research, American Gas Association, Catalog No. L30000, November 1965, pp. 43–82.

18. W A Maxey, J F Kiefner, R J Eiber, A R Duffy, 'Ductile Fracture Initiation, Propagation and Arrest in Cylindrical Vessels', ASTM STP 514, American Society for Testing and Materials, Philadelphia, 1972, pp. 70–81.

19. J F Kiefner, W A Maxey, R J Eiber, A R Duffy, 'Failure Stress Levels of Flaws in Pressurised Cylinders', American Society for Testing and Materials. ASTM STP 536, pp. 461–481, 1973.

20. R W E Shannon, 'The Failure Behaviour of Linepipe Defects', Int. J Press Vessel & Piping, (2), pp. 243–255, 1974.

21. G D Fearnehough, D G Jones, 'An Approach to Defect Tolerance in Pipelines', IMechE, Conference on Defect Tolerance of Pressure Vessels, May 1978. London, UK.

22. W Kastner et al., 'Critical Crack Sizes In Ductile Piping', International Journal of Pressure Vessels and Piping, Vol. 9, 1981, pp. 197–219.

23. A Cosham, P Hopkins, 'The Effect of Dents in Pipelines - Guidance in the Pipeline Defect Assessment Manual', Proceedings ICPVT-10. July 7–10, 2003, Vienna, Austria.

24. T Muntinga, 'Acceptance Criteria for Damaged Pipes', Paper 26, Volume II, Proceedings of EPRG/NG-18 Eighth Biennial Joint Technical Meeting on Line Pipe Research, Paris, France, May 14–17, 1991.

25. B N Leis, F W Brust P M Scott, 'Development and Validation of a Ductile Flaw Growth Analysis for Gas Transmission Line Pipe', Final Report to American Gas Association, NG-18, Catalog No, L51543, 1991.

26. B N Leis, N D Ghadiali, 'Pipe Axial Flaw Failure Criteria - PAFFC', Version 1.0 User's Manual and Software, Topical Report to American Gas Association, NG-18, Catalog No. L51720, 1994.

27. G Knauf, P Hopkins, 'The EPRG Guidelines on the Assessment of Defects in Transmission Pipeline Girth Welds', 3R International, 35, Jahrgang, Heft 10-11/1996.

28. J F Kiefner et al., 'Further Revalidation of RSTRENG', Pipeline Research Council International Catalogue Number L51749e, December 1996.

29. P Hopkins, I Corder, P Corbin, 'The Resistance of Gas Transmission Pipelines to Mechanical Damage, Paper VIII-3, International Conference on Pipeline Reliability, Calgary, Canada, June 1992.

30. P Hopkins, 'Some Experiences of Applying the Pressure Systems Regulations to Older Systems', IMechE Paper C454/006/93, Institute of Mechanical Engineers, London, 1993.

31. I Corder, 'The Application of Risk Techniques to the Design and Operation of Pipelines', Paper No.: C502/016/95, Proceedings of International Conference on Pressure Systems: Operation and Risk Management, Institution of Mechanical Engineers, London, UK, pp. 113–125, 1995.

32. G Re, V Pistone, G Vogt, G Demofonti, D G Jones, 'EPRG Recommendation for Crack Arrest Toughness for High Strength Line Pipe Steels', 3R International, Vol. 34, Jahrgang, Heft October-November 1995, pp. 607–611.

33. D R Stephens, R B Francini, 'A Review and Evaluation of Remaining Strength Criteria for Corrosion Defects in Transmission Pipelines', Proceedings of ETCE/OMAE 2000, ASME. 2000.

34. K McDonald, S Maddox, P Haagensen, 'Guidance For Fatigue Design and Assessment of Pipeline Girth Welds', 20th International Conference on Offshore Mechanics and Arctic Engineering', June 2001, Rio de Janeiro, Brazil, OMAE01/MAT-3331.

35. A Cosham, P Hopkins, 'The Pipeline Defect Assessment Manual', Proceedings of IPC 2002: International Pipeline Conference. 29 September–3 October, 2002; Calgary, Alberta, Canada. Paper IPC02-27067. 2002.

36. J F Kiefner, 'Procedure anlayzes low-frequency ERW, flash welded pipe for HCA integrity assessments', Oil and Gas Journal, 08/05/2002.

37. M J Rosenfeld, 'Factors To Consider When Evaluating Damage On Pipelines', Oil & Gas Journal, September 9, 2002.

38. M J Rosenfeld et al., 'Basis of New Criteria in ASME B31.8 for Prioritisation and Repair of Mechanical Damage', 4th International Pipeline Conference, October 2002.

39. M J Rosenfeld et al., 'Development Of Acceptance Criteria For Mild Ripples In Pipeline Field Bends', 4th International Pipeline Conference, September 29–October 3, 2002, Calgary, Alberta, Canada, Paper 27124.

40. G B Rogers et al., 'Integrity Evaluation of an Older Vintage ERW Pipeline', Proceedings of IPC 2002: International Pipeline Conference. September 29–October 30, 2002 Calgary, Alberta, Canada. IPC2002-27052. 2002.

41. M J Rosenfeld, 'Factors To Consider When Evaluating Damage On Pipelines', Oil & Gas Journal, September 9, 2002.

42. J F Kiefner, 'Dealing With Low-frequency-welded ERW Pipe And Flash - Welded Pipe With Respect To HCA-related Integrity Assessments', ETCE 2002. ASME Engineering Technology Conference on Energy. February 4–6, 2002. Houston, Texas. Paper No. Etce2002/pipe-29029.

43. C E Jaske, J A Beavers, 'Development and Evaluation of Improved Model for Engineering Critical Assessment of Pipelines', Proceedings of IPC 2002, 4th International Pipeline Conference, September 29–October 3, 2002, Calgary, Alberta, Canada.

44. P Hopkins, 'The Structural Integrity of Oil and Gas Transmission Pipelines'. In 'Comprehensive Structural Integrity', Editors I. Milne, R. O. Ritchie and B. Karihaloo. Elsevier Publishers. Amsterdam. Volume 1. 2003.

45. Anon., 'Integrity Management Program Delivery Order DTRS56-02-D-70036: Dent Study', Final Report. M Baker Jr. Inc. November 2004.

46. Anon., 'Pipe Wrinkle Study: Final Report', Michael Baker Jr., Inc. Report OPS TTO11. October 2004.

47. A Cosham, P Hopkins, 'The effect of dents in pipelines—guidance in the pipeline defect assessment manual', International Journal of Pressure Vessels and Piping, Vol. 81, 2004, pp. 127–139.

48. M J Rosenfeld, J F Kiefner, 'Basics of Metal Fatigue in Natural Gas Pipeline Systems—A Primer for Gas Pipeline Operators', Pipeline Research Council International, Report No. L52270. 2006.

49. C Jaske, B Hart, W Bruce, 'PRCI Pipeline Repair Manual', PRCI Report PR-186-0324. 2006.

50. K A Macdonald, A Cosham, C R Alexander, P. Hopkins, 'Assessing mechanical damage in offshore pipelines - Two case studies', Engineering Failure Analysis 14 (2007) 1667–1679. Elsevier.

51. C Alexander, K Brownlee, 'Methodology for Assessing the Effects of Plain Dents, Wrinkle Bends, and Mechanical Damage on Pipeline Integrity', NACE International Corrosion & Expo. March 2007. Nashville, USA. NACE2007-07139.

52. C Alexander, S Kulkarni, 'Evaluating the Effects of Wrinkle Bends on Pipeline Integrity', Proceedings of IPC2008 7th International Pipeline Conference. September 29–October 3, 2008, Calgary, Alberta, Canada. IPC2008-64039.

53. R J Eiber, 'Fracture Propagation-1: Fracture-arrest prediction requires correction factors', Oil and Gas Journal, October 20, 2008. Vol. 106, Issue 39.

54. C Alexander, 'Evaluating damage to on- and offshore pipelines using data acquired using ILI', Journal of Pipeline Engineering, 1st Quarter, 2009, pp. 35–46.

55. Anon., 'Corroded Pipelines', DNV-RP-F101. Det Norske Veritas, Norway. October, 2010.

56. M J Rosenfeld, 'A Synthesized Approach to Pressure Reduction for Investigating Mechanical Damage', 8th International Pipeline Conference. Canada. ASME Paper IPC2010-31245.

57. R Amend, 'Early Generation Pipeline Girth Welding Practices and Their Implications for Integrity Management of North American Pipelines'. 8th International Pipeline Conference IPC2010, September 27–October 1, 2010, Calgary, Alberta, Canada. IPC2010-31297.

58. J Kiefner, K Leewis, 'Pipeline Defect Assessment: A Review and Comparison of Commonly-used Methods', Pipeline Research Council International. Report No. L52314. May 2011.

59. Anon., 'Manual for Determining the Remaining Strength of Corroded Pipelines. A Supplement to ASME B31 Code for Pressure Piping', ASME B31G-2012. The American Society of Mechanical Engineers, New York, 2012.

60. A Cosham. P Hopkins, B Leis, 'Crack-like Defects in Pipelines: The Relevance of Pipeline-specific Methods and Standards', Proceedings of the 9th International Pipeline Conference. IPC 2012. September 24–28, 2012, Calgary, Alberta, Canada. IPC2012-90459.

61. Anon., 'Managing System Integrity for Hazardous Liquid Pipelines', American Petroleum Institute. API Recommended Practice 1160. Second Edition. September, 2013.

62. Anon., 'Welding of Pipelines and Related Facilities', Twenty-First Edition. September, API Standard 1104. American Petroleum Institute, USA, 2013.

63. R J Eiber, 'Drop-weight tear test application to natural gas pipeline fracture control', Journal of Pipeline Engineering. 3rd Quarter. 2013. p. 175.

64. B Leis, 'The Charpy impact test and its applications', Journal of Pipeline Engineering. 3rd Quarter. 2013. p. 183.

65. Yong-Yi Wang et al., 'Assessment of vintage girth welds and challenges to ILI tools', Journal of Pipeline Engineering. 2nd Quarter. 2013.

66. G Wilkowski, Y Hioe, D J Shim, 'Methodology for Brittle Fracture Control in Modern Line Pipe Steels', Pipeline Technology Conference. Ostende, Belgium, 2013.

67. Anon., 'Managing System Integrity of Gas Pipelines', ASME B31.8S-2014. American Society of Mechanical Engineers. New York, USA, 2014.

68. P Hopkins, 'Assessing the significance of corrosion in onshore oil and gas pipelines', Edited by M. Orazem. Woodside Publishing Ltd., UK, 2014.

69. R M Andrews, R M Denys, G Knauf, M Zarea, 'EPRG guidelines on the assessment of defects in transmission pipeline girth welds - Revision 2014', Pipelines International, June 2015.

70. M F Ashby, D R H Jones, 'Engineering Materials 1. An Introduction to their Properties and Applications', Pergamon. International Library. England, 1986.

71. J R Lund, J P Byrne, 'Leonardo da Vinci's Tensile Strength Tests: Implications for the Discovery of Engineering Mechanics', Civil. Eng. and Env. Syst., Vol. 00, pp. 1–8.

72. C E Inglis, 'Stresses in a plate due to the presence of cracks and sharp corners', Trans. Inst. Nav. Archit. London 55, 1913, pp. 219–230.

73. A A Griffith, 'Phenomena of rupture and flow in solids', Philos. Trans. R. Soc. London Ser. A221, 1920, pp. 163–198.

74. G R Irwin, 'Fracture dynamics', Trans. Am. Soc. Met. 40A, 1948, pp. 147–166.

75. G R Irwin, 'Analysis of stresses and strains near the end of a crack traversing in a plate', J. Appl. Mech. 24, 1957, pp. 361–364.

76. D S Dugdale, 'Yielding of Steel Sheets Containing Slits', J. Mech. Phys. Solids, vol. 8, pp. 100–104, 1960.

77. A Wells, 'Application of Fracture Mechanics at and Beyond General Yielding', British Welding Journal, Vol. 11, pp. 563–570, 1961.

78. J R Rice, 'A Path Independent Integral and the Approximate Analysis of Strain Concentration by Notches and Cracks', Journal of Applied Mechanics, 35, 1968, pp. 379–386, 1968.

79. O H Bjornoy, M Marley, 'Assessment of Corroded Pipelines: Past, Present and Future', Proceedings of tile Eleventh (2001) International Offshore and Polar Engineering Conference, Stavanger, Norway, June 17–22, 2001.

80. O H Bjørnøy, G Sigurdsson, and M J Marley, 'Background and Development of DNV-RP-F101 Corroded Pipelines'. ISOPE 2001, Paper no 2001-HM-10. 2001.

81. B N Leis, T A Bubenik, 'Topical Report Periodic Re-Verification Intervals for High-Consequence Areas', Gas Research Institute Report GRI-00/0230, January 2001.

82. B Leis, X Zhu, 'Corrosion Assessment Criteria: Rationalising their use for Vintage vs Modern Pipelines', Final Report. US Department of Transportation. Research and Special Projects Agency. September 2005.

83. B N Leis, T A Thomas, 'Linepipe Property Issues in Pipeline Design and Re-establishing MAOP', Paper ARC-17, PEMEX International Congress on Pipelines, Merida, Mexico, November 2001.

84. J F Kiefner, P H Vieth, 'A Modified Criterion for Evaluating the Strength of Corroded Pipe', Final Report for Project PR 3-805 to the Pipeline Supervisory Committee of the American Gas Association, Battelle, Ohio, 1989.

85. J F Kiefner. P H Vieth, I Roytman, 'Continued Validation of RSTRENG', Pipeline Research Council International, Report Number L51749e. USA, December 1996.

86. B Fu, A D Batte, 'Advanced Methods for the Assessment of Corrosion Defects in Line Pipe', Health & Safety Executive Summary Report, OTO 1999-051, HSE Books, 1999.

87. A Cosham, P Hopkins, 'The Assessment of Corrosion in Pipelines - Guidance in the Pipeline Defect Assessment Manual (PDAM)', Pipeline Pigging and Integrity Management Conference. 17–18th May 2004—Amsterdam, Netherlands, 2004.

88. Anon., 'Specifications and requirements for intelligent pig inspection of pipelines', Pipeline Operators Forum, Version 2009. http://www.pipelineoperators.org/publicdocs/POF_specs_2009.pdf. 2009.

89. Anon., 'Oil and gas pipeline systems', CSA Z662-15, Canadian Standards Association. 2015.

90. S Westwood, P Hopkins, 'Smart Pig Defect Tolerances: Quantifying the Benefits of Standard and High Resolution Pigs'. Proceedings of IPC 2004: International Pipeline Conference. Calgary, Canada. IPC04-0085. October 4–8, 2004.

91. B N Leis, 'Hydrostatic Testing of Transmission Pipelines: When It Is Beneficial and Alternatives When It Is Not', Pipeline Research Committee International PRCI Final Report PR 3-9523, 2002.

92. B Rothwell, R Coote, 'A Critical Review of Assessment Methods for Axial Planar Flaws', Ostende Pipeline Conference. Ostende. Belgium. 2009.

93. A Hosseini et al., 'Experimental Testing and Evaluation of Crack Defects in Line Pipe', 8th International Pipeline Conference, IPC2010. IPC2010. September 27–October 1, 2010, Calgary, Alberta, Canada. Paper IPC2010-31158.

94. J F Kiefner, E B Clark, 'History of Line Pipe Manufacture in North America', ASME Research Report. CRTD Vol 43. ASME 1996.

95. Anon., 'Selective Seam Weld Corrosion Literature Review', Det Norske Veritas Report for Pipeline and Hazardous Materials Safety Administration U.S. Department of Transportation. Report No.: ANEUS 811CSEAN120106, Rev. 2. November 11, 2012.

96. K Masamura, I Matsushima, "Grooving Corrosion of Electric Resistance Welded Steel Pipe in Water - Case Histories and Effects of Alloying Elements", Paper No. 75, Corrosion '81, NACE International. 1981.

97. Anon., 'Rupture of Hazardous Liquid Pipeline with Release and Ignition of Propane Carmichael, Mississippi. November 1', USA NTSB Accident Report: NTSB/PAR-09/01. PB2009-916501. 2009.

98. B N Leis, B A Young, J F Kiefner, J B Nestleroth, J A Beavers, G T Quickel, C S Brossia, 'Final Summary Report and Recommendations for the Comprehensive Study to Understand Longitudinal ERW Seam Failures - Phase One', Final Report - Task 4.5. U.S. Department of Transportation Pipeline and Hazardous Materials Safety Administration 1200 New Jersey Ave., SE Washington DC 20590. October 23, 2013.

99. J F Kiefner et al., 'Track Record of In-Line Inspection as a Means of ERW Seam Integrity Assessment', Battelle Final Report No. 12-180. November 15, 2012.

100. J F Kiefner, K M Kolovich, 'Models for Predicting Failure Stress Levels for Defects Affecting ERW and Flash-Welded Seams', Battelle Final Report No. 13-002. January 3, 2013.

101. J Kiefner, K M Kolovich, 'ERW and Flash Weld Seam Failures', U.S. Department of Transportation Battelle Final Report No. 12-139. September, 2012.

102. E Clark, B Leis, R J Eiber, 'Integrity Characteristics of Vintage Pipelines', The INGAA Foundation Inc., Report E-2002-50435. 2005.

103. Anon., 'Low Frequency ERW and Lap Welded Longitudinal Seam Evaluation', Final Report. Revision 3. Michael Baker Jr., Inc. April 2004. http://primis.phmsa.dot.gov/gasimp/docs/TTO05_Low FrequencyERW_FinalReport_Rev3_April2004.pdf.

104. P Hopkins, 'Some Experiences of Applying the Pressure Systems Regulations to Older Systems', IMechE Paper C454/006/93, Institute of Mechanical Engineers, London, 1993.

105. B N Leis, R J Eiber, 'Stress-Corrosion Cracking on Gas-Transmission Pipelines: History, Causes, and Mitigation', First international Business Conference on Onshore Pipelines. Berlin, December 2007.

106. A Batte et al., 'A new joint-industry project addressing the integrity management of SCC in gas transmission pipelines', The Journal of Pipeline Engineering. 2nd Quarter. 2012. p. 63.

107. Anon., 'Stress Corrosion Cracking Study', Final Report, Office of Pipeline Safety, USA. OPS TT08, January, 2005. https://primis. phmsa.dot.gov/iim/docstr/SCC_Report-Final_Report_without_ Database.pdf.

108. J Beavers, 'Pipeline Stress Corrosion Cracking: Detection and Control', Pipeline and Gas Journal. July 2015, Vol. 242, No. 7.

109. J A Beavers, N G Thompson, 'External Corrosion of Oil and Natural Gas Pipelines', ASM Handbook, Volume 13C, Corrosion: Environments and Industries (#05145). 2006.

110. J Paviglianti et al., 'An Analysis of 'significant' SCC data reported to the National Energy Board', Journal of Pipeline Engineering, 4th Quarter. 2007. p. 197.

111. R Fessler, 'Pipeline Corrosion: Final Report', Michael Baker Jr., Inc. Report for U.S. Department of Transportation Pipeline and Hazardous Materials Safety Administration, November 2008.

112. Anon., 'Stress Corrosion Cracking on Canadian Oil and Gas Pipelines', National Energy Board of Canada, Report MH-2-95, December 1996.

113. R Fessler, 'Predicting the Failure Pressure of SCC Flaws in Gas Transmission Pipelines', Proceedings of the 2012 9th International Pipeline Conference IPC2012 September 24–28, 2012, Calgary, Alberta, Canada. Paper IPC2012-90236.

114. Anon., 'Stress Corrosion Cracking: Recommended Practices', Second Edition. Canadian Energy Pipeline Association (CEPA). Canada, 2007.

115. J Dawson et al., 'Emerging Techniques for Enhanced Assessment and Analysis of Dents', International Pipeline Conference, Calgary, Canada. 2006. ASME. Paper IPC 2006-10264. Also, in Global Pipeline Monthly, Vol 4. Issue 5. July 2008.

116. http://www.phmsa.dot.gov/pipeline/library/data-stats/pipeline incidenttrends.

117. Anon., 'Mechanical Damage Final Report', Michael Baker Jnr Inc. DoE PHMSA OPS, April 2009.

118. J F Kiefner, 'Effectiveness of Various Means of Preventing Pipeline Failures from Mechanical Damage', Gas Research Institute Report GRI-99/0050. August 26, 1999.

119. P Hopkins et al., 'Recent Studies of Significance of Mechanical Damage in Pipelines', AGA and EPRG Research Seminar, San Francisco, USA, September 1983.

120. K C Wang, E D Smith, 'The Effect of Mechanical Damage on Fracture Initiation in Linepipe Part III—Gouge in a Dent', Canadian Centre for Mineral and Energy Technology (CANMET), Canada, Report ERP/PMRL 85-69 (TR). 1985.

121. K C Wang, E D Smith, 'The Effect of Mechanical Damage on Fracture Initiation in Linepipe: Part II—Gouges', Canadian Centre for Mineral and Energy Technology (CANMET), Canada, Report ERP/PMRL 88-16 (TR). 1988.

122. W R Tyson, K C Wang, 'Effects of External Damage (Gouges and Dents) on Performance of Linepipe. A Review of Work at MTL', CANMET, Canadian Centre for Mineral and Energy Technology (CANMET), Canada, Report MTL 88-34. 1988.

123. Anon., 'Pipeline Investigation Report P09H0084', Transportation Safety Board of Canada. Canada. September 2009. http://www.tsb. gc.ca/eng/rapports-reports/pipeline/2009/p09h0084/p09h0084.asp.

124. A Cosham, M Kirkwood, 'Best Practice in Pipeline Defect Assessment', Proceedings of IPC 2000: International Pipeline Conference October 2000; Calgary, Alberta, Canada. IPC00-0205. 2000.

125. G Malatesta et al., 'Extension of Current Defect Assessment Methods for Gouge and Corrosion Defects in X80 Grade Pipeline', Proceedings of the 2012 9th International Pipeline Conference IPC2012. September 24–28, 2012, Calgary, Alberta, Canada. IPC2012-90613.

126. B N Leis et al., 'Leak versus Rupture Considerations for Steel Low-Stress Pipelines', Gas Research Institute, USA. Report No. GRI-00/0232. January 2001.

127. M Rosenfeld, R Fassett, 'Study of pipelines that ruptured while operating at a hoop stress below 30% SMYS', Pipeline Pigging and Integrity Management Conference. Houston, USA. 13–14 February, 2013.

128. J R Fowler, C R Alexander, P J Kovach, L M Connelly, 'Cyclic pressure fatigue life of pipelines with plain dents, dents with gouges, and dents with welds'. AGA Pipeline Research Committee, Report PR-201-927 and PR-201-9324, June 1994.

129. I Corder, P Chatain; 'EPRG Recommendations for the Assessment of the Resistance of Pipelines to External Damage', Proceedings of the EPRG/PRC 10th Biennial Joint Technical Meeting on Line Pipe Research, Cambridge, UK, April 1995.

130. C R Alexander, J F Kiefner, 'Effects of Smooth and Rock Dents on Liquid Petroleum Pipelines', API Publication 1156, First Edition, November 1997.

131. Anon., 'Pipeline Accident Report. Pipeline Rupture and Subsequent Fire in Bellingham, Washington June 10, 1999', National Transportation Safety Board. Washington, D.C.NTSB/PAR-02/02. PB2002-916502. 1999.

132. P Hopkins, 'The Application of Fitness for Purpose Methods to Defects Detected in Offshore Transmission Pipelines, Conference on Welding and Weld Performance in the Process Industry, London, 1992.

133. C. de Larminat et al., 'Assessment of bending wrinkles - A review of available acceptance criteria', 17th EPRG/PRCI Joint Technical Meeting. Milan, 2009.

134. Anon., 'Safety Study: Integrity Management of Gas Transmission Pipelines in High Consequence Areas', National Transportation Safety Board, USA. NTSB/SS-15/01. PB2015-10273. January, 2015.

CONFORMANCE AND COMPLIANCE VERIFICATION AND ACTION PLANS

Chapter 13 describes the terminology, approaches, elements, process, and methods for conducting compliance and conformance verification of a Pipeline Integrity Management System (PIMS). The chapter also includes some available enforcement tools as well as processes and examples of Corrective and Preventive Action Plans (CPAP). The chapter closes with some fundamentals about incident investigation of pipeline integrity incidents or accidents. Some of the processes explained in this chapter may help as an input for company own oversight of PIMS depending on the regional framework and requirements [1–3].

13.1   INTRODUCTION

Regulators and company's regulatory groups are typically independent, reporting to senior levels of accountability (e.g., ministerial, parliament, company's Chief Executive Officer or Board of Directors). They can be assigned to promoting, verifying, and enforcing laws and regulations to prevent pipeline incidents, non-compliances, and non-conformances. Their directives may cover the entire pipeline life cycle overseeing the design, operation, inspection, testing and maintenance, deactivation/reactivation and abandonment of hydrocarbon transmission pipelines. Some proactive regulators participate in the development of consensus standards. These standards can be adopted as regulation as is and can be complemented or enhanced to meet the jurisdictional needs. Governments promote and encourage improving safety, environmental protection and sometimes the security and efficiency of the pipeline infrastructure.

13.2   CONFORMANCE AND COMPLIANCE CONCEPTS

13.2.1   Compliance

It is a state of being in accordance with an established official requirement such as a pipeline act, law, regulation, adopted industry and/or company own standards. Compliance action or activity meets the requirements of legislation, rules and regulations and adopted consensus standards.

Example: Compliance with Regulations
A pipeline company demonstrated its compliance with regulations by presenting the most recent internal audit including all elements of the IMP.

13.2.2   Non-Compliance

Non-compliance is a failure to adhere to legislation, rules and regulations, and adopted consensus standards. Non-compliance can be classified as high, medium, and low risk.

Example: Non-Compliance
A pipeline company had not developed and/or implemented the elements of a Management System supporting the integrity management program. This finding does not meet the requirements of the regulation requiring a pipeline company to submit a Corrective and Preventive Action Plans (CPAP) for regulator's approval.

A high-risk non-compliance is one where the nature of the non-compliance makes it more likely than other non-compliances to result in a major loss, such as life-altering injury or fatality, immediate loss of containment of a system, or other similar losses. Where non-compliance is considered high risk, corrective and/or preventive actions should be immediately undertaken.

Example: High-Risk Non-Compliance
A pipeline company did not have an overpressure protection system on a high-pressure pipeline susceptible to cracking in a highly populated area.

13.2.3   Conformance

Conformance is being in accordance with a requirement, standard, code or procedure defined by the company, but not required in the act, law, regulation nor adopted consensus standard by any corresponding jurisdiction.

13.2.4   Non-Conformance

Non-conformance is a failure to comply with a requirement, standard, code or established procedure defined by the company, but not required in the act, law, regulation nor adopted by any legal entity.

Example: Non-Conformance
A pipeline company has not conducted/updated the risk assessment during the last two (2) years exceeding the company own requirement interval. This finding does not meet the pipeline company requirement (conformance), but does meet the act, law, and regulation of the jurisdiction as it is within the five (5) year interval.

13.3 COMPLIANCE AND CONFORMANCE OVERSIGHT APPROACHES

The following four (4) distinctive compliance and conformance oversight approaches can be used by regulators and companies to oversee compliance and conformance.

1. Prescriptive Approach
2. Goal Setting or Performance-Based Approach
3. Goal-Oriented Approach
4. Management-Based Approach

These four (4) compliance and conformance oversight approaches are explained below by defining their focus, advantages, and limitations.

13.3.1 Prescriptive Oversight Approach

The prescriptive approach creates pipeline integrity requirements with a high level of detail for assisting the pipeline companies with the specifics of *What-To, When-To, and How-To Do* in order to comply. Hence, the prescriptive approach defines methods for oversight in terms of protocols, forms, and questionnaires providing the companies with clear expectations for adequacy (i.e., input, process) and implementation (i.e., process and output) to be reached.

The oversight is expected to be *easier to an extent* than the other two (2) approaches due to this methodic and almost all-written expectation approach. Consequently, comprehensive verification should be obtaining results in the short-term as to whether the requirements have been met.

This methods assumes that "*one size fits all*" evaluating all pipeline companies and associated conditions to be the same. Even though "Special Permits" can be created as an option to deal with specific conditions, integrity of pipelines are driven by multiple combination of conditions such as manufacturing, environment, operations, and external forces that make it difficult to apply one size fits all.

For instance, a requirement can be a given re-inspection interval whether the number is low or high (e.g., 3 to 20 years); a pipeline exposed to high corrosivity product may require a shorter metal loss re-inspection interval than a sweet natural gas pipeline. Conversely, a pipeline manufactured, constructed, and operated with high quality standards may require a longer cracking re-inspection interval than a pre-1980 vintage pipeline.

13.3.2 Goal Setting or Performance-Based Oversight Approach

The Goal Setting oversight approach creates pipeline integrity requirements focused on *outcomes* (e.g., people safety, environmental protection) reflecting effectiveness with a language stipulating:

a. Goal or expected end-result *(e.g., pipeline operator to ensure operation under safe limits)*
b. Requirement condition *(e.g., no fluid conveyed in the pipeline unless)* and;
c. If applicable, acceptability method *(e.g., so far as is reasonably practicable)*

The Goal Setting oversight approach focuses on outcomes such as safe work conditions, health and safety and protecting the environment [4]. This approach recognizes the differences among pipeline companies expecting all to achieve the expected outcomes or goals as well as creating opportunities for innovation and new technologies to meet the intent. This approach typically requires pipeline companies to implement management systems (e.g., safety, environmental, integrity, security) in order to build a platform with senior management commitment and leadership, processes and procedures and internal verifications for continuous improvement.

For pipelines with potential for major hazards and/or consequences, Goal Setting oversight approaches typically establish specific requirements, inspection and audit intervals, and performance measures. For approvals, these pipelines may also be required to demonstrate through a *Safety Case* that their hazards, threats, consequences, and risks have been identified and properly mitigated to As Low as Reasonably Practicable (ALARP).

The oversight is expected to be more challenging as the requirements may or may not be interpreted to the same extent by both the verifier and the verified pipeline company. Pipeline companies have the liberty to reasonably choose the method(s) to meet the requirement, and for the verifier to reasonably evaluate whether their methods and results can demonstrate effectiveness in reaching the goal. The verifiers are required to be well trained in both legal and integrity areas as well as keeping themselves up-to-date on new technologies, knowledge and methodologies.

13.3.3 Goal-Oriented or Combined Oversight Approach

The goal-oriented oversight approach combines the prescriptive with the performance-based approaches creating pipeline integrity requirements focused on *process, output, and outcomes* reflecting adequacy, implementation, and effectiveness. The prescriptive components come from adopting consensus industry standards, which are minimum requirements that could be applied accounting for specific pipeline design, manufacturing, construction, and operational conditions [5].

The prescriptive requirements can be used as a reference for assessing adequacy or completeness of the IMP processes; however, management system elements (e.g., senior management leadership, planning, verification) are also used in conjunction with IMP for determining the levels of implementation of an integrated PIMS, which can be translated into outputs or results (e.g., mitigation, prevention, releases). Furthermore, effectiveness is also evaluated by determining if the end-goals were achieved.

The oversight is expected to be less challenging than Goal Setting as it counts on the prescriptive standard as the reference to verify concrete requirements for adequacy and implementation. However, Goal-Oriented oversight approach requires a well-trained inspector, auditor, or officer in assessing whether the company has been effective in achieving the goals and objectives of PIMS. Inspectors, auditors, or officers need to continue learning new technologies and methodologies as they would be assessing a pipeline industry continuously improving.

13.3.4 Management-Based Oversight Approach

The management-based oversight approach would keep pipeline integrity requirements focused on improving internal management planning processes (e.g., risk assessment, inspection, and mitigation, prevention and monitoring) to ensure adequacy or completeness of all required plans for reaching integrity goals. It is a means (plan) to an end (outputs and outcomes) [6].

13.4 OVERSIGHT VERIFICATION PROGRAM ELEMENTS AND PROCESS

The external (e.g., regulators) and internal (e.g., company's own regulatory group or third party) oversight entities expect pipeline companies to adequately implement and effectively achieve the goal of the requirements (e.g., regulations, standards and practices) within the day-to-day operation activities. Oversight entities develop processes and methods for verifying and monitoring pipeline operators for compliance with regulatory requirements and/or conformance with adopted industry standards and procedures. Internal oversight entities may trigger enforcement processes from regulators; if the company's own regulatory group identifies non-compliance, Self-Disclosure Non-Compliance should be reported to the regulator.

As illustrated in Figure 13.1, a verification program may follow the framework of a Quality Management System process such as "PLAN-DO-CHECK-ACT" (PDCA).

13.4.1 Planning of the Verification (Plan)

The rationale and objectives of the verification should first be clearly defined and substantiated on authority to conduct the verification. The plan should discuss the requirements to be covered. The criteria for the entities selection should also be documented. Thus, the selection of the verification method, qualified personnel, and tools can be aligned with the objectives.

Verifier's notification to the company entity/department should be formally written (e.g., letter/e-mail) specifying the objectives and authority under which the request for verification is made with a tentative schedule. In urgent cases, communication via phone or in-person is acceptable followed by a letter/e-mail request. Sending the request for verification in advance ensures the attendance of the right personnel as well as the agenda agreed by both parties and consensus of the schedule and locations for the verification.

A verification plan conceptually covers the following:

- *Why* a verification program is required (i.e., pipeline incident occurrence and trends, risk exposure, corrective action plan, whistleblower, lack of clarity and transparency, deficiency notice, correction process, or audit)
- *What is the verification objective* (e.g., get-to-know for the first time, adequacy, implementation, effectiveness, confirm next steps) based on assessment, scoring model, analyses, audits, implementation, follow-ups, operational events, pipeline incidents, complaints, and other considerations
- *Which department(s) or company (ies)* is (are) the focus of the verification (e.g., senior management, operations, integrity, safety, damage prevention)
- *How* is the verification going to be conducted by selecting the *method(s)* is (are) adequate (i.e., through investigation, inspection or audit) for the achieving the verification objective
- *Who* is qualified to perform the verification method (e.g., inspector, auditor, multi-disciplinary team)
- *What* tools and associated authority may be required (e.g., stop-work by inspection officer, corrective action order by audit team)
- *When* the selected method is to be carried out as agreed with the verified entity

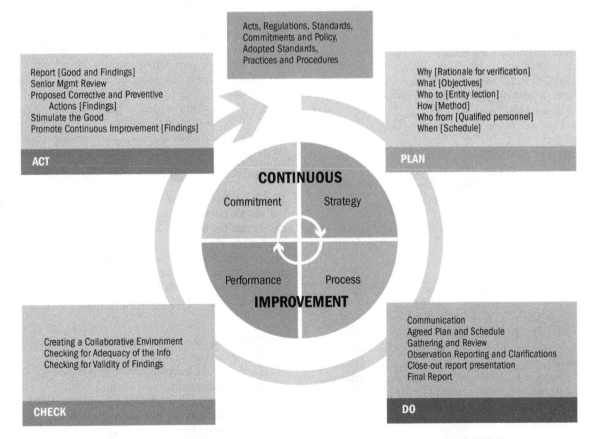

FIG. 13.1 THE OVERSIGHT VERIFICATION ELEMENTS AND PROCESS

13.4.2 Implementation of the Verification (Do)

The implementation step requires a procedure for conducting the verification method. The implementation of a verification method conceptually covers the following:

1. *Verification Preparation* including the verifier to communicate the verification scope, expectations, protocols, and initial data needed as well as a proposed schedule.
2. *Agreed Plan and Schedule* identifying locations, personnel responsible and selected for interview, and verification duration
3. *Gathering and review* of documentation at the selected location visits and personnel *interviews*
4. *Reporting* daily or weekly *observations* identified during the review
5. *Clarification of observations* with the verified party for confirming whether they are *findings* (i.e., natural justice): additional information or clarification may be beneficial
6. *Close-out report* to verified party's senior management and further clarifications
7. *Final report of the verification* and request of the Corrective and Preventive Action Plans process

---

Example: Verification Implementation Activities

- Verification activity preparation
  - Verifier's own in-house preparation
  - Scope, expectations, protocols and initial data requirements communicated to the party to be verified
  - Meeting with the party to be verified for clarifications and schedule review
  - Initial documentation review by the verifier
- Kick-off verification meeting with senior management of the party to be verified
- Site visits to A, B and C locations and interviews
  - Daily summaries with observations (not yet findings) shared with the party being verified
- Close Out verification meeting to senior management of the verified party
  - Summary of the verification findings
  - Clarifications
- Draft of the verification report for verified party's review
  - Clarifications and resolution
- Final report of the verification
  - Request for a Corrective and Preventive Action Plans

Example: Verification of Documentation and Field Activities

- A verifier conducted a review of the Non-Destructive Examination (NDE) process identifying the need for essential roles and responsibilities missing in the evaluation of the NDE results
- A verifier conducted a field visit to a pipeline excavation finding that the safe operating pressure to protect the workers did not get identified in the dig sheet nor implemented during the installation of a sleeve

13.4.3 Checking the Verification: Information and Findings (Check)

13.4.3.1 Creating a Transparent and Collaborative Environment The transparency and collaboration from both the parties (i.e., verifier and verified) make a huge difference during the *checking stage* by promoting an environment of *trust* and *reasonableness* in the understanding of the observations and findings. The lack of those checking attributes can induce potentially unnecessary scrutiny and legally defensive reactions that diminish the end goal of verification: *acknowledgment of the good* and the *need for continuous improvement.*

Example: Steps for building a trust and reasonableness environment in the verification process

a. Let's start with the ancient practice of "putting yourself in the other person's shoes" by asking what is someone's objectives and how to remove barriers
b. Identify the good practices and acknowledge them: *creating good takes effort*
c. Use well-known principles to act fairly such as Natural Justice (see 13.6.1): *rule against bias*
d. Expect and accept that the verification outcome will have areas for improvement: *no company is perfect*
e. Ensure that you have understood the information (received by you) and you have resolved your questions: *read and understand; otherwise ask instead of assuming*
f. Verify that the information (provided by you) has been understood: *check if your intended objective is shared by the receiver; otherwise, please clarify*
g. Rephrase the agreed action items requesting confirmation of a common understanding on what and when they will be done; if how-to is also agreed, please discuss: *all participants on the same page*
h. Ensure that the integrity regulations and industry standards used are applicable: *clear rules of the verification*
i. If you committed to or promised an action you should make sure it is completed. If there is a risk of not fulfilling the promised action, communicate that risk before the deadline: *Walk the talk* and *do not wait until are called for it to provide an excuse (too late).*

13.4.3.2 Checking for Adequacy of the Information A verifier should conduct a test for *adequacy of the information* to ensure that the sampled information is applicable, complete, representative of the quality and quantity of information available, and credible and timely. The sampled information received via documentation and interviews leading to observations or findings require checking or testing for adequacy by applying the following criteria:

- *Applicable*: the sampled information is in context of the verification scope
- *Complete*: the sampled information covers the extent requested by the requirement
- *Representative*: the sampled information contains the expected elements in quantity and quality representing the entire "population" information

- *Credible*: the sampled information is likely to reflect a phenomenon, reality, process or event
- *Timely*: the sample information is valid within the timeframe being evaluated

---

Example: Adequacy Checking of the Metal Loss In-Line Inspection Information

Metal loss in-line inspection (ILI) results from two (2) years ago have been received as sampled information of the integrity condition of the pipeline. The verifier has used the following *test* to determine whether the metal loss ILI information is adequate to be used for the verification process:

- Metal loss ILI data is *applicable* for the scope and objectives of the verification as the metal loss is considered an identified or susceptible integrity threat to the pipeline
- ILI data produced by a tool is complete as it has properly covered the intended length
- Metal loss ILI data is *representative* as the frequency (how many/quantity) and severity (how large/quality) of the integrity threats expected in the pipeline
- Metal loss ILI data is *credible* as the ILI field verification confirmed the metal loss phenomena (anomaly classification, sizing and detection) reported by the ILI
- ILI data is *timely* as the anomaly growth is very confirmed to be slow and then, not expected to significantly change the integrity criticality condition during the last two (2) years

The *verifier has concluded that the metal loss ILI information is adequate* for the verification process related to assessing the integrity condition of the pipeline.

A good method to check whether the information is verifiable is to request a third party to ask probing questions requesting supporting documentation. Furthermore, verification of previous or repetitive occurrences may provide context in defining whether the finding is reasonable.

13.4.3.3 Checking for Validity of Findings If a potential finding is identified, the verifier should conduct a test of the *validity of the findings* to ensure that the verification outcome is balanced, substantiated, transparent, and reasonable. The findings require checking or testing for validity by applying the following criteria:

- *Balanced:* findings should describe both the good practices (useful context) and the areas identified for improvement.
- *Substantiated:* findings should be supported by the evidence in the form of documentation or interviews indicating the analysis made and/or rationale used.
- *Transparent:* findings should not be a surprise for the verified party. They require mutual discussion, but not agreement.
- *Reasonable:* findings should be able to be concluded by a third party reviewer (e.g., peer review) based on the written document without verbal explanation. Otherwise, the finding is NOT reasonable (lack of basis), well supported (not sufficient documentation) or well explained (lack of written quality).

Example: Validity Test of a Finding about Integrity Threat Assessment

A verifier has identified a potential pipeline integrity finding result through a verification process of a Pipeline Integrity Management Program of a company. The verifier has used the following test to determine whether the integrity finding is valid:

- *[Balanced approach with context of the good work]* The integrity team has conducted annual threat assessments for corrosion and cracking with effective results.
- *[Substantiated with evidence of the finding]* The documentation and interviews indicated that the assessment for the other threats (e.g., incorrect operations, 3rd party damage) is conducted by their own corresponding departments (e.g., operations, Right-Of-Way), but results have not been integrated.
- *[Finding]* The company does not have an adequate and fully implemented process for conducting the integrity threat assessment for all the threats and integrating the input from all relevant departments.
- *[Transparency describing the finding basis]* This finding is considered a *non-conformance* with the company's own process and procedures described in PIMS section 5 as well as a *non-compliance* with the regulation section C.10.
- *[Reasonableness Test]* Another auditor was consulted to review the write-up indicating that the finding is reasonable based on the evidence shown.

13.4.4 Senior Management Review of Verification and Action Plan (Act)

The verifier should produce a draft report of the verification with the objective of giving an opportunity to the verified party to know, review and provide feedback prior to the issuance of the final report. Typically, the verification reports are addressed to the senior management responsible for the processes verified. Verification findings are ultimately senior management's responsibility as to their expected capacity to lead, resource and follow-up on the effective resolution of the findings.

A verification report conceptually covers the following:

- *Purpose and Objectives* of the Verification (high level expectation at the end of the process)
- Verification *Terminology* (e.g., findings, non-conformance, non-compliance)
- *Scope* of the Verification (e.g., facilities inspected, detailed regulatory, industry and company's own practice framework used, focus applied)
- Verification *Process* (e.g., flowchart)
- *Plan and Schedule* conducted (e.g., activities, location and dates)
- Verification Activity Team (e.g., verified and verifier)
- Verification *Analysis and Results* (e.g., what good areas and areas for improvement were found and why)
- Verification *Conclusion* (e.g., summary) *and Recommendations* (e.g., future deliverables)

Each finding should be identified with the most appropriate corrective and preventative action with the associated timing for resolution. Non-compliance would typically take a higher priority over non-conformance; however, the timing for findings with potential for affecting the safety and environment in the near term should be prioritized first.

The corrective and preventative actions should be defined in the plan identifying the expected date of completion. They should be documented, followed-up, and closed-out with the approval of senior management.

13.5 VERIFICATION METHODS

Verification methods are used for checking, verifying, and identifying good practices, non-compliances, non-conformance and areas for improvement [7, 8]. The selection of the verification method depends on the level of information known by the regulator and the level of scrutiny required at a given time. External and internal oversight entities assure compliance and conformance, as part of their own verification programs using methods such as the following:

a. Information Exchange
b. Screening
c. Field Inspections
d. Audits
e. Show case

13.5.1 Information Exchange (IE)

Information Exchange (IE) describes a two-way method intended to get-to-know better both the verifier and the verified party related to the conformance and compliance verification (Figure 13.2). Practically, the IE provides both parties with each other's understanding of their culture, processes, implementation, and senior management commitment approaches.

Concisely, an Information Exchange should be provided at the end of the process to achieve:

a. The verified party understands better the expectations from the verifier

b. The verifier gets specific information of interest about the verified party (i.e., Integrity Management System and Operation and Maintenance procedures)
c. Both parties understand the context of the information provided via Q&As
d. The verified party has an opportunity to demonstrate their compliance and conformance

Subsequently, the IE results would allow the verifier to rank at a higher level their expected conformance and/or conformance risk. As IE process is used as the first interaction between verifier and verified, its ranking typically drives the verification schedule towards other more in-depth verification programs based on the level of knowledge of the verified party (e.g., finding history or repeatability, lack of collaboration, pipeline leak or program implementation). On the other hand, the IE would allow the verified party to understand verifier's processes and expectations in the near future.

13.5.2 Screening Verification (SV)

Screening verification is a thorough process that comprises a systematic examination, evaluation, investigation, or assessment performed especially to detect and identify non-compliances and non-conformances. The main purpose of the screening verification is to assess and diagnose the pipeline integrity activities against pre-defined criteria and expectations related to adequacy and implementation. The systematic assessment may have a large number of subjects to ensure compliance and conformance. A Checklist methodology (Figure 13.3) can be used as a screening verification by inquiring the verified party using specific questions about the integrity management program.

In some regulatory jurisdictions, a requirement may be associated to establish, implement and maintain a Pipeline Integrity Management System (PIMS). Thus, guidance questions may include inquiries about whether PIMS elements have been developed, implemented, and measured.

This verification method enables the verified and the verifier with the understanding of the areas requiring improvement via a high-level systematic and focused interaction.

FIG. 13.2 INFORMATION EXCHANGE (IE): A TWO-WAY COMMUNICATION

FIG. 13.3   SCREENING VERIFICATION (SV): CHECKLIST IS ONE OF THE SYSTEMATIC METHODS

Example: Screening Verification

A screening verification meeting was scheduled to assess the integrity program implementation. Eight (8) elements from PIMS were selected to evaluate at a high level the implementation of integrity management program pursuant current pipeline regulation, industry standards and company's own practices.

The verified party demonstrated that it performs pipeline integrity activities to ensure the integrity of the pipeline system is maintained and no immediate threat to safety and environment exist. However, the screening method revealed that verified party did not have a fully developed, implemented and performance measured section of its PIMS. As a result, the verifier encountered five (5) non-compliances and non-conformances, but no safety nor environmentally critical findings.

Subsequently, the Table 13.1 with the commitments from both parties was jointly developed to ensure resolution and tracking of all requirements. The corrective and preventive action plans (CPAP) addressed both non-compliances and non-conformances identified by the verification screening method.

TABLE 13.1   EXAMPLE OF CORRECTIVE AND PREVENTIVE ACTION PLANS FROM A SCREENING VERIFICATION

		Corrective and Preventive Action Plans	
Element	**Report finding reference numeral**	**Description**	**Completion date**
Risk assessment	3.1	• Conduct risk assessment for line ABC and XYZ • Develop a mitigation plan based on the results of risk assessment	30 June 2016
Internal audit	5.2	• Develop Internal Audit program including PIMP Program Audit Protocol • Perform Internal Audit program for line ABC	31 December 2016 31 December 2018
Performance measurement	5.3	• Develop performance metrics and targets for integrity program • Conduct gap analysis to identify gaps on performance metrics and targets, and then formalize them	31 December 2015
Evaluation the PIMP	5.4	• Conduct a formal management program review of the PIMP • Assess the effectiveness of the PIMP	30 March 2016 31 December 2016
Continuous improvement plan	6.1	• Develop a formal continuous improvement plan for the entire pipeline system • Implement the continuous improvement plan for its pipeline system	29 April 2017 31 December 2018

13.5.3 Field Inspection

Field inspections are conducted by the verifier for the main purpose of verifying the implementation of compliance and conformance requirements. Inspections in the field would have the advantage of being able to trigger stop work (prevent incidents), if the severity of the finding is critical or poses a threat to safety and/or the environment (Figure 13.4).

Inspecting assets transporting hydrocarbon fluids contributes toward a safe, reliable, and environmentally secure operation along the pipeline. Inspection verification involves two (2) main tasks:

1. Inspection of activities to evaluate the adequacy and implementation of integrity programs
2. Development of corrective and preventive actions, if findings are identified

Field inspection is typically a programmatic review of the pipeline system construction, operation, maintenance and integrity activities, procedures and processes; however, inspections may occur at any time during the life of the pipeline (e.g., construction, operation, abandonment).

FIG. 13.4   INSPECTION: FOCUS ON IMPLEMENTATION AND PREVENTION

Example: Corrective and Preventive Action Plans

A pipeline company gathering and transporting petroleum hydrocarbon product was in the process of finalizing the expansion of segments of a pre-1970 vintage pipeline terminal to deliver products to an existing storage tank farm.

Table 13.2 describes an example of a corrective and preventive action plans resulting from inspecting of a pipeline company.

TABLE 13.2   EXAMPLE OF CORRECTIVE AND PREVENTIVE ACTION PLANS FROM A INSPECTION

| | | **Compliance Corrective Action Plan** | |
No	Element	Description	Completion date
1	IMP Monitoring	Run In-Line-Inspection tool(s) for pipeline anomalies on the pipeline to detect, locate, size, and characterize metal loss, mechanical deformation and cracking type of anomaly by 2016 including validation and assessment of the performance specifications for the ILI tool(s).	31 December 2016
2	IMP Mitigation	Repair the soil-to-air (S/A) interface on the two risers for the pipelines, the coating was deteriorated with indication of external corrosion.	30 June 2016

13.5.4 Audits

An audit is a systematic and documented verification of the adequacy and effectiveness of the management systems, programs, processes, procedures, plans, manuals, records, and activities used by the verified party including the policy, commitment and review practices from senior management. Pipeline integrity activity processes can be audited within the International Organization for Standardization (ISO) management system framework. The protocols should be focused on regulatory requirements, industry standards and company's own process and procedures.

The *National Energy Board of Canada* describes "Audits" as an effective compliance verification method to promote compliance and proactively detect and correct non-compliances before they become issues." In the USA regulators have used the term "inspection" to describe some "audit" activities associated with the in-depth review of adequacy and effectiveness.

The main differences between audits and inspection are the focus and the mechanisms of the verification. Audits are focused on the systemic practices of the verified party across the entire pipeline system to verify the adequacy and effectiveness of the integrity programs (Figure 13.5); however, some localized/sampled field visits may encounter findings associated with the company not implementing the regulations or own procedures. Inspections are focused on specific facilities visited to verify the implementation of the integrity program including their records, personnel competency, and processes.

FIG. 13.5 AUDITS: ADEQUACY AND EFFECTIVENESS FOCUS WITH SAMPLED IMPLEMENTATION

Audits are typically conducted in the headquarters or main offices interacting with all levels of the organization and reviewing the processes, procedure and effectiveness of the implementation to identify systemic findings. Field inspections may detect adequacy (completeness) and effectiveness (reaching the goal) findings that may be characterized as local cases unless proven systemic by additional inspections (e.g., headquarters, similar facilities).

13.5.4.1 Internally-Driven Audit Process: Self-Assessment The main purpose of the company's own auditing process is to define a Self-Assessment approach for Pipeline Integrity Management System (PIMS) verification. A company's

own audit should verify principles, elements, and procedures. The main goal of the self-assessment plan is to promote continual improvement to be achieved by recognizing the good practices and developing a corrective and preventive action plans for areas needing improvement.

As illustrated in Figure 13.6, the ISO 19011:2011 standard provides some guidance to conduct audits of management systems or manage an audit program [9]. The overall approach proposed by ISO to managing audit programs is shown in the following figure. An audit manual document should describe the audit scope, objectives, and criteria as well as the audit team's competency, roles and responsibilities, protocols and checklists. Company's own audits

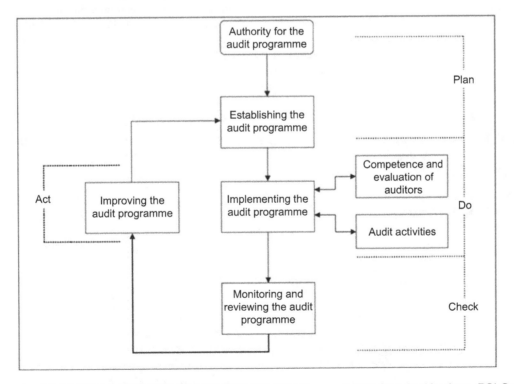

FIG. 13.6 MANAGING AUDIT PROGRAMS (BSI ISO 55) (*Source:* British Standard Institute, BSI Group)

can be conducted directed by internal or external parties depending on the objectives and resources.

The auditing program process can follow the PLAN-DO-CHECK-ACT structure and activities of ISO 19011. This proposed formal approach can help ensure your company's own audits are effective and consistent. Figure 13.7 provides an example of the steps that may assist in verifying parties in conducting company's own audits.

13.5.4.2 Externally-Driven Audit Process: Understanding Auditor's

13.5.4.2.1 Initiating an Audit. *Establish and Initiate the Audit*: To start, the auditor should initiate the audit by contacting the process owner to be audited and ensuring the audit will be feasible. Initial contact for conducting the audit may be a formal or informal contact and is made by the audit team leader. It is a good idea to make sure someone is available to present evidence when you want to audit, rather than try to surprise the verified party.

Request the Documents and Records: The main purpose of this task is to request access to the relevant information, document, and records. Auditors then need to review the documents and records. This will help auditors to plan how extensive the audit will be, whether it might take a whole day or week(s). This knowledge is critical for the next step. At this time the auditor should confirm any disclosure and confidentiality agreement, and make arrangement for the audit including schedule and dates.

Determine Feasibility of the Audit: the feasibility analysis should be considered the following minimum factors:

- Availability of sufficient and appropriate information for planning and conducting the audit,
- Adequate cooperation from the company being audited,
- Adequate time and resources.

13.5.4.2.2 Preparing for the Audit. *Review the Documentation*: Documentation such as policies, procedures, work instructions, forms, checklists, photographs, and other written information would help build understanding of the verified party practices. Conducting document review helps to prepare audit activities such as questions and drill-down focus. If the documentation is deficient and unsatisfactory, auditors may change the focus of the audit increasing the level of scrutiny as to why the verified party did not perform well in this activity.

Develop Audit Plan: The purpose of the document review and determination of the feasibility of the audit is to develop the audit plan:

- What will be audited,
- Why is it being audited,
- Who will do the auditing,
- When it will happen and,
- Who will be invited for an interview.

The audit team leader decides how the audit will be split-up whether more than one auditor will be used, and how much time will be dedicated to each process in the audit. Practically, preparation of the audit plan is based on information contained in the audit program and documentation provided (or lack of) by the verified party.

Assign Work to Auditors per Plan: Larger audits may assign work amongst several auditors, with each taking more than one process to audit. In this way, each auditor can focus on specific processes. Specific audit processes, activities, functions, and location would be assigned.

Prepare Working Plan: The assigned auditor then prepares the working plan that will identify

- What the auditor wants to verify,
- What questions to ask, and
- What they expect as evidence.

The working plan may include checklists and audit sampling plans and forms for recording information such as supporting evidence, audit findings, and records of meetings.

FIG. 13.7 EXAMPLE OF A MANAGEMENT SYSTEM AUDIT PROCESS USING ISO 19011 GUIDELINES

13.5.4.2.3 Conducting the Audit. *Kick-off Meeting*: The audit begins with an opening or kick-off meeting. The purpose is to make sure that everyone understands the scope and extent of this particular audit, to confirm the agreement of all parties on the audit plan, to introduce the audit team, and to ensure that all planned audit activities can be performed within the timeline and resources from both parties.

Determine the Audit Sequence: The next step is to determine the sequence of audit from the kick-off meeting up to presenting the audit findings to the verified party senior management. If done right, the sequence of process audits can help to make the audit flow easier. In this chapter, an example of a large audit is described.

Review of Additional Documents Collected during the Audit: The purpose is to determine conformity to audit criteria, gather information typically following the audit protocol to understand verified party's practices and potential questions and personnel for interviews.

Carry out the Audit: A beneficial practice is to maintain a good and open communication throughout the audit. Formal arrangements may be necessary for communication with the audit team, verified party and potential external bodies. Audit team meetings are needed to

- Communicate audit progress to verifier's main office
- Discuss observations
- Reassign work, and
- Redirect focus of the audit

The auditor asks the questions, and collects the records, determining observations following the audit protocol and drilling down where concerns may appear. Again, it is important to remember that an auditor is there to try to verify that a process conforms to the requirements set out, not to dig until fault is found. Verification can be achieved with

- Interviews
- Observations of activities
- Review of records, and/or
- Asking the five (5) Whys [10]

Generate Audit Findings: After the auditor finishes the verification, they must generate any audit findings and prepare any audit conclusions to be presented. Evaluation of audit findings against audit criteria and evidence may indicate whether compliance and conformance have been achieved or opportunities for improvement are recommended. It is equally important to highlight best practices, as it is to identify any shortcomings. Some companies also use a process of having company's own audits to identify opportunities for improvement (OFIs) that are initially reviewed by the process responsible for review whether they accept it and then senior management.

Audit Conclusions: Before the closing meeting, the audit team should review the audit findings and any other relevant information, agree on the audit conclusions, prepare recommendations, and discuss audit follow-up, as applicable.

Audit Close-out Meeting: In the close-out meeting, findings and conclusions are communicated via a summary (e.g., PowerPoint with key points) to senior management of the verified party; however, other members can be invited. The meeting is recommended to be formal with action items and attendance recorded.

13.5.4.2.4 Reporting of the Audit. *Draft Audit Report and Feedback Period*: With verified party's feedback from the audit close-out meeting, the draft audit report is drafted as a more complete, accurate, concise and clear record of the audit. The report should include or refer to everything listed in the audit plan. The good practices, findings, and opportunities for improvement are then presented in context to the overall process for the verified party to comment and provide feedback to the verifier on the report and verifier's audit process. The draft report also includes a statement of the confidential nature of the contents.

Final Audit Report: With verified party's feedback on the draft audit report, the final audit report is prepared. Report should be dated, approved, and distributed according to the audit program procedures. This provides a record of the outcome of the audit.

13.5.4.2.5 Completing the Audit Cycle. *Lesson Learned*: The feedback related to both audit process and findings should be analyzed with the continual improvement process for both the verifier and verified parties. This analysis may derive into a plan about what can be done differently at a systemic level (e.g., prevention focus).

Corrective and Preventive Action Plans Preparation: Probably the most important part of an audit is for the verified party to prepare an action plan for senior management and auditor's review. Later on, a designated verifier of the CPAP would follow up on any actions, as a way of ensuring remedial action is taken and completing the audit. Without follow up of corrections and corrective actions, the same problems could be found continually during subsequent audits, which defeat the purpose of the audit.

13.5.4.3 Audit Protocols From Regional Jurisdictions

13.5.4.3.1 Canada. The National Energy Board (NEB) is an independent federal regulatory body in Canada that is responsible for regulating pipelines that cross international borders or provincial borders. The NEB Act empowers the NEB to make regulations governing the design, construction, operation and abandonment of a pipeline in these areas. Pipelines transporting hydrocarbons solely within the borders of a single province are regulated by that province's regulatory body. For instance, the Alberta Energy Regulator (AER) is the regulator for pipelines within the province of Alberta. AER's mandate is to provide for the efficient, safe, orderly, and environmentally responsible development of energy resources in Alberta through the Regulator's regulatory activities.

Both the federal and provincial regulators use protocols for assessing compliance with their regulatory requirements while auditing pipeline companies. The protocols verify that companies develop and implement integrity management programs to anticipate, prevent, mitigate and manage conditions that may adversely affect the safety and security of the company's pipelines, employees, and the public, as well as property and the environment. The Canadian Standard CSA-Z662 has been adopted by all regulators in Canada as a regulation and in some provinces; further additions and modifications have been introduced during the adoption of the standard [11, 12].

13.5.4.3.2 United States. PHMSA has inspection protocols for verifying compliance with the Integrity Management Program requirements for gas (CFR 49 Part 192 Subpart O) and hazardous liquid (CFR 49 Part 195.452) pipelines in High Consequence Areas (HCA) illustrated in Figure 13.8. These regulations apply to all regulated transmission pipelines and certain gathering lines. At the time of writing this book, PHMSA is proposing new rulemaking for pipeline in non-High Consequence Areas (non-HCA).

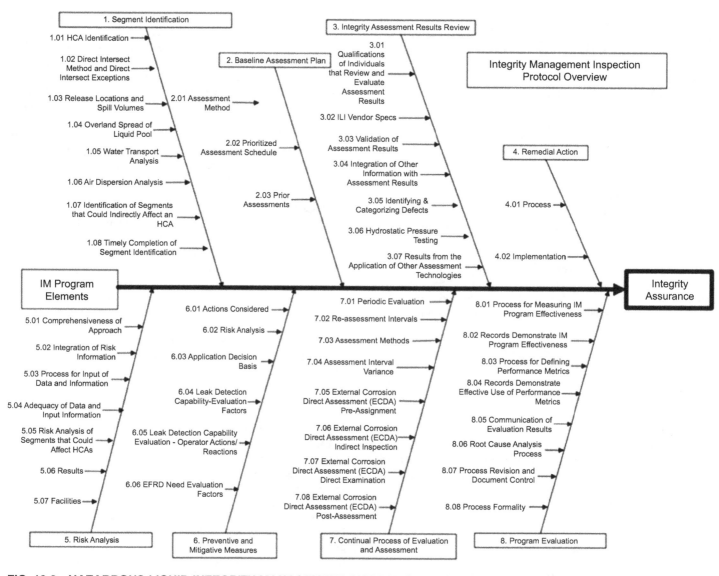

FIG. 13.8 HAZARDOUS LIQUID INTEGRITY MANAGEMENT INSPECTION PROTOCOL OVERVIEW—REVISION: 2013 (*Source:* USA Department of Transportation, Pipeline Hazardous and Material Safety Administration)

The IM Rule specifies how the USA pipeline operators must identify, prioritize, assess, evaluate, repair and validate the integrity of gas and oil pipelines that could, in the event of a leak or failure, affect High Consequence Areas (HCAs) within the United States to reduce the risk of injuries and property damage from pipeline failures.

Some states in the United States have requirements that go beyond the requirements set by the federal government. Operators are encouraged to check with the local state governments to determine applicable regulations [13, 14].

13.5.5 Show-Cause (SC)

A Show-Cause is a formal process used as the last resource for enforcing compliance when the previous enforcement methods have been exhausted in trying to reach effectiveness from the company. Show-cause seeks to provide the last opportunity for the company to be heard (i.e., hearing) and before the authority suspends or revokes the license or permit of operation, or other granted certificate. A Show-Cause hearing requires a case, argument, evidence, cause or reasons to justify, prove or explain how the company can resolve the serious non-compliance that may notably affect the public and/or environment [15].

Authorities commonly enforce the Show-cause process through an Order (i.e., Show Cause Order) or Notification Letter. An Order to Show Cause requires abrupt attention and solution from companies. The company would be requested to provide key information before deciding whether to issue conditions or take a proposed action to obtain fully compliance.

The company is expected to provide adequate reasons, evidences or causes to demonstrate (i.e., hearing process) why the authority should not take actions or enforcement; alternatively, what senior management commitment and corrective actions will be immediately taken to effectively resolve the non-compliance.

Example: Show-Cause

Following a number of ruptures of pipeline segment, a screening verification meeting and Emergency Response Exercise evaluation were conducted to assess and validate the capacity to an emergency response of a product release. The Emergency Response Program implementation review was conducted through a screening verification meeting.

The validation and evaluation of the response capacity was conducted through an Emergency Response exercise review comprised of the following activities:

a. Emergency Response Program and specific plan evaluation,
b. Test of the integration of different teams of the emergency response,
c. Evaluation and test of the Incident Command Post (ICP), mutual partners, industry, municipalities, and public first responders, and
d. Evaluation of communication within the ICP.

The observation and assessment showed that the verified party could demonstrate no familiarity with the Emergency Response, and the Emergency program has deficiencies and not developed appropriately. The Authority party concluded that the company cannot response efficiently to an emergency in a high-populated area. Additionally, the authority issued a Show-Cause letter directing the company to development a comprehensive Emergency program following regulation and standards within 60 days of the issuance of the letter to avoid a temporary suspension of the line operation till company demonstrate and prove through an exercise fully familiarity with their Emergency Response Program.

13.6 REGULATORY ENFORCEMENT PROCESS AND TOOLS

Oversight entities (internal or external) have developed processes and tools for enforcing compliance of companies. If findings are identified by a regulatory entity, the authority may enforce the requirements to obtain compliance by using the most suitable enforcement tool (e.g., warning, non-compliance notification, stop-work, monetary penalties). If findings are identified by a company's own verifier, senior management should be informed to follow the company procedures in the implementation of a resolution including timely self-disclosing to the authorities, if required.

Regulatory agencies hold toolkits to enforce compliance. The main goal is to enforce actions by regulated company to achieve compliance quickly, efficiently and effectively. The objective of achieving compliance is to reduce the hazard and threat likelihood and consequence in a timely manner so the safety of employees and the public as well as property and the environment are protected.

In Canada, Inspection Officers have the authority to take immediate enforcement action if they encounter non-compliances that could affect the safety of the employees and/or public safety and the environment [16].

In the USA, PHMSA has a full range of enforcement tools to ensure that an operator takes appropriate and timely corrective actions for violations, responds appropriately to incidents, and that they take preventive measures to preclude future failures or non-compliant operation [17]. The Pipeline Enforcement Program has a number of different mechanisms to assure operator compliance and safe operation.

The following are some of the North American and European regulatory authority enforcement tools:

Canada

• Corrected Non-compliance (immediate)
• Notice of Non-compliance

• Inspection Officer Order
• Inspection Officer Direction
• Board Orders (i.e., Safety Order, Amendment Order, or Miscellaneous Order)
• Administrative Monetary Penalty (fines)
• Revoke Authorization to operate pipeline
• Prosecution
• Imposition of terms and conditions

USA

• Corrective Action Order
• Notice of Amendment
• Notice of Probable Violation
• Warning Letter
• Safety Order

Europe

• Verbal or written information
• Advice
• Letters
• Notice (i.e., Improvement or Prohibition Notice)
• Prosecution

13.6.1 Natural Justice

Natural justice or procedural fairness ensures a fair decision is reached by an objective decision maker. These two basic legal safeguards govern all decisions by judges and government officials (i.e., inspection officer) when they make quasi-judicial or judicial decisions [18]. Three common law rules are referred to in relation to natural justice.

1. *The hearing rule*, a regulated company has a right to be heard prior to making any decision. They must have a fair opportunity to present their case whenever their interests might be adversely affected by a decision maker;
2. *The evidence rule*, this rule is that a decision must be based upon logical proof or evidence material, not on mere speculation or suspicion, and the decision must be communicated in a way that makes clear what evidence was used in making the decision. Relevant information and evidence should be considered prior to making any decision; and
3. *The bias rule*, the ruling must be made by somebody free of bias.

Natural justice is a principle used for the regulatory authority to provide the opportunity to the regulated company to make representations prior to any administrative decision being made. Typically, the administrative decision is related to a possible non-compliance requiring the use of an enforcement tool that can significantly affect the regulated company. Natural justice can also be named as regulators *"duty to act fairly"* and *"fair play in action."*

Relevant evidence such as arguments, allegations, documents, photos, data, reports, and engineering assessments can be presented by one party. This evidence must be disclosed to the other party, who may then subject it to scrutiny.

Natural justice principles and rules should be used when the regulator is performing enforcement duties. Being an independent and impartial decision maker without making assumptions prior to a verification activity ensures fairness and equal treatment for all pipeline companies. Additionally, the regulator must ensure that affected pipeline company must be given the opportunity to explain non-compliance or otherwise show that the company is in compliance.

13.6.2   Corrected Non-Compliance (CNC)

Some regulators define a CNC as the non-compliances observed in the field that were addressed while the inspection officer was on-site. This tool is used when there is minimal safety risk or environmental harm because of the non-compliance and it is corrected immediately in the field. The non-compliance will be noted in the inspection report along with the corrective measures that were taken to return to compliance.

It is used when the probability and consequence of non-compliance is low and it needs to be addressed, but there is a preventive plan to prevent reoccurrence. The Non-compliance activity observed during a compliance verification activity should be recorded and documented even though it has been corrected. The outcome of this enforcement option could be the issuance of an immediate resolution of the non-compliance, a report documenting the corrected non-compliance and/or the notification of the non-compliance to the company.

13.6.3   Notice of Non-Compliance (NNC)

Some regulators defines a NNC as a written undertaking issued by the inspection officer to a regulated company or third party when non-compliance with a low probability of harm to people or the environment is observed, and time is required to address the issue. It is intended to bring compliance issues to the attention of the company/individual, in order to generate the necessary action to return to compliance. The inspection officer will determine an effective corrective action, consult with the regulated company, and propose a reasonable timeline to correct the non-compliance. It is not a finding of guilt or civil liability, but will form part of the records when planning its compliance verification activities and determining what enforcement action to take for future or reoccurring non-compliances.

This enforcement tool may be used based on the following criteria:

* When harm to people or the environment may (low probability) occur if not corrected
* When a root cause assessment with a corrective action plan is required from the company
* When the regulator needs to receive a commitment to resolve a non-compliance that will extend beyond the time of the inspection (i.e., days to several months)

The Notice of Non-Compliance option requires all non-conformities be addressed. A corrective action plan with a completion date should be prepared with regulator's approval.

13.6.4   Inspection Officer Order

An inspection officer order is used when a high probability and/or high consequence non-compliance is observed, and either a hazard to the safety of the public or employees is being or will be caused or high potential detriment to property or the environment is being or will be caused. A corrective and preventive action plan to prevent reoccurrence is issued. This Inspection Order may be issued by an official representative of the regulatory body while conducting a compliance verification activity.

If a company refuses to either comply with requirements or sign a Notice of Non-Conformance where a hazard or detriment exists, an inspection officer order may be used.

13.6.5   Inspection Officer Direction (IOD)

The Inspection Officer Direction purpose is to gain access to property and information to fulfill the regulatory requirement, and provide information that may lead to evidence of non-compliance. Additional enforcement tools may be applied in the event that a company does not comply with the Inspection Officer Direction and if they refuse to meet them. A report may be issued to document uncooperative company behavior.

13.6.6   Monetary Penalties (MP)

Some regulators define a MP as a financial penalty imposed on individuals or companies in response to contravention of legislative requirements to promote safety and environmental protection.

13.6.7   Safety Order

Regulators have the authority to enforce certain requirements or restrict operations using Orders. Orders that restrict operations are commonly referred to as Safety Orders. In addition, designated Officers have the authority to direct and, in some situations, order parties to correct non-compliance [1]. Considerations may be taken into issuing a Board Order as to whether the immediate hazard or detriment has been resolved and other non-compliances require direction.

Example: Regulatory Safety Order Considerations

A class location designation has changed due to the increase of population in a pipeline area. The company did not comply with this requirement. This non-compliance action has been identified and notice sent to the company, but it has taken extended periods to address it. Immediate resolution of the non-compliance and long-term corrective action is needed by a specific date. The objective of the Safety Order is to prevent harm to the public and obtain compliance with the regulations.

13.6.8   Revoke Authorization

Some regulators may consider revoking an authorization that enables the company's operation when it has lost confidence in the ability of a company to operate safely or protect the environment. Suspension or cancelation of operations is a serious consequence.

13.6.9   Prosecution

If serious offences are discovered, some regulators may refer the details of the offence to the Office of the Attorney General for prosecution. An Order can also be made an order of the Court for the purpose of enforcement.

13.7   CORRECTIVE AND PREVENTIVE ACTION PLANS (CPAP)

A corrective and preventive action plans is developed to eliminate the cause of an identified non-conformance and/or non-compliance. A corrective action is taken to prevent recurrence; whereas, a preventive action is taken to eliminate the cause of a potential nonconformity or other undesirable situation. All nonconformities may not necessarily result in CPAP.

A corrective action plan can be the outcome of any verification method such as inspection, audit or assessment of the integrity program implementation. The CPAP is typically developed by the verified party, and then revised and approved by the verifier.

The following practices are recommended in managing CPAP:

- Evaluate the significance of nonconformities on risk to safety of employees and public and the environment; operating costs; cost of correction; company performance; dependability; any other risks
- Keep appropriate records of all CPAP steps
- Ensure timely completion of any open CPAP actions documenting evidence to justify the closure of open action items during the operation
- Develop Performance indicators to measure the effectiveness of the CPAP process to reduce time for correction, problem reoccurrence, costs, and impact on productivity
- Maintain a summary report for management or regulator review

As illustrated in Figure 13.9 a roadmap can be developed for supporting continual improving of corrective and preventive action plans using a progression workflow. The illustrated approach can help to ensure the action plan is effective and consistent, and builds the continual improvement of the pipeline integrity system.

13.7.1 Corrective and Preventative Action Identification

This task defines the corrective and preventative actions to be taken for resolving the integrity or operational issue leading to the finding. If an independent third party can say what the outcome should be (or is expected to be), then you may have identified an action to likely resolve the nonconformity.

13.7.2 CPAP Scope Development

Defining the scope of the action plan is conducted by integrating and prioritizing the identified actions in the previous step.

Nonconformities that affect the public, environment, or a property need more attention and a more comprehensive action plan.

In the interim, containment actions may need to be undertaken while fixing the root-cause of the nonconformity. These actions become immediate measures to decrease the probability of failure for a given period of time, but not resolving it permanently.

Finally, the CPAP should identify measures to eliminate the root cause of the nonconformity within a timeline. Sensitivity analysis may be used for evaluating alternatives determining their benefit/cost ratio related to time for implementation. Regulators usually skip this step when safety and environmental protection are at the core of the issue by directing the regulated party to act within a defined term.

> *Example:* Plan Development
>
> A pre-1970 vintage gas pipeline has not a baseline integrity assessment, but experienced several leak incidents in highly populated areas. The risk associated with the nonconformity is higher requiring a higher priority for resolution to be reflected in the CPAP. A containment action may include measures such as a 20% pressure reduction, an increase to weekly right-of-way surveillance and leak detection interval assessment reduction. The regulator has also directed the regulated company to conduct dual crack in-line inspections followed by a confirmatory spike test of short duration to ensure effectiveness within a year.

13.7.3 CPAP Implementation

Implementation of the Corrective Action is following through on your plan and making it happen. It may require implementing a preventive maintenance program or an overprotection system because the old one could no longer provide the confidence

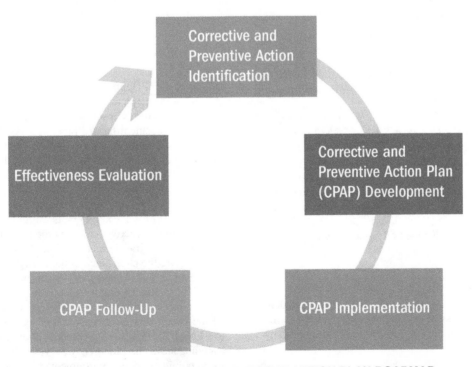

FIG. 13.9 CORRECTIVE AND PREVENTIVE ACTION PLAN ROADMAP

required. The tasks must be well defined in scope ensuring that they can be achievable within the schedule and effectiveness required.

13.7.4 CPAP Follow-Up

The follow-up is a very important step when the verified parties may struggle the most. Often, the verified party wants to close out the paperwork quickly to demonstrate timeliness and completeness, but the objective may not be achieved. The follow-up should be documented providing updates to verifiers and management.

13.7.5 Effectiveness Evaluation

The verified party should conduct periodic evaluation of the actions taken to determine whether the implemented measures have met the intent of the CPAP.

13.8 INCIDENT INVESTIGATION FUNDAMENTALS

13.8.1 Pipeline Incident, Accident, Damage, and Failure Definitions

An incident is an unplanned and undesired event that could adversely affect society, environment and company's operation, resources, focus and reputation. Pipeline integrity incidents may include employee and public injury and fatality, property damage and environmental damages. Incidents commonly result from a chain of changes, failures and/or errors rather than just a single mistake.

Example: Canada Definitions
In Canada, an "incident" is defined in section 1 of the OPR-99 as an occurrence that results in:

a. *The death of or serious injury to a person;*
b. *A significant adverse effect on the environment;*
c. *An unintended fire or explosion;*
d. *An unintended or uncontained release of LVP hydrocarbons in excess of 1.5 m³;*
e. *An unintended or uncontrolled release of gas or HVP hydrocarbons; or*
f. *The operation of a pipeline beyond its design limits as determined under CSA Z662 or CSA Z276 or any operating limits imposed by the Board.*

The 2015 version of the Canadian Standard CSA-Z662 distinguishes the difference between a damage incident and a failure incident, as follows:

• Damage incident—an event that results in damage to a pipe, component, tank, or coating without release of service fluid
• Failure incident—an unplanned release of service fluid

Example: USA Definitions
In the USA, an "accident" as defined in §195.50. An accident is a failure in a pipeline system in which there is a release of the hazardous liquid or carbon dioxide transported resulting in any of the following:

a. Explosion or fire not intentionally set by the operator;
b. Release of 5 gal (19 L) or more of hazardous liquid or carbon dioxide, except that no report is required for a release of less than 5 barrels (0.8 cubic meters) resulting from a pipeline maintenance activity if the release is:
   1. Not otherwise reportable under this section;
   2. Not one described in §195.52(a) (4);
   3. Confined to company property or pipeline right-of-way; and
   4. Cleaned up promptly.
c. Death of any person;
d. Personal injury necessitating hospitalization;
e. Estimated property damage, including cost of cleanup and recovery, value of lost product, and damage to the property of the operator or others, or both, exceeding $50,000.

In the USA, an "incident" as defined in 49 CFR §191.3,

1. An event that involves a release of gas from a pipeline or of liquefied natural gas, liquefied petroleum gas, refrigerant gas, or gas from an LNG facility, and that results in one of the following consequences:
   i. A death, or personal injury necessitating in-patient hospitalization;
   ii. Estimated property damage of $50,000 or more, including lost to the operator and others, or both, but excluding cost of gas lost;
   iii. Unintentional estimated gas loss of three million cubic feet or more;
2. An event that results in an emergency shutdown of an LNG facility. Activation of an emergency shutdown system for reasons other than an actual emergency does not constitute an incident.
3. An event that is significant in the judgment of the operator, even though it did not meet the criteria of paragraphs (1) or (2) of this definition.

Example: UK Definitions
In UK, a "major accident" in The Pipelines Safety Regulations 1996 means death or serious injury involving a dangerous fluid. A "major accident hazard pipeline" has the meaning given by regulation 18(1). A "major incident" can be defined as any emergency that requires the implementation of special arrangements by one or more of the Emergency Services, NHS or the Local Authority(s).
UKOPA [19] defines levels of pipeline emergencies:
Level 1: *Minor emergency* is a minor incident that involves checks and corrective action by the Pipeline Operator only, has no immediate impact on the public or the environment and does not require the attendance of the emergency services.
Level 2: *Local emergency* is an incident being investigated by the Pipeline Operator, has no immediate impact on the public or the environment but may require the attendance of the Emergency Services to ensure it is dealt with safely.
Level 3: *Pipeline emergency* is an incident requiring the attendance of the Emergency Services, but does not put the general public or wider environment at risk. The effects can be seen to be contained with no expectation of escalation.

Level 4: *Pipeline Major Emergency* is a major incident that requires the implementation of the Local Authority Emergency Plan. This type of incident will fit with the Government definition of a major incident in "Dealing with a Disaster":

- Affects a large number of people
- Causes significant public disruption
- Results in many injuries
- Causes major environmental damage
- Requires a significant response from many agencies

Regardless of the regulatory definition, any failure of primary containment should be investigated to understand the cause(s). A proper understanding of the root causes of pipeline incidents is an essential process to the development of suitable measures to prevent their recurrences. The incident investigation process is one of the main elements of the integrity management system.

13.8.2 Types of Pipeline Incident Causes

Incident causes include both

- Immediate, direct or primary causes and sub-causes
- *Basic or root causes* and sub-causes

13.8.2.1 Immediate, Direct, or Primary Causes   Immediate, direct, or primary causes correspond to the circumstances immediately or directly preceding a given incident and usually represent the situation that can be seen. They are the circumstances that precede the pipeline event and are typically observable by one or more of the five senses (e.g., see, hear, smell, feel, or taste).

Example of Pipeline Integrity Immediate Causes (Table 13.3 describes an example of causes adapted from NEB and CSA-Z662-15 Annex H)

TABLE 13.3   EXAMPLE OF PIPELINE INTEGRITY IMMEDIATE OR TECHNICAL CAUSES

Immediate/Primary/Direct cause	Immediate/Direct sub-cause
Metal Loss Wall thickness reduction due, for example but not exclusively, to corrosion or erosion	External Metal Loss Internal Metal Loss
External Interference External activities causing damage to the pipe or component	Company Employee Damage (1st party) Company Contractor Damage (2nd party) Other or Unknown Non-Company Damage (3rd party) Vandalism
Cracking Mechanically driven or environmentally assisted cracking of the pipe or component	Corrosion Fatigue Fatigue Cracking Hydrogen-Assisted Cracking Mechanical Damage Delayed Cracking Sulfide Stress Cracking Stress Corrosion Cracking Other
Geohazard	Mass Wasting: Landslides, Slow Ground Movement, Avalanches Hydro-geotechnical: River scour, Bank Erosion and Channel Migration Seismic Specialized Geohazards: man-made, thaw off, residual and sensitive soils, desert mechanisms including dune migration, volcanic overburden, geochemical karst, and acid rock drainage
Material, Manufacturing, or Construction Material or manufacturing defect in the failed pipe or component Construction defect, damage, or deficiency in the failed pipe or component	Defective Circumferential Weld Defective Helical Seam Weld Defective Long Seam Weld Defective Pipe Body or Component Diameter Deformation Other
Equipment/Component Failure	Control Equipment/Electronics failure Measurement Device Failure Pipe Body Failure Piping Component/Fitting Failure Pig Barrel/Receiver Failure Prime Mover Failure Rigging Failure Valve Failure Weld Failure
Incorrect Operations	Equipment/Control System Malfunction Improper Operation Unknown

13.8.2.2 Basic or Root Causes Basic or root causes are the causes behind immediate/direct causes and are often referred to as real, indirect, underlying, or contributing causes such as:

- Management systems
- Job factors
- Personal factors
- Environmental factors
- Material and Equipment Factors

When all root causes of an incident are corrected, long-term mitigation and prevention of similar incidents will likely be minimized.

Example of Pipeline Integrity Management System Causes including Basic or Root causes is described in Table 13.4

TABLE 13.4   EXAMPLE OF INTEGRITY MANAGEMENT SYSTEM BASIC OR ROOT CAUSES

Management system basic/ Root-cause	Management system basic/ Root sub-cause
Policy and Commitment	Policy and Commitment
Plan	Hazard Identification, Risk Assessment, and Control
	Legal and Other Requirements
	Goals, Targets, and Objectives
Do	Organizational Structure, Roles, and Responsibilities
	Training, Competence, and Evaluation
	Communication and Awareness
	Documentation and Document Control
	Operational Control—Normal Operations
	Operational Control—Upset or Abnormal Operations
Check	Inspection, Measurement, and Monitoring
	Corrective and Preventive Actions
	Records Management
	Internal Audit
Act	Management Review

Example of Pipeline Integrity Basic Causes (adapted from NEB Incident Investigation)

TABLE 13.5   EXAMPLE OF BASIC CAUSES RELATED TO THE PEOPLE, MATERIAL, EQUIPMENT, OPERATION, AND ENVIRONMENT

Basic cause	Basic sub-cause
Personal Factors	Improper Motivation
	Inadequate Mental/Psychological Capability
	Inadequate Physical/Physiological Capability
	Lack of Knowledge

(continued)

TABLE 13.5   (CONTINUED)

Basic Cause	Basic sub-cause
	Lack of Skill
	Mental or Psychological Stress
	Physical or Physiological Stress
Materials	Inappropriate Material for Conditions/Usage
	Manufacturing Anomaly
	Material Anomaly
Operating Conditions	Excessive Cold Temperatures
	Excessive Dynamic Load
	Excessive Hot Temperatures
	Excessive Static Load
	Excessive Wear and Tear
	Normal Wear and Tear
Environment	Adverse Weather
	Earth or Slope Movement
	Earthquake
	Flooding
	Freezing
	Frost Heave
	Lightning
	Snow Loading
	Wash-out Erosion
	Wild land Fires

13.8.2.3 Incident Finding Analysis Using a Management System Approach: Case Study
Excerpt from ASME IPC 2012-90045 paper

The Figure 13.10 shows the distribution of the MS elements (or causes) identified from the 2005–2009 NEB integrity-related incidents. The top two MS causes were *implementation* (do) and *planning* (plan) representing over 65% of the causes identified in the incident analysis.

The Figure 13.11 shows the distribution of the MS sub-elements (or sub-causes) identified from the 2005–2009 NEB integrity-related incidents.

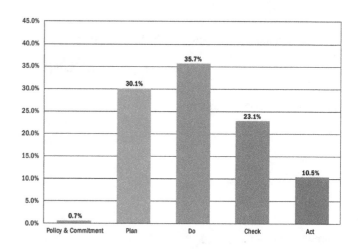

FIG. 13.10   THE 2005–2009 INTEGRITY-RELATED INCIDENT CAUSES BY MS ELEMENTS (ASME IPC 2012-90046 PAPER)

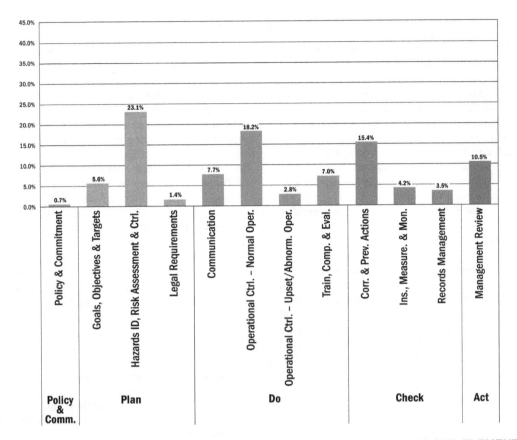

FIG. 13.11 THE 2005–2009 INTEGRITY-RELATED INCIDENT CAUSES BY MS SUB-ELEMENTS
(*Source:* ASME International Pipeline Conference IPC 2012-90046 Paper)

13.8.3   Pipeline Incident Investigation Process

The purposes of the incident investigation analysis of a pipeline failure are:

- Identify and describe the facts of the incident (what, where, and when)
- Identify the immediate and root causes/contributing factors of the incident (why)
- Identify pipeline integrity risk reducing measures to prevent future and comparable incidents
- Create and discover alternative possibilities for action plans
- Identify lessons learned associated with the investigation process and the incident systemic issues

As illustrated in the Figure 13.12, the process comprises:

- Data Gathering
- Analysis and determination of the causes
- Report of Findings
- Recommended Actions

Pipeline leaks and ruptures are the most serious incidents that require in-depth investigation. The investigation will require dispatching independent reviewers and/or regulatory personnel to the incident site to monitor the activities, to assess whether appropriate actions are being taken, and to advise decision-maker (e.g., senior management, regulator) whether and when the pipeline system may be safely returned to service.

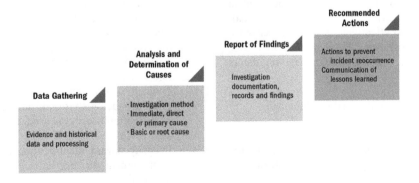

FIG. 13.12   INCIDENT INVESTIGATION PROCESS

Collecting and storing data in a systematic manner, developing a custom database, and cause analysis are essential tasks in the incident analysis process. Date should be arranged in a way that allows the company to be able to analyze it quickly and to update it easily. A systematic method of processing the information, as described in the previous figure, obtained from examination of the evidence leads to identification of the issue, identification of factors that make it an issue, definition of its significance, and development of the suitable corrective or remedial measures.

The type of method for analyzing causes is a crucial decision and it should be chosen after knowing the incident as much as possible. An investigative analysis may require numerous rounds of collecting evidences, facts, forming hypotheses, and testing hypotheses. Investigation step is fundamentally an iterative or repetitive process.

The Root Cause Analysis Checklist is a methodology by answering the questions in the checklist. The checklist incident analysis provides an easy-to-follow process to reveal primary and root-causes. Incident investigator can use this method as an initial evaluation against pre-established criteria in the form of one or more checklists.

The checklist questions can be developed based on historical information, knowledge from previous incidents and from other sources of information. It is a quick and simple method for incident analysis. The advantages of this Checklist incident analysis are:

- Simple technique
- Rapid to implement and develop
- Easy to implement, and
- Require few resources

Relationships between database record types and information sources may or may not play a role to confirm incident causes by identifying similarities and differences with previous incidents. Furthermore, the analysis of historical incident data compared to the actual incident data also helps identifying both the adequacy and effectiveness of proposed prevention and mitigation measures.

The incident report should provide the background, methodology, data gathering, analysis, conclusions, and recommendations. The recommendations would drive the corrective and preventive action plans.

13.8.4 Incident Investigation Documentation and Records

Documentation and records are parts of the information management system for better communication, management of the pipeline operations, decision-making process, keeping records and regulatory purposes.

Over the years, pipeline companies have been wisely working to strengthen database to collect accurately their pipeline incident data. Pipeline incident data gathering and reporting have challenges to increase efficiency and effectiveness of the process for analyzing, investigating, documenting, and reporting incidents. Challenges facing the pipeline companies and even regulators including:

- Difficulties receiving timely and detailed incident data;
- Low incident analysis and close out rates;
- Deficiencies in determining accurately the root causes;
- An aging Pipeline Incident Database (PID) that no longer meets pipeline companies and regulatory body information requirements;

- Lack of an effective incident report system that increases workload;
- Data entry manually and data standardization; and
- Difficulties delivering incident report.

A database-driven incident data analysis should be considered to review and categorize the amount of pipeline incident records according to their causes, dynamics, and consequences. Using an automated data-mining process followed by a peer-review of the record data, the pipeline flaws hazards, and deficiencies in the database should be identified and extracted.

The use of incident information should be to prepare cost estimates, forecast, incident trending analysis, and company performances. The documentation and records should be organized, stored, issued, revised, approved, distributed, and controlled with maintenance of revision histories.

13.8.5 Incident Investigation Learning

One of the main purposes of the incident investigation process is to determine what we have learned from the incident. A number of examples of "Lessons Learned" are described below:

- Having a formal incident investigation process allows pipeline companies investigate systematically incident to determine root causes to properly identify preventive and mitigative actions.
- Prevention action practically should become the ultimate goal of an incident such actions can be taken for minimizing their reoccurrence.
- Learning from pipeline deviations and investigations encourages employees to both report these instances and provide truthful feedbacks.
- Definitely, incidents depict real occurrences that generate logic activities to logical process analysis with final recommendations as outcomes.
- Outcomes (i.e., Identified hazards, preventive measures, etc.) have facilitated operators to build historical information including all potential hazards or measures that possibly were not identified previously.
- Incident investigation process also has contributed to developing the continuous improvement process and then effectiveness of the pipeline integrity activities.
- Identification of hazards through the incident investigation process allows determining improvements of process in any pipeline lifecycle. For instances, San Bruno incident investigation identified deficiencies in pipe section fabrication, Yield Strength and Welding but it also identified Integrity Management deficiencies [20].
- Hazards identified in the incident investigation can often be addressed in more than one pipeline system.
- The maximum preventive measure is achieved by implementing corrective actions of the pipeline conditions that led to the pipeline incident and may lead pipeline failure at other pipeline systems.
- Typically, the immediate cause is identified as the main contributing factor to the incident. When a detailed incident investigation process is used, multiple root causes are usually uncovered.
- Integrity hazard identification process used as a continuous improvement process helps to prevent or anticipate to the incident.

13.9 REFERENCES

1. NEB, *Regulatory Framework*, National Energy Board, Canada http://www.neb-one.gc.ca/sftnvrnmnt/prtctng/index-eng.html.

2. DOT PHMSA, *Pipeline Inspections 101*, Pipeline and Hazardous Materials Safety Administration, USA http://phmsa.dot.gov/pipeline/inspections.

3. HSE, *Health and Safety Inspection*, UK Health and Safety Executive (HSE), United Kingdom http://www.hse.gov.uk/stress/furtheradvice/hseinspectors.htm.

4. Penny, J., Eaton, A., 2001, *The Practicalities of Goal-based Safety Regulation*, Adelard, London, United Kingdom.

5. Bolzon, G., Boukharouba, T., Gabetta, G., Elboujdaini, M., 2010, *Integrity of Pipelines Transporting Hydrocarbons*, Springer, The Netherlands.

6. Coglianese, C., Lazer, D., 2003, Management-Based Regulation: Prescribing Private Management to Achieve Public Goals, Law & Society Review, Harvard University, USA.

7. DOT PHMSA, *Pipeline Safety Inspections*, Pipeline and Hazardous Materials Safety Administration, USA http://primis.phmsa.dot.gov/comm/Inspection.htm.

8. NEB, *Compliance Verification*, National Energy Board, Canada http://www.neb-one.gc.ca/sftnvrnmnt/prtctng/index-eng.html#s3.

9. ISO, 2011, *Guidelines for Auditing Management Systems*, ISO 19011:2011, International Organization of Standardization, Geneva, Switzerland.

10. Sondalini, M., Understanding How to Use The 5-Whys for Root Cause Analysis, Lifetime Reliability—Solutions, WA, Australia http://www.lifetime-reliability.com/tutorials/lean-management-methods/How_to_Use_the_5-Whys_for_Root_Cause_Analysis.pdf.

11. NEB, *Management System and Protection Program Audit Protocol*, National Energy Board, Canada https://www.neb-one.gc.ca/bts/ctrg/gnnb/nshrppln/dtprtcl-eng.html.

12. BC OGC, *IMP Self Assessment Protocol*, British Columbia Oil and Gas Commission (BC OGC), Canada https://www.bcogc.ca/content/imp-self-assessment-form.

13. DOT PHMSA, *Gas Integrity Management Inspection Protocol*, Pipeline and Hazardous Materials Safety Administration, USA https://primis.phmsa.dot.gov/gasimp/docs/Visio-GasIMP_October2013.pdf.

14. DOT PHMSA, *Hazardous Liquid Integrity Management Inspection Protocol*, Pipeline and Hazardous Materials Safety Administration, USA https://primis.phmsa.dot.gov/iim/docsp/HazLiquidIMProtocolForm_10_21_13.doc.

15. TSB, *Crude Oil Pipeline Rupture IPL Line 3 Mile Post 506.6830*, Pipeline Investigation Report P96H008, Transportation Safety Board of Canada (TBS), Canada.

16. NEB, *Enforcement Toolkit*, National Energy Board, Canada http://www.neb-one.gc.ca/bts/nws/fs/nbnfrcmnttlktfctsht-eng.html.

17. DOT PHMSA, Enforcement, Pipeline and Hazardous Materials Safety Administration (PHMSA), USA http://phmsa.dot.gov/pipeline/enforcement.

18. Cody, P., *Natural Justice*, Canada http://www.justice4you.org/natural%20_justice.php.

19. UKOPA—United Kingdom Onshore Pipeline Operator's Association. www.ukopa.co.uk.

20. CUPC, *September 9, 2010 PG&E Pipeline Rupture in San Bruno, California—Investigation Report*, Consumer Protection & Safety Division, California Public Utilities Commission, California, USA http://www.cpuc.ca.gov/NR/rdonlyres/28720A78-1DC7-4474-B51F-00C5E8BB5069/0/AgendaStaffReportreOIIPGESanBrunoExplosion.pdf.

PIMS MANAGEMENT REVIEW (ACT): PERFORMANCE ASSESSMENT AND KPIs

Chapter 14 provides a process for conducting performance assessments as well as developing, implementation and benchmarking KPIs to enable management in reviewing the adequacy, implementation, and effectiveness of PIMS.

The chapter provides terminology differentiating the management review areas of focus as well as integrity performance indicators, integrity performance measures, criteria and track symbols. As illustrated in Figure 14.1, the performance assessment process oriented to the management review is guided a PLAN-DO-CHECK-ACT management system approach, as follows:

- goals, objectives and targets; and planning (Plan)
- implementation (Do)
- verification and action plans (Check)
- continual improvement (Act)

Some key areas where to look for developing integrity performance indicators are recommended. The integrity performance cycle is described in four (4) stages from which KPIs can be measured and evaluated. Cumulative indices are also explained providing the trend interpretation.

The chapter provides guidance about how to review the adequacy, implementation and effectiveness of PIMS. Residual risk assessed after contributions made by the integrity programs is discussed. Some examples to initiate the management review are also provided.

The chapter closes presenting a benchmarking methodology by referencing your own performance to external entities' performance published by industry associations, regulatory agencies at the Federal, State and Provincial level, and industry standards. The benchmarking methodology compares KPI integrity approach, cycle stage and measure type.

14.1 INTRODUCTION

Management review purpose is to assess the progress of both the management system and IMP in achieving the goals, objectives, and targets established for PIMS. Performance assessments for management review should portrait the level of

- Adequacy of integrity processes and procedures (e.g., development);
- Implementation within the plans (e.g., completion), and;
- Effectiveness in reducing probability, consequence, and risks (e.g., goal achievement).

Societal and professional responsibility with a focus onto *"doing the right thing"* can be applied in assessing the level of PIMS performance. Furthermore, management is also encouraged to review the performance assessment in context (e.g., societal, environment, market) for understanding their contribution and/or impact to stakeholders (i.e., phronesis).

The review allows management to clarify or adjust the focus (i.e., center of attention), prioritize, or re-prioritize (i.e., order) and re-define expected outcomes (i.e., final result) in consideration to the ongoing internal and/or external business environment changes. Management review outcomes may trigger changes in the pipeline system, technology, process, and organizational structure.

Management system review can assist to:

- Combat confirmation bias (e.g., *If I fulfill my plans or regulatory requirements, my system is safe as demonstrated in the past*)
- Uncover management system weaknesses (e.g., *effectiveness verification*)
- Identify emerging integrity threats (e.g., *seam weld cracking of pre-1980 vintage pipe*), and
- Recognize the absence of adequate and effective barriers (e.g., *pressure cycling management*)

Management reviews the results presented as KPIs that represent or anticipate integrity conditions, events and processes (e.g., technical/technological, management) within a period of time linked to an objective or goal. Conceptually, KPIs are built-in with evaluation status, time and expectation, which applied periodically, will trigger the need to reaffirm or change the path of action taken or to be taken.

Managers measure performance of their departments, management systems, and activities using very detailed KPIs for management review. However, this information may either partially answer the questions or not be transformed in terms of "big picture" benefits/impacts needed to be known by senior management and informed outsiders such as regulators, Boards of Directors, industry associations, public and media.

The development of pipeline KPIs has been historically activated by unsuccessful events motivating the organization (e.g., company, regulator, association) to identify the performance drivers capturing the past and anticipating the near-future.

KPIs can also be calculated segmenting by regions, processes, and teams. Pipeline sections can also be identified with either specific characteristics and operating conditions or specific applied technology for their differentiation within the performance analysis

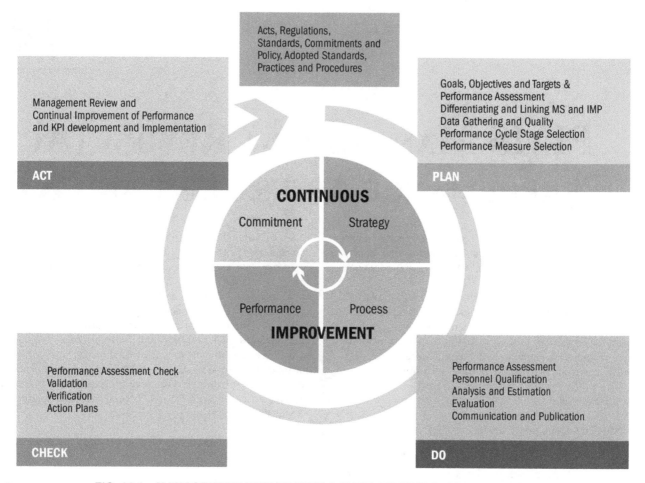

FIG. 14.1   MANAGEMENT REVIEW WITH A MANAGEMENT SYSTEM APPROACH

to be trended and analyzed for future actions. The absolute value of KPI related to their benchmark or reference can be compared as long as their specific conditions and environment in which they were measured in are acknowledged.

14.2   TERMINOLOGY

Adequacy
Adequacy is achieved when the stakeholders' *intent, company's responsibilities, and PIMS completeness* of processes, procedures, roles, and responsibilities and associated organization meet specific requirements from either regulations or adopted industry standards.

Implementation
Implementation is achieved when *plans* are conducted in a safe, quality, and timely manner following approved processes and procedures.

Effectiveness
Effectiveness is achieved when the *PIMS goals and objectives* are met in a safe and environmentally responsible manner in context of the stakeholders (i.e., phronesis).

Integrity Performance Indicator
An integrity performance indicator represents an integrity focus, objective, or outcome to be measured and compared against stakeholders' goals, targets, or expectations. A condition indicator would reflect the level of degradation of the pipeline related to an integrity threat *(e.g., corrosion)*, group of threats *(e.g., time-dependent)* or overall pipeline condition; whereas, the integrity management indicator would reflect the level of adequacy, implementation or effectiveness of a given activity *(e.g., pipeline cleaning)*, process, or program *(e.g., corrosion inhibition)*.

Integrity Performance Measure
An integrity performance measure quantifies one (or the key) factor or component *(e.g., maximum depth of corrosion anomalies)* of an integrity performance indicator within a time period *(e.g., monthly, quarterly, or annually)*. A collection of selected measures *(e.g., corrosion identification and mitigation)* can be used to build up a performance indicator *(e.g., corrosion performance)*. Hence, integrity performance indicators are comprised of one or more integrity performance measures.

Integrity Performance Criteria and Symbols
The integrity performance criteria provides the acceptability of the integrity performance measure based on a pre-defined expectation, benchmark or standard *(e.g., <80% deep corrosion)*, which can be graded in multiple levels from acceptance to rejection *(e.g., <50%, 50% to 70%, >70%)*. The results of the comparison can also be expressed as a three-color code *(e.g., green, yellow, red)*, binary response *(e.g., Yes/No)*, percentage and qualitative measure *(e.g., low, medium, high)*.

Integrity Management Performance Cycle

The integrity management performance cycle describes the objectives and interrelationship of the performance elements of integrity management driven by continuous improvement. The performance cycle is divided in four (4) stages: input or *integrity feed*, process or *integrity progress*, output or *system response to integrity actions*, and outcome or *integrity end-results towards goals*.

14.3 INTEGRITY PERFORMANCE ASSESSMENT PROCESS: MANAGEMENT REVIEW

14.3.1 PIMS Goals, Objectives and Targets and Performance Assessment

The written PIMS goals, objectives, and targets for managing integrity of pipelines should be directed to minimizing the residual risk to acceptable levels in accordance with the goals. The objectives and targets defined in PIMS should also be used for the performance assessment.

If PIMS goals are focused on *protecting our employees, public and environment while providing a reliable service to our shippers or customers*, the prevention of a cracking failure or its reoccurrence should be part of the performance assessment.

Example: Objective, Target, and KPIs

- MS Objective: Adequate Procedures for Implementing Programs
  KPI: *Percentage of completion of procedures for running crack in-line inspection tools*
- IMP Objective: Identify the pipeline sections susceptible to cracking and inspect them.
  KPI1: *Number of Crack ILIs conducted in susceptible pipeline sections this year*
  KPI2: Percentage of *Crack ILIs completed versus planned in susceptible pipeline sections*

14.3.2 Performance Assessment Planning (Plan)

14.3.2.1 Differentiating and Linking MS and IMP Approaches: MS and IMP Integrity performance assessment should be planned using two (2) interrelated integrity approaches:

1. Integrity Management Program (IMP), and;
2. Management Systems (MS).

The MS approach to performance assessment is management-oriented and can be used following the ISO operating principle by qualitatively or quantitatively monitoring elements against organization's expectations or industry benchmarks. MS-driven performance assessment tracks management processes such as planning, implementation, checking, and verification.

The IMP approach to performance assessment is technical and can be applied by using elements such as integrity health or condition (i.e., identification and monitoring), assessment or verification, mitigation including incident investigations, monitoring, and prevention results. These approaches are explained in detailed later in this chapter.

Performance measures should be representative of the system or systemic as well as systematic or repeatable. The systematic

approach would rely on the consistency of the data gathering, analysis and interpretation to derive into comparable conclusions. The value of each developed performance measure will be known during their use. Some of the drawbacks (i.e., lack thereof) that an organization should avoid during the planning of a performance assessment are:

- Measuring tactic or operational processes without accounting for senior management needs
- Providing definitions of the without the quality and consistency required by the end-users (MS)
- Not having long term focus on performance assessments (MS)
- Local focus without accounting for system or corporation-wide efforts (MS)
- Not understanding the causes of failure on their system (IMP)
- Not having a process for identifying barriers are required to prevent failure (IMP)

Example: Guidance for the development and implementation of KPIs with both a management system and integrity management program approach:

KPIs DNV RP-F116 *Integrity Management of Submarine Pipeline Systems* standard

- Management System Approach
- Uses Barriers to create KPIs
  - Design (e.g., Wall thickness)
  - Protection Systems (e.g., CP)
  - Control Systems
  - Processes and Procedures (e.g., Inspect, Mitigate, assess, monitor)
- Management-driven implementation

API RP 1173 *Pipeline Safety Management Systems Requirements*

- Management System Approach
- Adequacy (Completion)
- Effectiveness (Risk Reduction)
- Improve pipeline safety performance
- Use of Lagging and Leading KPIs
- Frequency to identify trends and corrective actions
- Management-driven implementation

14.3.2.2 Data Gathering and Quality Performance assessments require capturing data from past events that could be used either to rear-view the actual outcomes (i.e., lagging) or to anticipate future outcomes (i.e., leading). Data should also be prepared and aligned such as design, construction, and inspection. Data associated with timeline can be useful such as immediate and scheduled repairs.

The quality of the data gathered needs to be assessed to understand the basis for the performance assessment. This may lead to provide context to the audience and to establish a level of confidence for decision making.

14.3.2.3 Performance Cycle Stage Selection: Input, Process, Output and Outcome The performance assessment may focus on one or multiple stages of the cycle based on the stakeholders' needs. Therefore, it is important to recognize the advantages and disadvantages of each cycle stage.

Example: Performance cycle stage selection

- The Integrity Feed (Input) and the Progress (Process) stages anticipate challenges and actions to be taken focusing on measuring efficiency or how to do it better.
- Integrity Response to Action (Output) and End-Results (Outcome) should be the stages to be selected for measuring effectiveness.

Sometimes performance measures are picked to assess how the things they did do worked out in the *process* (e.g., program completion) and not to assess if negative *outputs* (e.g., anomaly growth) and *outcomes* (e.g., reputation) occurred due to absent or bypassed integrity barriers (e.g., cleaning pigs, inhibition, recoating). Near-misses, abnormal events, lack of operational control, reliability, and tools like benchmarking may assist in anticipating hidden negative outcomes.

The performance cycle stages are explained in detailed later in this chapter.

14.3.2.4 Performance Measure Selection: Lagging, Leading, and Qualitative
There might be consistent common factors between the past, present, and future associated to the nature of the phenomenon being measured that may allow for trending. Lagging measures are useful as long as they are analyzed in time context. For instance, a lagging measure such as "number of releases" may provide a trend looking backwards identifying an increase or decrease in the performance. Conversely, the next period to be measured may not follow the trend due to change in conditions. In the other hand, lagging measures from system responses (output) and end-results (outcome) stages can be compared to the objectives, targets, and goals providing a high value to the organization.

Leading measures are useful in providing a trending comparison in terms of the level of integrity feed provided to a process or the percentage of completion, which both allow for anticipating (not predicting) the next results.

During unexpected events (e.g., emergencies), qualitative measures may offer an alternative for assessing performance during the event.

The performance measure types are explained in detailed later in this chapter.

14.3.3 Performance Assessment Implementation (Do)

14.3.3.1 Qualification of Personnel
Performance assessment should be conducted by personnel with experience in the management and technical processes as well as aware of the external stakeholders (e.g., regulators) views and perceptions. Understanding management systems, integrity management programs and public relationships with some qualification in data analysis and trending can be an asset in conducting performance assessments.

14.3.3.2 Analysis and Estimation
The analysis should confirm that the end results can provide the level of clearness, practicality, and usefulness of the performance assessment required by the stakeholders.

The data analysis should be comprised of data comparison, identification of outliers and confirmation bias as well as normalization of data for proper interpretation. The calculation process should be checked ensuring that the formulation and units are correct.

The analysis should also determine how the performance assessment represents the issues (e.g., reality checks) and how much confidence can provide to triggers management actions. Conversely, erratic or no changes in estimated performance measures over time would pose a challenge in determining a representative trend for

decision making. Otherwise, improvements should be made prior to the management review.

14.3.3.3 Evaluation
The performance assessments typically establish criteria for acceptability of given measures. The acceptability can be highlighted with symbols (e.g., green, yellow, orange, red) for which ranges of acceptability (e.g., <50%, 50% to 70%, 70% to 80%, >80%) should be previously defined.

14.3.3.4 Communication and Publication
The creation of a pilot project for first communicating the newly created performance assessment indicators provide a flexible atmosphere for searching and meeting the needs of multiple management levels. The pilot project should be nurtured by Subject Matter Expert (SME) expertise and end-user feedback. The pilot project may have multiple iterations until management approval is reached.

There are three (3) types of performance publication methods that may serve as a means for communicating to different stakeholders:

1. *Company-own or internal KPIs publication for tracking*
   Company-own integrity performance indicators should be shared with operations, right-of-way, and senior management.
2. *Industry association Aggregate KPIs publication for benchmarking*
   Industry associations typically gather performance data under non-disclosure agreements from their members calculating aggregate performance indicators and providing an average industry performance benchmark. Industry associations periodically publish on the public domain lagging measures (e.g., output and outcome) such as incident frequencies and releases. Occasionally, industry associations publish leading progress (process) measures tracking the fulfillment of regulatory requirements such as integrity baseline assessments (e.g., in-line inspection, hydrostatic testing).
3. *Company-own external KPIs publication for safety assurance to the public*
   Companies can also publish their own KPIs on the public domain typically intended to increase public safety and regulatory confidence and/or demonstrate transparency and good will; these public reporting can be voluntary or enforced by a regulatory order.

14.3.4 Performance Assessment Validation, Verification and Action Plans (Check)

Performance assessment results should be *validated by operational and integrity SME* to determine a proper representation of the phenomena or reality. Furthermore, verification of calculations factoring all representative variables should be made. Benefits and drawbacks for decision making should also be identified.

Performance measures should be tested to decide whether they can be accepted or require modification or removal ensuring that measures met their intent. *Senior management review* would provide feedback ensuring that the measures are aligned with company goals and expectations for measuring integrity performance.

14.3.5 Performance Continual Improvement (Act)

Continual improvement process is a twofold process in the performance assessment, as follows:

1. Improvement of the Performance Assessment Process and Measures
2. Improvement of the Integrity Performance

The improvement of process and measures would be focused on what and how-to assess the performance increasing the efficiency; whereas, the improvement of the integrity performance of the pipeline system would be focused on what-to-do for increasing the effectiveness of the PIMS.

14.4 WHERE TO LOOK FOR INTEGRITY KPIs?

One of the difficult steps in developing performance indicators and measures is to know where to start looking. Every pipeline system, station-to-station section, segment or pipe joint has their own special characteristics (e.g., design, manufacturing, and construction) and conditions (e.g., age, operation/reactivation/ deactivation, MOP, consequences) that may require a specific differentiation associated with integrity performance. The following are some sources from which indicators and measures can be developed:

1. Your Stakeholders' Needs
2. Historical Issues and Challenges
3. Integrity Practices
4. Threat@time: Fault Trees for Intermediate and Basic Causes
5. Threat-in-Scenario: Event Trees for Initiating Events and Barrier Identification
6. Human Factors: Human, Workplace and Operational

14.4.1 Your Stakeholders' Needs

Your stakeholders (e.g., internal and external audience) are of great guidance in providing areas for measuring performance. Their input should be considered the primary focus in communicating performance information needed by your stakeholders. However, other focuses (i.e., secondary) should be also explored to communicate emerging trends or challenges that may not be in stakeholders' radar.

14.4.2 Historical Issues and Challenges

Pipeline incident databases from regulators (e.g., USA PHMSA, Canada NEB, UK HSE), industry associations (e.g., CEPA, UK OPA, EGIG) and individual companies provide variables associated with time, conditions, previous inspections (e.g., ILI), geographical (e.g., coordinates, references), consequences (e.g., fatalities, environmental), and root-causes (e.g., cracking under overpressure) and contributing factors (e.g., pressure cycling).

As illustrated in Figure 14.2, identifying variables in the pipeline incident databases such as pipe manufacturing year and their incident frequency allows their comparison against a company's own pipeline characteristics and conditions establishing similarities in the integrity threat susceptibility and capacity to cause an incident. As illustrated in Figure 14.3, stress level at the failure and seam weld type of the incidents showed a correlation with incidents. Hence, the need for assessing (e.g., correlation), monitoring (e.g., ILI, pressure changes) and mitigating (e.g., repairs, pipe replacements) via performance measures can serve as identifiers for pipeline susceptibility and conditions that may require improvement.

14.4.3 Integrity Practices

Integrity practices should be also monitored to identify areas for improvement in integrity performance. Performance in terms of planned and actual time, resources, and results of integrity practices can be tracked down. Integrity practices such as in-line inspection, cathodic protection, integrity assessments, pipe repairs, internal and external inspection or audit findings and training and competency of personnel are areas typically used in developing performance indicators. The performance indicators will be narrowed-down to performance measures for following up on the current and future pipeline integrity condition as well as the management adequacy, implementation or effectiveness of PIMS.

14.4.4 Threat@time: Fault Tree for Immediate and Intermediate Events and Root-Causes

A fault tree describes multiple paths to root-cause, namely cut sets of a potential pipeline integrity threat incident encompassing

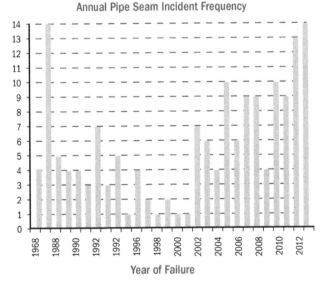

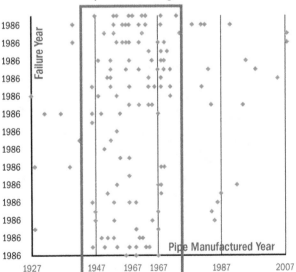

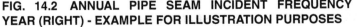

FIG. 14.2 ANNUAL PIPE SEAM INCIDENT FREQUENCY (LEFT) AND MANUFACTURING YEAR VERSUS FAILURE YEAR (RIGHT) - EXAMPLE FOR ILLUSTRATION PURPOSES

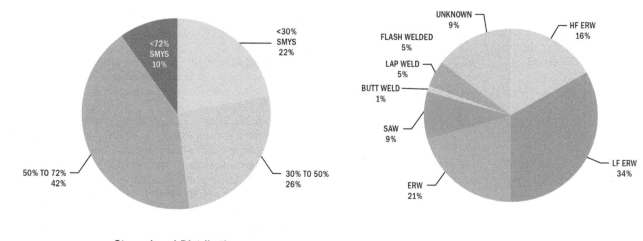

FIG. 14.3 PIPE STRESS LEVEL AT THE FAILURE (LEFT) AND THEIR SEAM WELD TYPE (RIGHT) - EXAMPLE FOR ILLUSTRATION PURPOSES

specific conditions at a given time. A fault tree contains top, immediate and intermediate event(s), and basic or root causes. Some of those events can be used as performance measures to track down how to reduce via barriers the chance of the *Top Event* [1] or integrity threat(s) failure to occur. Hence, some of immediate and intermediate events may become barriers or lack of protection that can be selected as integrity performance measures for assessing their effectiveness in mitigating or preventing an incident.

As illustrated in Figure 14.4, the *Pipeline Fails on Seam weld During Operations* fault tree has events showing multiple cut sets such as lack of manufacturing quality and inadequate operations could fail the pipeline on a seam weld during in-service. Low toughness from legacy pre-1980's pipe manufacturing technology with critical pressure cycling resulting from either intermittent supply or non-integrity sound hydraulic design can also be translated into a performance measure shown in (a.). Similarly, pipeline remaining ILI-reported anomalies, even though non-critical, in conjunction with Operations error could trigger an incident, for which accounting for ILI accuracy and growth (e.g., fatigue, corrosion fatigue)

assist in regularly tracking the current and future condition of the pipeline as shown in measure (b.) below [2].

Performance measures can lead to proactive actions such as crack In-Line Inspection (ILI), pressure change monitoring, crack growth and remaining life assessments, non-critical anomaly mitigation and even changes in the operation for a smoother pressure cycling as shown in (c).

Example: Integrity performance measures to track susceptibility to, criticality of, or operational effects to seam weld anomalies

a. *Number of pipeline sections with lower toughness and non-light pressure cycling [susceptibility]*
b. *Number of remaining crack anomalies expected to reach or exceed 40% depth and/or 1.25 Safety Factor within 1, 2, and 5 years [identified criticality]*
c. *Number of pipeline sections with pressure changes producing higher pressure cycling than average [Changes]*

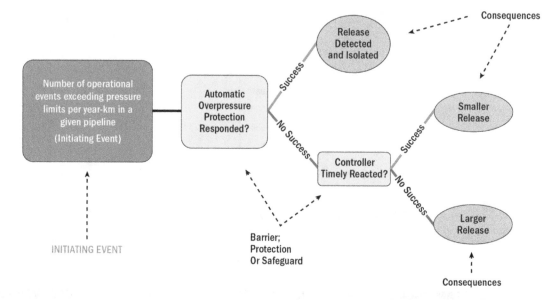

FIG. 14.4 ELEMENTS (POTENTIAL BARRIERS) FOR A FAULT TREE (PIPELINE FAILS ON SEAM WELD DURING OPERATIONS)

The status (i.e., integrity threat in a pipeline) changes from "susceptible" to "identified" when the integrity threat is found in the field. As part of the reliability metrics, ongoing monitoring of pressure provides an understanding of growth and timing for re-inspection and further mitigation.

14.4.5 Threat-in-Scenario: Event Trees for Initiating Events, Barrier ID, and PoF and Consequence

An event tree describes potential scenarios or sequence of events started from an integrity deviation or *initiating event* that may result in unwanted effects or consequences with a given probability of occurrence or failure. Either initiating or intermediate events provide examples of performance measures for tracking that can be minimized or strengthened in reducing the chance for an incident.

As illustrated in Figure 14.5, barriers with potential for avoiding or creating (if absent) effects "initiated from" events and affecting the integrity of a pipeline can be monitored through performance measures. The reliability of overpressure protection systems, quality of the pipe manufacturing, and controllers' competency provide barriers or resistance to affecting the integrity of the pipeline. Conversely, the frequency and severity of the pressure spikes can affect the integrity of the pipeline requiring their monitoring, identification and assessment.

Example: Integrity performance measures to track initiating and intermediate events from an event tree aimed to mitigating, minimizing, and preventing unwanted effects from inadequate operational systems or incorrect operations

a. *Number of pressure events exceeding 110% MOP per kilometer of affected pre-1980's vintage pipe*
b. *Number of controller working hours "after hours" (fatigue susceptibility)*
c. *Reliability of the overpressure protection system achieved during testing (%)*

Event trees may capture some of the key intermediate events required for monitoring integrity performance and identifying barriers, controls or safeguards to avoid unwanted consequences. For quality assurance purposes, Fault Trees can be used to determine any outstanding intermediate events overlooked in the development of event trees.

14.4.6 Human Factors: Human, Workplace and Operational in the context of Control Rooms

Reason, J. [3–5] studied the *Human Error* in industrial incidents correlating unsafe (human) acts with human performance, local workplace, and operational factors. These *Human Factors* can be detailed into *Performance Factors* [6] for analyzing *Control Room* factors playing an essential role in pipeline operations control and monitoring. Performance factors can help in minimizing or eliminating upset or abnormal operations, which may lead to worsen an existing pipeline integrity condition (threat). In addition, Performance factors can help in timely response to an incident (i.e., damage, leak and rupture) reducing its consequence (e.g., fluid volume and extent).

In the context of *Control Rooms*, McCallum, M. and Richard, C. built the taxonomy of human factors for determining the type and level of risk as well as their prioritization for mitigation. The human factor taxonomy has been defined in a three (3) breakdown structure (i.e., Areas, Topics, and Performance Factors):

- Human Factor **Areas**: 11 *(e.g., 1. Task Complexity and Workload)*
  - Human Factor **Topics**: 29 *(e.g., 1.1 Task Design)*
    - **Performance Factors**: 138 *(e.g., 1.1.1 Routine Activities (e.g., batch cutting) are too complex)*

Incorrect Operations is one of the integrity threats influenced by *Human Factors* that can be categorized by:

a. Pipeline incidents or near-losses from incorrect operations *(e.g., improper valve selection or operation, overpressurization, or improper selection or installation of equipment)*; or
b. Pipeline susceptibility to incorrect operations even if non-related incidents have been reported *(e.g., Performance Factors resulting from controllers lack of experience, and fatigue; workspaces with inadequate display monitors or lighting; inadequate processes for batch follow-up and control).*

Example: Incorrect Operations performance measures related to *Human Factors*

a. *Number of pipeline incidents associated to overpressurization*
b. *Number of pipeline incidents contributed by controllers fatigue, schedule and rest, slow work periods, management practices (Controller Alertness)*
c. *Number of pipeline operations in the control room with a very small margin of error (Task Design)*

TOP EVENTS	Level 1	Pipeline Fails on Seam Weld During Operations												
INTERMEDIATE	Level 2	Lack of Manufacturing Quality		Construction Issues			Inadequate Operations							
	Level 3	Legacy Pre-1980's Technology	Post-1980's Technology Quality Control	Inadequate Pipe Transport	Pipe Deformation or Damage		Overpressure		Critical Pressure Cycling		Human Error			
BASIC or ROOT CAUSE	Level 4	Low Toughness	Remaining crack anomalies susceptible to growth	Inadequate Manufacturing Process and QA/QC	Incorrect Pipe Stacking [Orientation] for Transport	Lack of Training Competency	Ineffective Pipe Bending Control or Protection	No Overpress. Protection	System Fault	Intermitt. Supply	Hydralic Design	Fatigue	No Adequate Console	Operation Training

FIG. 14.5 ELEMENTS OF AN OVERPRESSURE INCIDENT EVENT TREE (SEE BOXES WITH RED OUTLINE)

The USA Department of Transportation (DOT) Pipeline and Hazardous Materials Safety Administration (PHMSA) developed some performance metrics for Control Room Management factoring Human Factors. Please refer to the website https://primis.phmsa.dot.gov/crm/index.htm.

14.5 CHARACTERISTICS OF INTEGRITY KPIs

Integrity performance measures can be characterized understanding their integrity management approach, stage within the cycle, and ability to record the past or anticipate performance. Those characteristics can be defined prior to or after the development of performance measures.

The following are the four (4) characteristics of a performance measure (KPI):

1. *Approach*: every KPI has a primary focus that can be categorized as either
   a. *Integrity Management Program* (e.g., inspection); and
   b. *Management System* (e.g., training)
2. *Performance Cycle Stage*: every KPI can be mainly measuring in one of the stages of the cycle (e.g., program):
   a. *Input* (e.g., hours utilized);
   b. *Process* (e.g., completion);
   c. *Output* (e.g., 300 defects), and;
   d. Outcome (e.g., people safety)
3. *Measure Type*: every KPI can be defined based on its objective as
   a. *Lagging* (e.g., measuring the past)
   b. *Leading* (e.g., anticipating the future)
   c. *Qualitative* (e.g., assessing an unintended change)

4. Measurement Units: every KPI can be expressed as a:
   a. *Magnitude* (e.g., kilometers/miles)
   b. *Percentage* (e.g., % of completion)
   c. *Frequency* (e.g., 0.3 millimeters or mils/year)
   d. *Cumulative* (e.g., rate of 3 incidents/km during 5 years)

14.5.1 Integrity Approach: IMP or MS

Every integrity management performance indicator or measure intrinsically has an approach centered on either the *Integrity Management Program* (IMP) or *Management System* (MS) described in Figure 14.6. IMP performance centers on threat frequency and severity (i.e., identification and monitoring), assessment or verification, potential consequences, risks and their mitigative, preventative, and monitoring strategies.

MS-driven performance measures look into elements such as senior management commitment, training and competency, planning, compliance verification results (e.g., audits, inspections) and corrective and preventive action plans, communications and continuous improvement. MS performance centers on the process adequacy, the timely and quality execution, and the effectiveness of the management processes through the review of the results as to how they achieved the goals, objectives, and targets set out by senior management.

Example: IMP-driven Objectives and KPIs

- Objective: Identification of susceptibility to external interference (i.e., 1st, 2nd, and 3rd party damage)
  KPI: *number of unauthorized activities on the Right-Of-Way*
- Objective: Effectiveness of corrosion programs such as cathodic protection, pipe recoating and repairs
  KPI: *number of external metal loss anomalies exceeding an criterion over two (2) last in-line inspections*

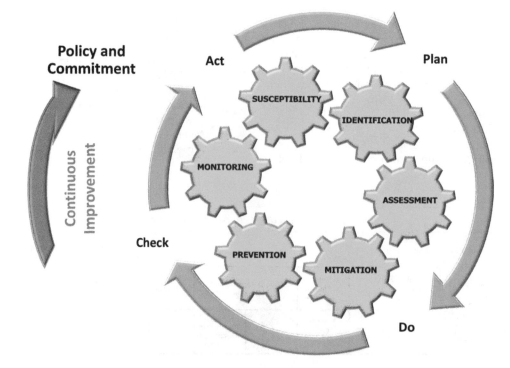

FIG. 14.6   INTEGRITY MANAGEMENT APPROACH: IMP (DENTED GEAR) AND MS (CIRCLE)

14.5.2 Performance Cycle Stage

The integrity performance cycle has four (4) stages in which a performance measure can be developed. Conversely, a performance indicator can be codified or supported by performance measures in multiple stages within the cycle.

As illustrated in Figure 14.7, the following are the four (4) stages of the integrity performance cycle:

1. Integrity Feed (Input)
2. Progress (Process)
3. Integrity Response to Action (Output)
4. End-Results (Outcome).

14.5.2.1 Integrity Feed (Input) An input measure shows the amount of *integrity feed* provided to the IMP and MS processes for achieving integrity objectives (e.g., threat, consequence, and risk reduction). The input can be supplied by operational, maintenance and integrity processes. The *input* stage provides the *fuel for performance* such as equipment and personnel utilization, training, qualification and competency years, inspection technology use, knowledge, and lessons learned translated into requirements and resources.

Example: Input or integrity feed performance measures

a. *Number of equipment [or man] hours [or additions] in a given activity per site or kilometer (e.g., Variable Speed Pump, Right-of-Way surveillance, landslide remediation, assessments)*

b. *Number of years of experience of a given job classification (e.g., pipeline controller, cathodic protection technician, integrity engineer, in-line inspection specialist, defect assessment professional)*

c. *Number of kilometers inspected by a given technology (e.g., Electro-Magnetic Acoustic Emission—EMAT-, Phased Array—PA-, Circumferential Magnetic Flux Leakage—CMFL-)*

14.5.2.2 Progress (Process) A process measure shows the progress (or lack thereof) or fraction of an *integrity process* completed. The percentage of completion and its tracking anticipates whether the process, procedure, activity, or task would be finalized as planned. The processes can be of operational, maintenance and integrity nature. Processes such as identification, assessment, mitigation, prevention, and monitoring can be measured against a benchmark, plan or reference. Sometimes this measure is captured as a fraction or portion of the overall system. In both cases, these measures describe a relative measurement of the entire plan or system.

Example: Process or progress performance measures

a. *Percentage (%) of inhibition program completed [Progress of Internal Corrosion Mitigation Program: Executed versus Planned/year]*

b. *Percentage (%) of pipeline surveillance patrols completed [Progress of Geotechnical Monitoring Program: Executed versus Planned/year]*

c. *Number of pipeline kilometers identified as susceptible to geohazards [Fraction of pipeline system with susceptibility]*

14.5.2.3 Integrity Response to Action (Output) The output shows how the pipeline system has responded to the integrity feed and progress achieving an integrity health of improving, same as before or *status quo ante*, or deteriorating. Sometimes the integrity health (or lack of) in a pipeline system results in a leak, damage or rupture falling under the *"Direct Integrity Measure"* category as named by ASME B31.8S [7].

The pipeline integrity health, being measured regularly, describes how the integrity of a pipeline is responding to conditions (e.g., anomalies) that could be deteriorating faster/slower than

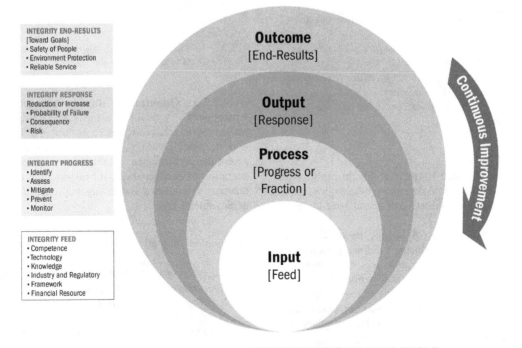

FIG. 14.7 INTEGRITY MANAGEMENT PERFORMANCE CYCLE

their mitigation (e.g., digs) or prevention (e.g., damage). The worst conditions at any given time would trigger short and midterm mitigation in reducing probability of failure. Furthermore, the rate of change in time of the condition (e.g., growth of existing anomalies, new anomalies) would trigger an adequate prioritization for risk reduction strategies until the deteriorating trend changes to an acceptable integrity condition.

Pipeline integrity condition changes in anomaly growth or frequency, cathodic protection potential, product water content and inhibition use, pressure cycling and spikes, non-authorized activities on RoW, rain precipitation, landslide movements, water flooding and operational upsets can be used as performance measures for tracking integrity health improvement, *status quo ante* or degradation.

Example: Integrity output performance measures

    a. *Number of releases (i.e., leaks and ruptures) on the pipeline (Direct Output/Response to Integrity Management)*
    b. *Percentage (%) of change in the minimum, average and maximum growth from the last 3 consecutive assessments [Output/Response to Anomaly Growth Mitigation]*
    c. *Percentage (%) of change in the number of non-authorized activities on the RoW over the last 3 years [Output/Response to Damage Prevention]*

14.5.2.4 End-Results (Outcome) An outcome measure shows how much has been achieved related to the integrity goals set out by the corporation. The integrity goals are expressed in terms of safety, environmental and reliability performance such as injuries or deaths, level of environmental affectation and service reductions or interruptions. The difference between output and outcome measures is the output is resulting from a process, but the outcome may require the integration and success of multiple outputs, processes and conditions to achieve a goal.

Example: Integrity outcome or end-result performance measures

    a. *Number of employee (or public) injuries (or deaths) resulting from pipeline incidents*
    b. *Volume of product released onto the facilities (or outside environment) from a pipeline incident*
    c. *Number of hours of interrupted service to customers or shippers this year*

14.5.3 Measure Types

Measures can be differentiated by their ability to report past events that may or may not anticipate the future (i.e., lagging, leading), which can be expressed in qualitative or quantitative terms.

14.5.3.1 Lagging—Quantitative A lagging measure reports the performance occurred in the past during a defined period (e.g., monthly, quarterly, annually). Lagging measures are commonly trended reflecting the performance over a past period of time, but their rear-view approach does not necessarily anticipate the next performance. Typical lagging measures associated to incidents and spills maybe trended for past-behavior analysis, but not guarantee of their increase or decrease in the future.

Example: Lagging performance measures

    a. *Number of employee (or public) injuries (or deaths) resulting from a pipeline incident—[Outcome/End-result]*
    b. *Number of defects reported exceeding the acceptable criteria (e.g., depth or Safety Factor)—[Output/Response]*
    c. *Number of kilometers inspected by a given technology (e.g., Electro-Magnetic Acoustic Emission—EMAT-, Phased Array—PA-, Circumferential Magnetic Flux Leakage—CMFL-) [Process/Progress-Fraction]*

14.5.3.2 Leading—Quantitative A leading measure reports the performance occurred in the past during a period of time, but it is intrinsically connected to the future providing a better chance to anticipate (not predicting). Integrity processes such as the execution of planned in-line inspections, repairs, RoW surveillance, and public awareness allow for anticipating their performance as they report progress along time.

Example: Leading performance measures

    a. *Percentage (%) of completion of the cracking In-Line Inspection Program [Process/Progress]*
    b. *Percentage (%) change in the number of non-authorized activities on the RoW over the last 3 years [Output/Response to Damage Prevention]*
    c. *Percentage (%) of completion of the pipeline surveillance program in the year [Process/Progress]*

Leading indicators can also be selected based on behaviors or the performance of ongoing tasks within defined tolerances, provided that those behaviors or tasks are known or suspected to be preventative barriers against an adverse integrity outcome.

Example: Leading performance measures based on behaviors or tasks within defined tolerances

    a. *First run success rate of ILI*
    b. *Console alarms/day within tolerable ranges*
    c. *Pressure cycle regime*

14.5.3.3 Qualitative—Rapid Change, Breakdown, or Emergency A qualitative measure reports the performance in non-numeric relative terms (e.g., low, medium high) for a given period; typically when quantitative measures cannot properly represent the performance. Qualitative measures provide better representation of performance over quantitative measures during rapid changes, breakdown and emergency capturing events, issues, and achievements [8].

Example: Qualitative performance measures

    a. *Level of susceptibility of a given integrity threat during a product change (e.g., corrosivity)*
    b. *Rank of pipeline controller response during the incident*
    c. *Extent of environmental consequence (e.g., land, water, fauna, flora) during a emergency response*

14.5.3.4 Predictive—Signal A predictive measure reports the conditions in anticipation of the event potentially expected to occur. Predictive measures act as a warning or advanced notice of conditions being created for an event that may occur. Predictive measures allow for monitoring processes and preventing occurrences increasing the effectiveness and efficiency of actions, if conducted on time.

Example: Predictive or signal measures

    a. *Number of monthly equivalent pressure cycles (e.g., anticipation of crack growth)*

    b. *Contaminant levels in crude oil pipelines (e.g., corrosivity, leak-dependent)*

    c. *Number of unauthorized activities in the right-of-way (e.g., pipeline damage)*

14.5.4 Measurement Units

Performance measures can be expressed in magnitude, frequency or ratio, and percentage. Those units can themselves become cumulative creating another performance measure. Furthermore, cumulative values can be transformed in cumulative rates graphed as changing slopes, which can express increase, *status quo ante* (same as before) or decrease in the performance.

The following are the types of measure units:

    a. *Magnitude: Numeric (e.g., number of leaks) or qualitative (e.g., controller response: low, medium, high)*

    b. *Frequency or ratio (e.g., kilometers per year)*

    c. *Percentage (e.g., % change in the minimum, average and maximum growth)*

    d. *Cumulative Rate (e.g., leak rate change between a given year and the year before)*

14.6 REVIEWING ADEQUACY: ENABLED BY PROCESSES AND SYSTEMS?

Management review for adequacy should be focused on understanding all processes, procedures and activities and their assurance to be safe and efficient. Adequacy is achieved when the level of completeness required meeting specific requirements from either regulations or adopted industry standards are met.

Managers can review for adequacy by following a risk approach on IMP including its supporting management system. This review can be initiated with the question.

Example: what we are doing for...?

- Reducing the chance of failure? All threats identified or deemed susceptible
- Assessing and minimizing the consequence of Failure? Additional valves needed
- Identifying risk mitigation, prevention, and monitoring strategies? Prioritized plans
- Assigning adequate resources aligned with PIMS priorities and timing? Enough and on track

---

The management system (MS) supporting IMP should also need to be reviewed for completeness. The core MS elements to be reviewed are policy and commitment, planning, implementation, checking and verifications, and management review: *Are all core MS elements defined as to what, how and when they are working?*

As managers perceived concerns, inconsistencies or issues, the method of *drill-down deeper* in those areas is to repeatedly asking *Why?* to get down to the core completeness findings. Managers may encounter during the adequacy review some of the following findings:

- *Processes or procedures not documented or incomplete;*
- *Multiple procedure versions of how to do things (e.g., headquarters versus field);*
- *Not all integrity threats and/or consequences being assessed and/or not periodically;*
- *Processes focused on probability of failure not accounting for changing consequence.*

14.7 REVIEWING IMPLEMENTATION: WALKED THE TALK BY DOING WHAT WE PLANNED?

Management review of the implementation of process and procedures should be focused on comparing plans and budgets against their execution, identifying differences and unplanned activities, and asking for their rationale. This review can be initiated with the question *Did we do what we planned....? Why not completed as planned/envisioned? What are the root-causes and corrective actions?*

14.8 REVIEWING EFFECTIVENESS: ACHIEVED GOALS, OBJECTIVES, AND TARGETS?

Management review for effectiveness should be focused on comparing the goals against what was achieved. This process does not look at how much work was done; it looks at what goals were achieved and in retrospective, what actions contributed in achieving those goals. This review can be initiated with the question *What goals were achieved? How did we achieve effective results? no releases? no environmental damage? no unplanned downtime in the service to shippers and customers?*

Furthermore, goals associated to system response (e.g., corrosion growth) can be compared against the results of mitigation (e.g., inhibition, cathodic protection). Similarly, goals associated to the management system (e.g., competency, audits) can be also compared against the results of management processes (e.g., training and qualification stats) for assessing the effectiveness of IMP and MS.

The analysis of the adequacy, implementation, and effectiveness findings should be aimed to develop Corrective and Preventive Action Plans (CPAP). However, Continuous Improvement (CI) plan would look at lessons learned, event repetition, prevention, system modifications, introduction of new technologies, and changes in the organizational structure. The conceptual difference from CPAP to CI is the core change nature of CI to break the trend of issues, re-engineer the process, or re-design the system instead of just correcting the issues, which could repeat.

14.9   BENCHMARKING YOUR PERFORMANCE

Benchmarking involves comparing the content, state of completion or maturity of the process to a desired standard, such as an industry code, regulation, or statement of best practice. It is often used to drive program improvement and capability.

Benchmarking process may utilize techniques such as independent audit, gap analysis, survey, interviews, focus groups or peer review and typically plots its results on a maturity scale. Often performance benchmarks are chosen based on the average, best quartile, or leading performers in a population.

Development of performance assessment based on company's own conditions provide the internal benefits of focusing on *what matters the most* related to your goals, objectives, and targets as well as motivating your team to implement them. However, companies need to know how their performance relates to the external environment or stakeholders.

Industry Associations, Private Research Firms, Process Safety Consultants, and Regulatory Agencies which collect integrity and safety metrics are common sources for information. A draw back to the use of industry association or regulatory data sources is the company may not be able to identify metrics for comparative analysis that ideally address their area of interest. Regarding to the performance assessment formulation, your KPIs may differ from your external environment or stakeholder formulation, which is acceptable; a subset KPI (external) can be created for comparing it to the regulatory or industry organization.

Benchmarking your own company's KPIs related to KPIs from industry associations, regulatory jurisdictions and consensus standards can be conducted by developing a table, as follows:

a. Headers: Institution, Approach, Performance Cycle stages, and Type of measure.
b. Start with your own company KPIs checking whether (Yes/No) they have the KPI characteristics described in the headers.
c. Collect the information from the institutions (e.g., industry standards, regulators, industry associations) identifying whether (Yes/No) they have the KPI characteristics.

Example: Data sources with KPIs from Industry Associations, Regulatory Agencies, and Consensus Standards

- API 1160-2013
- ASME B31.8S-2014
- Colombia NTC 5901-2011
- DNV RP-F116
- API RP-1173
- Alberta Energy Regulator
- British Columbia Oil and Gas Commission Regulator
- Ontario Technical Standards and Safety Agency Regulator
- USA DOT PHMSA
- CANADA National Energy Board
- Canada Transmission Pipeline Associations
- European Onshore Pipeline Associations
- North America Liquids Pipeline Associations

d. Compare benchmarking how your own KPIs do (Yes/No) related to the selected institutions.

A balanced approach of looking both inside and outside of the organization is healthy as long as it is focusing on the positive continuously improvement of your own performance.

14.10   REFERENCES

1. Marcogaz, 2005, *Guideline for the definition of Performance Indicators for Safety Management System*, GI-TP-05-24, Technical Association of the European Natural Gas Industry, http://www.marcogaz.org/index.php/publications.

2. Kiefner, J. F., 2002, *Dealing with Low Frequency Welded ERW Pipe and Flash Welded Pipe with Respect to HCA-Related Integrity Assessments*, ASME ETCE2002/PIPE-29029, Proc. of the ASME Engineering Technology Conference on Energy, American Society of Mechanical Engineers, Houston, Texas, USA.

3. Reason, J., 1990, *Human Error*, Cambridge University Press, Cambridge, UK, ISBN 0 521 31419 4.

4. Reason, J., 1995, *Ergonomics, A Systems Approach to Organizational Error*, Volume 38, Taylor & Francis, London, England, doi:10.1080/00140139508925221.

5. Reason, J., 1995, Understanding adverse events: human factors, *Quality in Health Care*, 4: 80–89, http://www.ncbi.nlm.nih.gov/pmc/articles/PMC1055294/pdf/qualhc00016-0008.pdf.

6. McCallum, M., Richard, C., Battelle, 2008, *Human Factors Analysis of Pipeline Monitoring and Control Operations*, PR-003-03405, PRCI/USA DOT PHMSA, USA, http://primis.phmsa.dot.gov/crm/docs/FinalTechnicalReportNov2008.pdf.

7. Anon., 2014, *Managing System Integrity of Gas Pipelines—ASME B31.8S*, ASME Press, American Society of Mechanical Engineers, New York, USA.

8. Leitão, P., Restivo, F., 2004, *The use of qualitative indicators for performance measurement in manufacturing control system*, IFAC, Faculdade de Engenharia da Universidade do Porto, Portugal, https://web.fe.up.pt/~fjr/mPIPS/incom04-PL-FR.pdf.

GLOSSARY

A

Adequacy

Adequacy is achieved when *completeness* of processes, procedures, roles, and responsibilities and associated organization meet specific requirements from either regulations or adopted industry standards.

ALARP

As Low As Reasonably Practicable.

Anode

The location in a corrosion cell where oxidation occurs. The anode is used to protect the cathode.

Anomaly

An unexamined deviation from the norm in pipe material, coatings, or welds, which may or may not be a defect.

Appurtenance

A component that is attached to the pipeline (e.g., valve, tee, casing, instrument connection).

Audit

Systematic, independent, and documented process for obtaining audit evidence and evaluating it objectively to determine the extent to which the audit criteria are fulfilled [1].

C

Cathode

The location in a corrosion cell where reduction occurs. The pipeline is desired to be the cathode.

Characteristic

Any physical descriptor of a pipeline (e.g., grade, wall thickness, manufacturing process) or an anomaly (e.g., type, size, shape).

Class Location

A Class Location is a geographical area classified according to the number of dwellings that are considered when designing

ASME B31.8S-2014

Fig. 13-1 Hierarchy of Terminology for Integrity Assessment

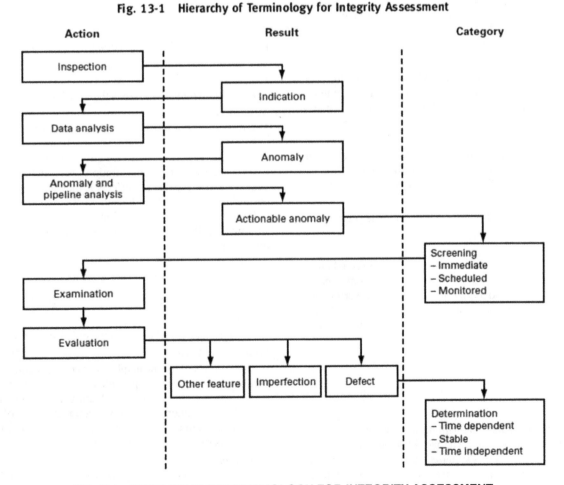

FIG. G.1   HIERARCHY OF TERMINOLOGY FOR INTEGRITY ASSESSMENT
(*Source:* American Society of Mechanical Engineers - ASME B31.8S-2014 Fig. 13-1)

and pressure testing pipelines to be located in the area. The class location-assessment area is comprised of a moving area of 200-m perpendicular to either side of the pipeline centerline with a length of 1600 m used for counting the number of dwellings with the purpose of designating a class ranging from one to four.

Reference: CSA-Z662-11

Combustible

Combustible liquids have the ability to burn at temperatures that are usually above normal working temperatures. Under the Workplace Hazardous Materials Information System (WHMIS), Combustible liquids have a flashpoint at or above 37.8°C (100°F) and below 93.3°C (200°F).

Reference: www.ccohs.ca/oshanswers/chemicals/flammable/flam.html

Compliance

It is a state of being in accordance with an established official requirement such as a pipeline act, law, regulation, adopted industry and/or company own standards. Compliance action or activity meets the requirements of legislation, rules and regulations and adopted consensus standards.

Compliance, Non-

Non-compliance is a failure to adhere to legislation, rules and regulations, and adopted consensus standards. Non-compliance can be classified as high, medium, and low risk.

A high-risk non-compliance is one where the nature of the non-compliance makes it more likely than other non-compliances to result in a major loss, such as life-altering injury or fatality, immediate loss of containment of a system, or other similar losses. Where non-compliance is considered high risk, corrective and/or preventive actions should be immediately undertaken.

Conformance

Conformance is being in accordance with a requirement, standard, code, or procedure defined by the company, but not required in the act, law, regulation nor adopted consensus standard by any corresponding jurisdiction.

Conformance, Non-

Non-conformance is a failure to comply with a requirement, standard, code, or established procedure defined by the company, but not required in the act, law, regulation nor adopted by any legal entity.

Consequence

Any unplanned effect on the environment, people, or pipeline operation.

Corrective Action Plan

A documented plan outlining activities and actions taken to eliminate and address the causes of an existing non-conformance or non-compliant findings identified in a compliance verification method to rectify or prevent recurrence. The Corrective Action Plan describes the methods and actions, which will be used to correct the non-compliance.

D

Defect

A physically examined anomaly with dimensions or characteristics that exceed acceptable limits. The word "defect" implies substandard. Substandard refers to a system that does not comply with its original fabrication standard. Hence, "defect" does not mean "failure": it means it is not acceptable to a standard.

DEM: Digital Elevation Model

DEM represents the 3D format-surface of the terrain. For consequence assessments, DEM assists in understanding terrain topography, geomorphology, and physical geography for developing models that can be used for areas such as farming, agriculture, forestry, engineering, and pipeline fluid dispersion and consequence assessments.

Detect

To sense or obtain a measurable indication from a feature.

Direct Assessment (DA)

Direct Assessment (DA) is a non-destructive assessment technique for classifying pipeline regions with common characteristics (i.e., *Pre-Assessment*) that may be experiencing the selected integrity threat (e.g., external corrosion, internal corrosion or stress corrosion cracking). Those regions are supplemented with field surveys (i.e., *Indirect Inspection*) and validated through excavation (i.e., *Direct Examination*) and results used for evaluating the effectiveness of the assessment and estimating the reassessment interval (i.e., *Post Assessment*).

Disbonded

Coating that is not adhered to the pipe surface.

E

Effectiveness

Effectiveness is achieved when the *PIMS goals and objectives* are met in a safe manner.

Efficiency

Efficiency is achieved when the *effort (e.g., time, resources) invested is optimal accomplishing the expected results* in a safe, quality, and time manner.

Electrolyte

A liquid medium that allows the transfer of ions in the corrosion cell.

Emergency Response Planning Guidelines (ERPG)

ERPGs are exposure guidelines designed to anticipate health effects from exposure to certain airborne chemical concentrations. ERPGs estimate the concentrations at which most people will begin to experience health effects if they are exposed to a hazardous airborne chemical for 1 hour.

Reference: response.restoration.noaa.gov/oil-and-chemical-spills/chemical-spills/resources/emergency-response-planning-guidelines-erpgs.html

F

Feature

Any physical object detected by an ILI system.

Fitness-For-Service

Fitness-for-service principle is that a structure is considered adequate for its purpose, provided the conditions to cause failure are not reached. It involves conducting a quantitative engineering evaluation to demonstrate the structural integrity of a component that may contain a flaw or damage. This evaluation is often called an 'engineering critical assessment' (ECA).

Flammable

Flammable liquids will ignite (catch on fire) and burn easily at normal working temperatures. Under the Workplace Hazardous

Materials Information System (WHMIS), flammable liquids have a flashpoint below 37.8°C (100°F).

Reference: www.ccohs.ca/oshanswers/chemicals/flammable/flam. html

Flammable and Explosion Limits

The range of fuel vapor-to-air ratio necessary for a concentration to fire or explode is called *fire* or *explosion limits*. They range between the *Lower Flammability/Explosion Limit* (LFL or LEL) and *Upper Flammability/Explosion Limit* (UFL or UEL), respectively.

Flashpoint

The flashpoint of a liquid is the lowest temperature at which the liquid gives off enough vapor to be ignited (start burning) at the surface of the liquid.

Reference: www.ccohs.ca/oshanswers/chemicals/flammable/flam. html

H

Hazard

Hazard means anything that can cause harm (e.g., chemicals, electricity, working from ladders, etc.), whereas risk is the chance, high or low, that somebody will be harmed by the hazard.

Hazards (Integrity)

Hazard is defined by Webster's dictionary as *"a source of danger"* and danger is explained as the *exposure or liability to harm or loss*. Thus, an *Integrity Hazard* can be defined as *any situation or event or condition* (e.g., coating damage) *able to initiate or grow an integrity treat (e.g., corrosion)*.

Hazardous Effects

Hazardous effects can be caused to humans, environment, and property resulting from the dispersion of a vapor cloud or fluid (e.g., spread velocity, density, or concentration) and/or its flammability, combustibility, explosion and toxicity characteristics.

Holiday

A small location where pipe steel is exposed through the coating.

I

Imperfection

An anomaly with characteristics that do not exceed acceptable limits.

Implementation

Implementation is achieved when *plans* are conducted in a safe, quality, and timely manner following approved processes and procedures.

Incident

It is an unexpected event or occurrence preceding the loss that is likely to cause an unintended impact, harm, or damages to the public, property, or environment, or lead to severe consequences.

Indication

A signal from an ILI system.

Individual risk

Individual risk is the measure of risk as perceived by a specific individual that might be located near of the pipeline during a pipeline incident assuming that the individual is present 100% of the time.

Injurious

A feature that could lead to failure of the pipeline.

In-Line Inspection (ILI)

ILI is a non-destructive inspection technique that can be used for integrity assessment of pipelines. The type of ILI survey performed is dependent upon the type of integrity threat that is being assessed. In-Line Inspection pigs are distinguished from other pigs in that they have sensors for collecting and recording information about the condition of the pipeline. In-Line Inspection is explained in detail in the chapter 8.

Inspection Test Plan (ITP)

Document that covers all of the steps for a paint or coating application to a pipeline.

L

Lethality Probability

Lethality probability specifies the change of having fatalities under a certain conditions (e.g., heat intensity threshold) or boundaries (e.g., open or confined space).

LIDAR: Light Detection and Ranging Technology

LiDaR is a remote sensing method used to capture topographic (with near-infrared laser) and/or bathymetric (with water-penetrating green light) information of the surface of the earth. This information can be used for producing Digital Elevation Models (DEM). LIDAR instruments mainly consist of a laser, a scanner, and a specialized GPS receiver.

Reference: oceanservice.noaa.gov/facts/lidar.html

Likelihood

The probability that an event occurs.

M

MAOP

Maximum Allowable Operating Pressure defined as per US 49 CFR DOT. Canada standard, CSA Z662, and regulation, NEB OPR, defines MOP (Maximum Operating pressure) to similarly United States regulation.

Mitigation

The act of removing a threat or reducing the likelihood of a threat causing a consequence on the pipeline.

Monitoring

The act of ensuring that known threats do not increase likelihood of.

N

Near misses

An event could have the potential to incur a loss under certain circumstances.

P

People Concentration

People concentration refers to locations with concentrations equal or higher than 20 persons that may be used infrequently such as weekly, monthly, or seasonal.

Performance Criteria and Symbols

The integrity performance criteria provides the acceptability of the integrity performance measure based on a pre-defined expectation, benchmark or standard *(e.g., <80% deep corrosion)*, which can be graded in multiple levels from acceptance to rejection *(e.g., <50%, 50% to 70%, >70%)*. The results of the comparison can also be expressed as a three-color code *(e.g., green, yellow, red)*, binary response *(e.g., Yes/No)*, percentage and qualitative measure *(e.g., low, medium, high)*.

Performance Cycle

The performance cycle describes the objectives and interrelationship of the performance elements of integrity management driven by continuous improvement. The performance cycle is divided in four (4) stages: input or *integrity feed*, process or *integrity progress/fraction*, output or *system response to integrity actions*, and outcome or *integrity end-results towards goals*.

Performance Indicator

An integrity performance indicator represents an integrity focus, objective, or outcome to be measured and compared against stakeholders' goals, targets, or expectations. A condition indicator would reflect the level of degradation of the pipeline related to an integrity threat *(e.g., corrosion)*, group of threats *(e.g., time-dependent)* or overall pipeline condition; whereas, the integrity management indicator would reflect the level of adequacy, implementation or effectiveness of a given activity *(e.g., pipeline cleaning)*, process or program *(e.g., corrosion inhibition)*.

Performance Measure

An integrity performance measure quantifies one (or the key) factor or component *(e.g., maximum depth of corrosion anomalies)* of an integrity performance indicator within a time period *(e.g., monthly, quarterly or annually)*. A collection of selected measures *(e.g., corrosion identification and mitigation)* can be used to build up a performance indicator *(e.g., corrosion performance)*. Hence, integrity performance indicators are comprised of one or more integrity performance measures.

Pipeline

Pipeline includes all components through which oil and gas industry fluids are conveyed including pipe, isolating valves and other appurtenances, but excluding pumping, compression, and metering stations. This definition makes the pipeline starting and ending at the station fence excluding components within the stations or facilities. Simply, pipeline definition would include pipe and accessories from fence to fence.

Pipeline Integrity

Pipeline Integrity is the status of a pipeline defined by its *structural reliability and availability to transport a fluid under safe conditions*.

Pipeline Integrity Management Program (IMP)

IMP can be defined as an engineering-sound process to manage the integrity a pipeline through the identification, susceptibility, assessment, prevention, mitigation and monitoring of risks to protect people and environment providing a reliable service to shippers and customers.

Pipeline Integrity Management System (PIMS)

Pipeline Integrity Management System (PIMS) can be defined as a *leadership driven-approach to direct plan, implement, verify, measure and continuously improve the integrity of a pipeline(s) to protect our people and environment providing a reliable service* to shippers and customers. PIMS is *supported at the core by an engineering-sound Integrity Management Program (IMP)*.

Pipeline Lifecycle

It is a series of changes in the life of a pipeline, covering the design, construction, operation, maintenance, discontinuation/deactivation, and abandonment.

Pipeline Systems

Pipeline System includes all components of a pipeline and facilities required to move oil, gas and products including measurement, non-formation storage (e.g., tank farms), transportation and distribution. Furthermore, industry standards typically define *Pipeline Systems* within their applicability flowcharts having in common the exclusion of production, underground formations, bulk plant, steam generation, tanker and barge loading/unloading, gas processing and refinery facilities.

Pipelines, Distribution

Distribution Pipelines transport treated fluids from transmission pipelines, distribution centers, or terminals to the end-user or customer. Distribution pipelines are also known as city or town networks. Distribution pipelines cross populated and environmentally sensitive areas to reach houses, factories, or dwellings with high exposure to excavation or drilling damage. Distribution pipelines are designed for low pressures (i.e. <30% SMYS) transporting smaller volumes for shorter distances with thinner pipe walls underground.

Pipelines, Gathering

Gathering Pipelines transport untreated fluids from onshore or offshore wells to battery/treatment, processing or refining facilities for extracting natural gas, natural gas liquids (e.g., propane, butane, and ethane) or crude oil. Gathering pipelines are also known as feeders.

Pipelines, Transmission

Transmission Pipelines transport treated fluids from processing, production, or refining facilities to distribution centers or terminals. Transmission pipelines are also known as trunk lines. Facility laterals or interconnecting pipelines can be also included within the transmission pipeline category, if the fluids have been treated.

Population Density

The population density at any point along the pipeline can be calculated as the number of occupants of all buildings and facilities within an assessment area centered on that point, divided by the size of the selected assessment area.

Reference: CSA-Z662-11

In the United States, census block data can also be used as an estimate for residential population estimates, but it does not represent daytime business population distributions. Expand (Robin).

Potential Impact Radius (PIR)

The radius of a circle within which the potential failure of a pipeline could have significant impact on people or property. PIR is determined by the formula $r = 0.69*$ (square root of $(p*d^2)$), where 'r' is the radius of a circular area in feet surrounding the point of failure, 'p' is the maximum allowable operating pressure (MAOP) in the pipeline segment in pounds per square inch and 'd' is the nominal diameter of the pipeline in inches.

Note: 0.69 is the factor for natural gas. This number will vary for other gases depending upon their heat of combustion.

Reference: http://www.phmsa.dot.gov/staticfiles/PHMSA/Pipeline/TQGlossary/

Pressure Reversal

A pressure reversal is a phenomenon where a pipeline fails at progressively lower pressures during depressurization/unloading or subsequent tests

Pressure testing

Pressure testing is a destructive testing technique to detect/eliminate (by failing) the largest defect in the pipeline at the time of the testing that can fail due to internal pressure. The test is performed at a pressure higher than the proposed pipeline operating pressure providing a pressure-dependent safety margin to the normal operation and leak-tightness at the time of the testing. Pressure testing is explained in detail in the Chapter 9.

Prevention

The act of stopping a threat from occurring on a pipeline during any phase of a pipeline's life cycle.

Probability of Detection (POD)

The probability of a feature being detected by an ILI tool.

Probability of Failure (PoF)

Probability of Failure (PoF) is the probability or likelihood that an event or condition will result in failure factoring their resistance and loading characteristics. PoF is mathematically defined as the Unitarian complement of reliability or R (i.e., PoF – R). The probability of failure is typically expressed as a dimensionless number (changes of failure); however, as the number reach very low magnitude, they can also be expressed in terms of kilometer or per year per mile (i.e., PoF/year-km, PoF/year-mile).

Probability of Identification (POI)

The probability that the type of anomaly or other feature can be correctly classified (e.g., as metal loss, dent).

Probit Analysis

Probit analysis is an approach for determining the effects of a received dose of either toxic substances or thermal radiation. The Probit functions calculate the fatality rate of personnel exposed to harmful agents over a given period.

Reference: www.hse.gov.uk/foi/internalops/hid_circs/technical_osd/spc_tech_osd_30/spctecosd30.pdf

Purge

Process to remove hydrocarbon fluids out of the pipeline typically using nitrogen or non-toxic liquids until the pipeline is free of potential pollutants or contaminants.

R

Receptors

Receptors are humans, water wells, surface water, livestock, vegetation, and wildlife subjected to chemical exposure.

Reference: www.neb-one.gc.ca/clf-nsi/rsftyndthnvrnmmt/nvrnmmt/rmdtnprcssgd/rmdtnprcssgd-eng.html

Reliability

Reliability is defined as the probability that a component of pipeline system will perform its required function without failure during a specified time interval (usually taken as one year). Reliability is expressed as 1 minus PoF, where the PoF is the probability that a segment, section, component part, or equipment associated to a pipeline system fails its required function under given operating conditions for a specific time period.

Risk

Risk is a measure of human injury, environment damage, or economic loss in terms of both the incident likelihood and the magnitude of the loss or injury (Centre for Chemical Process Safety, 2000). In pipeline integrity, is a measure of the safety of the structural integrity of a pipeline system, which considers the probability of failure associated to its consequence of an undesirable event (i.e., leak or rupture).

Risk, Individual

Individual risk is the probability that an individual at a specific pipeline location will be a casualty as a function of distance from a pipeline because of pipeline rupture assuming that the individual is present 100% of the time.

Risk, Societal

Societal risk is a measure of the risk where the consequence considered is a function of the expected number of fatalities (N) or injured persons occurring due to a pipeline failure, and the probability incident causing number of fatalities (i.e., # of Fatalities/km-year).

Risk-based design

Risk-based design is a process that enables the pipeline design to minimize risk in a cost-effective manner and to demonstrate safe and reliable pipeline operations.

Risk Management

The systematic application of management policies, procedures, and practices to the tasks of analyzing, assessing, and controlling risk in order to protect employees, the general public, and the environment as well as company assets while avoiding business interruptions [2].

S

Service Provider

Any organization or individual providing services (e.g. in-line inspection, cathodic protection, depth of cover, above ground surveys, laboratory analysis) to operators.

Shielding

Prevention of cathodic protection current from reaching the pipe.

Sizing Accuracy

The accuracy with which an anomaly dimension or characteristic is reported.

Societal risk

Societal risk measures the overall risk where consequences considered is a function of the expected of fatalities occurring due to a pipeline failure. Societal risk is defined as the relationship between frequency of number of fatalities (F) and the number of fatalities (N) suffering from a specified level of harm in a given population from the consequence of hazardous fluids being released from a pipeline.

T

Test, Leak

This test is used to determine that a pipeline segment does not show evidence of leakage.

Test, Spike

This test is used to verify the integrity of a pipeline containing "time dependent" defects.

Test, Strength

This test establishes the operating pressure limit of a pipeline segment.

Threats, Integrity

An Integrity Threat is defined as an *abnormal state affecting a pipeline that may have the capacity to cause a failure.* Abnormal states or threats (e.g., Stress Corrosion Cracking) are typically created, made active, or grown by one or more hazards (e.g., tape coating damage, stress levels, pipe manufacturing thermal treatment).

Toxic and Toxicity

Toxic means able to cause harmful health effects to receptors. Toxicity is the ability of a substance to cause harmful health effects. Under the *Canadian Controlled Products Regulations* and the *U.S. OSHA HAZCOM Standard*, there are specific technical criteria for identifying a material as toxic for the purpose of each regulation.

Reference: http://www.ccohs.ca/oshanswers/chemicals/glossary/msds_gloss_n.html#_1_67

Reference: en.m.wikipedia.org/wiki/Digital_elevation_model

V

Vapor Cloud

A vapor cloud can be formed from the discharged pipeline fluid by

a. Being a fluid in gaseous state (i.e., gas);
b. Vaporizing fluid (i.e., liquid or liquefied gas) while escapes from the pipeline, and/or;
c. Evaporating from the spills (i.e., liquids) shorter distances with thinner pipe walls underground.

REFERENCES

1. ISO 19011-2011—Guidelines for Auditing Management System, Section 3, Audit Definitions.

2. Guidelines for Developing Quantitative Safety Risk Criteria by Center for Chemical Process Safety, August 2009.

INDEX

Page numbers followed by f and t indicate figures and tables, respectively.

CPSIA information can be obtained
at www.ICGtesting.com
Printed in the USA
LVHW010552130722
723323LV00005B/113

9 780791 861110